T0325672

Wireless Receiver Architectures and Design

To Toby, John, and Hend
and
in memory of Carry, George, Rose, Jean, and Zouzou.
But those who hope in the LORD will renew their strength.
They will soar on wings like eagles;
they will run and not grow weary, they will walk and not grow faint.
Isaiah 40: 31.

Wireless Receiver Architectures and Design

Antennas, RF, Synthesizers, Mixed Signal, and Digital Signal Processing

Tony J. Rouphael

AMSTERDAM • BOSTON • HEIDELBERG • LONDON
NEW YORK • OXFORD • PARIS • SAN DIEGO
SAN FRANCISCO • SINGAPORE • SYDNEY • TOKYO
Academic Press is an Imprint of Elsevier

Academic Press is an imprint of Elsevier
The Boulevard, Langford Lane, Kidlington, Oxford OX5 1GB, UK
225 Wyman Street, Waltham, MA 02451, USA

First edition 2014

British Library Cataloguing in Publication Data
A catalogue record for this book is available from the British Library

Library of Congress Cataloging-in-Publication Data
A catalog record for this book is availabe from the Library of Congress

ISBN–13: 978-0-12-378640-1

For information on all Academic Press publications
visit our web site at store.elsevier.com

Printed and bound in the United States of America
14 15 16 17 18 10 9 8 7 6 5 4 3 2 1

Contents

Preface

Background

The advent of modern wireless devices, such as smart phones and MID[1] terminals, has revolutionized the way people think of personal connectivity. Such devices encompass multiple applications ranging from voice and video to high-speed data transfer via wireless networks. The voracious appetite of twenty-first century users for supporting more wireless applications on a single device is ever increasing. These devices employ multiple radios and modems that cover multiple frequency bands and multiple standards with a manifold of wireless applications often running simultaneously. This insatiable requirement for multistandard multimode devices has ushered in new innovations in wireless transceiver designs and technologies that led and will continue to lead the way to highly integrated radio solutions in the form of SoCs.[2] These solutions include several radios and digital modems all designed on a single chip.

The design of such futuristic wireless radios and modems, capable of supporting multiple standards and modes, is drastically different from traditional designs supporting a single standard. The complications stem due to the high level of integration, low power, and low cost requirements imposed on the chip manufacturers. Until recently, most designers and architects designed for one particular application or another, which in turn meant familiarity with certain modem or radio architecture targeted for a certain wireless standard. However, designing future wireless devices, which entail designing for multimode and multistandard wireless applications, will require the radio engineer to be familiar with all facets and nuances of various transceiver architectures. This in essence will help maximize the efficiency of the design, lower the overall cost of the device, and allow for coexistence and simultaneous operations of various radios.

The intent of this book is to address the various designs and architectures of wireless receivers in the context of modern multimode and multistandard devices. The aim is to present a unique coherent and comprehensive treatment of wireless receivers from analog front ends to mixed signal design and frequency synthesis with equal emphasis on theory and practical design. Throughout the text, the design process is characterized by a close interaction between architecture on the one hand and algorithm design on the other with particular attention paid to the whole design process. This is in contrast to most other wireless design books that focus either on the theory or implementation. The focus will be on architecture and partitioning between mixed signal, digital and analog signal processing, analysis and design

[1]MID stands for Mobile Internet Device

[2]SoC stands for system on chip.

trade-offs, block level requirement, and algorithms. The reader will learn how certain design trade-offs in the analog back end, for example, can positively or negatively impact the design in the analog front end and vice versa. Furthermore, the reader will learn how the choice of certain algorithms for a given architecture can reduce the cost and design cycle. Conversely, how the choice of certain parameters or blocks in the lineup, such as filters, mixers, and amplifiers can significantly ameliorate the performance of the wireless receiver.

Scope

The intended audience of this book is engineers, researchers, and academics working in the area of wireless communication receivers. More precisely, the book is targeted toward engineers and researchers focused on designing low-power low-cost multimode multistandard receivers of the twenty-first century. It targets professionals who seek a comprehensive treatment and deep practical and theoretical understanding of wireless receiver design.

The book is divided into eight chapters:

- Chapter 1 focuses on antenna theory, transmission lines, and matching networks. The chapter provides an overview of antenna design and analysis. Topics such as radiation fields, gain, directivity and efficiency, matching circuits and networks, and transformers are addressed. The chapter concludes with a discussion on the most common types of antennas notably the monopole, the dipole, the patch, and the helical antennas.
- Chapter 2 addresses microwave network design and analysis. The chapter is divided into three major sections. In Section 2.1, we address the various circuit network models and their advantages and disadvantage particularly the impedance and admittance multiport network model, the hybrid network model, and the scattering parameters network model. Section 2.2 presents the topics of signal flow graphs, power gain equations and analysis, and stability theory. In Section 2.3, we discuss the most common types of two-port and three-port devices found in low-power wireless receivers.
- Chapter 3 is focused on noise in wireless receivers and how it manifests itself in circuits and components. The first section discusses thermal noise and how it appears in electronic components. Topics such as noise figure at the component and system level and how it impacts the performance of the receiver are discussed in some detail. Next, we introduce the concept of phase noise and present both Leeson's model and the Lee-Hajimiri model. The impact of phase noise on performance is also addressed. Finally, external noise sources due to interferers and blockers are discussed. The impact of coexistence of multiple radios on a single chip and its impact on the desired received signal is also discussed.
- Chapter 4 is concerned with system nonlinearity and its impact on receiver performance and design. The chapter focuses mainly on weakly nonlinear

systems with a brief mention strongly nonlinear systems. The various nonlinear phenomena are first classified. The chapter then delves into a discussion on memoryless nonlinearities using power series methods. The chapter concludes with a discussion on nonlinear systems with memory and the Volterra series techniques used to model them. The chapter contains several appendices.

- Chapter 5 is concerned with signal sampling and distortion. It starts with describing the sampling and reconstruction processes of baseband and bandpass signals and then examines the various degradations caused by the signal conversion and sampling imperfections. Topics such as quantization noise, sampling clock jitter, impact of phase noise on the sampling clock, signal overloading and clipping are discussed. The anti-aliasing filtering requirements and their impact on signal quality are also examined. An exact formulation of the quantization noise and a re-derivation of signal-to-quantization-noise ratio that takes into account the statistics of the input signal are also presented.
- Chapter 6 is concerned with analog to digital conversion and the architecture of the various data converters. The various building blocks of the ADC such as track and hold amplifiers and comparators are studied. The impact of aperture time accuracy clock feedthrough, and charge injection and their impact on signal SNR (signal to noise ratio) are also discussed. Next the Nyquist converter as a class of data converters is discussed. We delve into the architectural details of the FLASH, pipelined, and folding ADC architectures. Key performance parameters such as dynamic range, SNR, and harmonic distortion among others are also discussed. Next, we discuss oversampled converters and contrast it with Nyquist converters. The basic loop dynamics are derived. The architectures of continuous-time and discrete-time $\Delta\Sigma$ modulators are presented and their advantages and disadvantages are also contrasted. The MASH converter architecture is then presented and its pros and cons versus a single loop converter are also discussed.
- Chapter 7 presents two topics that deal with loops: the automatic gain control (AGC) loop algorithm and frequency synthesis phase-locked loop (PLL) design. While discussing the AGC loop, we discuss topics such as receiver lineup and gain, noise and noise figure, linear optimization, and dynamic range. We also discuss the digital signal processing implementation of the AGC loop. Closed loop dynamics are also presented in some detail. In the second portion of the chapter we discuss frequency synthesis. The Linear PLL model is discussed and performance parameters pertaining to error convergence, order and type, stability, and operating range are presented. The various blocks that constitute the PLL such as the frequency and phase detector, loop filter, and VCO are discussed. Finally, we focus on synthesizer architectures relevant to integer-N and fractional-N PLLs where we discuss various programmable digital counters such as the dual modulus prescalar and counters based on $\Delta\Sigma$ – modulators. The performance of both architectures is evaluated especially with respect to spurs, loop bandwidth, and resolution.
- Chapter 8 is focused on receiver architecture. In particular, we study the direct conversion architecture, the superheterodyne architecture, and the low-IF

architecture. We present the nuances and pros and cons of each of the architectures and discuss the various performance parameters. Two appendices concerning DC offset compensation and IQ imbalance compensation are also provided.

Tony J. Rouphael
San Diego, California
October 2013

Acknowledgments

I am very grateful to the teachers, colleagues, friends, and the editorial staff at Elsevier Academic Press whom without their support and encouragement this book would not have been written. I also would like to thank the wonderful engineers who reviewed various parts of the manuscript, in particular Dr Hui Liu and Mr Ricke Clark of RFMD, Mr Joe Paulino of MaxLinear, and Mr Tony Babaian of the University of California at San Diego. I am also indebted to Miss Charlie Kent and Mr Tim Pitts of Elsevier Academic Press for their patience and encouragement during the entire writing and editing process. Finally, I especially would like to thank my wonderful wife Toby. Toby thank you for encouraging me and making me believe I can complete this book. Finally, I would like to thank my son John for waking me up early in the morning and challenging me to finish the manuscript a section at a time. Toby and John I love you with all my heart. I also would like to thank my mother for her kind words and prayers. DJ and Tom you helped in your own unique ways as well!

CHAPTER

1

Antenna Systems, Transmission Lines, and Matching Networks

CHAPTER OUTLINE

Wireless Receiver Architectures and Design. http://dx.doi.org/10.1016/B978-0-12-378640-1.00001-9

In wireless communication, the antenna represents a fundamental component of the radio and exerts a major influence on its link budget. Therefore the choice of a proper antenna system becomes of paramount importance to ensure satisfactory connectivity amongst the various communicating devices. On the other hand, a poorly designed antenna impacts both the transmit and receive chains of the radio. In the uplink for example, an inefficient antenna could force the power amplifier along with the other gain stages in the transmit path to operate at higher gain, causing, amongst other things, excessive current consumption. In the downlink, a poorly designed antenna impacts the sensitivity of the receiver and hence the area of coverage under which the wireless device should be able to operate.

When designing an antenna, there are many factors and specifications that have to be taken into consideration. The antenna gain and directivity, the signal bandwidth and data throughput, the frequency band or bands of operation, path loss and radio frequency (RF) propagation environment, the interfaces to the antenna, and the matching networks are just a few of the key parameters that impact the performance of the radio as a whole via the antenna. Other factors involved in the design that must be well understood are the antenna form factor, the interaction between the antenna and its environment such as the package in which the antenna is housed, or the interaction between the antenna and the human body and other nearby objects. Given the intricacies of modern digital communication systems, it is then imperative that the antenna be designed as part of the whole radio rather than as an individual element.

The aim of this chapter is to provide an overview of antenna system design and analysis. The chapter is divided into three major sections. In Section 1.1, we discuss basic transmission line theory and antenna parameters, including topics ranging from radiation fields, gain, directivity, and efficiency to circuit realization. Section 1.2 discusses matching circuits. The section starts out by describing the impact of load mismatch on antenna performance. The discussion then further investigates transmission lines followed by an in-depth discussion of matching networks and transformers. Finally, Section 1.3 presents an overview of common antenna types. Antennas such as the dipole, monopole, small loop, patch antenna, and helical antenna are discussed.

1.1 Basic parameters

The purpose of this section is to discuss basic antenna parameters. Topics such as radiation density, radiation fields, antenna gain and antenna directivity, antenna efficiency, and load mismatch are discussed. The intent is to provide the reader with an overview of antenna theory and to point out critical design parameters.

1.1.1 Radiation density and radiation intensity

Simply put, an antenna is a transducer that changes the energy from magnetic (inductive) or electric (capacitive) to electromagnetic and vice versa. For example, on the receiver side, the electromagnetic energy received by the antenna is converted to electric or magnetic energy to be processed by the analog front end. The instantaneous power associated with an electromagnetic wave can be obtained via the Poynting[a] vector as [1]:

$$\vec{P} = \vec{E} \times \vec{H} \text{ Watts/m}^2 \tag{1.1}$$

where again $\vec{P}$ designates the instantaneous Poynting vector denoting the power density or energy flux, $\vec{E}$ is the instantaneous electric field intensity in Volt per meter, and $\vec{H}$ is the instantaneous magnetic field intensity in Ampere per meter. The time-averaged Poynting vector or average power density can be defined over a certain period T as

$$\vec{P}_{average} = \frac{1}{T} \int_{t}^{t+T} \vec{P} \mathrm{d}t \tag{1.2}$$

where $T = 2\pi/\omega$ is the period. A field is said to be monochromatic if the said field vector changes sinusoidally in time with respect to a single angular frequency ω. In a Cartesian coordinate system, we can relate the complex electric and magnetic fields to $\vec{E}$ and $\vec{H}$ as

$$\vec{E}_{x,y,z,t} = \mathrm{Re}\{E_{x,y,z,t} e^{j\omega t}\} \quad \text{and} \quad \vec{H}_{x,y,z,t} = \mathrm{Re}\{H_{x,y,z,t} e^{j\omega t}\} \tag{1.3}$$

The Cartesian and time indices in Eqn (1.3) will be eliminated in consequent equations for the sake of simplifying the notation. Substituting Eqn (1.3) into Eqn (1.1), we obtain the relationship for the instantaneous Poynting vector as

$$\vec{P} = \vec{E} \times \vec{H} = \frac{1}{2}\mathrm{Re}\{E \times H^*\} + \frac{1}{2}\mathrm{Re}\{E \times H^* e^{-j2\omega t}\} \tag{1.4}$$

Substituting Eqn (1.4) into Eqn (1.2), we obtain time-averaged Poynting vector over a period of time T as

$$\vec{P}_{average} = \frac{1}{T} \int_{t}^{t+T} \vec{P} \mathrm{d}t = \frac{1}{2}\mathrm{Re}\{E \times H^*\} \tag{1.5}$$

The relationship in Eqn (1.5) is often referred to as the radiation density of the antenna in its respective far field. The average power radiated by an antenna over a certain area can be expressed as the contour integral over the surface $\widehat{A}$ as

$$P_{radiated} = \oint_{\widehat{A}} \vec{P}_{average} \mathrm{d}\widehat{A} = \frac{1}{2} \oint_{\widehat{A}} \mathrm{Re}\{E \times H^*\} \mathrm{d}\widehat{A} \tag{1.6}$$

[a]Named after the English physicist John Henry Poynting.

At this point, it is befitting to define the radiation intensity of an antenna. The radiation intensity in the far field region of the antenna is simply the radiation density multiplied by the square of the distance

$$P_{Watts/\text{unit solid angle }(\theta,\phi)} = \vec{P}_{Watts/m^2} R^2_{m^2} \qquad (1.7)$$

The radiation intensity can be also related to the electric field of the antenna in the far field region as defined in the following section.

1.1.2 Radiation fields

The far-field region of radiation of the antenna, also known as the Fraunhofer region, is the region of radiation situated at a distance $R > 2l^2/\lambda$ where l is the overall dimension of the antenna and λ is the wavelength defined as the ratio of the speed of light $c = 3 \times 10^8$ m/s to the frequency F or simply $\lambda = c/F$. In this region, the radiation pattern of the antenna has the same field distribution regardless of distance.

On the other hand, the near field of radiation of the antenna can be classified as either the reactive near field region or the Fresnel region. The reactive near field region is within the immediate vicinity of the antenna where the distance R is such that $R < 0.62\sqrt{l^3/\lambda}$. In this region the electric and magnetic fields are 90° out of phase, unlike the far field region where both fields are orthogonal but in phase.

Finally, the Fresnel region or the radiating near field region is typically defined within the distance R from the antenna and is expressed as

$$0.62\sqrt{\frac{l^3}{\lambda}} < R < 2\frac{l^2}{\lambda} \qquad (1.8)$$

In this region, the shape of the radiating fields varies with distance. As the distance increases, the shape of the fields becomes more like far field waves and less like reactive near field waves.

1.1.3 Antenna gain, directivity, and efficiency

The directivity of an antenna is a dimensionless quantity defined as the ratio of the radiation intensity in a given direction to the average radiation intensity given out by the antenna in all directions. In other words, it is given with respect to an isotropic element in terms of a solid angle

$$D_{\theta,\phi} = \frac{P_{Watts/\theta,\phi}}{P_{Watts/isotropic/\theta,\phi}} \qquad (1.9)$$

where θ is the elevation angle and ϕ is the azimuth angle. The directivity of an antenna, when it is not specified in terms of elevation angle or azimuth, refers to the ratio of maximum radiation intensity to the radiation intensity of an isotropic element with respect to a solid angle as

$$D_{\max} = \frac{\max\left\{P_{Watts/\theta,\phi}\right\}}{P_{Watts/isotropic/\theta,\phi}} \tag{1.10}$$

The directivity can also be expressed as the ratio of maximum radiation intensity to the total radiated power by modifying Eqn (1.10) as

$$D_{\max} = \frac{4\pi\max\left\{P_{Watts/\theta,\phi}\right\}}{P_{\text{total radiated power of isotropic element}}} \tag{1.11}$$

Note that the directivity of an isotropic element according to Eqns (1.10) and (1.11) is unity. The reason is that an isotropic source radiates power in equal amounts in all directions. For all other radiating sources, the maximum directivity in Eqn (1.11) will be greater than unity or in 0 dB in the log scale.

Define the normalized radiation intensity of an antenna as

$$F_{\theta,\phi} = \frac{1}{c_o} P_{Watts/\text{unit solid angle }(\theta,\phi)} \tag{1.12}$$

where c_o is a certain normalizing constant representing $\max\{P_{Watts/\text{unit solid angle }(\theta,\phi)}\}$. Then the relationship in Eqns (1.10) and (1.11) can be expressed respectively as

$$D_{\theta,\phi} = 4\pi \frac{F_{\theta,\phi}}{\int_0^{2\pi}\int_0^{\pi} F_{\theta,\phi} \sin\theta d\theta d\phi} \tag{1.13}$$

and

$$D_{\max} = 4\pi \frac{\max\left\{F_{\theta,\phi}\right\}}{P_{Total} = \int_0^{2\pi}\int_0^{\pi} F_{\theta,\phi} \sin\theta d\theta d\phi} \tag{1.14}$$

Note that the denominator of Eqn (1.14) is the power per unit solid angle integrated over a spherical surface. The *average* power per unit angle can then be obtained by dividing the P_{total} by 4π.[b] The *absolute gain* of the antenna with 100% efficiency is defined as the ratio of antenna intensity $P_{Watts/\theta,\phi}$ in a given direction (θ, ϕ) to an equivalent isotropic radiating source. In both cases, the power delivered to the directive antenna and the power delivered to the isotropic source is assumed to be the

[b]Recall that there are 4π steradians on the surface of a sphere.

same after all losses are taken into account. Hence, the *absolute gain* in a given direction is

$$G_{absolute/\theta,\phi} = \frac{4\pi P_{Watts/\theta,\phi}}{P_{input}} \tag{1.15}$$

where P_{input} is the power at the antenna after all the losses have been accounted for.

The *relative gain* with 100% efficiency, on the other hand, refers to the power gain of a given antenna in a direction (θ, ϕ) to the power gain of a reference antenna in the same direction. For example, in reference to an ideal lossless isotropic source,[c] the ratio may be defined as

$$G_{relative/\theta,\phi} = \frac{4\pi P_{Watts/\theta,\phi}}{P_{isotropic}} \tag{1.16}$$

However, a more practical and reasonable definition of gain must include the efficiency of the antenna. When accounting for losses due to impedance mismatch or polarization loss, for example, the gain of the antenna can then be expressed as

$$G_{\eta/\theta,\phi} = \eta \frac{4\pi P_{Watts/\theta,\phi}}{P_{isotropic}} \tag{1.17}$$

where η is the antenna efficiency defined as the ratio of the radiated power out of the antenna to the total input power into the antenna. This definition of antenna gain is a more common figure of merit in the wireless engineering circles. It is also common to specify the gain of the antenna in terms of directivity and maximum directivity as

$$\begin{aligned} G_{\eta/\theta,\phi} &= \eta D_{\theta,\phi} \text{ relative to a solid angle}(\theta, \phi) \\ G_{\eta/\max} &= \eta D_{\max} \text{ relative to maximum directivity} \end{aligned} \tag{1.18}$$

1.1.4 Circuit realization

While transmitting, the antenna acts as a load impedance that mainly consists of a resistive element R_{ant} and a reactive element X_{ant} as depicted in Figure 1.1. In simplest terms, the resistive portion of the load can be attributed to the power losses in the antenna circuit and its surroundings, whereas the reactive part signifies the energy stored in fields in the vicinity of the antenna. To the transmitter, the antenna then appears as a load impedance. The transmitter itself can be modeled as a voltage generator V_{TX} with a load impedance of its own consisting of a resistive element R_{TX} and a reactive element X_{Tx}.

[c]The reference antenna or source could also be a monopole or a dipole, since they are more practical antennas.

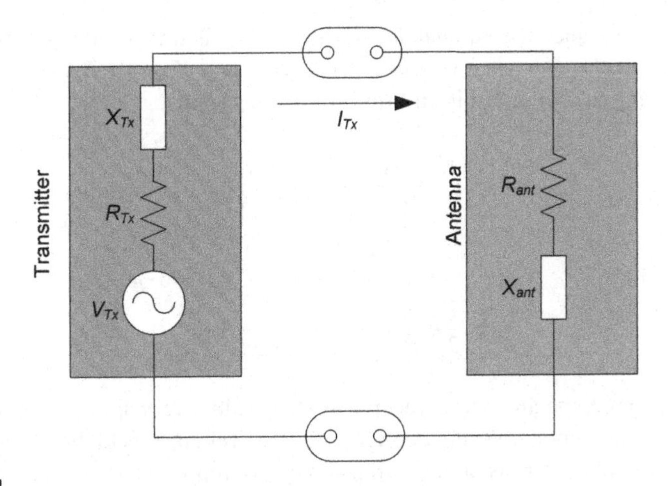

FIGURE 1.1

Transmitter with antenna represented as a circuit element. (For color version of this figure, the reader is referred to the online version of this book.)

Using Kirchhoff's current law and Figure 1.1, we can determine the current at the antenna input terminals as $I_{Tx} = V_{Tx}/(Z_{Tx} + Z_{ant})$ where Z_{Tx} is the impedance of the transmitter $R_{TX} + X_{Tx}$ and Z_{ant} is the impedance of the antenna $R_{ant} + X_{ant}$. The power lost in the transmitter due to its resistance R_{TX} can also be determined in a similar manner as

$$P_{\text{transmitter}} = \frac{1}{2}|I_{Tx}|^2 R_{Tx} = \frac{1}{2}\frac{R_{Tx}}{|Z_{Tx} + Z_{ant}|^2}|V_{Tx}|^2 \tag{1.19}$$

The total power delivered to the antenna can then be simply determined as

$$P_{ant} = \frac{1}{2}|I_{Tx}|^2 R_{ant} = \frac{1}{2}\frac{R_{ant}}{|Z_{Tx} + Z_{ant}|^2}|V_{Tx}|^2 \tag{1.20}$$

However, not all the power delivered to the antenna will be radiated. Some of the power delivered to the antenna will be lost in the antenna itself. These resistive-type losses are known as ohmic losses. The rest of the power is radiated. The resistance associated with the radiated power is known as *radiation resistance*. The ratio of radiated power, say $P_{radiated}$ to the power delivered to the antenna P_{ant}, is the efficiency factor η defined earlier in Eqn (1.17). In order to maximize the power delivered to the antenna, and hence prolong the battery life of the device, it is then essential to conjugate match the transmitter load Z_{Tx} to the antenna load Z_{ant}, thus minimizing the power lost in the transmitter. Conjugate matching of the load impedances implies that $Z_{Tx} = Z^*_{ant}$, which in turns implies that the transmitter resistance and antenna resistance are equivalent, or $R_{TX} = R_{ant}$, and that the reactance of the transmitter and the reactance of the antenna are such $X_{Tx} = -X_{ant}$. With this in mind, revisiting the relationship in Eqn (1.20), we obtain

$$P_{ant,\max} = \frac{1}{2}\frac{R_{ant}}{|Z_{Tx}+Z_{ant}|^2}|V_{Tx}|^2\Bigg|_{\substack{R_{ant}=R_{Tx}\\ Z_{Tx}=R_{Tx}+jX_{Tx}\\ Z_{ant}=R_{Tx}-jX_{Tx}}} = \frac{1}{2}\frac{R_{Tx}}{4R_{Tx}^2}|V_{Tx}|^2 = \frac{1}{8R_{Tx}}|V_{Tx}|^2 \tag{1.21}$$

On the other hand, the antenna in receive mode acts in a very similar manner to the antenna in transmit mode. While receiving, the receiver appears to the antenna at the output of its connector as a load impedance as represented by the circuit in Figure 1.2. Using the reciprocity principle, and given for the sake of simplicity that the transmit frequency is the same as the receive frequency, the load of the antenna in receive mode is identical to the load of the antenna in transmit mode Z_{ant}. Let the receiver load be given as $Z_{Rx} = R_{Rx} + jX_{Rx}$, then the current flowing into Z_{Rx} is $I_{ant} = V_{ant}/(Z_{Rx} + Z_{ant})$. The power delivered to the receiver can then be given in a manner similar to Eqn (1.19) as

$$P_{\text{receiver}} = \frac{1}{2}|I_{ant}|^2 R_{Rx} = \frac{1}{2}\frac{R_{Rx}}{|Z_{Rx}+Z_{ant}|^2}|V_{ant}|^2 \tag{1.22}$$

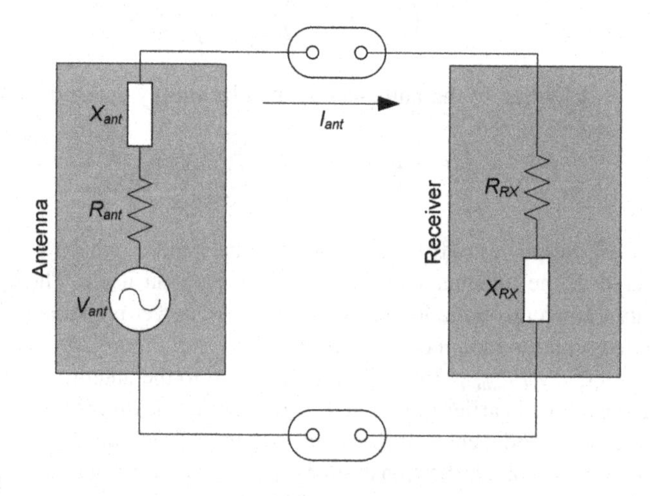

FIGURE 1.2

Receiver with antenna represented as a circuit element. (For color version of this figure, the reader is referred to the online version of this book.)

Furthermore, when the antenna load is matched to that of the receiver load, that is $Z_{ant} = Z_{Rx}^*$, the maximum power is delivered to receiver

$$P_{Rx,\max} = \frac{1}{8R_{Rx}}|V_{ant}|^2 \tag{1.23}$$

1.1.5 Friis link budget formula

The effective aperture or effective area of the receive antenna is defined as the ratio of the power absorbed by the receiving antenna to the incident power density of the transmit antenna at a given distance R

$$A_{Rx} = \frac{P_{Rx}}{\rho_{Tx}} \tag{1.24}$$

The effective isotropic radiated power of the antenna is defined as the product of the transmit gain and transmit power of the antenna or simply $P_{EIRP} = P_{Tx}G_{Tx}$, where P_{Tx} is the power of the transmitting antenna and G_{Tx} is gain of the transmitting antenna. The free-space power density at a distance R can then be obtained:

$$\rho_{TX} = \frac{\delta P_{Tx}}{\delta S(R,\theta,\phi)} = \frac{P_{Tx}G_{Tx}}{4\pi R^2} \tag{1.25}$$

Combining Eqns (1.24) and (1.25), we obtain the well-known Friis formula

$$P_{Rx} = A_{Rx}\rho_{Tx} = \frac{P_{Tx}G_{Tx}A_{Rx}}{4\pi R^2} \tag{1.26}$$

The transmit and receive gains G_{TX} and G_{RX} of the transmitting and receiving antennas can be further expressed in terms of the effective area as

$$\begin{aligned} G_{Tx} &= 4\pi\frac{A_{Tx}}{\lambda^2} \\ G_{Rx} &= 4\pi\frac{A_{Rx}}{\lambda^2} \end{aligned} \tag{1.27}$$

where λ is the free-space wavelength of the transmitted signal defined as the ratio of the speed of light c in meters per second to the carrier frequency F_c in Hz or $\lambda = c/F_c$ where $c = 2.9979 \times 10^8$ m/s. Substituting Eqn (1.27) into Eqn (1.26), we obtain the free-space Friis formula

$$\begin{aligned} P_{Rx} &= \left(\frac{\lambda}{4\pi R}\right)^2 G_{Tx}G_{Rx}P_{Tx} \\ P_{Rx,dBm} &= 20\log_{10}\left(\frac{\lambda}{4\pi R}\right) + G_{Tx,dB} + G_{Rx,dB} + P_{Tx,dBm} \end{aligned} \tag{1.28}$$

The inverse of the square term in Eqn (1.28) is known as the free-space path loss and is often expressed in equation form as $10\log_{10}(16\pi^2R^2/\lambda^2)$.

EXAMPLE 1.1 ANTENNA GAIN AND PATH LOSS

Consider a UWB MBOA signal transmitted at the carrier frequencies of 4488 MHz. The transmit power is set at −14 dBm EIRP at the output of the antenna. What is the received signal power at 9.5 m assuming the gain of the antenna is −2 dBi? Assume a free-space model.

The free-space path loss at 9.5 m can simply be computed as

$$\text{Path Loss in dB} = 20\log_{10}\left(\frac{4\pi R}{\lambda}\right) = 20\log_{10}\left(\frac{4\pi \times 9.5\text{ m}}{0.0688\text{ m}}\right) = 65.0434\text{ dB} \quad (1.29)$$

Where $\lambda = \frac{c}{F_c} = \frac{2.9979 \times 10^8 \text{m/s}}{4488 \times 10^6 \text{Hz}} = 0.0668$ m.

The transmit power at the output of the antenna of −14 dBm simply implies that $G_{Tx,dB} + P_{Tx,dBm} = -14$ dBm. Hence using Eqn (1.28), the received signal power at the receiving antenna is

$$\begin{aligned} P_{Rx,dBm} &= 20\log_{10}\left(\frac{\lambda}{4\pi R}\right) + G_{Tx,dB} + G_{Rx,dB} + P_{Tx,dBm} \\ &= -65.0434\text{ dB} - 14\text{ dBm} - 2\text{ dB} = -81.0434\text{ dBm} \end{aligned} \quad (1.30)$$

1.1.6 Electrical length

In antenna theory, engineers differentiate between the physical length of the antenna and its electrical length. This differentiation is particularly important when analyzing an antenna. Two antennas can have the same physical length L but different electrical lengths. Recall that in free space, the wavelength is related to the *velocity* of light as

$$\lambda = \left.\frac{U}{F}\right|_{U=c \text{ in free-space}} = \frac{c}{F}, c = 3 \times 10^8 \text{m/s} \quad (1.31)$$

The velocity in general depends on the electric permittivity and magnetic permeability, defined as the ratio

$$U = \frac{1}{\sqrt{\mu\varepsilon}} \quad (1.32)$$

where the magnetic permeability and electric permittivity are simply defined as

$$\begin{aligned} \mu &= \mu_0\mu_r, \quad \text{where } \mu_0 = 4\pi \times 10^{-7}(\text{H/m}) \\ \varepsilon &= \varepsilon_0\varepsilon_r, \quad \text{where } \varepsilon_0 = \frac{1}{36\pi} \times 10^{-9}(\text{F/m}) \end{aligned} \quad (1.33)$$

Substituting Eqn (1.33) into Eqn (1.32), we obtain

$$U = \frac{3 \times 10^8 \text{m/s}}{\sqrt{\mu_r \varepsilon_r}} = \frac{c}{\sqrt{\mu_r \varepsilon_r}} \text{m/s}, \mu_r \geq 1, \varepsilon_r \geq 1, U \leq c \tag{1.34}$$

where ε_r is the dielectric constant. The maximum attainable velocity as presented in Eqn (1.34) is the speed of light. In RF components and antennas, U never attains the speed of light, and hence the wavelength in the medium $\lambda_{medium} = U/F$ is always less than the free-space wavelength $\lambda = U\sqrt{\mu_r \varepsilon_r}/F$. Note that the electric permittivity can be defined as the ratio of the electric flux density D_ε to the electric field E

$$\varepsilon = \frac{D_\varepsilon}{E} \tag{1.35}$$

Similarly the magnetic permeability can be defined as the ratio of magnetic flux density B_μ to the magnetic field H or

$$\mu = \frac{B_\mu}{H} \tag{1.36}$$

For example, a short vertical antenna has capacitive impedance with a small resistive impedance.[d] Capacitive impedance stores the energy and does not radiate or dissipate power.

1.2 Matching circuits

In order to avoid high voltage standing wave ratio (VSWR) losses, a source is matched to a load, possibly an antenna via a matching circuit. This section starts out by elaborating on the impact of load mismatch on power transfer and then discusses the various circuit matching schemes.

1.2.1 Antenna load mismatch and reflection coefficient

Thus far, we have treated the antenna as a matched load to the transceiver in both transmit and receive modes. In this section, we address the important issue of load mismatch. In Eqn (1.21), for example, we assumed that the maximum power was delivered to the antenna. However, in reality, only a fraction of the power delivered to the antenna is radiated mainly due to ohmic and dielectric losses. Therefore, in order to address antenna matching and the effect of load mismatch on the circuit, consider the relationship given in Eqns (1.20) and (1.21). Then we can relate

$$P_{\text{out}} = \eta_m P_{ant,\max} \tag{1.37}$$

[d]In reality, the small radiation impedance for a length $L < \lambda/20$ has a radiation resistance approximately given as $R_{rad} = 20(2\pi L/\lambda)^2$.

where η_m is the mismatch efficiency, as opposed to the total efficiency defined in Eqn (1.17), expressed as the ratio

$$\eta_m = \frac{\dfrac{1}{2}\dfrac{R_{ant}}{|Z_{Tx}+Z_{ant}|^2}}{\dfrac{1}{8}\dfrac{|V_{Tx}|^2}{R_{Tx}}} = \frac{4R_{Tx}R_{ant}}{|Z_{Tx}+Z_{ant}|^2} \tag{1.38}$$

When Z_{Tx} and Z_{ant} are purely resistive loads, that is when $X_{Tx} = X_{ant} = 0$, the mismatch efficiency can be written as

$$\eta = \frac{4R_{Tx}R_{ant}}{R_{Tx}^2 + R_{ant}^2 + 2R_{Tx}R_{ant}} \tag{1.39}$$

From Eqn (1.39), we obtain

$$1 - \eta = \frac{R_{Tx}^2 + R_{ant}^2 - 2R_{Tx}R_{ant}}{R_{Tx}^2 + R_{ant}^2 + 2R_{Tx}R_{ant}} = \frac{(R_{Tx} - R_{ant})^2}{(R_{Tx} + R_{ant})^2} = |\Gamma|^2 \tag{1.40}$$

where Γ is termed as the reflection coefficient and expressed in terms of load impedances as

$$\Gamma = \frac{Z_{ant} - Z_{Tx}}{Z_{ant} + Z_{Tx}} \tag{1.41}$$

1.2.2 Transmission lines and the wave equation

In this section we will use transmission line theory and the Telegraph Equations[e] to determine the voltage and current relationship in a transmission line with respect to two parameters: time t and distance Δz. Consider the circuit representation of an

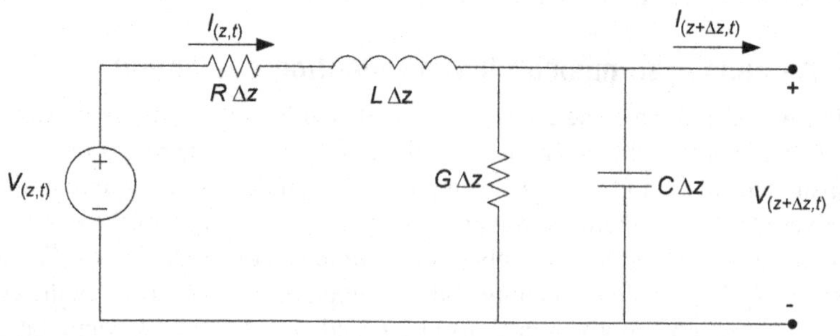

FIGURE 1.3

Circuit representation of an elementary component of a transmission line.

[e]Oliver Heaviside is credited with the transmission line model and the discovery of the telegraph or the telegrapher equations.

elementary transmission line component as depicted in Figure 1.3. The resistance R, the inductor L, the capacitance C, and the conductance G are specified per unit distance Δz. The steady state voltage $V(z, t)$ given as a function of time and distance can be expressed as

$$\begin{aligned} -V(z,t) + \Delta z(R + j\Omega L)I(z,t) + V(z+\Delta z,t) &= 0 \\ V(z+\Delta z,t) - V(z,t) &= -\Delta z(R + j\Omega L)I(z,t) \end{aligned} \quad (1.42)$$

where Ω is the frequency. Divide both sides of Eqn (1.42) by Δz and we obtain

$$\frac{V(z+\Delta z,t) - V(z,t)}{\Delta z} = -(R + j\Omega L)I(z,t) \quad (1.43)$$

Similarly, consider the current $I(z, t)$ in Figure 1.3 as a function of time and distance, then $I(z, t)$ can be expressed as

$$\begin{aligned} I(z,t) - I(z+\Delta z,t) + \Delta z(G + j\Omega C)V(z,t) &= 0 \\ I(z,t) - I(z+\Delta z,t) &= -\Delta z(G + j\Omega C)V(z,t) \end{aligned} \quad (1.44)$$

Again, divide both sides of Eqn (1.44) by Δz and we obtain

$$\frac{I(z,t) - I(z+\Delta z,t)}{\Delta z} = -(G + j\Omega C)V(z,t) \quad (1.45)$$

As Δz approaches zero, the relation in Eqns (1.43) and (1.45) can be expressed as differential equations by taking the partial derivatives:

$$\begin{aligned} \frac{\partial V(z,t)}{\partial z} &= \lim_{\Delta z \to 0} \frac{V(z+\Delta z,t) - V(z,t)}{\Delta z} = -(R + j\Omega L)I(z,t) \\ \frac{\partial I(z,t)}{\partial z} &= \lim_{\Delta z \to 0} \frac{I(z+\Delta z,t) - I(z,t)}{\Delta z} = -(G + j\Omega C)V(z,t) \end{aligned} \quad (1.46)$$

From Eqn (1.46), we can deduce that the change of voltage is proportional to the resistance and inductance parameters, whereas the change in current is proportional to the conductance and capacitance parameters. Furthermore, taking the second derivative of Eqn (1.46) and dropping the time variable t, we obtain the second order differential equations:

$$\begin{aligned} \frac{d^2V(z)}{dz^2} - (R + j\Omega L)(G + j\Omega C)V(z) &= 0 \\ \frac{d^2I(z)}{dz^2} - (R + j\Omega L)(G + j\Omega C)I(z) &= 0 \end{aligned} \quad (1.47)$$

It is interesting to note that for small values of R and G, the transmission line of Figure 1.3 becomes lossless. Define the complex propagation constant γ as

$$\gamma = \alpha + j\beta = \sqrt{(R + j\Omega L)(G + j\Omega C)} \quad (1.48)$$

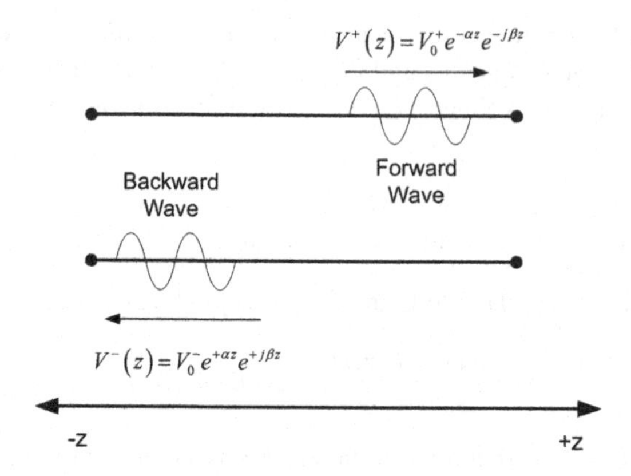

FIGURE 1.4

Forward and backward wave propagation in a transmission line.

Obviously, for a lossless system, $\alpha = 0$ and $\beta = \Omega\sqrt{LC}$ rad/m. Decomposing Eqn (1.48) results in α and β as

$$\begin{aligned} \alpha &= \sqrt{\frac{1}{2}\left(\sqrt{(R^2+\Omega^2L^2)(G^2+\Omega^2C^2)} - (\Omega^2LC - RG)\right)} \\ \beta &= \sqrt{\frac{1}{2}\left(\sqrt{(R^2+\Omega^2L^2)(G^2+\Omega^2C^2)} + (\Omega^2LC - RG)\right)} \end{aligned} \tag{1.49}$$

The mathematical solutions to the second order differential equations presented in Eqn (1.47) are:

$$\begin{aligned} V(z) &= V_0^+e^{-\gamma z} + V_0^-e^{+\gamma z} \\ I(z) &= I_0^+e^{-\gamma z} + I_0^-e^{+\gamma z} \end{aligned} \tag{1.50}$$

where $\{V_0^+, V_0^-, I_0^+, I_0^-\}$ are constants defined by the boundary conditions.

Further investigation of Eqn (1.50) indicates that there are two possible waves: one wave propagating forward (incident wave) and one wave propagating backwards (reflection wave) as illustrated in Figure 1.4. The incident wave is propagating in the positive z direction and is given as

$$\begin{aligned} V^+(z) &= V_0^+e^{-\gamma z} = V_0^+e^{-z(\alpha+j\beta)} = V_0^+e^{-\alpha z}e^{-j\beta z} \\ I^+(z) &= I_0^+e^{-\gamma z} = I_0^+e^{-z(\alpha+j\beta)} = I_0^+e^{-\alpha z}e^{-j\beta z} \end{aligned} \tag{1.51}$$

whereas the reflection wave is propagating in the negative z direction and is given as

$$\begin{aligned} V^-(z) &= V_0^-e^{+\gamma z} = V_0^-e^{z(\alpha+j\beta)} = V_0^-e^{\alpha z}e^{j\beta z} \\ I^-(z) &= I_0^-e^{+\gamma z} = I_0^-e^{z(\alpha+j\beta)} = I_0^-e^{\alpha z}e^{j\beta z} \end{aligned} \tag{1.52}$$

Examining the relationship in Eqns (1.51) and (1.52) reveals that the phase velocity is simply the ratio

$$v_{phase} = \frac{\Omega}{\beta} \tag{1.53}$$

From the relationships in Eqns (1.46), (1.47), and (1.50), the characteristic impedance of the transmission line is given as

$$\begin{aligned} \frac{V_0^+}{I_0^+} &= Z_{char} \\ \frac{V_0^-}{I_0^-} &= -Z_{char} \end{aligned} \tag{1.54}$$

where the characteristic impedance is found to be

$$Z_{char} = \sqrt{\frac{R + j\Omega L}{G + j\Omega C}} \tag{1.55}$$

The negative sign in the second equation in Eqn (1.54) is due to the direction of propagation (in the negative z direction) of the backward wave.

At this point it is interesting to note that for a low-loss transmission line, that is when $\Omega L \gg R$ and $\Omega C \gg G$, the characteristic impedance and its corresponding parameters α and β can be approximated as

$$\begin{aligned} &\alpha \approx \frac{1}{2}\left(R\sqrt{\frac{C}{L}} + G\sqrt{\frac{L}{C}}\right) \\ &\beta \approx \Omega\sqrt{LC} \\ &Z_{char} \approx \frac{1}{2}\sqrt{\frac{L}{C}}\left(2 - j\underbrace{\left(\frac{R}{\Omega L} - \frac{G}{\Omega C}\right)}_{\approx 0}\right) \approx \sqrt{\frac{L}{C}} \end{aligned} \tag{1.56}$$

In a similar manner, when the transmission line experiences high loss, that is when $\Omega L \ll R$ and $\Omega C \ll G$, the characteristic impedance and its corresponding parameters α and β can be approximated as

$$\begin{aligned} &\alpha \approx \sqrt{RG} \\ &\beta \approx 0 \\ &Z_{char} \approx \sqrt{\frac{R}{G}} \end{aligned} \tag{1.57}$$

In this case, the electromagnetic wave attenuates rapidly across the line.

1.2.3 Impedance and standing wave ratio

Consider the transmission line depicted in Figure 1.5 connected to a load Z_{Load}. The load could be any microwave element including an antenna. The load itself can be determined by the ratio

$$Z_{Load} = \frac{V_{Load}}{I_{Load}} \tag{1.58}$$

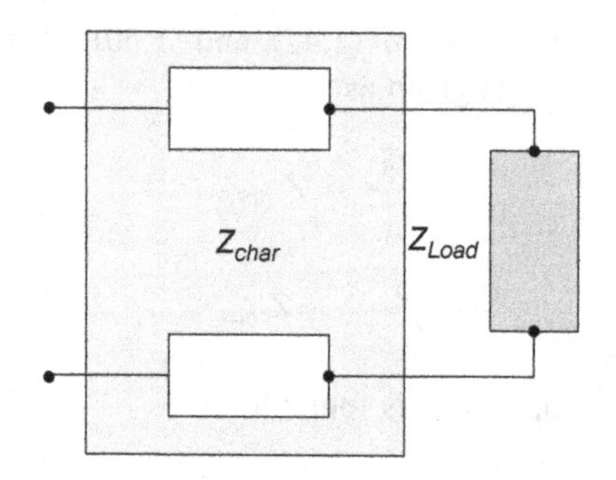

FIGURE 1.5

A transmission line Z_{char} connected to a load Z_{Load}. (For color version of this figure, the reader is referred to the online version of this book.)

The voltage across the load is the sum of the incident and reflected voltages

$$V_{Load} = V^+ + V^- \tag{1.59}$$

Furthermore, the current passing through the load is the sum of the incident and reflected currents

$$I_{Load} = I^+ + I^- = \frac{V^+}{Z_{char}} - \frac{V^-}{Z_{char}} \tag{1.60}$$

The ratio of the reflected voltage to the incident voltage can be found via simple manipulations of Eqns (1.58), (1.59), and (1.60) to be

$$\frac{V^-}{V^+} = \frac{Z_{Load} - Z_{char}}{Z_{Load} + Z_{char}} = \Gamma \tag{1.61}$$

Not surprisingly, this ratio in Eqn (1.61) is the same reflection coefficient defined in Eqn (1.41). A further manipulation of Eqns (1.51), (1.52), and (1.61) results in the lossless reflection coefficient as a function of location z being expressed as the complex ratio

$$\Gamma(z) = \frac{V^-}{V^+} = \frac{V_0^- e^{+j\beta z}}{V_0^+ e^{-j\beta z}} = \frac{V_0^-}{V_0^+} e^{+j2\beta z} = \Gamma_{Load} e^{+j2\beta z} \tag{1.62}$$

where Γ_{Load} is the reflection coefficient at the load. Observe that Eqn (1.62) implies that the origin of the location z-axis is taken at the load. For a lossless transmission line, the magnitude of the reflection coefficient does not change as a function of z, that is $|\Gamma(z)| = \Gamma_{Load}$.

Given the reflection coefficient as defined in Eqn (1.62) and the relation in Eqns (1.51) and (1.52), we can rewrite Eqn (1.59) as

$$V_{Load} = V^+ + V^- = V_0^+ e^{-j\beta z} + V_0^- e^{+j\beta z} = V_0^+ \left(e^{-j\beta z} + \Gamma_{Load} e^{+j\beta z}\right) \tag{1.63}$$

And by the same token, we could also rewrite the current load as

$$I_{Load} = \frac{V_0^+}{Z_{char}}\left(e^{-j\beta z} - \Gamma_{Load}e^{+j\beta z}\right) = I_0^+\left(e^{-j\beta z} - \Gamma_{Load}e^{+j\beta z}\right) \tag{1.64}$$

At this point, it is interesting to examine the average power at the load. Recall that the average power is defined as the product $P_{average} = 0.5\text{Re}\{V_{Load}I_{Load}^*\}$, then the average power at the load is given as

$$P_{Load} = \frac{1}{2}\frac{|V_0^+|^2}{Z_{char}}\left(1 - |\Gamma_{Load}|^2\right) = P_{Load}^+ - P_{Load}^- \tag{1.65}$$

where P_{Load}^+ and P_{Load}^- are the incident and reflected powers defined as

$$\begin{aligned} P_{Load}^+ &= \frac{1}{2}\frac{|V_0^+|^2}{Z_{char}} \\ P_{Load}^- &= \frac{1}{2}|\Gamma_{Load}|^2\frac{|V_0^+|^2}{Z_{char}} \end{aligned} \tag{1.66}$$

In general, the total voltage $V(z)$ and total current $I(z)$ at any location z on the transmission line can be expressed with the aid of Eqn (1.62) as

$$\begin{aligned} V(z) &= V^+(1 + \Gamma(z)) \\ I(z) &= I^+(1 - \Gamma(z)) \end{aligned} \tag{1.67}$$

And hence, the input impedance can then be expressed based on the ratio of $V(z)$ over $I(z)$ as

$$\begin{aligned} Z_{input} &= \frac{V(z)}{I(z)} = \frac{V^+(1 + \Gamma(z))}{I^+(1 - \Gamma(z))} = Z_{char}\frac{1 + \Gamma_{Load}e^{+j2\beta z}}{1 - \Gamma_{Load}e^{+j2\beta z}} \\ &= Z_{char}\left.\frac{1 + \Gamma_{Load}e^{-j2\beta l}}{1 - \Gamma_{Load}e^{-j2\beta l}}\right|_{l=-z} = Z_{char}\frac{Z_{Load} + jZ_{char}\tan(\beta l)}{Z_{char} + jZ_{Load}\tan(\beta l)} \end{aligned} \tag{1.68}$$

The relationship expressed in Eqn (1.68) is not constant across the transmission line. This relationship forms the basis of the Smith chart as shown by way of an example in Figure 1.6.

A further important observation concerning Eqn (1.63) is in relationship to its absolute maximum and minimum. For a lossless transmission line, we have already established that $|\Gamma(z)| = \Gamma_{Load}$, then a reexamination of Eqn (1.63) reveals that

$$\begin{aligned} \max\{V_{Load}\} &= |V_{Load}|_{\max} = |V_0^+|(1 + |\Gamma(z)|) \\ \min\{V_{Load}\} &= |V_{Load}|_{\min} = |V_0^+|(1 - |\Gamma(z)|) \end{aligned} \tag{1.69}$$

Hence we define the voltage standing wave ratio as the ratio

$$VSWR = \frac{|V_{Load}|_{\max}}{|V_{Load}|_{\min}} = \frac{1 + |\Gamma(z)|}{1 - |\Gamma(z)|} \tag{1.70}$$

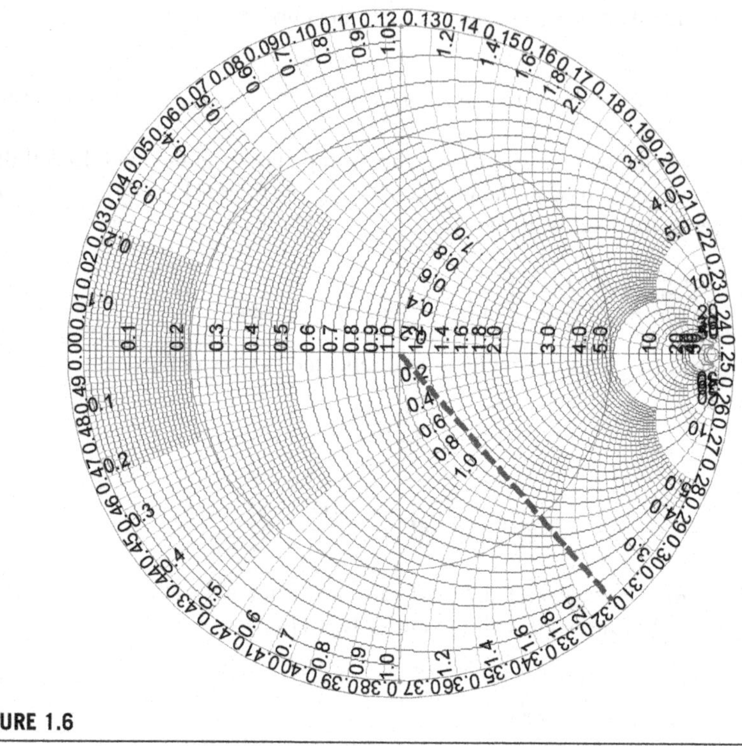

FIGURE 1.6

Smith chart of simple load $Z_{Load} = 50 - j30$ and $Z_{char} = 75 + j50$. (For color version of this figure, the reader is referred to the online version of this book.)

Note that there are two extreme cases and one general case that are of interest. The first two result in either a pure standing wave or a pure traveling wave along the transmission line, while the last case results in a mixed wave.

1.2.3.1 Standing wave case

In this case, the load is either open or short. In either scenario, the load does not consume any energy. To start out describing the standing wave case, we assume that the load is open, that is $Z_{Load} = \infty$, then according to Eqn (1.61) we have

$$\Gamma_{open} = \lim_{Z_{Load} \to \infty} \left\{ \frac{1 - \dfrac{Z_{char}}{Z_{Load}}}{1 + \dfrac{Z_{char}}{Z_{Load}}} \right\} = 1 \tag{1.71}$$

And in the second case, we assume that load is short, that is $Z_{Load} = 0$, and again according to Eqn (1.61) we have

$$\Gamma_{short} = \lim_{Z_{Load} \to 0} \left\{ \frac{Z_{Load} - Z_{char}}{Z_{Load} + Z_{char}} \right\} = -1 \tag{1.72}$$

Note that the VSWR is ∞ for both open and short loads. At this point it is instructive to look at the voltage and current for both open and short loads. For an open load, the voltage and current along the line can be expressed as

$$V(z) = V_0^+\left(e^{-j\beta z} + e^{+j\beta z}\right) = 2V_0^+ \cos(\beta z) = 2V_0^+ \cos\left(\frac{2\pi}{\lambda}z\right)\bigg|_{\beta=\frac{2\pi}{\lambda}}$$

$$I(z) = \frac{V_0^+}{Z_{char}}\left(e^{-j\beta z} - e^{+j\beta z}\right) = -\frac{2jV_0^+}{Z_{char}}\sin(\beta z) = -\frac{2jV_0^+}{Z_{char}}\sin\left(\frac{2\pi}{\lambda}z\right)\bigg|_{\beta=\frac{2\pi}{\lambda}} \tag{1.73}$$

The load $Z(z)$, the ratio of $V(z)$ over $I(z)$, is always *reactive* in this case, that is

$$Z(z) = jZ_{char}\cot(\beta z) = -jZ_{char}\cot(\beta l)|_{l=-z} \tag{1.74}$$

In this case the load does not absorb any energy and the transmission line is in *pure standing wave state*. The phase difference between the current and the voltage is 90° as shown in Figure 1.7. Likewise, for a shorted transmission line, the voltage and current along the line can be expressed in a manner similar to Eqn (1.73) as

$$V(z) = V_0^+\left(e^{+j\beta z} - e^{-j\beta z}\right) = j2V_0^+ \sin(\beta z) = j2V_0^+ \sin\left(\frac{2\pi}{\lambda}z\right)\bigg|_{\beta=\frac{2\pi}{\lambda}}$$

$$I(z) = \frac{V_0^+}{Z_{char}}\left(e^{-j\beta z} + e^{+j\beta z}\right) = \frac{2V_0^+}{Z_{char}}\cos(\beta z) = \frac{2V_0^+}{Z_{char}}\cos\left(\frac{2\pi}{\lambda}z\right)\bigg|_{\beta=\frac{2\pi}{\lambda}} \tag{1.75}$$

It can be easily shown that the impedance of a shortened transmission line is also purely reactive. This is done by taking the ratio of $V(z)$ over $I(z)$ or

$$Z(z) = -jZ_{char}\tan(\beta z) = jZ_{char}\tan(\beta l)|_{l=-z} \tag{1.76}$$

The impedance for both open and shorted transmission lines alternate along the z axis to periodically appear as inductive reactance or capacitive reactance as shown for the open transmission line case of Figure 1.7.

1.2.3.2 Traveling wave case

In this case, the reflection coefficient is equal to zero, that is $\Gamma = 0$, and there is no wave reflecting back from the load. The input impedance is equivalent to the characteristic impedance, and the total voltage along the line is simply the incident wave voltage

$$\begin{aligned} V_{Load} &= V^+ = V_0^+ e^{-j\beta z} \\ Z_{input} &= Z_{char} \end{aligned} \tag{1.77}$$

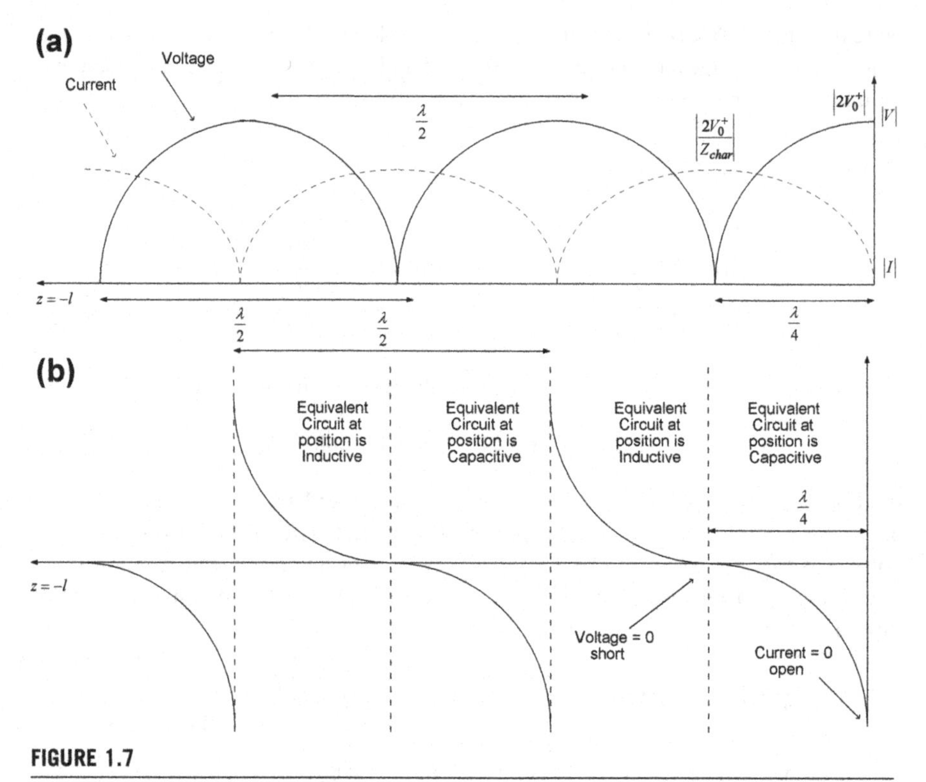

FIGURE 1.7

The standing wave of an open transmission line. (a) Voltage amplitude and current amplitude, and (b) impedance distribution along the transmission line. (For color version of this figure, the reader is referred to the online version of this book.)

As a matter of fact, the input impedance at any given cross-section along the transmission line is equal to the characteristic impedance of the line. Furthermore, the load impedance is also equivalent to the characteristic impedance of the transmission line resulting in a matched load. It can be seen that in this case, the matching load to the transmission line absorbs all the energy from the wave generator, say in our case a transmitter. That is

$$\begin{aligned} P_{Load}^{+} &= \frac{1}{2}\frac{\left|V_0^+\right|^2}{Z_{char}} \\ P_{Load}^{-} &= 0 \end{aligned} \tag{1.78}$$

1.2.3.3 Mixed wave case

The mixed wave is the most prominent case that is encountered in practical systems. Here, some of the energy will be absorbed by the load and some of the energy will be reflected back along the transmission line, thus resulting in a mixed wave. A mixed wave then is a combination of a traveling wave and a standing wave. The voltage and current along the line are given as

$$\begin{aligned} V(z) &= V_0^+ e^{j\beta l}\left(1+\Gamma_{Load}e^{-j2\beta l}\right) \\ I(z) &= \frac{V_0^+}{Z_{char}} e^{j\beta l}\left(1-\Gamma_{Load}e^{-j2\beta l}\right) \end{aligned} \tag{1.79}$$

Note that $\Gamma_{Load}e^{-j2\beta l}$ in Eqn (1.79) is periodic, which implies that the voltage and current on a transmission line in the mixed wave case are distributed periodically with a period equal to $\lambda/2$.

EXAMPLE 1.2 TRANSMISSION LINE AND ANTENNA MATCHING

Consider the transmitter circuit with internal impedance of 50 ohm connected to an antenna via a transmission line and matching circuit as shown in Figure 1.8. Assume for the sake of this example that the transmission line is also resistive with resistance of 50 ohm and antenna load of 10 ohm. Determine the VSWR without any matching element. Determine the load impedance of the matching circuit to obtain a VSWR of 1.4:1. Note that this is not a realistic scenario, however, it will serve to illustrate certain aspects of antenna matching.

First, in order to obtain the VSWR of the circuit, we must determine the reflection coefficient of the circuit according to Eqn (1.61)

$$\Gamma = \frac{Z_{Load}-Z_{char}}{Z_{Load}+Z_{char}} = \frac{R_{TL}-R_{ant}}{R_{TL}+R_{ant}} = \frac{10-50}{10+50} = -0.667 \tag{1.80}$$

And consequently, the VSWR, expressed in Eqn (1.70), implies

$$VSWR = \frac{1+|\Gamma|}{1-|\Gamma|} = \frac{1+0.667}{1-0.667} = 5 \tag{1.81}$$

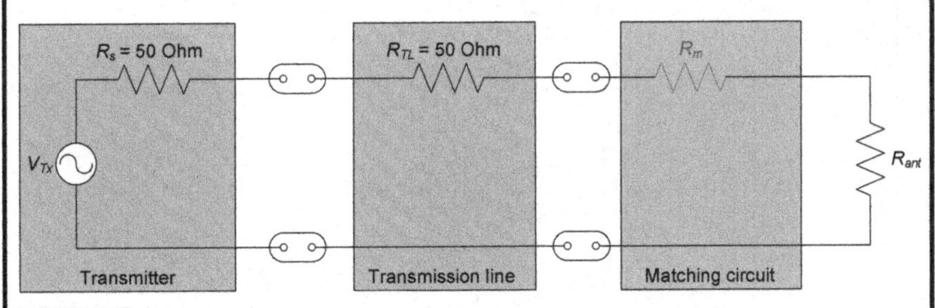

FIGURE 1.8

Resistive matching circuit between antenna and transmission line. (For color version of this figure, the reader is referred to the online version of this book.)

Continued

EXAMPLE 1.2 TRANSMISSION LINE AND ANTENNA MATCHING—cont'd

Obviously, a VSWR of 5 is far away from the required VSWR of 1.4. Therefore, we can determine the required reflection coefficient from the VSWR, that is

$$VSWR = \frac{1+|\Gamma|}{1-|\Gamma|} \Rightarrow |\Gamma| = \frac{VSWR-1}{VSWR+1} = \frac{1.4-1}{1.4+1} = 0.1667$$
$$\Gamma = \pm 0.1667 \tag{1.82}$$

Now, let's determine the required load that needs to be added. According to Eqn (1.61), we have

$$\Gamma = \frac{Z_m - Z_{TL}}{Z_m + Z_{TL}} = \frac{R_m - R_{TL}}{R_m + R_{TL}} \Rightarrow R_m = R_{TL}\frac{1+\Gamma}{1-\Gamma}$$
$$R_m = 50\frac{1 \pm 0.1667}{1 \mp 0.1667} = \begin{cases} 70 \text{ Ohm} & \Gamma = +0.1667 \\ 35.7143 \text{ Ohm} & \Gamma = -0.1667 \end{cases} \tag{1.83}$$

In terms of power consumption, choosing $R_m = 35.7143$ Ohm makes more sense.

1.2.4 Antenna bandwidth and quality factor

The antenna bandwidth is the frequency band for which the antenna will allow the transmitted or received signal to pass through with minimal attenuation. In essence, the antenna in this case has the passband characteristics of a filter. Define the antenna bandwidth as the difference between the 3-dB frequencies F_{High} and F_{Low} as illustrated in Figure 1.9.

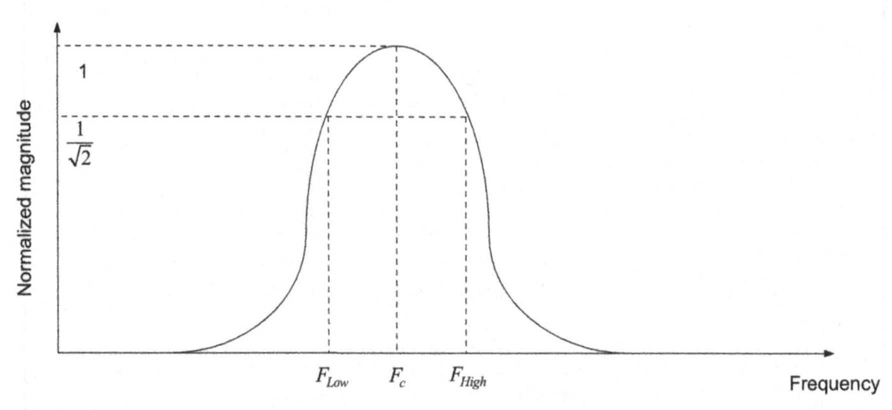

FIGURE 1.9

The 3-dB bandwidth of an antenna.

Another figure of merit that is used to define the performance of an antenna is the bandwidth factor or fractional bandwidth defined in terms of resonant frequency as

$$BW_{fractional} = \frac{F_{High,VSWR} - F_{Low,VSWR}}{F_{center}} \times 100\% \tag{1.84}$$

where $F_{High,VSWR}$ and $F_{Low,VSWR}$ are the upper and lower frequencies at a given antenna VSWR. The center or resonant frequency F_c is defined as the square root of the product

$$F_c = \sqrt{F_{High,VSWR} \times F_{Low,VSWR}} \tag{1.85}$$

It is the frequency where the antenna load appears purely resistive,[f] that is, the impedance is neither capacitive nor inductive, and therefore the imaginary part of the impedance is zero. Recall that in a series or parallel *RLC*-circuit, resonance occurs when $\Omega_c L - 1/\Omega_c C = 0$ or simply $\Omega_c = 2\pi F_c = 1/\sqrt{LC}$.

The quality factor Q could simply be defined as the ratio of maximum energy stored in an inductor L or capacitor C over the energy dissipated in a given cycle. In a series RLC circuit, for example, the quality factor is given as

$$Q = \frac{2\pi F_c L}{R} = \frac{2\pi F_c}{RC} \tag{1.86}$$

The relationship in Eqn (1.86) reveals that the Q of an *RLC*-circuit increases as the resistance decreases and decreases as the resistance increases. This quality factor is known as the unloaded Q. Similarly, for a parallel *RLC*-circuit, the quality factor is given as

$$Q = \frac{R}{2\pi F_c L} = 2\pi F_c RC \tag{1.87}$$

This quality factor is typically called the loaded Q. For an antenna, a more general definition of Q is needed. The general expression for Q is

$$Q \triangleq 2\pi \frac{\text{Energy stored in capacitor or inductor}}{\text{Energy lost per cycle}} \tag{1.88}$$

In more practical terms, and in the context of this chapter, the quality factor is defined as the ratio

$$Q = \frac{F_{center}}{F_{High,VSWR} - F_{Low,VSWR}} = \frac{F_{center}}{BW} \tag{1.89}$$

which is the inverse of the fractional bandwidth. Equation (1.89) implies that a narrowband antenna has a high quality factor whereas a broadband antenna has a low quality factor.

---

[f]That is, the resistance in this case is a combination of loss resistance and radiation resistance.

EXAMPLE 1.3 ANTENNA Q

This is a hypothetical example. Determine the radiation resistance of an antenna operating in the 698–746 MHz band. Assume that the capacitive reactance is in series and is equal to $j200$, $j20$, and $j2$ ohm as shown in Figure 1.10. What is the inductance in each case? What is the impact of the various capacitive reactances on the transmit power?

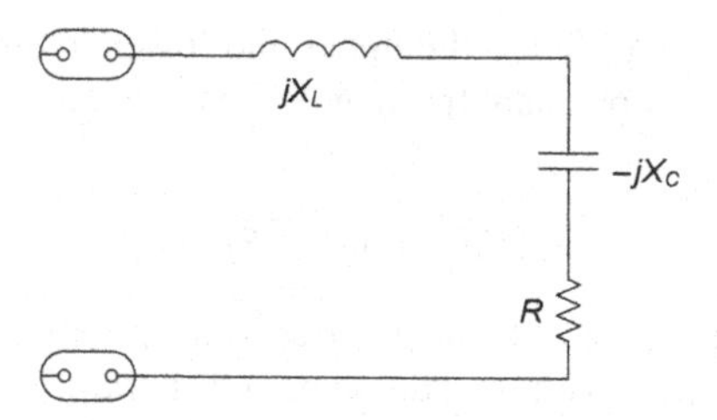

FIGURE 1.10

Schematic for capacitive load antenna.

Assume that the center frequency of the band is the resonant frequency of the antenna, that is, $F_C = (746\text{ MHz} + 698\text{ MHz})/2 = 722\text{ MHz}$ and that the bandwidth of the antenna is $BW = (746\text{ MHz} - 698\text{ MHz}) = 48\text{ MHz}$, the quality factor can then be simply determined from Eqn (1.89) as

$$Q = \frac{F_{center}}{BW} = \frac{722\text{ MHz}}{48\text{ MHz}} = 15.04 \tag{1.90}$$

Furthermore, for a series *RLC*-circuit, the radiation resistance is the ratio of the reactance to the quality factor

$$R_{rad} = \frac{X_L}{Q} = \frac{200\text{ ohm}}{15.04} = 13.3\text{ ohm} \tag{1.91}$$

Note that in order for the circuit to resonate, the inductance must be $+j200$ ohm. Using Eqn (1.91) the radiation resistance for the capacitive reactance of $j20$ ohm and $j2$ ohm cases at resonance is $R = 1.33$ ohm and 0.133 ohm, respectively.

It is important to keep in mind that this is *a hypothetical example*, especially when answering the last question. The antenna circuit depicted in Figure 1.10 obviously is not realistic since it does not exhibit any resistive loss. Having said that, recall that the efficiency of an antenna is the ratio of the radiated power over the total power:

$$\eta = \frac{P_{radiated}}{P_{radiated} + P_{loss}} \tag{1.92}$$

The total power is the sum of the radiated power plus the power lost in the antenna resistance. The radiated power is directly proportional to the radiation resistance, that is $P_{radiated} = |I|^2 R_{Rad}$, and hence higher radiation resistance simply implies higher transmit power.[g]

[g]Note that this is only valid with current-source transmitter.

Another circuit configuration that is of importance is a mixed series—parallel circuit as shown in Figure 1.11. In practice, the coil has a certain resistance associated with it, say R_{coil}. In order to compute the resonance of the circuit depicted in

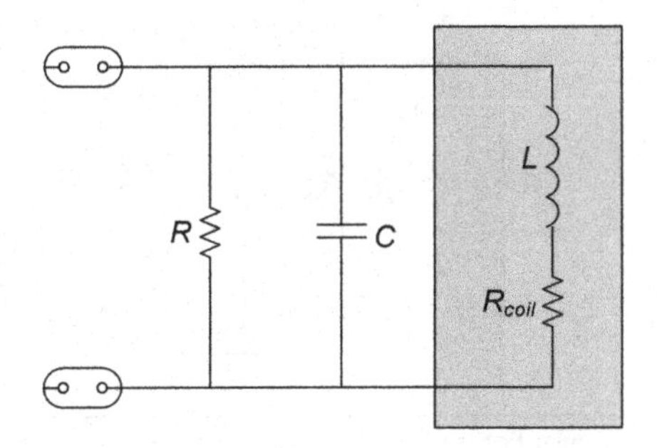

FIGURE 1.11

A mixed series-parallel *RLC*-circuit. (For color version of this figure, the reader is referred to the online version of this book.)

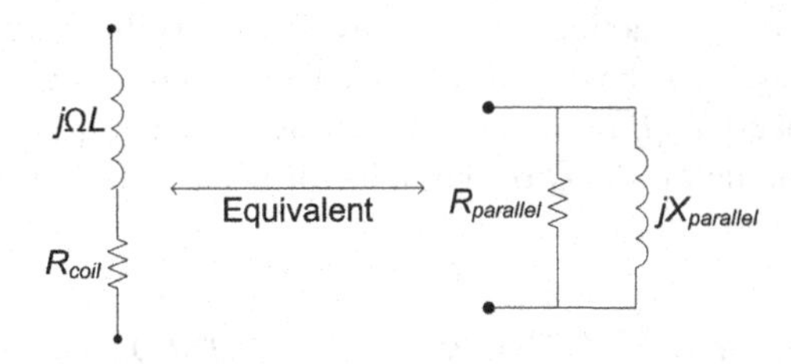

FIGURE 1.12

Parallel equivalent circuit of the model series *L*-*R* circuit of the coil.

Figure 1.11, we first represent the shaded series *L*-*R* circuit of the coil with a parallel equivalent as shown in Figure 1.12. Using the admittance transformation, we can write

$$\frac{1}{R_{parallel}} + \frac{1}{jX_{parallel}} = \frac{1}{R_{coil} + j\Omega L} \tag{1.93}$$

which simply implies that

$$\begin{aligned} R_{parallel} &= \frac{R_{coil}^2 + \Omega^2 L^2}{R_{coil}} \\ X_{parallel} &= \frac{R_{coil}^2 + \Omega^2 L^2}{\Omega L} \end{aligned} \tag{1.94}$$

The equivalent parallel circuit realization of Figure 1.11 is the circuit depicted in Figure 1.13. From circuit theory, we know that the resonant frequency of the circuit in Figure 1.13 is

$$\Omega_c = \sqrt{\frac{1}{LC} - \frac{R_{coil}^2}{L^2}} \tag{1.95}$$

This can be obtained by realizing that the input admittance of Figure 1.13 is

$$Y = \frac{1}{R} + \frac{1}{R_{parallel}} + j\left(\frac{1}{X_C} - \frac{1}{X_{parallel}}\right) \tag{1.96}$$

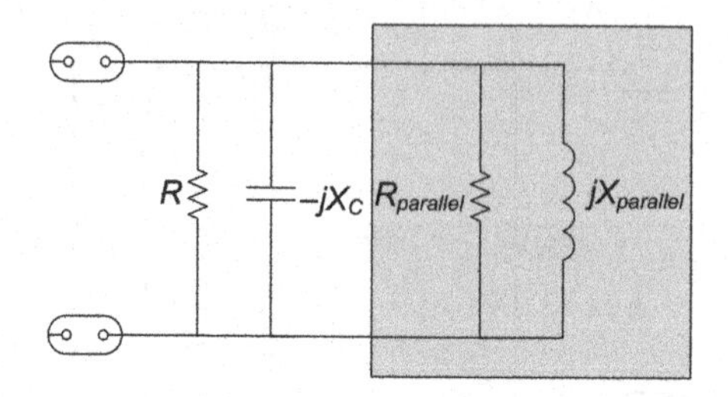

FIGURE 1.13

Equivalent parallel circuit realization of Figure 1.11. (For color version of this figure, the reader is referred to the online version of this book.)

Note that capacitors in practice are also lossy and exhibit a certain amount of resistance. The losses in these practical elements can always be expressed in terms of their respective Q. In general, if a lossless capacitor is connected in parallel with a lossy inductor, the Q of the inductor is the Q of the resulting parallel circuit[h] given as

$$Q_{inductor} = 2\pi F_c CR = \frac{R_{inductor}}{2\pi F_c L} \Rightarrow R_{inductor} = 2\pi F_c L Q_{inductor} = X_L Q_{inductor} \tag{1.97}$$

Similarly, when a lossless inductor is connected to a lossy capacitor that exhibits a certain amount of resistance, the latter, that is the resistance, can be described in terms of the capacitor's Q.

In general, both capacitors and inductors have resistive losses associated with them. Given a parallel RLC circuit, where the inductor and its corresponding resistance, the capacitor and its corresponding resistance, and a load resistance R_{Load} are all placed in parallel, the circuit's total resistance can be expressed as

$$\begin{aligned}\frac{1}{R_{Total}} &= \frac{1}{Q_{inductor}X_L} + \frac{1}{Q_{capacitor}X_C} + \frac{1}{R_{Load}} \\ R_{Total} &= \frac{1}{\frac{1}{Q_{inductor}X_L} + \frac{1}{Q_{capacitor}X_C} + \frac{1}{R_{Load}}}\end{aligned} \tag{1.98}$$

At resonance, the capacitive and inductive loads are equal $X_L = X_C$ and the relationship in Eqn (1.98) becomes

[h]That is placing the inductor and its corresponding resistor and the capacitor in parallel to form a parallel *RLC* circuit.

$$\frac{1}{R_{Total}} = \frac{1}{X_L}\left(\frac{1}{Q_{inductor}} + \frac{1}{Q_{capacitor}}\right) + \frac{1}{R_{Load}} \text{ or}$$
$$\frac{X_L}{R_{Total}} = \left(\frac{1}{Q_{inductor}} + \frac{1}{Q_{capacitor}}\right) + \frac{X_L}{R_{Load}} \tag{1.99}$$

According to Eqn (1.99), the unloaded Q is the Q related to the reactive elements only without the load and, hence, can be found as

$$\frac{1}{Q_{unloaded}} = \frac{1}{Q_{inductor}} + \frac{1}{Q_{capacitor}} \tag{1.100}$$

1.2.5 Matching networks

Thus far, we have limited our discussion of matching circuits to single components. In reality, matching circuits over an entire frequency band or frequency bands coupled with certain VSWR requirements tend to be complex circuit problems. There are several basic circuit topologies that are usually considered by designers. In this section we will examine some of the more common ones. Note that the objective of all matching networks is to ensure the maximum transfer of power to the load over a certain frequency or frequency band. This, in essence, implies that the resistive part of the load is made to match (or equal) that of the source while neutralizing the reactance part. The various matching techniques at the designer's disposal vary in complexity, cost, availability of parts, size, and certainly bandwidth. So in choosing an appropriate matching technique, one would have to take into account these parameters.

1.2.5.1 L-matching networks

We start our discussion with the *L*-matching networks shown in Figure 1.14 and Figure 1.15. The letters S and P designate the components that are in series and in parallel, respectively. In Figure 1.14, the parallel-series or shunt network has

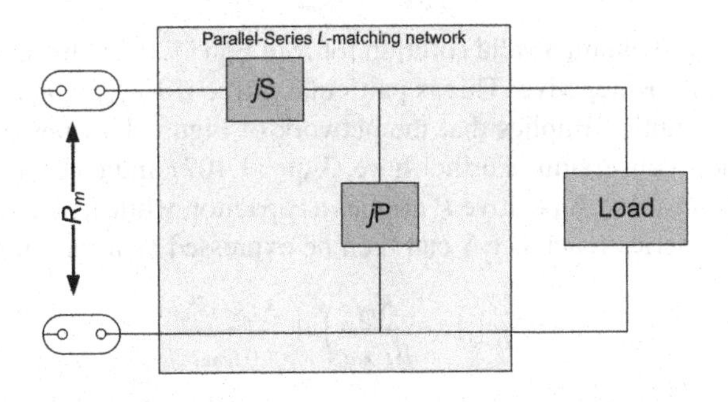

FIGURE 1.14

Parallel-series or shunt *L*-matching network. (For color version of this figure, the reader is referred to the online version of this book.)

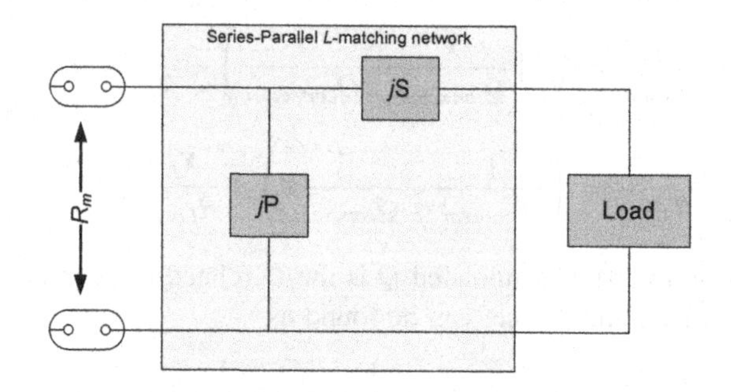

FIGURE 1.15

Series-parallel or simply series L-matching network. (For color version of this figure, the reader is referred to the online version of this book.)

the component P in parallel with the load. On the other hand, the series-parallel or simply series network displayed in Figure 1.15 shows the S component to be placed in series with the load. The components S and P are either capacitors or inductors. To gain an intuitive understanding of these circuits, consider the impedance looking into the matched L-network and load impedance Z_{Load} in Figure 1.14. The load impedance in this discussion refers to the *antenna load.* If the circuits are matched, it implies that the resistance (or real part) of the input impedance R_m is [2]

$$R_m = jS + \frac{1}{jP + 1/Z_{Load}} = jS + \frac{Z_{Load}}{jPZ_{Load} + 1} = \frac{-PSZ_{Load} + jS + Z_{Load}}{jPZ_{Load} + 1} \tag{1.101}$$

Let Z_{Load} be a complex load of the form $Z_{Load} = R_{Load} + jX_{Load}$, then separating the real from the imaginary, we obtain two equations and two unknowns P and S. Hence, solving for P and S in Eqn (1.101), we obtain

$$P = \frac{X_{Load} \pm \left(\sqrt{R_{Load}^2 + X_{Load}^2 - R_m R_{Load}}\right)\left(\sqrt{\frac{R_{Load}}{R_m}}\right)}{R_{Load}^2 + X_{Load}^2} \tag{1.102}$$

Note that in order to obtain a valid solution for P in Eqn (1.102), the argument in the square root cannot be negative. This is particularly true if $R_{Load} > R_m$ for any value of X_{Load}. This condition implies that the network of Figure 1.14 performs a downward impedance conversion. Furthermore, Eqn (1.102) indicates that there are two valid solutions for P. A positive P implies a capacitor, while a negative P implies an inductor. The series reactance S can then be expressed as a function of P as

$$S = \frac{1}{P}\left(1 - \frac{R_m}{R_{Load}}\right) + \frac{X_{Load} R_m}{R_{Load}} \tag{1.103}$$

Similarly, S can take either a positive or a negative value. A positive value implies that S is an inductor, while a negative value implies that S is a capacitor. So there are four possible configurations for Figure 1.14.

Next, consider the series-parallel L-matching network of Figure 1.15. Again, the impedance looking into the matched L network and load impedance Z_{Load} is matched to the circuit if the resistance (or real part) of the input impedance R_m is given as [2]

$$\frac{1}{R_m} = jP + \frac{1}{Z_{Load} + jS} = \frac{-PS + 1 + jPZ_{Load}}{Z_{Load} + jS} = \frac{-PS + 1 - PX_{Load} + jPR_{Load}}{R_{Load} + j(X_{Load} + S)} \tag{1.104}$$

Similar to the previous analysis, separating the real from the imaginary in Eqn (1.104) we obtain two equations and two unknowns P and S

$$S = -X_{Load} \pm \sqrt{R_m R_{Load} - R_{Load}^2} \tag{1.105}$$

The relation in Eqn (1.105) provides two valid solutions for S if $R_m > R_{Load}$. Solving for P as a function of S, we obtain

$$P = \pm \frac{\sqrt{\frac{R_m}{R_{Load}} - 1}}{R_m} \tag{1.106}$$

Again, we observe that the condition $R_m > R_{Load}$ implies that the network of Figure 1.15 performs an upward impedance conversion. Similar to the previous case, S and P can take either positive or negative values. A negative S value implies a capacitor, while a positive S value implies an inductor. Likewise, a negative P value implies an inductor, while a positive P value implies a capacitor. Consequently, there are eight different configurations of Ld-matching networks for Figure 1.15.

The L-matching network provides for two degrees of freedom, namely P and S. For a given center frequency, the circuit Q automatically determines the bandwidth of the system. Furthermore, the transformation ratio can also be determined. The beauty of the L-matching network is its simplicity. However, L-matching networks are limited by the matching frequency and have been effectively used up to 1.3 GHz.

EXAMPLE 1.4 *L*-TYPE MATCHING NETWORK FOR CAPACITIVE ANTENNA LOAD

Assume that the antenna impedance at 900 MHz is $Z_L = 0.3 - j20$ ohm. The antenna is connected to a 50 ohm transmission line. Design an L-network matching circuit.

From our previous discussion, it is obvious that the L-matching network needs to perform an upward impedance conversion. Therefore, given the capacitive nature of the antenna load, we design an L-matching network as depicted in Figure 1.16. In order to obtain the values X_C and X_L we use the relationships expressed in Eqns (1.105) and (1.106). That is

$$\begin{aligned} X_L &= -X_{Load} \pm \sqrt{R_m R_{Load} - R_{Load}^2} = -X_{ant} \pm \sqrt{Z_m R_{ant} - R_{ant}^2} \\ &= 20 \pm \sqrt{50 \times 0.3 - 0.3^2} = \begin{cases} 16.138 \\ 23.86 \end{cases} \text{Ohm} \end{aligned} \tag{1.107}$$

Continued

EXAMPLE 1.4 *L*-TYPE MATCHING NETWORK FOR CAPACITIVE ANTENNA LOAD—cont'd

Note that both solutions are mathematically valid. Next compute X_C according to Eqn (1.106), that is

$$X_C = \frac{\sqrt{\frac{R_m}{R_{Load}} - 1}}{R_m} = \frac{\sqrt{\frac{Z_m}{R_{ant}} - 1}}{Z_m} = \frac{\sqrt{\frac{50}{0.3} - 1}}{50} = 0.2574\,\text{Ohm} \tag{1.108}$$

Now, compute the inductance based on the value of X_L

$$L = \frac{X_L}{2\pi F_c} = \begin{cases} \dfrac{16.138}{2\pi \times 900 \times 10^6} = 2.85\text{nH} & X_L = 16.138\,\text{Ohm} \\ \dfrac{23.86}{2\pi \times 900 \times 10^6} = 4.22\text{nH} & X_L = 23.86\,\text{Ohm} \end{cases} \tag{1.109}$$

Similarly, the capacitance may be computed as

$$C = \frac{1}{2\pi F_c X_C} = \frac{1}{2\pi \times 900 \times 10^6 \times 0.2574} = 687\text{ pF} \tag{1.110}$$

What is the VSWR of this circuit? Hint: use Eqn (1.104) to compute Z_m.

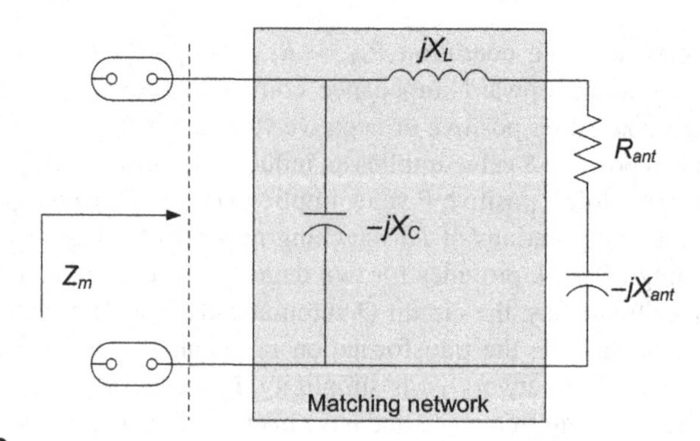

FIGURE 1.16

L-matching network with capacitive antenna load.

EXAMPLE 1.5 *L*-TYPE MATCHING PARALLEL-SERIES RESONANT FREQUENCY AND INPUT RESISTANCE

Consider the parallel-series network depicted in Figure 1.17. Find the resonant frequency of the network and the corresponding input resistance R_m. Does the load resistance impact the resonant frequency?

The input admittance to Figure 1.17 is given as

$$\frac{1}{Z_m} = j\Omega C + \frac{1}{R_{Load} + j\Omega L} = \frac{R_{Load}}{R_{Load}^2 + \Omega^2 L^2} + \underbrace{j\Omega\left(C - \frac{L}{R_{Load}^2 + \Omega^2 L^2}\right)}_{jX} \tag{1.111}$$

EXAMPLE 1.5 *L*-TYPE MATCHING PARALLEL-SERIES RESONANT FREQUENCY AND INPUT RESISTANCE—cont'd

Resonance frequency occurs when the imaginary portion of Eqn (1.111) is zero, that is, the shunt susceptance is $X = 0$ and hence the resonant frequency can be found as

$$\frac{L}{R_{Load}^2 + \Omega_{resonant}^2 L^2} = C \Rightarrow \Omega_{resonant} = \sqrt{\frac{1}{LC} - \frac{R_{Load}^2}{L^2}} \tag{1.112}$$

From Eqn (1.112), the load resistance R_{Load} directly impacts the resonant frequency. Note that at resonance, the reciprocal of the conductance of Eqn (1.111) is none other than the input resistance, that is

$$R_m = \frac{R_{Load}^2 + \Omega_{resonance}^2 L^2}{R_{Load}} = R_{Load} + \frac{\Omega_{resonance}^2 L^2}{R_{Load}} \tag{1.113}$$

Substituting Eqn (1.86) into Eqn (1.113), we obtain

$$R_m = R_{Load}\left(1 + Q^2\right) \tag{1.114}$$

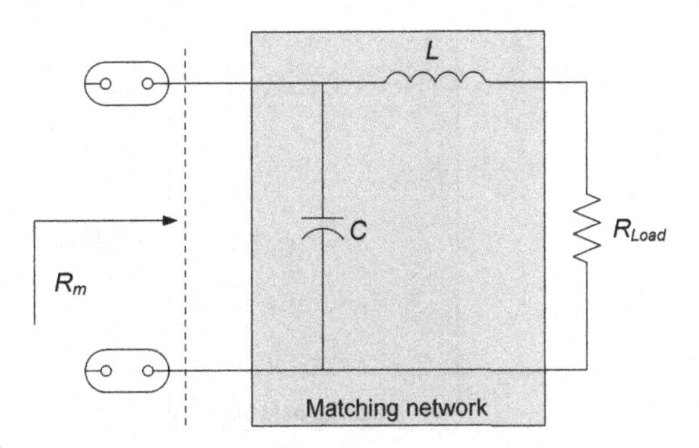

FIGURE 1.17

L-matching parallel-series network.

EXAMPLE 1.6 *L*-TYPE MATCHING SERIES-PARALLEL RESONANT FREQUENCY AND INPUT RESISTANCE

Now consider the shunt network depicted in Figure 1.18. In a manner similar to the previous example, find the resonant frequency of the network and the corresponding input resistance R_m.

The input admittance to Figure 1.18 is given as

$$Z_m = j\Omega L + \frac{1}{1/R_{Load} + j\Omega C} = \frac{1/R_{Load}}{1/R_{Load}^2 + \Omega^2 C^2} + \underbrace{j\Omega\left(L - \frac{C}{1/R_{Load}^2 + \Omega^2 C^2}\right)}_{jX} \tag{1.115}$$

Continued

EXAMPLE 1.6 *L*-TYPE MATCHING SERIES-PARALLEL RESONANT FREQUENCY AND INPUT RESISTANCE—cont'd

In order to achieve resonance, the series reactance would have to equal to zero, or $X = 0$, and the resonant frequency can then be obtained as

$$L = \frac{C}{1/R_{Load}^2 + \Omega^2 C^2} \Rightarrow \Omega_{resonant} = \sqrt{\frac{1}{LC} - \frac{1}{R_{Load}^2 C^2}} \quad (1.116)$$

Substituting Eqn (1.116) into the input admittance relation in Eqn (1.115), we obtain the input resistance as

$$R_m = \frac{R_{Load}}{1 + \Omega^2 C^2 R_{Load}^2} \quad (1.117)$$

Again, to obtain the input resistance in terms of the matching circuit Q, substitute Eqn (1.86) into Eqn (1.117) we obtain

$$R_m = \frac{R_{Load}}{1 + Q^2} \quad (1.118)$$

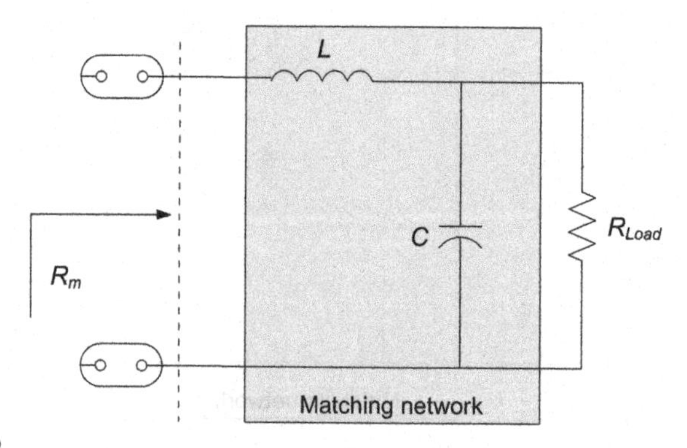

FIGURE 1.18

L-matching parallel-series network.

1.2.5.2 π-matching networks

As mentioned in the previous section, despite the fact than the *L*-matching network can match any arbitrary load to any arbitrary source, the *Q*-factor and matching bandwidth are uniquely determined by the matching elements. The addition of a third element to the matching network, as is the case of the π-matching network[i] depicted in Figure 1.19, provides the necessary flexibility of controlling the bandwidth. As a matter of fact, the matching network can be made arbitrarily narrow.

A π-matching network can be thought of as a shunt *L*-matching network connected to another shunt *L*-matching network joined via the series elements as shown by way

[i]The π-matching network is sometimes referred to as the Δ-matching network.

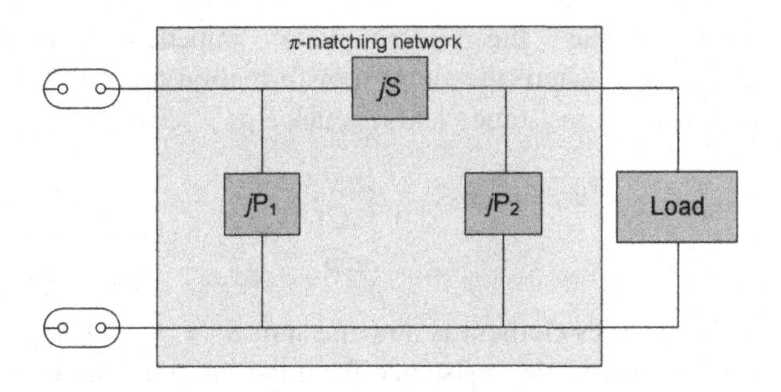

FIGURE 1.19

The π-matching network. (For color version of this figure, the reader is referred to the online version of this book.)

of an example in Figure 1.20. The load resistance undergoes two transformations. The first transformation lowers the load resistance to a certain intermediate resistance at the connection of the two inductances. A second transformation performed by L-matching circuit farthest from the load transforms the intermediate resistance up to a certain input resistance. The importance of the intermediate resistance is simply the fact that it decouples the matching circuit Q from the transformation ratio.

Next, we will show how in this circuit configuration, the Q of the matching network can be chosen as a design parameter. To do so, let us reexamine Figure 1.19. The elements S and P_2 perform the impedance transformation, while P_1 is used as a compensation element to further tune the reactance of S and P_2, respectively. According to the previous discussion on L-matching networks, the shunt element P_2 and the series element S, or at least part of it, serve to reduce the resistance by a factor of $1/(1+Q_2^2)$ as indicated in Eqn (1.118) where Q_2 is the Q-factor with respect to P_2. Likewise, the shunt element P_1 and the series element S serve to increase the resistance by a factor of $1+Q_1^2$ as indicated in Eqn (1.114), where Q_1 is the Q-factor with respect to P_1. The final transformation results in an input resistance R_m that, by design, is either larger or smaller than the load resistance R_{Load}. The answer of course depends on the values of Q_1 and Q_2. For R_m less than R_{Load}, Q_1 must then be less than Q_2. In this case, Q_2 will to a great extent determine the bandwidth of the matching circuit.

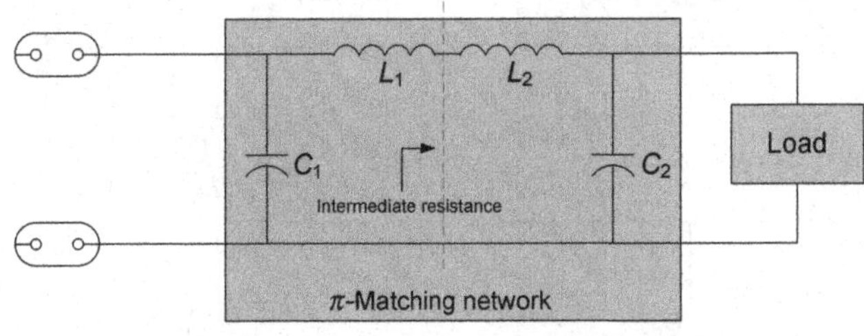

FIGURE 1.20

A π-matching network made-up of two L-matching networks. (For color version of this figure, the reader is referred to the online version of this book.)

In this case, define the intermediate impedance $Z_{intermediate} = R_{intermediate} + jX_{intermediate}$ where the shunt transformation due to P_2 results in the values for the intermediate resistance and reactance as

$$\begin{aligned} R_{intermediate} &= \frac{R_{Load}}{1+Q_2^2} \\ X_{intermediate} &= -Q_2 R_{intermediate} \end{aligned} \tag{1.119}$$

Note that the resulting series element is now the sum $\widehat{S} = S - Q_2 R_{intermediate}$. This series reactance will increase the resistance from the intermediate resistance value according to the relation

$$\begin{aligned} R_m &= R_{intermediate}\left(1+Q_1^2\right) \\ X_m &= -\frac{R_m}{Q_1} \end{aligned} \tag{1.120}$$

where

$$Q_1 = \frac{R_{intermediate}\left(\frac{S}{R_{Load}} - Q_2\right)}{R_m} \tag{1.121}$$

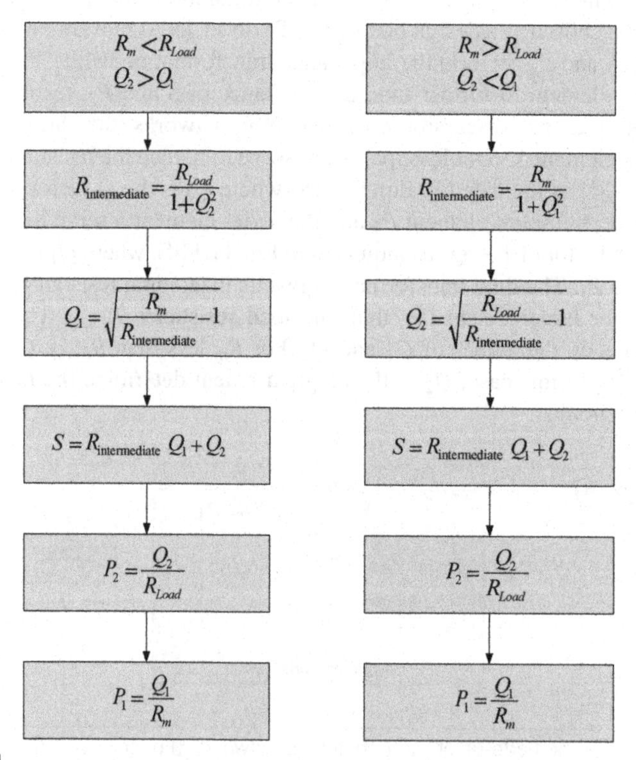

FIGURE 1.21

Design procedure for π-matching networks. (For color version of this figure, the reader is referred to the online version of this book.)

Note that the intermediate resistance can be expressed as

$$R_{\text{intermediate}} = \frac{S}{Q_1 + Q_2} \tag{1.122}$$

The design procedures for $R_m < R_{Load}$ or $R_m > R_{Load}$ are summarized in Figure 1.21. Finally, the π-matching network is ideal when the source and termination parasitics are capacitive, thus enabling them to become part of the matching network.

1.2.5.3 T-matching networks

The *T*-matching network, sometimes referred to as the *Y*-matching network, is shown in Figure 1.22. The matching circuit has a dual topology of the π-matching network, that is, all the parameters of the *T*-matching network can be derived from the π-matching network and vice versa. The *T*-matching network can be thought of as a cascade of two *L*-matching networks. However, in this case, the load resistance is transformed by the series reactance S_2, thus raising the intermediate resistance value $R_{\text{intermediate.}}$ The value of $R_{\text{intermediate}}$ is then lowered with the aid of the shunt susceptance to its final input matching resistance R_m as illustrated in Figure 1.23. In theory then, the *T*-matching network is equivalent to the π-matching network. However, this topology is preferred when the source and termination parasitics are inductive. In this manner, their effect can be absorbed into the network. The design procedure for implementing *T*-matching networks is outlined in Figure 1.24.

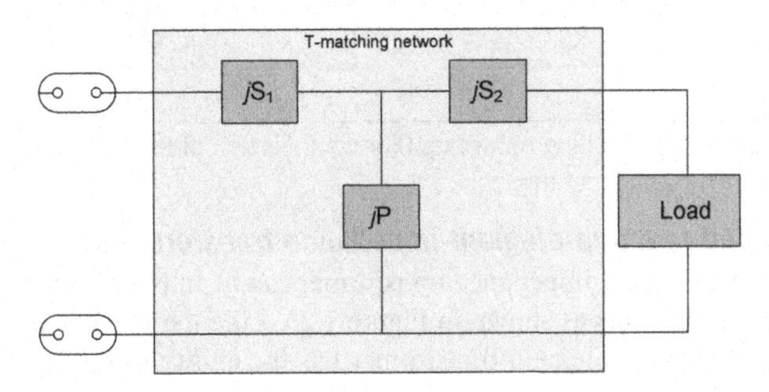

FIGURE 1.22

The *T*-matching network. (For color version of this figure, the reader is referred to the online version of this book.)

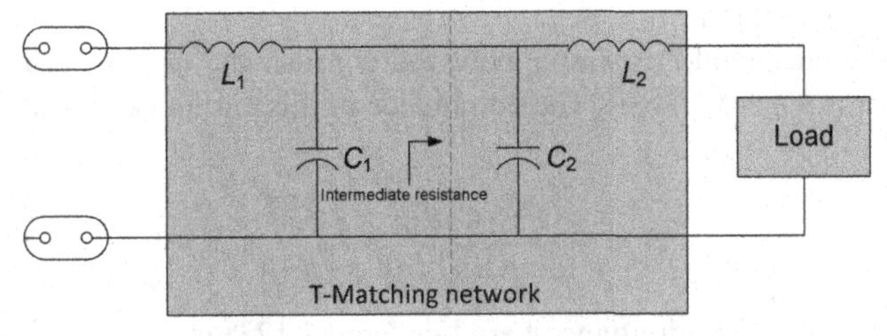

FIGURE 1.23

A *T*-matching network made-up of two *L*-matching networks. (For color version of this figure, the reader is referred to the online version of this book.)

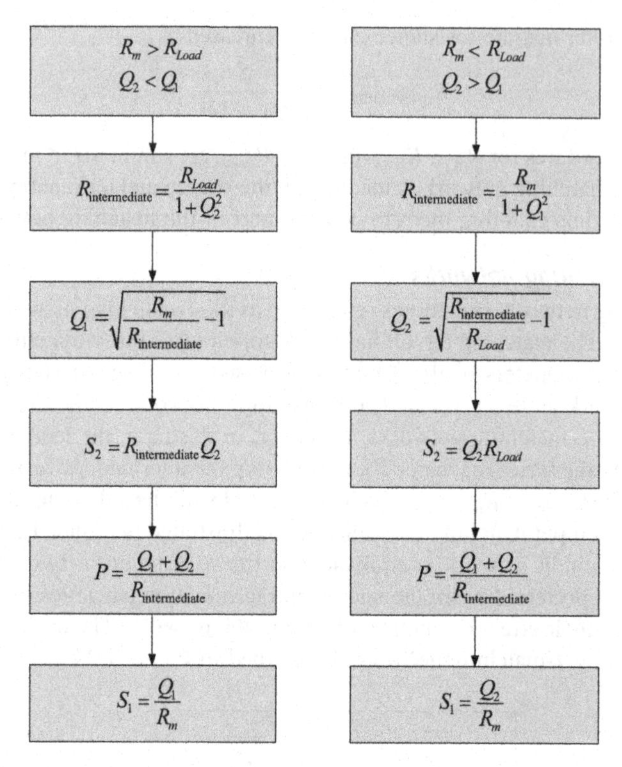

FIGURE 1.24

Design procedure for *T*-matching networks. (For color version of this figure, the reader is referred to the online version of this book.)

1.2.5.4 Tapped reactive-element impedance transformers

A tapped reactive-element impedance transformer can be implemented using capacitive or inductive elements as shown in Figure 1.25. Like the π and *T*-matching networks, a tapped reactive-element transformer has the ability to set the circuit Q as well as the transformation ratio at a given RF. Ideally, this impedance transformer can transform impedances without any loss as opposed to a purely resistive voltage divider implemented using resistors. The latter is especially not desirable at RFs. In reality, however, all capacitive and inductive components have certain resistive losses associated with them.

To further understand this simple network, consider the tapped capacitor transformer shown in Figure 1.25(a). The admittance of this circuit can be simply found as

$$Y = \frac{-\Omega^2 C_1 C_2 R_{Load} + j\Omega C_1}{1 + j\Omega R_{Load}(C_1 + C_2)} \tag{1.123}$$

The resistive part of the admittance based on Eqn (1.123) is

$$\mathrm{Re}\{Y\} = \frac{\Omega^2 C_1^2 R_{Load}}{1 + \Omega^2 R_{Load}^2 (C_1 + C_2)^2} \tag{1.124}$$

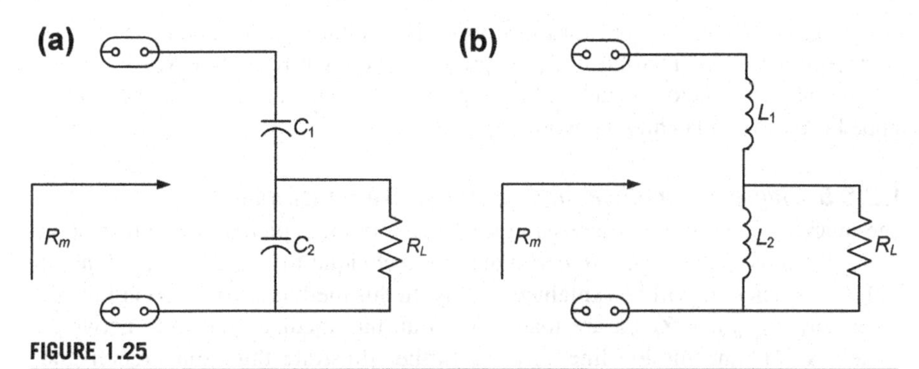

FIGURE 1.25

Tapped reactive-element impedance transformer: (a) tapped capacitor transformer, and (b) tapped inductor transformer.

While the reactive part can be found as

$$\mathrm{Im}\{Y\} = \frac{\Omega C_1 + \Omega^3 R_{Load}^2 C_1 C_2 (C_1 + C_2)}{1 + \Omega^2 R_{Load}^2 (C_1 + C_2)^2} \tag{1.125}$$

For a large value of Ω that is at very high frequency, the expression in Eqns (1.124) and (1.125) can be reduced to

$$\begin{aligned}\mathrm{Re}\{Y\} &\approx \frac{\Omega^2 C_1^2 R_{Load}}{\Omega^2 R_{Load}^2 (C_1 + C_2)^2} \qquad \text{for } \Omega^2 R_{Load}^2 (C_1 + C_2)^2 \gg 1 \\ &= \frac{1}{R_{Load}} \frac{C_1^2}{(C_1 + C_2)^2} = R_m\end{aligned} \tag{1.126}$$

And

$$\begin{aligned}\mathrm{Im}\{Y\} &\approx \frac{\Omega C_1 + \Omega^3 R_{Load}^2 C_1 C_2 (C_1 + C_2)}{\Omega^2 R_{Load}^2 (C_1 + C_2)^2} \qquad \text{for } \Omega^2 R_{Load}^2 (C_1 + C_2)^2 \gg 1 \\ \mathrm{Im}\{Y\} &= \frac{C_1 + \Omega^2 R_{Load}^2 C_1 C_2 (C_1 + C_2)}{\Omega R_{Load}^2 (C_1 + C_2)^2} \\ &= \Omega \frac{C_1 C_2}{(C_1 + C_2)} \qquad \text{for } \Omega^2 R_{Load}^2 C_1 C_2 (C_1 + C_2) \gg C_1\end{aligned} \tag{1.127}$$

The expected resistive impedance transformation according to Eqn (1.126) is then

$$R_m = \left(1 + \frac{C_2}{C_1}\right)^2 R_{Load} \tag{1.128}$$

In summary, the tapped capacitor transformer is a voltage divider and hence serves to *transform* the load resistance R_{Load} upward to the input resistance R_m.

The tapped inductor–matching network works in an analogous manner to the tapped capacitor–matching network.

1.2.5.5 Single and multisection quarter-wave transformers

The quarter-wave or $\lambda/4$ transmission line transformer, depicted in Figure 1.26 is by far the most popular *distributed*-matching technique that enables matching at high frequencies, as will be explained shortly. In this method, two different impedances, say Z_{input} and Z_{Load}, are matched via an intermediate matching network, namely a $\lambda/4$ transmission line $Z_{\lambda/4}$. To further illustrate this point, recall that the input impedance of a lossless transmission line of length $\lambda/4$ can be found as given in Eqn (1.68) as

$$\begin{aligned} Z_{input} &= \lim_{\beta l \to \frac{\pi}{2}} Z_{\lambda/4} \frac{Z_{Load} + jZ_{\lambda/4}\tan(\beta l)}{Z_{\lambda/4} + jZ_{Load}\tan(\beta l)}\Bigg|_{\beta l = \frac{2\pi}{\lambda}\frac{\lambda}{4} = \frac{\pi}{2}} \\ &= \lim_{\beta l \to \frac{\pi}{2}} Z_{\lambda/4} \frac{Z_{Load} + jZ_{\lambda/4}\tan\left(\frac{\pi}{2}\right)}{Z_{\lambda/4} + jZ_{Load}\tan\left(\frac{\pi}{2}\right)} = \lim_{\beta l \to \frac{\pi}{2}} Z_{\lambda/4} \frac{\dfrac{Z_{Load}}{\tan\left(\frac{\pi}{2}\right)} + jZ_{\lambda/4}}{\dfrac{Z_{\lambda/4}}{\tan\left(\frac{\pi}{2}\right)} + jZ_{Load}} = \frac{Z_{\lambda/4}^2}{Z_{Load}} \end{aligned} \tag{1.129}$$

This simply implies that

$$Z_{\lambda/4} = \sqrt{Z_{input} Z_{Load}} \tag{1.130}$$

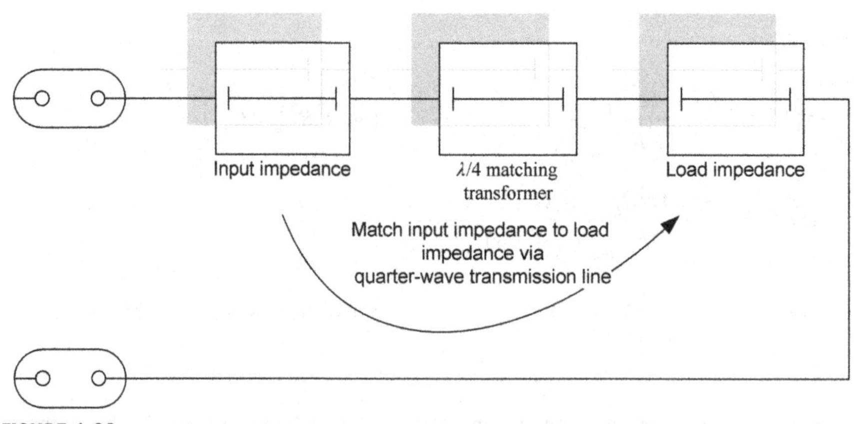

FIGURE 1.26

Single section quarter-wave matching transformer.

where $Z_{\lambda/4}$ is the characteristic impedance of the matching circuit. Note that the purpose of the matching circuit is to match the load impedance Z_{Load} to the input impedance Z_{input}, thus implying that $\Gamma = 0$ looking into the quarter-wave–matching circuit. This further implies, according to Eqn (1.129), that the load impedance gets transformed into its reciprocal value normalized by the square of the matching circuit impedance. This is true, however, at only one frequency. At this one frequency where $\beta l = \pi/2$, Z_{input} is equal to the characteristic impedance Z_0 of the transmission line. And hence, there is no reflection beyond this point toward the generator. However, at frequencies where $\beta l \neq \pi/2$, the reflections toward the generator are not zero, and the reflection coefficient can then be computed using Eqn (1.129) as

$$\Gamma_{in} = \frac{Z_{input} - Z_0}{Z_{input} + Z_0} = \frac{Z_{\lambda/4}\dfrac{Z_{Load} + jZ_{\lambda/4}\tan(\beta l)}{Z_{\lambda/4} + jZ_{Load}\tan(\beta l)} - Z_0}{Z_{\lambda/4}\dfrac{Z_{Load} + jZ_{\lambda/4}\tan(\beta l)}{Z_{\lambda/4} + jZ_{Load}\tan(\beta l)} + Z_0}$$

$$= \frac{Z_{\lambda/4}(Z_{Load} - Z_0) + j\tan(\beta l)\left(Z_{\lambda/4}^2 - Z_{Load}Z_0\right)}{Z_{\lambda/4}(Z_{Load} + Z_0) + j\tan(\beta l)\left(Z_{\lambda/4}^2 + Z_{Load}Z_0\right)} \tag{1.131}$$

From Eqn (1.131), we can compute the magnitude of the reflection coefficient as

$$|\Gamma_{in}| = \frac{Z_{Load} - Z_0}{\sqrt{(Z_{Load} + Z_0)^2 + 4Z_{Load}Z_0\tan^2(\beta l)}} \tag{1.132}$$

Assume that F_m is the frequency for which $\beta l = \pi/2$, then for a given frequency F_c such that $\beta l = (\pi/2)(F_c/F_m)$, we can write based on Eqn (1.132)

$$\frac{F_c}{F_m} = \frac{2}{\pi}\tan^{-1}\left\{\sqrt{\frac{(Z_{Load} - Z_0)^2 - |\Gamma_{in}|^2(Z_{Load} + Z_0)^2}{4Z_{Load}Z_0}}\right\} \tag{1.133}$$

The relationship in Eqn (1.133) is a design equation from which one can determine a set of frequencies where the magnitude of the reflection coefficient does not exceed a certain given value. Note that the impedance matching method presented above is limited to real load impedances only. One such load is a resonant antenna. A resonant antenna is typically a quarter-wavelength long or a multiple thereof where the reactance of the antenna itself is zero. This is not practical for most mobile and portable devices where the antenna is electrically small. These types of antennas are typically capacitive with low radiation resistance. In this case, an inductive element is added to cancel the capacitive reactance.

An extension of the single quarter-wave transformer is the multisection transformer comprised of say N-quarter-wave sections connected between the load and a transmission line with characteristic impedances Z_{Load} and Z_0, respectively. All

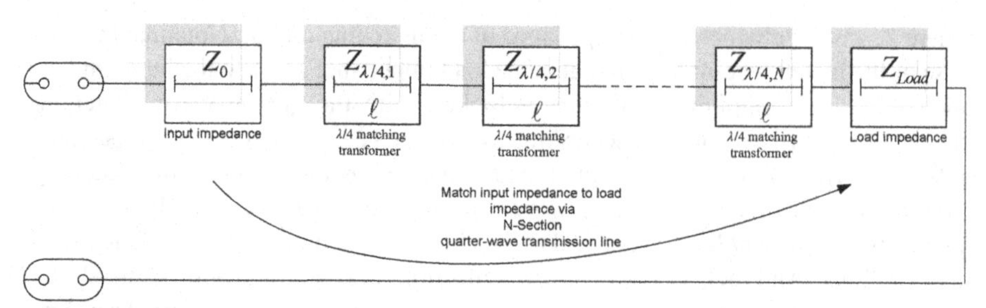

FIGURE 1.27

Multisection quarter-wave—matching transformer.

individual sections have the same length but with varying characteristic impedances, as depicted in Figure 1.27. The input impedance of the N-section quarter-wave lossless transmission line transformer can be found according to the simplified relationship

$$Z_{input}^{N} = Z_{N,\lambda/4}\frac{e^{j\beta l} + \Gamma_N e^{-j\beta l}}{e^{j\beta l} - \Gamma_N e^{-j\beta l}} \tag{1.134}$$

where

$$\Gamma_N = \frac{Z_{Load} - Z_{N,\lambda/4}}{Z_{Load} + Z_{N,\lambda/4}} \tag{1.135}$$

The reflection coefficient between the Nth and $(N-1)^{th}$ section is given as the ratio

$$\Gamma_{N-1} = \frac{Z_{N,\lambda/4} - Z_{N-1,\lambda/4}}{Z_{N,\lambda/4} + Z_{N-1,\lambda/4}} \tag{1.136}$$

Then the reflection coefficient as seen from the $(N-1)^{th}$ section is simply

$$\begin{aligned}\gamma_{N-1} &= \frac{Z_{input}^{N} - Z_{N-1,\lambda/4}}{Z_{input}^{N} + Z_{N-1,\lambda/4}}\\ &= \frac{\left(Z_{N,\lambda/4} - Z_{N-1,\lambda/4}\right)e^{j\beta l} + \Gamma_N\left(Z_{N,\lambda/4} + Z_{N-1,\lambda/4}\right)e^{-j\beta l}}{\left(Z_{N,\lambda/4} + Z_{N-1,\lambda/4}\right)e^{j\beta l} + \Gamma_N\left(Z_{N,\lambda/4} - Z_{N-1,\lambda/4}\right)e^{-j\beta l}}\\ &= \frac{\Gamma_{N-1} + \Gamma_N e^{-j2\beta l}}{1 + \Gamma_{N-1}\Gamma_N e^{-j2\beta l}}\end{aligned} \tag{1.137}$$

It is interesting to note that if $1 \gg \Gamma_{N-1}\Gamma_N$, that is Γ_N and Γ_{N-1} are small, then the denominator in Eqn (1.137) can be approximated simply as $1 + \underbrace{\Gamma_{N-1}\Gamma_N}_{1\gg\Gamma_{N-1}\Gamma_N} e^{-j2\beta l} \approx 1$, and hence

$$\gamma_{N-1} \approx \Gamma_{N-1} + \Gamma_N e^{-j2\beta l} \tag{1.138}$$

Note that the assumption that Γ_N and Γ_{N-1} are small implies that Z_{Load} is close to both $Z_{N,\lambda/4}$ and $Z_{N-1,\lambda/4}$. This approximation can be applied to subsequent sections if the same assumption concerning the reflection coefficients can be made.

EXAMPLE 1.7 QUARTER-WAVE TRANSFORMER

Match a 50 ohm transmission line to a 40.5 ohm antenna using a $\lambda/4$ impedance matching transformer. Assume that the frequency for which the matching circuit is $\lambda/4$ is F_m. What is the magnitude of the reflection coefficient at the frequency $F_c = F_m/2$?

The characteristic impedance of the matching circuit can be found using the relationship obtained in Eqn (1.130)

$$Z_{\lambda/4} = \sqrt{Z_0 Z_{Load}} = \sqrt{50 \times 40.5} = 45 \text{ ohm} \tag{1.139}$$

where Z_0 is the impedance of the transmission line. Recall that βl in Eqn (1.129) is related to the matching frequency as

$$\beta l = \left(\frac{2\pi}{\lambda}\right)\left(\frac{\lambda_m}{4}\right) = \left(\frac{2\pi F}{C_{light}}\right)\left(\frac{C_{light}}{4F_m}\right) = \frac{\pi F}{2F_m} \tag{1.140}$$

where $\lambda_m = C_{light}/F_m$ and C_{light} is the speed of light. Now for $F = F_c = F_m/2$, the βl in Eqn (1.140) becomes $\pi F_c/2F_m = \pi F_m/4F_m = \pi/4$, then the input impedance, which is *dependent on the frequency*, according to Eqn (1.129) becomes

$$\begin{aligned} Z_{input} &= Z_{\lambda/4}\left.\frac{Z_{Load} + jZ_{\lambda/4}\tan(\beta l)}{Z_{\lambda/4} + jZ_{Load}\tan(\beta l)}\right|_{\beta l=\frac{\pi}{4}} = Z_{\lambda/4}\frac{Z_{Load} + jZ_{\lambda/4}\tan\left(\frac{\pi}{4}\right)}{Z_{\lambda/4} + jZ_{Load}\tan\left(\frac{\pi}{4}\right)} = Z_{\lambda/4}\frac{Z_{Load} + jZ_{\lambda/4}}{Z_{\lambda/4} + jZ_{Load}} \\ &= 44.7514 + j4.7238 \end{aligned} \tag{1.141}$$

The magnitude of the reflection coefficient then becomes

$$|\Gamma_{in}| = \left|\frac{Z_{input} - Z_0}{Z_{input} + Z_0}\right| = \left|\frac{44.7514 + j4.7238 - 50}{44.7514 + j4.7238 + 50}\right| = 0.0744 \tag{1.142}$$

And consequently, the VSWR is

$$VSWR = \frac{1 + |\Gamma|}{1 - |\Gamma|} = 1.1608 \tag{1.143}$$

Next, we examine a scenario where the matching circuit happens to be less than $\lambda/4$. If the load impedance is an open circuit, that is if $Z_{Load} = \infty$, the relationship in Eqn (1.129) becomes

$$\begin{aligned} Z_{input} &= \lim_{Z_{Load}\to\infty} Z_{\lambda/4}\frac{Z_{Load} + jZ_{\lambda/4}\tan(\beta l)}{Z_{\lambda/4} + jZ_{Load}\tan(\beta l)} \\ &= \lim_{Z_{Load}\to\infty} Z_{\lambda/4}\frac{1 + j\frac{Z_{\lambda/4}}{Z_{Load}}\tan(\beta l)}{\frac{Z_{\lambda/4}}{Z_{Load}} + j\tan(\beta l)} = -jZ_{\lambda/4}\cot(\beta l) \end{aligned} \tag{1.144}$$

There are two characteristics that we must note according to Eqn (1.144) for a transmission line that is less than a quarter wave. First, the relation in Eqn (1.144) is imaginary and hence Z_{input} is a pure reactance. And second, the transmission line is capacitive. In a like manner, assume now that the load impedance is a short circuit that is $Z_{Load} = 0$, then Eqn (1.129) becomes

$$Z_{input} = Z_{\lambda/4} \frac{Z_{Load} + jZ_{\lambda/4}\tan(\beta l)}{Z_{\lambda/4} + jZ_{Load}\tan(\beta l)}\bigg|_{Z_{Load}=0} = jZ_{\lambda/4}\tan(\beta l) \tag{1.145}$$

Again, from Eqn (1.145) we deduce that for a transmission line that is less than a quarter wave, the Z_{input} is a reactance and that it is inductive.

EXAMPLE 1.8 HALF-WAVE TRANSFORMER

What is the input impedance of a lossless half-wave transmission line transformer?

The answer is simple. For $\beta l = \left(\frac{2\pi}{\lambda}\right)\left(\frac{\lambda}{2}\right) = \pi$ the relationship in Eqn (1.129) becomes

$$Z_{input} = Z_{\lambda/4} \frac{Z_{Load} + jZ_{\lambda/4}\tan(\beta l)}{Z_{\lambda/4} + jZ_{Load}\tan(\beta l)}\bigg|_{\beta l=\pi} = Z_{\lambda/4} \frac{Z_{Load} + jZ_{\lambda/4}\tan(\pi)}{Z_{\lambda/4} + jZ_{Load}\tan(\pi)} = Z_{Load} \tag{1.146}$$

This is an interesting transformer since it can match the input to the load (called 1:1 transformer). However, this transformer has a limited bandwidth capability. Increasing the bandwidth can be done via a multisection quarter-wave transformer.

1.2.6 Transformers

Transformers in general are used for various applications in RF circuits. Some of these applications are impedance transformation, power combining for amplifiers, power splitting, phase inversion and phase shifting, converting single-ended circuits to differential circuits and vice versa (baluns), and balanced mixers.

A transformer, by definition, is an RF device that uses magnetic coupling to couple energy from one circuit to another. Coupling implies that a current running through one wire will induce a current in the second wire. The coupling strength depends on the medium separating them. One type of medium may have a lower resistance or reluctance than another. In order to better understand the transformer and its parameters, consider the ideal transformer circuit depicted in Figure 1.28(a). The mutual inductance M is the coupling measure between two inductors. It is present on the primary (input) side of the transformer as well as the secondary (output) side of the transformer. The self inductances are L_1 on the primary side and L_2 on the secondary side. The input and output voltages as referred to in Figure 1.28(b) can be found as

$$\begin{aligned} V_{in}(\Omega) &= j\Omega L_1 \vec{I}_{in}(\Omega) \pm j\Omega M \vec{I}_{out}(\Omega) \text{ Self Voltage} \\ V_{out}(\Omega) &= j\Omega L_2 \vec{I}_{out}(\Omega) \pm j\Omega M \vec{I}_{in}(\Omega) \text{ MutualVoltage} \end{aligned} \tag{1.147}$$

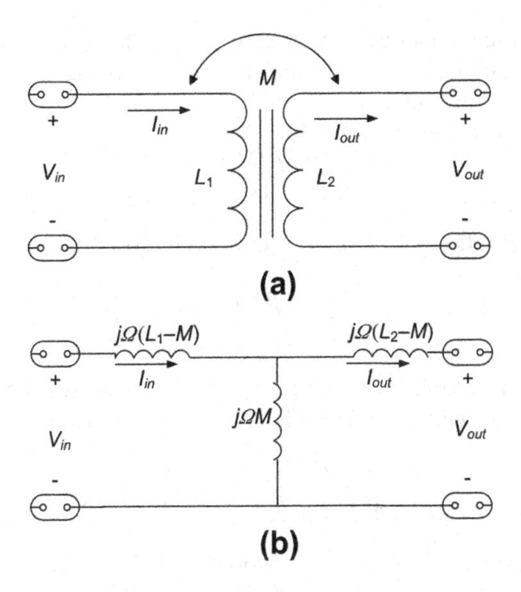

FIGURE 1.28

The ideal transformer (a) circuit symbol, and (b) equivalent *T*-type circuit model.

Similarly, the relationship in Eqn (1.147) can be expressed in the more traditional time-domain expression

$$\begin{aligned} v_{in}(t) &= L_1 \frac{\mathrm{d}\vec{i}_{in}(t)}{\mathrm{d}t} \pm M \frac{\mathrm{d}\vec{i}_{out}(t)}{\mathrm{d}t} \\ v_{out}(t) &= L_2 \frac{\mathrm{d}\vec{i}_{out}(t)}{\mathrm{d}t} \pm M \frac{\mathrm{d}\vec{i}_{in}(t)}{\mathrm{d}t} \end{aligned} \tag{1.148}$$

Note that the sign $\pm$ in front of the mutual inductance terms in Eqns (1.147) and (1.148) is dependent on the direction of the windings in a typical transformer. Conventionally, if the currents $\vec{I}_{in}$ and $\vec{I}_{out}$ go into the coils L_1 and L_2, that is $\vec{I}_{out}$ flows in the opposite direction in Figure 1.28(a) while $\vec{I}_{in}$ stays the same, the sign due to the mutual term in Eqn (1.147) is positive. If the currents flow in the same direction as depicted in Figure 1.28(a), that is $\vec{I}_{in}$ flows in the direction of the coil L_1 whereas $\vec{I}_{out}$ flows away from it, then the sign due to the mutual term in Eqn (1.147) is negative. To denote this convention, the currents in Eqn (1.147) will be expressed in vector notation.

Next, define the coupling coefficient k as the ratio

$$k = \frac{M}{\sqrt{L_1 L_2}} \tag{1.149}$$

The coupling coefficient can be simply derived as follows. First, let us obtain the input coupling coefficient k_i from Eqn (1.147) by forcing the input current to zero, that is, $\overrightarrow{I}_{in} = 0$ we obtain

$$\begin{aligned} V_{in} &= j\Omega M \overrightarrow{I}_{out} \\ V_{out} &= j\Omega L_2 \overrightarrow{I}_{out} \end{aligned} \tag{1.150}$$

And hence k_i is the ratio of V_{in} over V_{out} in Eqn (1.150), or

$$k_i = \frac{V_{in}}{V_{out}} = \frac{j\Omega M \overrightarrow{I}_{out}}{j\Omega L_2 \overrightarrow{I}_{out}} = \frac{M}{L_2} \tag{1.151}$$

Similarly, the output coupling coefficient k_o can be obtained from Eqn (1.147) by forcing the output current to zero, that is, $\overrightarrow{I}_{out} = 0$ and hence we obtain

$$\left.\begin{aligned} V_{in} &= j\Omega L_1 \overrightarrow{I}_{in} \\ V_{out} &= j\Omega M \overrightarrow{I}_{in} \end{aligned}\right\} \Rightarrow k_o = \frac{V_{out}}{V_{in}} = \frac{M}{L_1} \tag{1.152}$$

Finally, the coupling coefficient is expressed as $k = \sqrt{k_i k_o} = M/\sqrt{L_1 L_2}$ as already given in Eqn (1.149). An ideal transformer has $k = 1$, that is $M = \sqrt{L_1 L_2}$. Realistically, in integrated circuits this figure could be as low as 0.54 and as high as 0.9.

The T-circuit model of Figure 1.28(b) is a simplified model of an actual transformer. This model does not feature any DC isolation, resistive or capacitive parasitics, or the transformer turn ratios. First we address the transformer turns ratio. In order to keep with traditional transformer notation, we define the transformer turn ratio as the ratio of two numbers. In an ideal transformer ($k = 1$), the turn ratio is proportional to the voltage as well as the inductances

$$\left|\frac{V_{out}}{V_{in}}\right| = \frac{N_2}{N_1} = \sqrt{\frac{L_2}{L_1}} = \frac{1}{\gamma} \tag{1.153}$$

and is inversely proportional to the current

$$\left|\frac{\overrightarrow{I}_{out}}{\overrightarrow{I}_{in}}\right| = \frac{N_1}{N_2} = \sqrt{\frac{L_1}{L_2}} = \gamma \tag{1.154}$$

The parameter γ is known as the turn ratio. Define the impedances $Z_{in} = V_{in}/\overrightarrow{I}_{in}$ and $Z_{out} = V_{out}/\overrightarrow{I}_{out}$, then using the relationships given in Eqns (1.153) and (1.154), we can simply relate Z_{in} to Z_{out}, that is

$$Z_{in} = \frac{V_{in}}{\overrightarrow{I}_{in}} = \frac{\frac{N_1}{N_2} V_{out}}{\frac{N_2}{N_1} \overrightarrow{I}_{out}} = \left(\frac{N_1}{N_2}\right)^2 \frac{V_{out}}{\overrightarrow{I}_{out}} = \left(\frac{N_1}{N_2}\right)^2 Z_{out} = \left(\frac{L_1}{L_2}\right)^2 Z_{out} = \gamma^2 Z_{out} \tag{1.155}$$

The result presented in Eqn (1.155) is intriguing since it implies that the reflected impedance from the secondary Z_{in} is directly proportional to Z_{out} and vice versa.

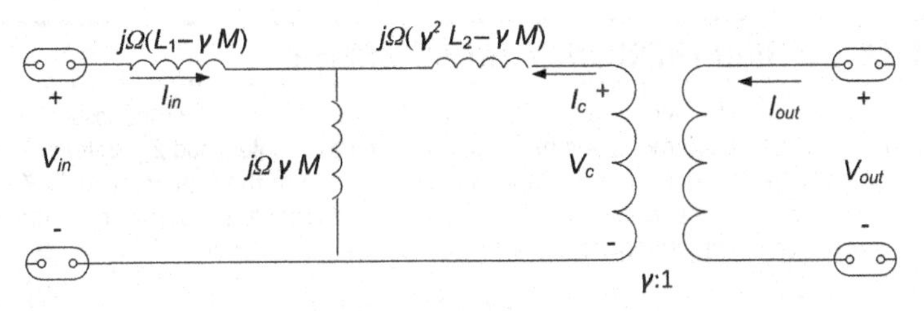

FIGURE 1.29

Ideal transformer model incorporating a turn ratio γ.

Thus, in RF systems, a transformer may be used to perform *impedance transformation*, thus manipulating impedances and matching circuits.

At this point, it is important to address the *T-type* circuit model depicted in Figure 1.28(b). Note that in this model, the turn ratio γ as well as the coupling coefficient k are not featured in the circuit but rather indirectly incorporated in the values of the components. Furthermore, the shared connection exhibited by the mutual inductance does not exhibit the inherent DC isolation in the actual component. A more realistic model that incorporates an ideal transformer showing the turn ratio γ is depicted in Figure 1.14. The voltage–current relationships in this case can be simply expressed as

$$\begin{aligned} V_{in}(\Omega) &= j\Omega L_1 \vec{I}_{in}(\Omega) + j\Omega M\gamma \vec{I}_C(\Omega) \\ V_C(\Omega) &= j\Omega L_2\gamma^2 \vec{I}_C(\Omega) + j\Omega M\gamma \vec{I}_{in}(\Omega) \end{aligned} \tag{1.156}$$

In this case, the $\pm$ sign was eliminated in Eqn (1.156) to reflect the direction of the current I_C shown in Figure 1.29. A more comprehensive model that incorporates a nonideal transformer with coupling ratio $k \neq 1$ is further shown in Figure 1.30.

At this point, it is important to mention the parasitic effects that exist in transformers. This topic may seem out of place in an antenna chapter, however, it is important to note that resistive and capacitive parasitics can greatly influence the transformer matching characteristics at high frequency and adversely affect performance.

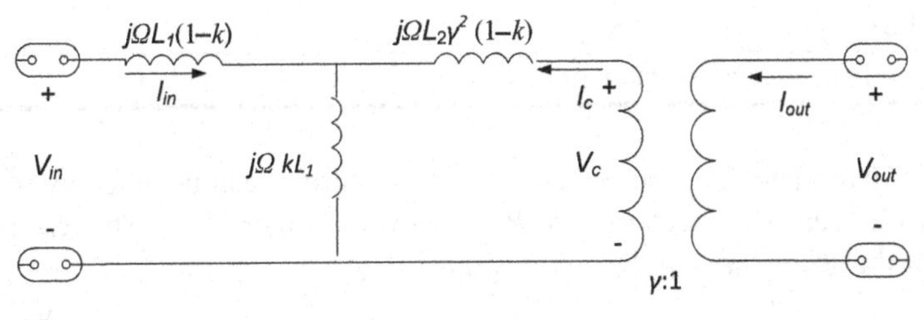

FIGURE 1.30

Nonideal transformer model incorporating a turn ratio γ and coupling factor k.

EXAMPLE 1.9 HIGH FREQUENCY TRANSFORMER

Consider the ideal transformer circuit depicted in Figure 1.31. Let the wiring losses be given as R_{in} and R_{out}. This transformer has a tuned primary and a complex load Z_L. Determine the input impedance. Next, determine the impedance reflected into the primary side. If Z_L is capacitive, would the reflective impedance into the primary appear capacitive or inductive?

First, consider the loop equations according to Eqn (1.147). Then

$$V_{in} = (R_{in} + j\Omega L_1)\vec{I}_{in} - j\Omega M\vec{I}_{out} \tag{1.157}$$

and

$$0 = (R_{out} + Z_L)\vec{I}_{out} - j\Omega M\vec{I}_{in} \tag{1.158}$$

Solve for $\vec{I}_{out}$ in Eqn (1.157), we obtain

$$\vec{I}_{out} = \frac{j\Omega M}{R_{out} + Z_L}\vec{I}_{in} \tag{1.159}$$

Substituting $\vec{I}_{out}$ in Eqn (1.158), we obtain

$$\frac{V_{in}}{\vec{I}_{in}} = Z_{in} = (R_{in} + j\Omega L_1) + \left(\frac{\Omega^2 M^2}{R_{out} + Z_L}\right) \tag{1.160}$$

The second term in Eqn (1.160) is the impedance reflected or transformed into the primary side, namely

$$Z_{in,reflected} = \frac{\Omega^2 M^2}{R_{out} + Z_L} \tag{1.161}$$

Let the load Z_L be capacitive, that is $Z_L = R_L - jX_L$, then the reflected load in Eqn (1.161) becomes

$$Z_{in,reflected} = \frac{\Omega^2 M^2}{(R_{out} + R_L) - jX_L} \tag{1.162}$$

which is obviously capacitive.

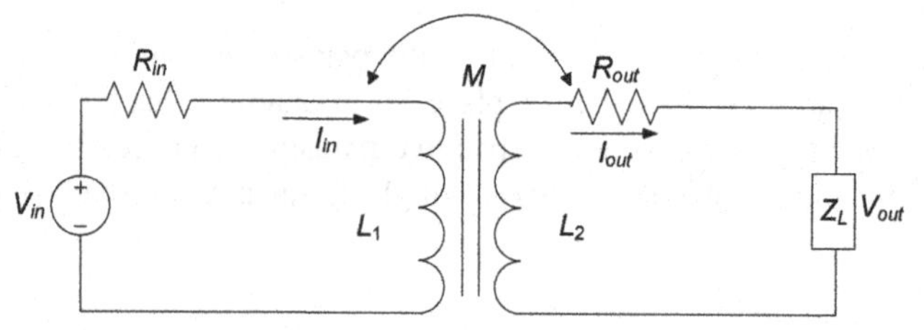

FIGURE 1.31

Single-frequency tuned transformer circuit.

In a similar manner, the secondary of the transformer can be tuned by placing a capacitor C_{out} across the load resistor R_L as shown in Figure 1.32. The resonant frequency of the secondary can be computed as

$$F_c = \frac{1}{2\pi\sqrt{L_2 C_{out}}} \tag{1.163}$$

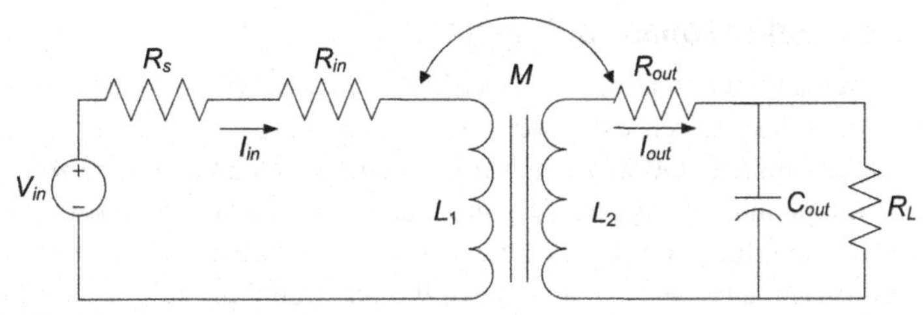

FIGURE 1.32

Secondary-tuned transformer circuit.

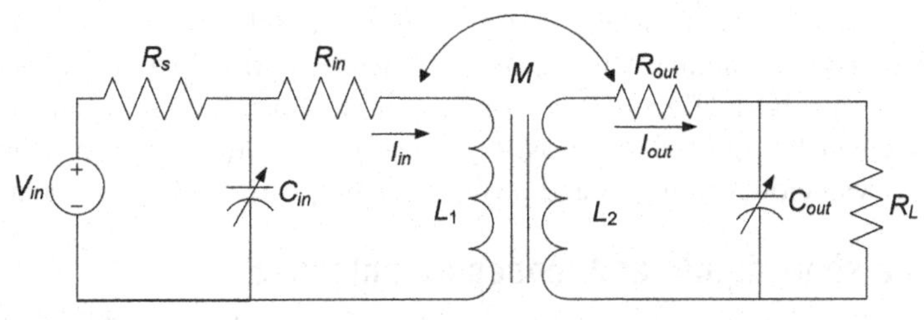

FIGURE 1.33

Primary and secondary-tuned RF transformer circuit.

The total impedance in the secondary is the sum of the reflected impedance due to the primary plus R_{out} or

$$R_{out} + Z_{out,reflected} = R_{out} + \frac{\Omega_c^2 M^2}{R_s + R_{in}} \tag{1.164}$$

where R_s is the resistance of the primary circuit. Finally, the Q of the secondary becomes

$$Q = \frac{2\pi F_c L_2}{R_{out} + Z_{out,reflected}} \tag{1.165}$$

A natural extension of this discussion is to look at transformers where both the primary and the secondary can be tuned as shown in Figure 1.33. The tuning capacitors are C_{in} and C_{out}. Each side of the transformer can be tuned separately with resonant frequency for the RLC circuit given as $F_c = 1/2\pi\sqrt{LC}$. Only when the primary and the secondary circuits are well isolated, their respective Qs become $Q_{\text{primary}} = \Omega_c L_1/R_{in}$ and $Q_{\text{secondary}} = \Omega_c L_2/R_{out}$.

1.3 Common antenna types

Thus far, we have treated the antenna in terms of circuits and general performance parameters. In this section, we will delve into specific antenna types that are commonly used in handheld and portable systems.

1.3.1 Small and resonant antennas

In order to understand the treatment of electrically small antennas as circuit elements, we must highlight the differences between long, short, and resonant antennas and how this classification impacts the way a certain antenna is treated and analyzed.

A resonant antenna, or resonant-length antenna, is an antenna whose length is a quarter of a wavelength, or multiples thereof, long. In this case, the antenna is purely resistive and its reactance is zero, and hence the maximum amount of current flows through the antenna. In most wireless applications, however, the antenna is much smaller than $\lambda/4$ in order to constrain its size. Due to this condition, an electrically small antenna has much lower radiation resistance than a resonant antenna and possesses a capacitive reactance thus resulting in multiple reflections at the load. In order to tune the antenna with capacitive reactance, an inductive reactance can be used. For an electrically small monopole, for example, the input impedance of the antenna can be approximated as $Z_{input} \approx -jZ_{char}\cot(\beta l)$, which is negative for frequencies below resonance and positive for frequencies above resonance.

1.3.2 The short dipole and monopole antennas

Consider a two-wire open-circuited transmission line as shown in Figure 1.34(a). From Eqn (1.68), the load impedance is infinite and hence

$$Z_{input} = Z_{char}\left.\frac{Z_{Load} + jZ_{char}\tan(\beta l)}{Z_{char} + jZ_{Load}\tan(\beta l)}\right|_{Z_{Load}\to\infty} = -jZ_{char}\cot(\beta l) \qquad (1.166)$$

Note that the input impedance Z_{input} is imaginary, that is a reactance, given the characteristic impedance Z_{char} that is real. Z_{input} is either positive or negative depending on βl. The open circuit depicted in Figure 1.34(a) has a zero current at the ends away from the source and thus causes a standing wave on the line. Note that at any cross-section on the line, the currents are of equal magnitude and opposite directions thus causing their respective radiated fields to be equal in magnitude and opposite in direction and thus do not radiate.

Next consider a scenario where the wires are bent at 90 degrees in opposite directions at their respective ends as shown in Figure 1.34(b). The currents in the bent portions of the wires as shown in Figure 1.34(b) are no longer flowing in the opposite direction, but rather in the same direction, causing the wires or *dipole antenna*, to radiate. The current distribution is depicted as decaying linearly away from the junction, which is a good approximation for $L < 0.1\lambda$. Note that the currents at the ends of the wires are still zero. Furthermore, the current distribution has also changed from that of the open-circuit transmission line. The current distribution can be found analytically as

$$I(l) = I_{junction}\frac{\sin\left(\frac{2\pi}{\lambda}\left(\frac{L}{2} - |l|\right)\right)}{\sin\left(\frac{\pi L}{\lambda}\right)}, \quad -\frac{L}{2} < l < \frac{L}{2} \qquad (1.167)$$

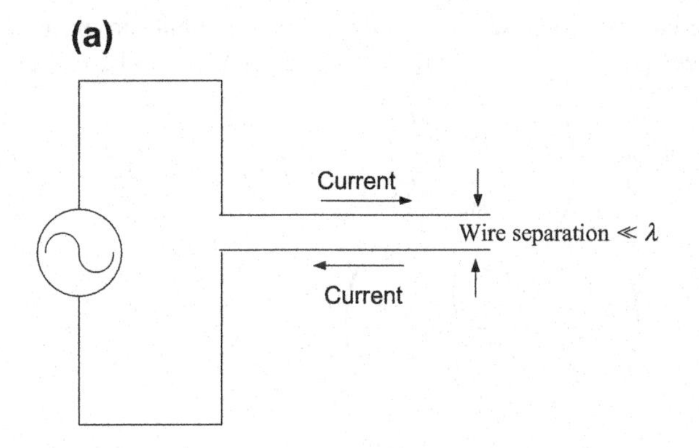

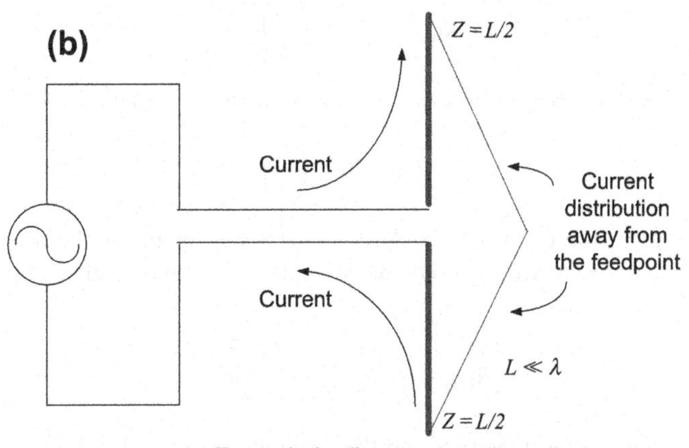

FIGURE 1.34

Current distribution on (a) open transmission line and (b) dipole antenna. (For color version of this figure, the reader is referred to the online version of this book.)

where $I_{junction}$ is the current at the dipole–transmission line junction, l is the current at given location on the dipole, and L is the length of the dipole. The total electric field radiating from the dipole is the sum of the individual fields

$$E_\theta = \frac{j\eta I_{junction} L e^{-j\beta r}}{2\lambda r} \frac{1 - \cos\left(\frac{\pi L}{\lambda}\right)}{\frac{\pi L}{\lambda} \sin\left(\frac{\pi L}{\lambda}\right)} \sin(\theta) \tag{1.168}$$

where θ is the elevation angle, r is the distance between an infinitesimally small point on the dipole to a point in the electric field, and $\eta = \sqrt{\mu/\varepsilon}$ as defined in Eqn (1.33).

A short dipole can be expressed as a simple three-element circuit as shown in Figure 1.35. In free space $\eta = 120\pi$ and the radiated power can be expressed as

$$\begin{aligned} P_{\text{total radiated power}} &= \frac{\eta I_{junction}^2}{8}\left(\frac{L}{\lambda}\frac{1-\cos\left(\frac{\pi L}{\lambda}\right)}{\frac{\pi L}{\lambda}\sin\left(\frac{\pi L}{\lambda}\right)}\right)^2 \int_0^{2\pi}\int_0^{\pi} F_{\theta,\phi}\sin\theta d\theta d\phi \Bigg|_{\eta=120\pi \text{ in free space}} \\ &= 80\left(\frac{L}{\lambda}\frac{1-\cos\left(\frac{\pi L}{\lambda}\right)}{\frac{\pi L}{\lambda}\sin\left(\frac{\pi L}{\lambda}\right)}\frac{I_{junction}}{\sqrt{2}}\right)^2 \end{aligned} \tag{1.169}$$

For a dipole with uniform current, the radiation resistance for L less than 0.1λ can be expressed as

$$R_{rad} = 80\pi^2\left(\frac{L}{\lambda}\right)^2 \tag{1.170}$$

On the other hand, the radiation resistance for an open-ended short dipole is given as

$$R_{rad} = 20\pi^2\left(\frac{L}{\lambda}\right)^2 \tag{1.171}$$

Note that the radiation resistance in Eqn (1.170) is dependent on the wavelength as well as the length of the antenna. The resistive loss in a dipole can be approximated simply as [3]

$$R_{loss} \approx \frac{L}{3d_{perimeter}}\sqrt{\frac{\pi\mu F_c}{\sigma}} \tag{1.172}$$

where $d_{perimeter}$ is the distance around the perimeter of the wire, and σ is the conductivity expressed in mhos per meter. The relationship in Eqn (1.172) assumes that the current diminishes linearly away from the junction as shown in

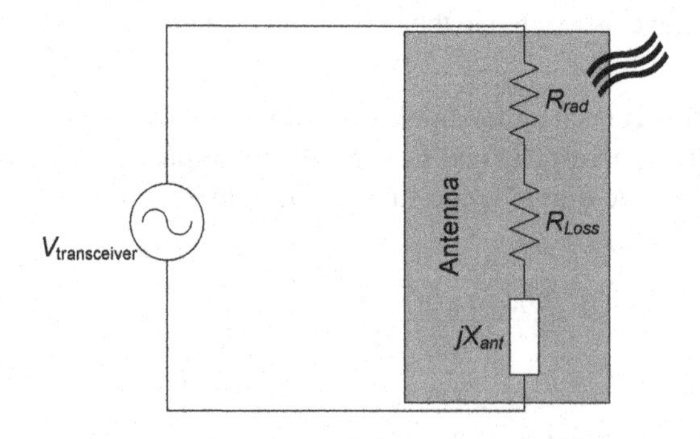

FIGURE 1.35

Circuit representation short dipole antenna. (For color version of this figure, the reader is referred to the online version of this book.)

Figure 1.34(b). The capacitance of a short dipole, on the other hand, can be also approximated as [4]

$$C = \frac{0.5\pi\varepsilon L}{\ln\left(\frac{L}{d_w}\right) - 1} \tag{1.173}$$

where d_w is the diameter of the wire.

The results discussed thus far are only true if the antenna is placed away from any large obstacles that can alter its radiation pattern. In practice, however, nearby objects can largely alter the radiation pattern of a transmitting antenna and can induce currents back into the antenna itself changing its impedance. While the former effect of radiation altering can be thought of as a far field effect, the change in impedance can be regarded as a near field effect. In order to understand this phenomenon, consider the impact of a vertical dipole placed a distance $z = d_x$ above a perfect ground plane. Due to this symmetry, the radiated fields due to the dipole can be expressed as

$$E_\theta = \frac{j\eta I_{junction} L e^{-j\beta r}}{2\lambda r} \frac{1 - \cos\left(\frac{\pi L}{\lambda}\right)}{\frac{\pi L}{\lambda} \sin\left(\frac{\pi L}{\lambda}\right)} \sin(\theta)\cos(\beta d_x \cos(\theta)) \tag{1.174}$$

Although not obvious, the relationship in Eqn (1.174) implies that the dipole image below the conducting plane is situated at a distance $z = -d_x$ and has the same fields as the dipole apart from for a phase delay thus adding to the field strength. This is only true for $d_x \leq 0.05\lambda$. For larger distances, the phase differences become more pronounced, thus reducing the field strength. The radiation resistance for the open-wire dipole over a perfect conducting plane is given as

$$R_{rad} = 40\pi^2 \left(\frac{L}{\lambda}\right)^2 \tag{1.175}$$

which is double the radiation resistance found in Eqn (1.171).

The monopole antenna is half of a dipole antenna placed on top of a ground plane, and hence its radiation pattern will be identical to that of the dipole in the upper hemisphere. Given a perfect conducting infinite plane, a monopole antenna is equivalent to a dipole antenna whose lower half is the exact image of its upper half as shown in Figure 1.36. The total radiated power of a monopole is half of that of a dipole, and hence the monopole radiation resistance is also half of that of the dipole, that is

$$\begin{aligned} P_{\text{total radiated power,monopole}} &= \frac{1}{2} P_{\text{total radiated power,dipole}} \\ R_{rad,\text{monopole}} &= \frac{1}{2} R_{rad,\text{dipole}} \end{aligned} \tag{1.176}$$

On the other hand, the directivity of the monopole is twice that of the dipole since $P_{Watts/isotropic/\theta,\phi}$ in the denominator of Eqn (1.10) is doubled.

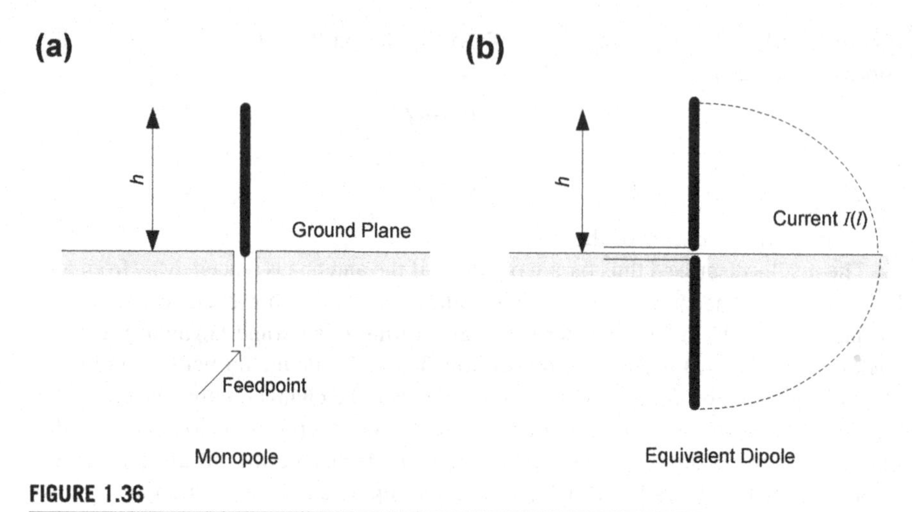

FIGURE 1.36

(a) Monopole above perfect ground plane and (b) its equivalent dipole.

1.3.3 The small loop

Another common antenna type is the small loop antenna depicted as a circular loop in Figure 1.37. A circular loop antenna is considered small if the radius is roughly less than or equal to $\lambda/20$ [5]. The electric field radiating from the circular antenna can be expressed in a manner similar to Eqn (1.168) and according to the coordinates shown in Figure 1.37 as [3]

$$\begin{aligned} E_\theta &= \frac{\eta\pi I_{\text{terminal}}\beta A e^{-j\beta r}}{\lambda^2}\vec{\phi}\sin(\theta) \\ \vec{\phi} &= -\vec{x}\sin(\phi)+\vec{y}\cos(\phi) \end{aligned} \tag{1.177}$$

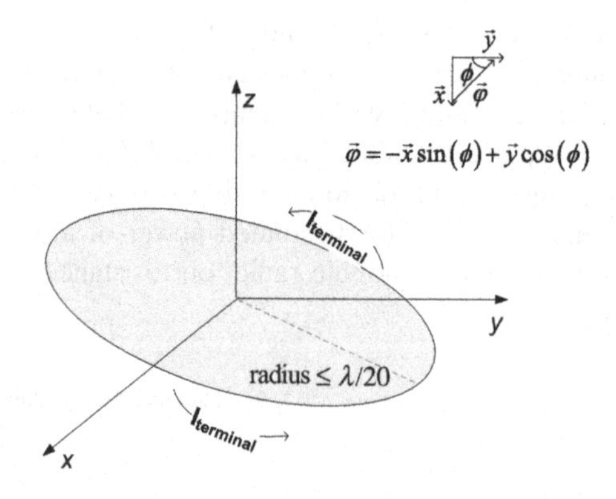

FIGURE 1.37

Circular loop antenna with uniform current in x–y plane. (For color version of this figure, the reader is referred to the online version of this book.)

where A is the area of the loop. It is important to note that the electric field of a given loop, unlike a dipole, scales by the wavelength squared, and the field strength varies in inverse proportion to λ^2. Furthermore, the relationship expressed in Eqn (1.177) is true of any loop regardless of geometry and rather depends on the total area A. The horizontal small loop is horizontally polarized and has the same magnitude pattern as the vertical dipole, which happens to be vertically polarized. Incidentally, this implies that in the near field, the short dipole stores its energy in the electric fields whereas the small loop stores its energy in the magnetic field.

The equivalent circuit for a small loop is shown in Figure 1.38. According to [6,7], the resonance of a small loop occurs at approximately $\pi D_{diameter}/\lambda = 0.49$, where $D_{diameter}$ is the loop diameter in meters. This number tends to vary even at frequencies where the loop is considered small. It is somewhat higher for thinner loops and somewhat lower for thicker loops, and thus this variation can be modeled by the shunt capacitor C shown in Figure 1.38. To resonate at $\pi D_{diameter}/\lambda = 0.49$, the capacitance value is found to be

$$C = \frac{D_{diameter}^2}{(0.98)^2 L} \tag{1.178}$$

where L is the inductance given in μH. The radiation resistance for the loop antenna can be expressed in terms of its area squared as

$$R_{rad} = \frac{8\eta\pi^3 A^2}{3\lambda^4} \tag{1.179}$$

In free space where $\eta = 120\pi$, the radiation resistance can be simply computed by substituting $\lambda = c/F$, where c is the speed of light as

$$R_{rad} = 0.346F^4A^2 \tag{1.180}$$

and the frequency F is in MHz.

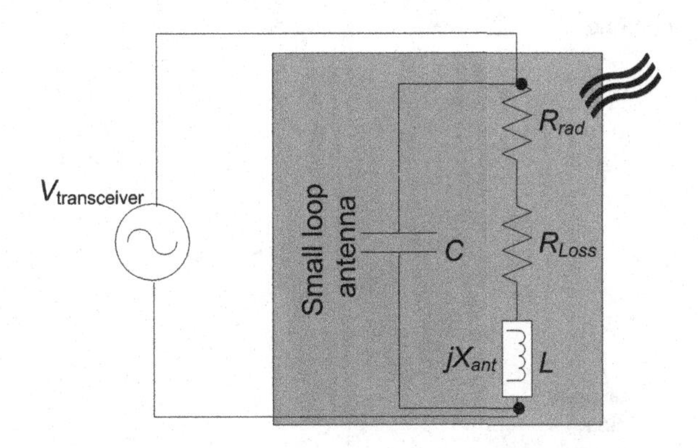

FIGURE 1.38

Circuit representation of small loop antenna. (For color version of this figure, the reader is referred to the online version of this book.)

The resistive loss per unit length is the same as that of the dipole, and hence the total resistive loss is given as

$$R_{loss} \approx \frac{1}{d_{perimeter}} \sqrt{\frac{\pi \mu F_c}{\sigma}} \tag{1.181}$$

where $d_{perimeter}$ is the distance around the perimeter of the wire, and σ is the conductivity expressed in mho per meter. The inductance L, on the other hand, can be found as [4]

$$L = \frac{\mu}{2} D_{diameter} \left[\ln\left(\frac{8D_{diameter}}{d_w} \right) - 2 \right] \tag{1.182}$$

Note that the same can be computed for a square loop or a rectangular loop. An excellent reference on this subject is found in [3]. Finally, it is worthy to note that loop antennas are less susceptible to nearby objects than wire antennas. Their biggest drawback lies in their inefficiencies.

1.3.4 Patch antennas

A microstrip patch antenna is comprised of a radiating metallic *patch* situated on one side of a nonconducting substrate panel with a metallic ground plane placed on the other side of the panel. A patch antenna, depicted in Figure 1.39, can take many geometric shapes, with the rectangle, square, and circle being the most common. The radiation of the patch is perpendicular to the board on which it is placed. The dimension L is slightly less than half the free-space wavelength divided by the square root

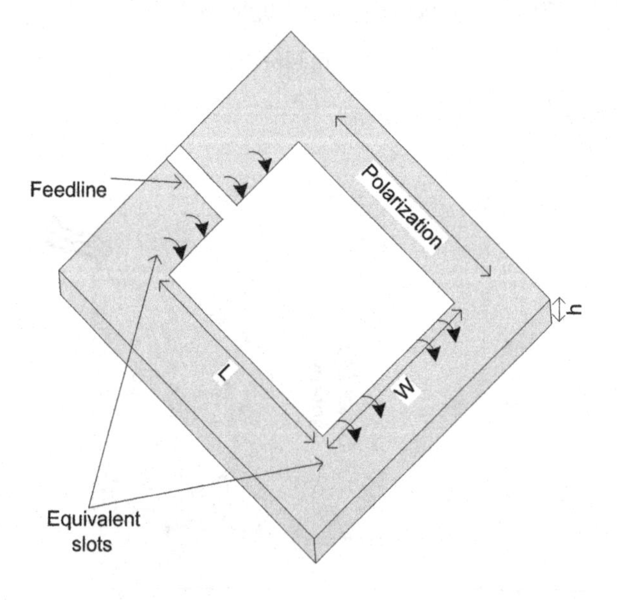

FIGURE 1.39

Patch antenna with two equivalent slots and microstrip feedline. (For color version of this figure, the reader is referred to the online version of this book.)

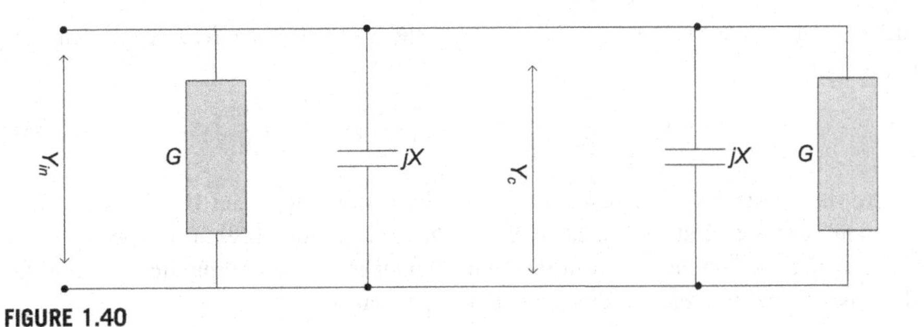

FIGURE 1.40

Equivalent circuit model of a patch antenna.

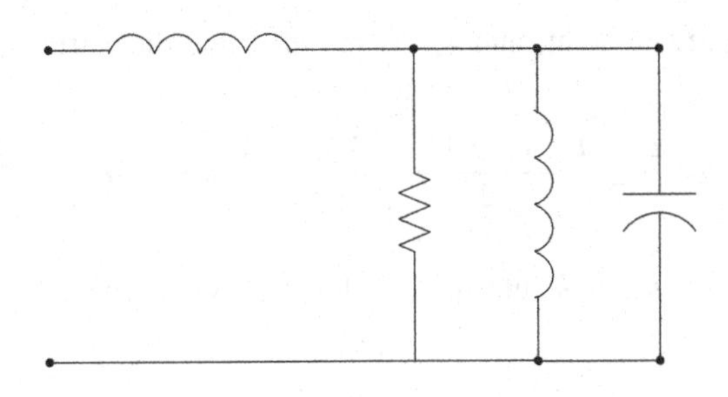

FIGURE 1.41

Equivalent circuit for the microstrip feedline.

of the effective dielectric constant of the board. The feedline, in this case a microstrip line, is etched alongside the patch at the center of its width as shown in Figure 1.39 for the rectangular patch. The equivalent circuit for the microstrip line is shown in Figure 1.41.

For a rectangular patch, the radiation is generated from the two edges with two equivalent slots as shown in Figure 1.39 [8,9]. The other two opposing edges that are W apart do not radiate so long as the feedline is at the center of the radiating edges. Thus, it can be concluded that a radiating patch can be modeled by two slots separated by a transmission line. Each slot can be represented by a parallel circuit of susceptance X and conductance G as shown by the equivalent circuit depicted in Figure 1.40. Note, however, that the circuit presented in Figure 1.40 does not model the mutual coupling present between the two radiating slots[j] nor does it account for the radiation due to the nonradiating edges of the patch. Due to these limitations, this model becomes unsuitable to analyze nonrectangular shapes and hence it is very limited in its application [9].

A rectangular microstrip radiator can be modeled as an open cavity bounded by the patch and the ground plane. Assume the relative dielectric constant to be ε_r and

[j]A slot is a radiating narrow aperture.

the substrate thickness to be h,[k] then the total electric field in the leaky cavity can be expressed as

$$E_z(x,y) = \sum_m \sum_n C_{m,n} \cos\left(\frac{m\pi}{L}\right)x \times \cos\left(\frac{n\pi}{W}\right)y \tag{1.183}$$

where the constant $C_{m,n}$ depends on the feed location and L and W are the patch's resonant length and nonresonant width as shown in Figure 1.39. The resonant frequency of a rectangular patch in the fundamental mode can be predicted using its dimensions and the relative dielectric constant, that is

$$F_c = \frac{c}{2(L+h)\sqrt{\varepsilon_{eff}}} \tag{1.184}$$

where c is the speed of light, and ε_{eff} is the effective dielectric constant defined as [10]

$$\varepsilon_{eff} = \frac{\varepsilon_r+1}{2} + \frac{\varepsilon_r-1}{2}\frac{1}{\sqrt{1+12\frac{h}{W}}}, \quad \text{where } W/h > 1 \tag{1.185}$$

Note that for a very small h, the relationship for the resonant frequency in Eqn (1.184) becomes

$$F_{resonant} = \frac{c}{2L\sqrt{\varepsilon_r}} \tag{1.186}$$

At this point it is important to elaborate on ε_{eff} in Eqn (1.185). A basic patch antenna is depicted in Figure 1.42. The distribution of the electric field of a rectangular patch when excited in its fundamental mode is also shown. The electric field is zero at the center of the patch and progresses to become maximum positive on one side and maximum negative on the opposite side. The polarization of the field interestingly enough depends on the instantaneous phase of the applied signal. Note, however, that the electric fields do not end immediately at the patch's edges but rather extend somewhat to the outer periphery of the patch. These field extensions are known as *fringing fields*. The electric field radiates along the z axis, whereas the magnetic field is present in the x, y plane. It is because of these fringing fields that the patch looks electrically greater than its physical dimension. This increase in dimension of the patch along its path, and denoted by ΔL, is a function of the effective dielectric constant ε_{eff} and the ratio W/h as [10]

$$\frac{\Delta L}{h} = 0.412\left(\frac{\varepsilon_{eff}+0.3}{\varepsilon_{eff}-0.258}\right)\left(\frac{\frac{W}{h}+0.264}{\frac{W}{h}+0.8}\right) \tag{1.187}$$

[k]The substrate is assumed to be very small and hence the fields radiate in the z direction.

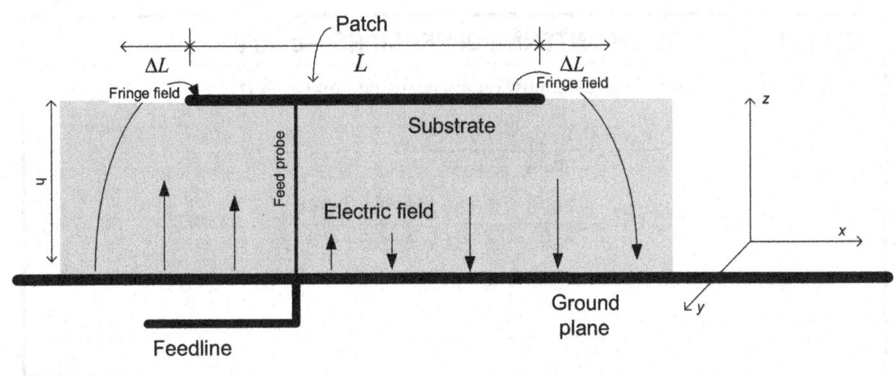

FIGURE 1.42

Basic patch antenna comprised of a flat plate over a ground plane and feedprobe.

Note that the resonant frequency in Eqn (1.186) does not account for the fringing effect. Accounting for fringing effects, the modified resonant frequency becomes

$$F_{resonant,fringing} = \frac{1}{2(L + 2\Delta L)\sqrt{\mu_0 \varepsilon_0 \varepsilon_{eff}}} \tag{1.188}$$

For an efficient radiator, the width can be predicted using the resonant frequency as

$$W = \frac{1}{2F_{resonant}\sqrt{\mu_0 \varepsilon_0}} \sqrt{\frac{2}{\varepsilon_r + 1}} \tag{1.189}$$

EXAMPLE 1.10 PATCH ANTENNA DIMENSIONS

Consider a rectangular patch antenna with a dielectric constant of 2.9, $h = 0.15$ cm. Compute the length and width of the patch for a resonant frequency of 2.5 GHz.

The width of the patch W can be computed using Eqn (1.189) as

$$W = \frac{1}{2F_{resonant}\sqrt{\mu_0 \varepsilon_0}} \sqrt{\frac{2}{\varepsilon_r + 1}} = \frac{1}{2\sqrt{4\pi \times 10^{-7} \times \frac{1}{36\pi} \times 10^{-9}} \times 2.5 \times 10^9} \sqrt{\frac{2}{2.9 + 1}}$$
$$= 4.2967 \tag{1.190}$$

Next, using the relationship in Eqn (1.185), the effective dielectric constant may be obtained as

$$\varepsilon_{eff} = \frac{\varepsilon_r + 1}{2} + \frac{\varepsilon_r - 1}{2} \frac{1}{\sqrt{1 + 12\frac{h}{W}}} = \frac{2.9 + 1}{2} + \frac{2.9 - 1}{2} \frac{1}{\sqrt{1 + 12\frac{0.15}{4.2967}}} = 2.7475 \tag{1.191}$$

Continued

EXAMPLE 1.10 PATCH ANTENNA DIMENSIONS—cont'd

To obtain the length L, first we must obtain ΔL using Eqn (1.187)

$$\Delta L = 0.412h\left(\frac{\varepsilon_{eff}+0.3}{\varepsilon_{eff}-0.258}\right)\left(\frac{\frac{W}{h}+0.264}{\frac{W}{h}+0.8}\right)$$
$$= 0.412 \times 0.15\left(\frac{2.7475+0.3}{2.7475-0.258}\right)\left(\frac{\frac{4.2967}{0.15}+0.264}{\frac{4.2967}{0.15}+0.8}\right) = 0.0743 \quad (1.192)$$

Using Eqn (1.188), we can obtain L as

$$L = \frac{1}{2F_{resonant,fringing}\sqrt{\mu_0\varepsilon_0\varepsilon_{eff}}} - 2\Delta L$$

$$L = \frac{1}{2\times 2.5\times 10^9\sqrt{4\pi\times 10^{-7}\times\frac{1}{36\pi}\times 10^{-9}\times 2.7475}} - 2\times 0.0743 = 3.4712 \quad (1.193)$$

Reexamining the equivalent patch antenna circuit presented in Figure 1.40, and assuming that each slot posses the parallel admittance and that both slots are identical, then the admittance for the slot closest to the feedline in Figure 1.39 is given as [11]

$$Y = G + jX$$
$$G = \frac{W}{120\lambda}\left(1-\frac{1}{24}\left(\frac{2\pi}{\lambda}h\right)^2\right), \quad \frac{h}{\lambda} < 0.1, W \text{ finite} \quad (1.194)$$
$$X = \frac{W}{120\lambda}\left(1-0.636\ln\left(\frac{2\pi}{\lambda}h\right)\right) \quad \frac{h}{\lambda} < 0.1, W \text{ finite}$$

Finally, it is important to note that patch antennas, in wireless portable applications, are largely used at high frequencies for narrowband applications with semispherical coverage. Wider bandwidth can also be supported using various bandwidth widening methods.[1] Patch antennas are inexpensive and can be integrated with circuit elements. Incidentally, it is interesting to note that patch antennas exhibit large ohmic losses.

1.3.5 Helical antennas

A helical antenna, depicted in Figure 1.43, is a broadband antenna that exhibits circular polarization. The antenna itself is connected to a conductor, the shaded circular portion, which itself is connected to a transmission line. Given a circular ground plane, the radius ρ is chosen such that $\rho \geq 0.75\lambda$. The helix itself consists of a

[1]One such technique uses thick and low permittivity substrates.

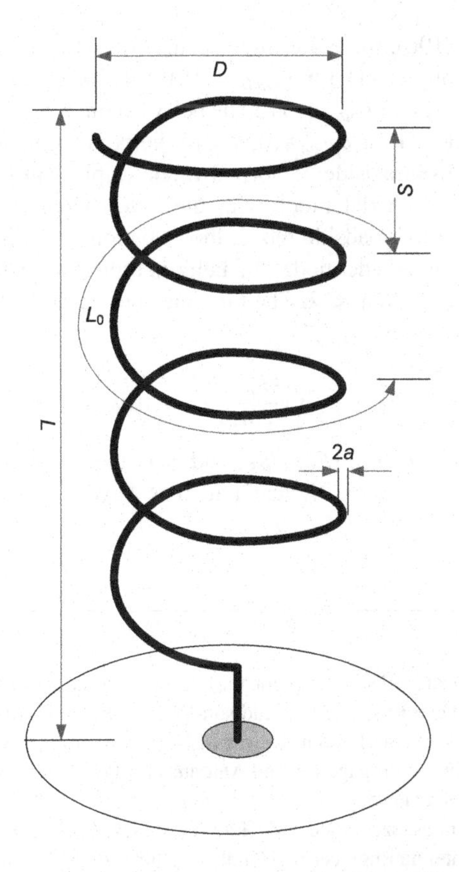

FIGURE 1.43

Helical antenna on a ground plane. (For color version of this figure, the reader is referred to the online version of this book.)

conducting wire wound N times. Each turn has a diameter D. The spacing between each turn is S, resulting in total length of the antenna being $L = NS$. The total length of the wire can be expressed as

$$L_{Total} = N\sqrt{S^2 + \pi^2 D^2} = NL_0 \tag{1.195}$$

where L_0 is the length of the wire between each turn as depicted in Figure 1.43 [10]. The geometric properties of the antenna along with the wavelength control its radiation characteristic.

Next, define the pitch angle as the angle that can be outlined between a tangent to the helix wire and a plane perpendicular to the helix axis

$$\theta_{pitch} = \tan^{-1}\left(\frac{S}{\pi D}\right), \quad 0 \leq \theta_{pitch} \leq \frac{\pi}{2} \tag{1.196}$$

According to Eqn (1.196), there are three scenarios that can ensue due to various values of the pitch angle: (1) for $\theta_{pitch} = 0$ the helix becomes a loop antenna comprised of N turns, (2) for $\theta_{pitch} = \pi/2$, the helix becomes a linear wire, and finally for all other scenarios (3) for $\theta_{pitch} \in (0, \pi/2)$ a helix is formed. A helical antenna operates principally in broadside or normal mode (typical in portable and mobile devices) or in endfire or axial mode with the latter being the most common in most applications. In the broadside mode, the radiation exists in a plane normal to the helix axis. In the broadside mode, the helix is typically small in size compared to the wavelength, that is $NL_0 \ll \lambda$. The far zone electric field is similar to that of a short dipole and is given as

$$E_{\theta_{pitch}} = j\eta \frac{kI_0 S e^{-jkr}}{4\pi r} \sin\left(\theta_{pitch}\right) \tag{1.197}$$

where I_0 is a constant, $k = 2\pi F_c/\sqrt{\mu\varepsilon}$, and η is the intrinsic impedance of the medium given as $\eta = \sqrt{\mu/\varepsilon}$. Helical antennas have been used extensively in cellular phones.

References

[1] W. Hayt, Engineering Electromagnetics, McGraw-Hill, New York, NY, 1981.
[2] D. Pozar, Microwave Engineering, Addison-Wesley, Reading, MA, 1993.
[3] D. Miron, Small Antenna Design, Newnes-Elsevier, Boston, MA, 2006.
[4] K. Siwiak, Radiowave Propagation and Antennas for Personal Communications, Artech House, Boston, MA, 1995.
[5] J.D. Kraus, Antennas, second ed., McGraw-Hill, New York, NY, 1988.
[6] R.C. Johnson, Antenna Engineering Handbook, third ed., McGraw-Hill, New York, NY, 1993.
[7] F.L. Dacus, J. Van Niekerk, S. Bible, Introducing loop antennas for short-range radios, Microwave RF (July 2002).
[8] A. Derneryd, A theoretical investigation of the rectangular microstrip antenna, IEEE Trans. Antennas Propag. 26 (July 1978) 532–535.
[9] H. Pues, A. Van de Capelle, Accurate transmission-line model for the rectangular microstrip antenna, Proc. IEE 131 (December 1984) 334–340.
[10] C.A. Balanis, Advanced Engineering Electromagnetics, John Wiley and Sons, New York, NY, 1989.
[11] R.F. Harrington, Time-Harmonic Electromagnetic Fields, McGraw-Hill, New York, NY, 1981.

Further reading

[1] I.J. Bahl, P. Bharita, Microstrip Antennas, Artech House, Dedham, MA, 1980.

CHAPTER

2 Microwave Network Design and Analysis

CHAPTER OUTLINE

In this chapter, we introduce a powerful set of tools to analyze microwave networks without having to resort to the more powerful but more complex Maxwell's equations. Maxwell's equations provide comprehensive information concerning the electromagnetic field distribution within a microwave network. Aside from the antenna, a microwave engineer may only be interested in how a microwave circuit reacts to external microwave signals. On the other hand, basic circuit analysis that is used typically at direct current (DC) or low frequency is far too simple to provide practical guidance for the design and analysis of microwave components. Circuits at microwave frequencies exhibit characteristics that cannot be easily delineated by mere application of Kirchhoff's voltage and current equations. Therefore, simply put, a set of analytical tools is needed to enable the engineer to obtain such simple, but realistic, parameters as current, voltage, and power through a microwave circuit or a set of interconnected microwave components without having to resort to very complex analysis.

One such tool is based on the scattering parameters (*S*-parameters) of the forward and reflected voltage and current waves derived in Chapter 1 from transmission line theory. Given a multiport network, *S*-parameters can be used to provide a thorough

Wireless Receiver Architectures and Design. http://dx.doi.org/10.1016/B978-0-12-378640-1.00002-0

description of the network behavior in terms of incident and reflected waves onto the various ports. *S*-parameters can be obtained either via network analysis or via measurements using a network analyzer. Once *S*-parameters are obtained, they can be related to more traditional network parameters such as impedance and admittance matrices. These parameters can also be related to a transmission (*ABCD*) network analysis matrix technique that is suitable for cascade analysis of multiport networks.

The chapter is divided into three major sections. While Sections 2.1 and 2.2 deal mostly with two-port network models, Section 2.3 addresses three-port and four-port network models. Section 2.1 addresses the various circuit network models and their advantages and disadvantages. Specifically, it discusses in some detail the impedance and admittance multiport network model, the hybrid network model, and scattering parameters network model. Section 2.2 presents an extremely useful analytical technique known as signal flow graphs. Topics ranging from power gain equations to stability theory are discussed. In Section 2.3, three-port and four-port network models are studied. However, rather than repeating the general theory already presented in the previous two sections, it presents two very specialized and very common three-port and four-port devices, namely power dividers and combiners and the directional coupler.

2.1 Network models

The purpose of this section is to give an overview of various popular circuit network models. It addresses the use of these models in front-end system analysis and circuit design.

2.1.1 The characteristic impedance revisited

Define the potential difference voltage V as the work done by a certain external source in moving a unit positive charge from an initial point $\ell_{initial}$ to a final point ℓ_{final}

$$V = -\int_{\ell_{initial}}^{\ell_{final}} \vec{E}\cdot d\vec{\ell} = \int_{\ell_{final}}^{\ell_{initial}} \vec{E}\cdot d\vec{\ell} \tag{2.1}$$

On the other hand, Ampère's law states that the total current flowing through a closed surface S is given as [1]

$$I = \oint_S \vec{H}\cdot d\vec{\ell} \tag{2.2}$$

The characteristic impedance for a traveling wave is then the ratio of the voltage in Eqn (2.1) to the current in Eqn (2.2) or [1,2]

$$Z_{char} = \frac{V}{I} \tag{2.3}$$

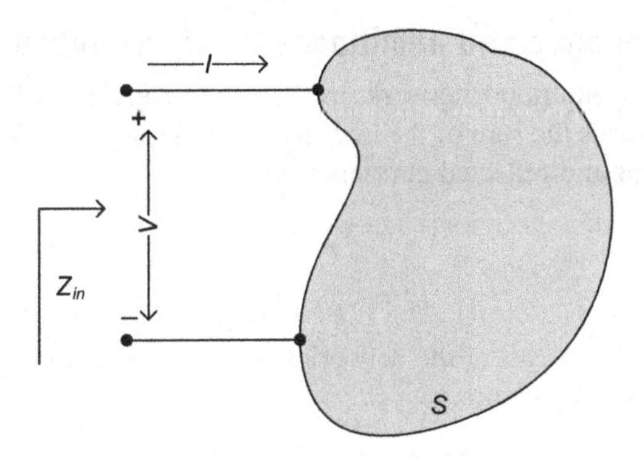

FIGURE 2.1

One-port network.

Given a one-port network as depicted in Figure 2.1 and recall from Chapter 1 that the complex power delivered to this network is $P = \frac{1}{2}\oint_S \vec{E}\cdot\vec{H}^* \cdot d\bar{s}$, then P is broken down into dissipated energy or the real value of P and stored energy in the form of electromagnetic energy or the imaginary value of P. The impedance at the input point of the network of Figure 2.1 can be expressed in the frequency domain as $Z(\Omega) = V(\Omega)/I(\Omega)$. For a real input voltage and current, that is $v(t) = v^*(t)$ and $i(t) = i^*(t)$, we know according to the Fourier transform that

$$\begin{aligned} V^*(\Omega) &= V(-\Omega) \\ I^*(\Omega) &= I(-\Omega) \end{aligned} \tag{2.4}$$

which implies for $Z^*(\Omega) = Z(-\Omega)$. This implies that for

$$Z(\Omega) = R(\Omega) + jX(\Omega) \tag{2.5}$$

$R(\Omega)$ is even in Ω and $X(\Omega)$ is odd in Ω.

Given the reflection coefficient $\Gamma(\Omega)$, defined as

$$\Gamma(\Omega) = \frac{Z(\Omega) - Z_{char}}{Z(\Omega) + Z_{char}} = \frac{R(\Omega) + jX(\Omega) - Z_{char}}{R(\Omega) + jX(\Omega) + Z_{char}} \tag{2.6}$$

The relationship in Eqn (2.6), in light of the characteristic of $Z(\Omega)$, also implies that

$$\begin{aligned} \Gamma^*(\Omega) &= \frac{Z^*(\Omega) - Z_{char}}{Z^*(\Omega) + Z_{char}} = \frac{R(\Omega) - jX(\Omega) - Z_{char}}{R(\Omega) - jX(\Omega) + Z_{char}} = \Gamma(-\Omega) \\ |\Gamma(\Omega)|^2 &= \Gamma(\Omega)\Gamma^*(\Omega) = \Gamma(\Omega)\Gamma(-\Omega) = |\Gamma(-\Omega)|^2 \end{aligned} \tag{2.7}$$

The relationship in Eqn (2.7) implies that the magnitude of the reflection coefficient as well as the magnitude squared are even functions of the frequency variable Ω.

2.1.2 The impedance and admittance multiport network model

At the nth port of a multiport network, as depicted in Figure 2.2, define the total voltage and currents as the sum of the incident and reflected voltages and the difference of the incident and reflected currents [3], or

$$\begin{aligned} V_n &= V_n^+ + V_n^- \\ I_n &= I_n^+ - I_n^- \end{aligned} \tag{2.8}$$

Then relate the total voltages of the network to the total currents via the impedance matrix Z defined via the relation

$$\underbrace{\begin{bmatrix} V_1 \\ V_2 \\ \vdots \\ V_N \end{bmatrix}}_{\overline{V}} = \underbrace{\begin{bmatrix} Z_{1,1} & Z_{1,2} & \cdots & Z_{1,N} \\ Z_{2,1} & Z_{2,2} & \cdots & Z_{2,N} \\ \vdots & \vdots & \cdots & \vdots \\ Z_{N,1} & Z_{N,2} & \cdots & Z_{N,N} \end{bmatrix}}_{\overline{Z}} \underbrace{\begin{bmatrix} I_1 \\ I_2 \\ \vdots \\ I_N \end{bmatrix}}_{\overline{I}} \tag{2.9}$$

$$\overline{V} = \overline{ZI}$$

where $\overline{Z}$ is the impedance matrix with elements defined as

$$Z_{m,n} = \left. \frac{V_m}{I_n} \right|_{I_l=0 \text{ for } l \neq n} \tag{2.10}$$

The relationship in Eqn (2.10) is obtained by setting the currents, the independent variable in this case, to zero at every port except at the driving port, that is for

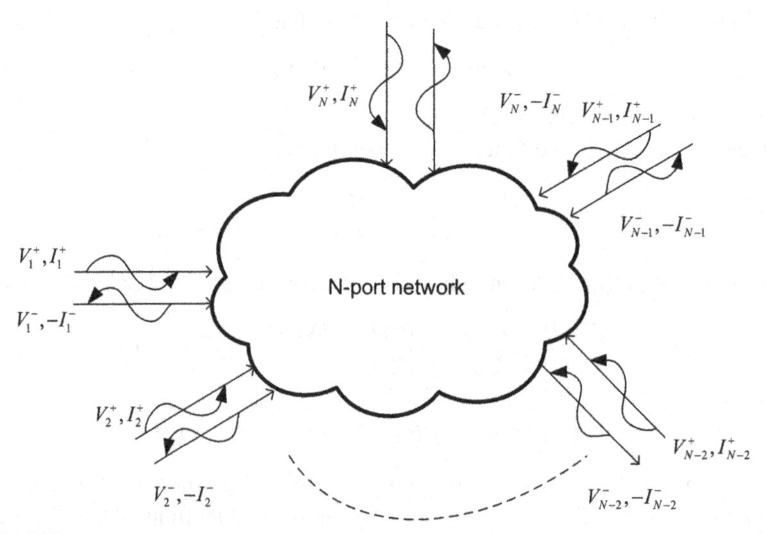

FIGURE 2.2

An abstract depiction of an N-port network.

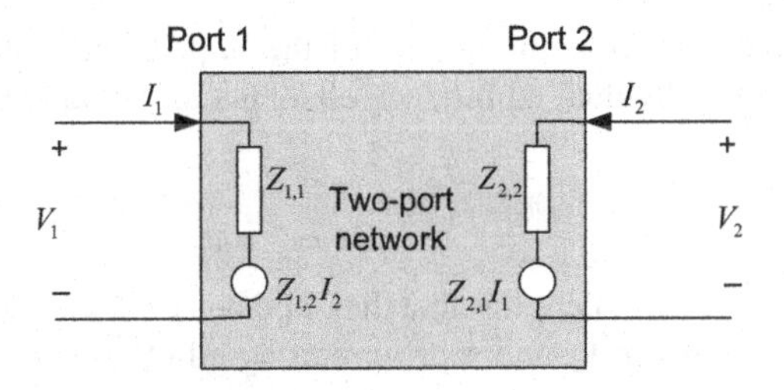

FIGURE 2.3

Two-port Thevenin impedance-equivalent network.

$I_l = 0$ and $l \neq n$, using open circuit termination. The parameters in Eqn (2.9) are therefore known as open circuit impedance parameters [4].

To further illustrate this point, consider the two-port network depicted in Figure 2.3. The impedance parameters are obtained via Eqn (2.9) as

$$\begin{bmatrix} V_1 \\ V_2 \end{bmatrix} = \begin{bmatrix} Z_{1,1} & Z_{1,2} \\ Z_{2,1} & Z_{2,2} \end{bmatrix} \begin{bmatrix} I_1 \\ I_2 \end{bmatrix} \tag{2.11}$$

or

$$V_1 = Z_{1,1}I_1 + Z_{1,2}I_2 \text{ and } V_2 = Z_{2,1}I_1 + Z_{2,2}I_2$$

To compute the input impedance $Z_{1,1}$ for example, set $I_2 = 0$, then we obtain $Z_{1,1} = V_1/I_1$. That is, $Z_{1,1}$ is the input impedance with the output port (port 2) terminated in an open circuit ($I_2 = 0$). In a similar manner, $Z_{2,2}$ is the output impedance obtained while the input port (port 1) is open circuited. Similarly, the forward transfer impedance $Z_{2,1}$ is computed with the output terminal open circuited while the reverse transfer impedance $Z_{1,2}$ is obtained with the input port terminated in an open circuit.

At this juncture, it is important to point out that the impedance model is most relevant at low frequencies. At high frequencies, however, fringing capacitances make use of an open circuit increasingly hard to implement. This is especially true when it comes to performing measurements on active devices. As we shall later see, this limitation is overcome by the use of the powerful "scattering parameters model".

Next we turn our attention to the admittance model. In this case, the voltages are the dependent variables. The admittance matrix relates the currents to the voltages according to

$$\underbrace{\begin{bmatrix} I_1 \\ I_2 \\ \vdots \\ I_N \end{bmatrix}}_{\overline{I}} = \underbrace{\begin{bmatrix} Y_{1,1} & Y_{1,2} & \cdots & Y_{1,N} \\ Y_{2,1} & Y_{2,2} & \cdots & Y_{2,N} \\ \vdots & \vdots & \cdots & \vdots \\ Y_{N,1} & Y_{N,2} & \cdots & Y_{N,N} \end{bmatrix}}_{\overline{Y}} \underbrace{\begin{bmatrix} V_1 \\ V_2 \\ \vdots \\ V_N \end{bmatrix}}_{\overline{V}} \tag{2.12}$$

$$\overline{I} = \overline{Y}\overline{V}$$

where the admittance matrix is the inverse of the impedance matrix or $\overline{Y} = \overline{Z}^{-1}$. From Eqn (2.12), the individual admittance elements are obtained such that

$$Y_{m,n} = \left.\frac{I_m}{V_n}\right|_{V_l=0 \text{ for } l \neq n} \tag{2.13}$$

The relationship in Eqn (2.13) implies that the impedance $Y_{m,n}$ is obtained by driving the nth port with the voltage V_n and short-circuiting all other ports such that $V_l = 0$ for $l \neq n$.

To further understand the impedance model, consider the two-port impedance network depicted in Figure 2.4. The admittance parameters are obtained via Eqn (2.12) as

$$\begin{bmatrix} I_1 \\ I_2 \end{bmatrix} = \begin{bmatrix} Y_{1,1} & Y_{1,2} \\ Y_{2,1} & Y_{2,2} \end{bmatrix} \begin{bmatrix} V_1 \\ V_2 \end{bmatrix}$$

or

$$I_1 = Y_{1,1}V_1 + Y_{1,2}V_2 \text{ and } I_2 = Y_{2,1}V_1 + Y_{2,2}V_2 \tag{2.14}$$

To estimate the input admittance $Y_{1,1}$, for example, set the output voltage $V_2 = 0$, then we obtain $Y_{1,1} = I_1/V_1$, that is, the input admittance is computed with the output short-circuited. In a similar manner, the output admittance $Y_{2,2}$ is obtained with the input voltage short-circuited, that is $V_1 = 0$, then we obtain $Y_{2,2} = I_2/V_2$. The forward $Y_{2,1}$ and reverse $Y_{1,2}$ transfer admittances are obtained by short-circuiting the output and input voltages, respectively, that is

$$Y_{2,1} = \left.\frac{I_2}{V_1}\right|_{V_2=0} \text{ and } Y_{2,1} = \left.\frac{I_1}{V_2}\right|_{V_1=0} \tag{2.15}$$

These parameters are in general very accurate and well represent the circuit behavior when used in higher impedance networks only. This is due to the difficulty in implementing a short circuit that is typically done using large capacitances.

At this point it is important to note that both the impedance and admittance parameters can be complex valued. Furthermore, for a purely lossless network, the admittances and impedances are purely imaginary values.

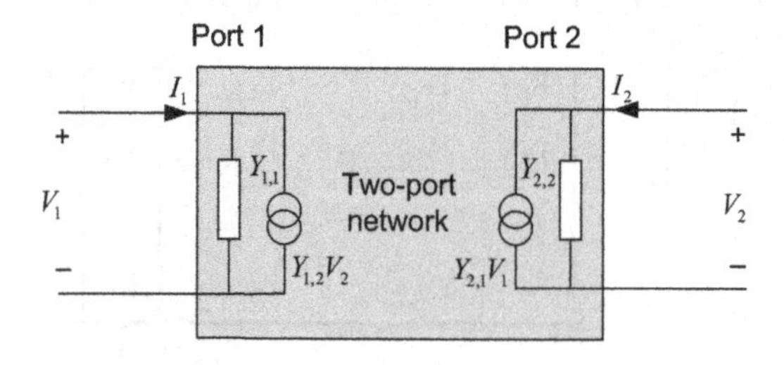

FIGURE 2.4

Two-port Thevenin admittance equivalent network.

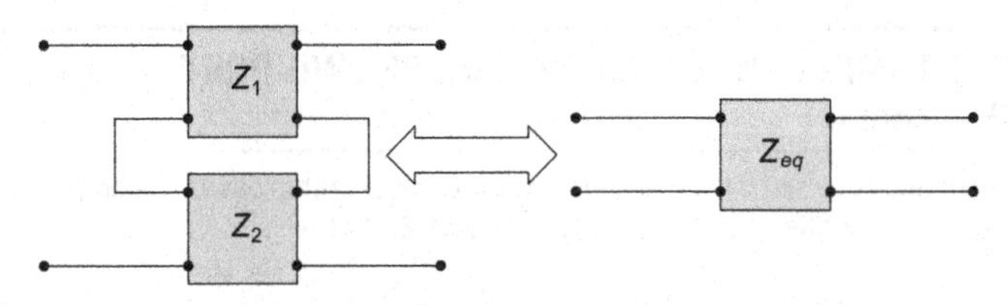

FIGURE 2.5

Series network represented by single impedance matrix.

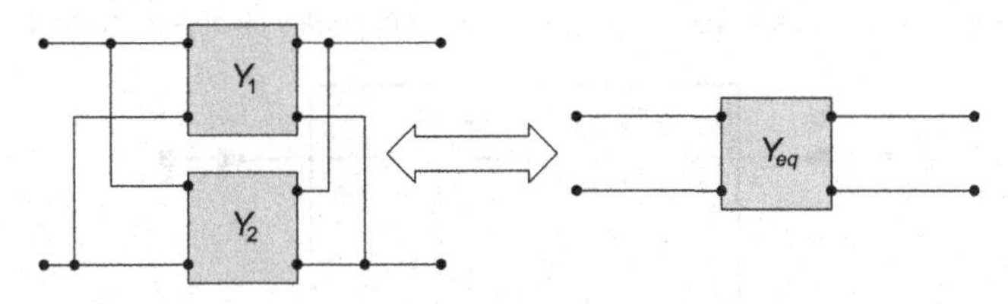

FIGURE 2.6

Parallel network represented by single admittance matrix.

Next, let us address the additive properties of impedance and admittance networks. Given two series networks with impedances Z_1 and Z_2 as depicted in Figure 2.5, it can be proven that the total impedance of the network is none other than the sum of the impedances Z_1 and Z_2 or

$$Z_{eq} = Z_1 + Z_2 \tag{2.16}$$

Similarly, given a parallel network comprised of two circuits characterized with admittance matrices Y_1 and Y_2, as depicted in Figure 2.6, then it can be shown that the total admittance of the network is the sum of the parallel circuit admittances, or

$$Y_{eq} = Y_1 + Y_2 \tag{2.17}$$

EXAMPLE 2.1 IMPEDANCE PARAMETERS OF TWO-PORT NETWORK

Consider the two-port network depicted in Figure 2.7. Compute the impedances Z_{11}, Z_{12}, Z_{21}, and Z_{22} for $R = 50$ ohm. What happens as R gets much larger?

The relationship in Eqn (2.11) implies that Z_{11} and Z_{22} can be found independently by leaving port 2 open and then port 1. That is, if we open the circuit at port 2, we find the input impedance as

$$\begin{aligned} V_1 &= Z_{1,1}I_1 + Z_{1,2}I_2\big|_{I_2=0} = Z_{1,1}I_1 \Rightarrow \\ Z_{1,1} &= \frac{V_1}{I_1} = R_1 + R = 10 + 50 = 60 \text{ ohm} \end{aligned} \tag{2.18}$$

In a likewise manner, if we open the circuit at port 1, we find the output impedance as

$$\begin{aligned} V_2 &= Z_{2,1}I_1 + Z_{2,2}I_2\big|_{I_1=0} = Z_{2,2}I_{21} \Rightarrow \\ Z_{2,2} &= \frac{V_2}{I_2} = R_2 + R = 50 + 50 = 100 \text{ ohm} \end{aligned} \tag{2.19}$$

Continued

EXAMPLE 2.1 IMPEDANCE PARAMETERS OF TWO-PORT NETWORK—cont'd

The forward and reverse transfer impedances are equal in this case. To compute Z_{12} allow port 1 to open circuit while applying a current at port 2, that is

$$V_1 = Z_{1,1}I_1 + Z_{1,2}I_2\big|_{I_1=0} = Z_{1,2}I_2 \Rightarrow Z_{1,2} = \frac{V_1}{I_2} = \frac{V_2}{I_2}\frac{R}{R+R_2}$$
$$Z_{1,2} = (R+R_2)\left(\frac{R}{R+R_2}\right) = R = 50\ \text{ohm} \tag{2.20}$$

Note that as R becomes larger and larger, all three impedances approach the value of R.

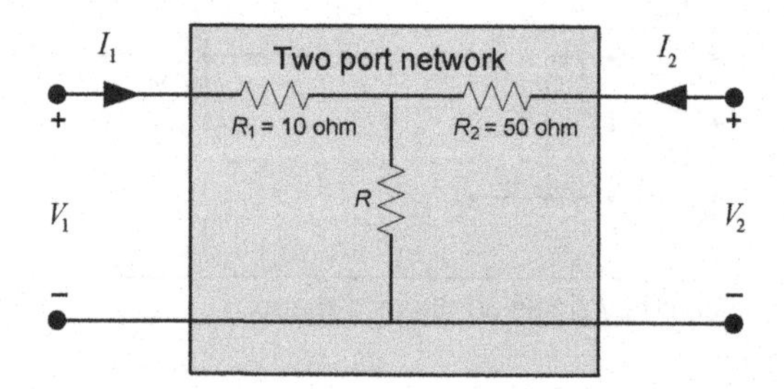

FIGURE 2.7

Two-port *T*-network.

2.1.3 The hybrid network model

Unlike the impedance or admittance matrices, the *ABCD* matrix is defined for a two-port model only [3,4]. A further difference can be seen by noticing that the output current in Figure 2.8 exits the network in comparison with the impedance network model shown in Figure 2.3.

$$\begin{bmatrix} V_1 \\ I_1 \end{bmatrix} = \begin{bmatrix} A & B \\ C & D \end{bmatrix}\begin{bmatrix} V_2 \\ I_2 \end{bmatrix} \tag{2.21}$$

or

$$V_1 = AV_2 + BI_2 \text{ and } I_1 = CV_2 + DI_2$$

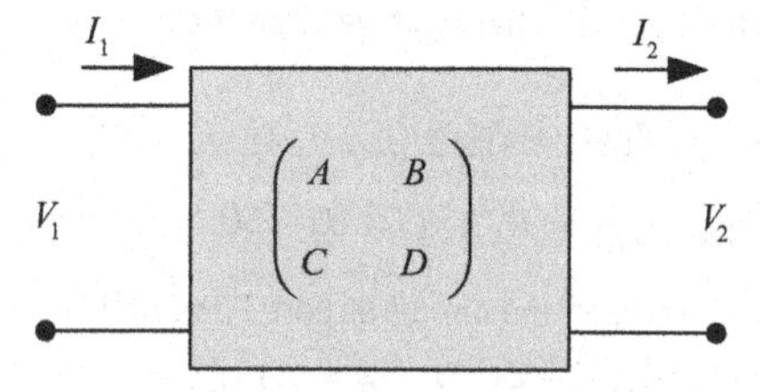

FIGURE 2.8

Voltage and current parameter representation for *ABCD* matrix.

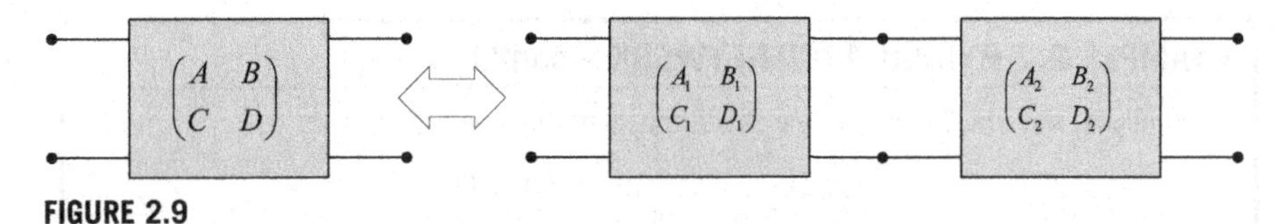

FIGURE 2.9

Cascade of *ABCD* matrices.

To obtain the *ABCD* parameters, both a short circuit termination ($V_2 = 0$) and an open circuit termination ($I_2 = 0$) are needed. Using the relationship expressed in Eqn (2.21), we obtain

$$\begin{aligned} A &= \left.\frac{V_1}{V_2}\right|_{I_2=0} \quad \text{and } B = \left.\frac{V_1}{I_2}\right|_{V_2=0} \\ C &= \left.\frac{I_1}{V_2}\right|_{I_2=0} \quad \text{and } D = \left.\frac{I_1}{I_2}\right|_{V_2=0} \end{aligned} \tag{2.22}$$

Observe that in the hybrid model, the parameters exhibit different units. Furthermore, the *ABCD* parameters can be complex similar to the impedance and admittance parameters.

A powerful feature of this network model can be seen when cascading several networks together in a chain. The overall parameters are found by multiplying the various *ABCD* matrices and obtaining a total *ABCD* matrix. By way of an example, consider the network depicted in Figure 2.9, the overall equivalent *ABCD* matrix is computed numerically as

$$\begin{aligned} \begin{bmatrix} A & B \\ C & D \end{bmatrix} &= \begin{bmatrix} A_1 & B_1 \\ C_1 & D_1 \end{bmatrix} \begin{bmatrix} A_2 & B_2 \\ C_2 & D_2 \end{bmatrix} \\ &= \begin{bmatrix} A_1A_2 + B_1C_2 & A_1B_2 + B_1D_2 \\ C_1A_2 + D_1C_2 & C_1B_2 + D_1D_2 \end{bmatrix} \end{aligned} \tag{2.23}$$

It is obvious that the matrix multiplication in Eqn (2.23) is not commutative. In other words, the order of the matrices is not interchangeable in the multiplication.

EXAMPLE 2.2 HYBRID *ABCD* NETWORK

Consider the *ABCD* network depicted in Figure 2.10. Assume that the network is made up of a cascade of two circuit elements. The first element can be modeled by a series impedance Z followed by a second element that can be modeled by a shunt admittance Y. Compute the parameters A, B, C, and D. What is the *ABCD* matrix if you swap the elements? Express the input impedance in terms of Z_L for the first cascade configuration.

The series and admittance elements are depicted in Figure 2.11. The corresponding *ABCD* matrix of the series impedance element is:

$$(ABCD)_{\text{series impedance}} = \begin{bmatrix} 1 & Z \\ 0 & 1 \end{bmatrix} \tag{2.24}$$

Continued

EXAMPLE 2.2 HYBRID *ABCD* NETWORK—cont'd

Similarly, the *ABCD* matrix of the shunt impedance can be expressed as

$$(ABCD)_{\text{shunt admittance}} = \begin{bmatrix} 1 & 0 \\ Y & 1 \end{bmatrix} \tag{2.25}$$

The cascaded *ABCD* matrix can then be computed according to Eqn (2.23) as

$$(ABCD)_{\text{cascade}} = \begin{bmatrix} 1 & Z \\ 0 & 1 \end{bmatrix}\begin{bmatrix} 1 & 0 \\ Y & 1 \end{bmatrix} = \begin{bmatrix} 1+ZY & Z \\ Y & 1 \end{bmatrix} \tag{2.26}$$

If we swap the elements, that is, if the shunt element precedes the series element in the cascade, the total *ABCD* matrix then becomes

$$(ABCD)_{\text{cascade}} = \begin{bmatrix} 1 & 0 \\ Y & 1 \end{bmatrix}\begin{bmatrix} 1 & Z \\ 0 & 1 \end{bmatrix} = \begin{bmatrix} 1 & Z \\ Y & YZ+1 \end{bmatrix} \tag{2.27}$$

Note that the swapping operation is not commutative, that is, the matrix operations in Eqn (2.26) and Eqn (2.27) result in a different cascade *ABCD* matrix.

Next, in order to obtain the input impedance, consider the ratio of V_1 by I_1 as given in Eqn (2.21), that is

$$Z_{in} = \frac{V_1}{I_1} = \frac{AV_2 + BI_2}{CV_2 + DI_2} = \frac{A\frac{V_2}{I_2} + B}{C\frac{V_2}{I_2} + D} = \frac{AZ_L + B}{CZ_L + D}\bigg|_{\frac{V_2}{I_2} = Z_L}$$
$$= \frac{Z_L + Z}{YZ_L + YZ + 1}\bigg|_{(ABCD) = \begin{bmatrix} 1 & Z \\ Y & YZ+1 \end{bmatrix}} \tag{2.28}$$

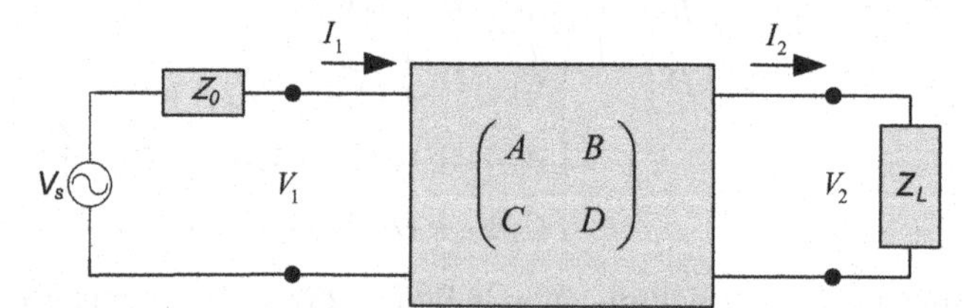

FIGURE 2.10

ABCD network.

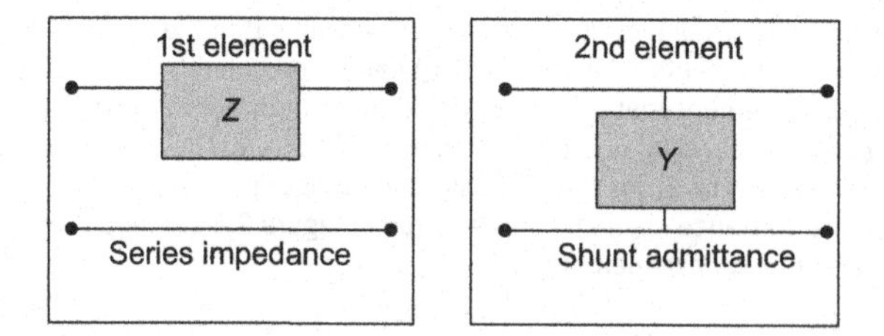

FIGURE 2.11

Series and shunt circuit elements.

EXAMPLE 2.3 INSERTION LOSS OF HYBRID *ABCD* NETWORK

Consider the *ABCD* network depicted in Figure 2.10. Compute the insertion loss of the network given that the impedance circuit element is cascaded with the shunt circuit element of Figure 2.11. Assume $Z_L = Z_0$ to be real.

The insertion loss of the network in Figure 2.10 is in reference to the voltage source V_s and hence can be expressed as

$$P_{\text{Insertion Loss}} = 10 \log_{10}\left(\frac{V_s}{V_2}\right)^2 \tag{2.29}$$

In order to compute the insertion loss, however, we must compute the *ABCD* matrix of the network including the impedance of the generator Z_0. That is,

$$\begin{aligned}(ABCD)_{\text{Total}} &= \begin{bmatrix} A_{\text{Total}} & B_{\text{Total}} \\ C_{\text{Total}} & D_{\text{Total}} \end{bmatrix} = \begin{bmatrix} 1 & Z_0 \\ 0 & 1 \end{bmatrix} (ABCD)_{\text{cascade}} \\ &= \begin{bmatrix} 1 & Z_0 \\ 0 & 1 \end{bmatrix} \begin{bmatrix} 1 & Z \\ Y & YZ+1 \end{bmatrix} = \begin{bmatrix} 1+Z_0Y & Z+Z_0(YZ+1) \\ Y & YZ+1 \end{bmatrix}\end{aligned} \tag{2.30}$$

According to Eqn (2.21), we have

$$\begin{aligned}\frac{V_s}{V_2} &= \frac{A_{\text{Total}}V_2 + B_{\text{Total}}I_2}{V_2} = \frac{A_{\text{Total}}V_2 + B_{\text{Total}}\frac{V_2}{Z_L}}{V_2} = A_{\text{Total}} + \frac{B_{\text{Total}}}{Z_L} \\ &= 1 + Z_0Y + \left.\frac{Z + Z_0(YZ+1)}{Z_L}\right|_{Z_L = Z_0} = 1 + Z_0Y + \frac{Z}{Z_0} + YZ + 1 \\ &= 2 + (Z_0 + Z)Y + \frac{Z}{Z_0}\end{aligned} \tag{2.31}$$

Hence, the insertion loss can be expressed as

$$P_{\text{Insertion Loss}} = 10 \log_{10}\left(\frac{V_s}{V_2}\right)^2 = 10 \log_{10}\left(2 + (Z_0 + Z)Y + \frac{Z}{Z_0}\right)^2 \tag{2.32}$$

EXAMPLE 2.4 *ABCD* NETWORK OF π-MATCHING NETWORK

Consider the π-matching network depicted in Figure 2.12. Determine the equivalent *ABCD* matrix.

In order to determine the overall *ABCD* matrix of the π-matching network of Figure 2.12, we must partition the network into three basic *ABCD* matrices and then compute the product to form the total *ABCD* matrix, that is

$$\begin{aligned}(ABCD)_{\text{Total}} &= \begin{bmatrix} A_{\text{Total}} & B_{\text{Total}} \\ C_{\text{Total}} & D_{\text{Total}} \end{bmatrix} = \begin{bmatrix} 1 & 0 \\ jX_{P_1} & 1 \end{bmatrix} \begin{bmatrix} 1 & jX_S \\ 0 & 1 \end{bmatrix} \begin{bmatrix} 1 & 0 \\ jX_{P_2} & 1 \end{bmatrix} \\ &= \begin{bmatrix} 1 & 0 \\ jX_{P_1} & 1 \end{bmatrix} \begin{bmatrix} 1 - X_SX_{P_2} & jX_S \\ jX_{P_2} & 1 \end{bmatrix} \\ &= \begin{bmatrix} 1 - X_SX_{P_2} & jX_S \\ j(X_{P_1} + X_{P_1}X_S + 1) & -X_{P_1}X_S + 1 \end{bmatrix}\end{aligned} \tag{2.33}$$

As an exercise to the reader, what is the total *ABCD* matrix of a *T*-matching network?

Continued

EXAMPLE 2.4 *ABCD* NETWORK OF π-MATCHING NETWORK—cont'd

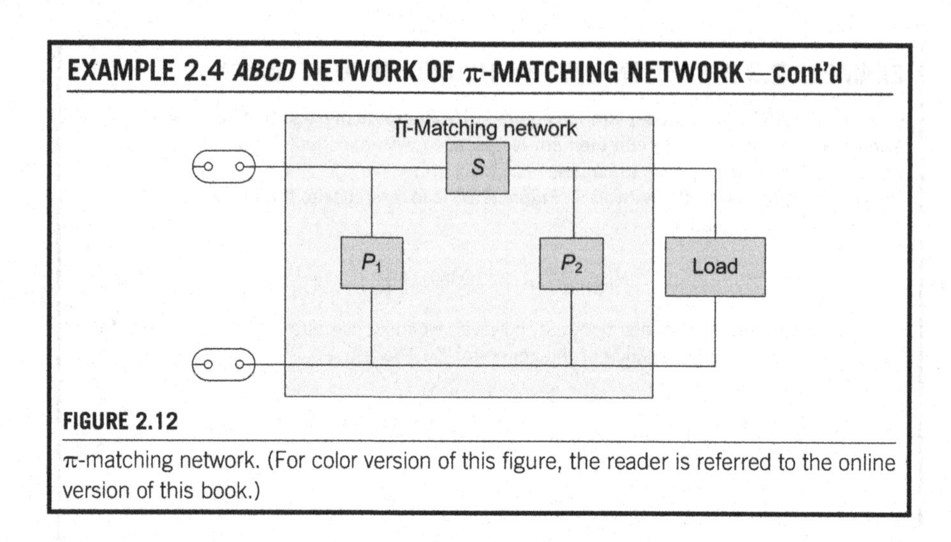

FIGURE 2.12

π-matching network. (For color version of this figure, the reader is referred to the online version of this book.)

2.1.4 Scattering parameters network model

At microwave frequency, implementing open and short circuits for accurate measurements is not practical mainly due to parasitic effects and the possible instability of the device under test. Further complications arise in the biasing requirements when open or short circuit loads are used. Consequently, a more robust method of defining the operating requirements for the various microwave devices than can be pragmatically and accurately measured is needed. The *S*-parameters, based on transmission line theory and traveling waves, fulfill this need [5].

Recall from Chapter 1 that the forward and reverse waves can be described in terms of incident and reflected waves as

$$\begin{aligned} V^+ &= V_0^+ e^{j(\Omega t-\beta z)} \text{ incident wave} \\ V^- &= V_0^- e^{j(\Omega t+\beta z)} \text{ reflected wave} \end{aligned} \tag{2.34}$$

where βz as defined in Chapter 1 is the propagation coefficient. Given the *N*-port network depicted in Figure 2.2, the scattering matrix can be defined in terms of the incident and reflected waves as

$$\underbrace{\begin{bmatrix} V_1^- \\ V_2^- \\ \vdots \\ V_N^- \end{bmatrix}}_{\overline{V}^-} = \underbrace{\begin{bmatrix} S_{1,1} & S_{1,2} & \cdots & S_{1,N} \\ S_{2,1} & S_{2,2} & \cdots & S_{2,N} \\ \vdots & \vdots & \cdots & \vdots \\ S_{N,1} & S_{N,2} & \cdots & S_{N,N} \end{bmatrix}}_{\overline{S}} \underbrace{\begin{bmatrix} V_1^+ \\ V_2^+ \\ \vdots \\ V_N^+ \end{bmatrix}}_{\overline{V}^+} \tag{2.35}$$

$$\overline{V}^- = \overline{S}\overline{V}^+$$

where the individual elements of the scattering matrix are given according to the relation

$$S_{m,n} = \left.\frac{V_m^-}{V_n^+}\right|_{V_l^+=0 \text{ for } l \neq n} \tag{2.36}$$

Each scattering element may be found by driving port n with an incident wave voltage V_n^+ while setting all other incident waves on the other ports to zero. Then the parameter is computed by measuring the reflected wave at the mth port V_m^- and then taking the ratio of the amplitudes V_m^-/V_n^+ as shown in Eqn (2.36). Note that all ports are terminated with matched loads in order to circumvent any reflections.

Define the reflection coefficient as the scattering matrix element according to Eqn (2.36) where $m = n$, or $S_{m,m}$. The reflection coefficient is found by driving port m with an incident wave and measuring the reflected wave. All other ports are terminated with matched loads to avoid reflections. In a similar vein, define the transmission coefficient $S_{m,n}$ where $m \neq n$ as the ratio of the reflected wave at port n due to an incident wave at port m and where all other ports are terminated with a matched load. For any matched port m, the reflection coefficient is zero or $S_{m,m} = 0$. Furthermore, for any passive circuit the reflection coefficient $|S_{m,m}| \leq 1$. A network is said to be reciprocal if

$$S_{m,n} \triangleq S_{n,m} \tag{2.37}$$

which implies that the S-matrix is equal to its transpose, or $S = S^T$. This in turn implies that a reciprocal network is a network that has identical transmission characteristics from port 1 to port 2 and vice versa.

A network is said to be lossless and reciprocal if

$$\sum_{m=1}^{N} |S_{m,n}|^2 = \sum_{m=1}^{N} S_{m,n} S_{m,n}^* = 1 \tag{2.38}$$

The relationship in Eqn (2.38) is true due to the energy conservation of the network, which further implies that the S-matrix is unitary, that is

$$SS^H = I \text{ or } I - SS^H = 0 \tag{2.39}$$

where $(\ )^H$ denotes hermitian or conjugate transpose. A further interpretation of a lossless network is the implication that the power incident on the network equals to the power reflected back from the network, which is exactly in line with the preservation of energy assumption. Finally, we note that lossless networks (or almost lossless) are typically used as matching networks between amplifier stages.

A lossy network on the other hand is a network where $I - SS^H$ is positive definite or

$$I - SS^H > 0 \tag{2.40}$$

The relationship in Eqn (2.40) implies that the eigenvalues of the S-matrix reside in the left-half plane, which in turn signifies that the impulse response of the network is made up entirely of decaying exponentials. The physical implication of a lossy

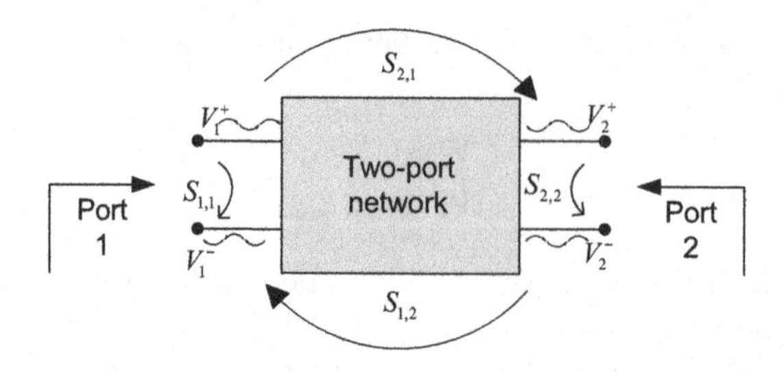

FIGURE 2.13

S-Parameters of two-port network.

network is that the net power reflected from the network is less than the incident power going into it.

To further illustrate the workings of *S*-parameters, let us consider the two-port network depicted in Figure 2.13. Using the relationship in Eqn (2.35), we obtain

$$\begin{bmatrix} V_1^- \\ V_2^- \end{bmatrix} = \begin{bmatrix} S_{1,1} & S_{1,2} \\ S_{2,1} & S_{2,2} \end{bmatrix} \begin{bmatrix} V_1^+ \\ V_2^+ \end{bmatrix} \tag{2.41}$$

$$V_1^- = S_{1,1}V_1^+ + S_{1,2}V_2^+ \text{ and } V_2^- = S_{2,1}V_1^+ + S_{2,2}V_2^+$$

The reflection coefficient at port 1 can be obtained by setting $V_2^+ = 0$ in Eqn (2.41) or

$$S_{1,1} = \left.\frac{V_1^-}{V_1^+}\right|_{V_2^+=0} = \Gamma_1 \tag{2.42}$$

Similarly, the reflection coefficient at port 2 can be obtained by setting $V_1^+ = 0$ in Eqn (2.41) or

$$S_{2,2} = \left.\frac{V_2^-}{V_2^+}\right|_{V_1^+=0} = \Gamma_2 \tag{2.43}$$

The transmission coefficients from port 1 to port 2 $S_{2,1}$ and from port 2 to port 1 $S_{1,2}$ can then be computed as

$$S_{2,1} = \left.\frac{V_2^-}{V_1^+}\right|_{V_2^+=0} \text{ and } S_{1,2} = \left.\frac{V_1^-}{V_2^+}\right|_{V_1^+=0} \tag{2.44}$$

At this point, it is important to introduce certain key design parameters. The gain defined as the ratio of output power P_{out} to the available power from the source P_s is given as

$$G_{dB} = 10\log_{10}\left(|S_{2,1}|^2\right) \tag{2.45}$$

The parameter $S_{2,1}$ also dictates the phase shift of the network, that is, $\angle S_{2,1}$. The insertion loss, on the other hand, is defined as the ratio of the power available from the source to the power delivered to the *matched* load, that is

$$IL|_{dB} = -10 \log_{10}\left(\frac{|S_{2,1}|^2}{1-|S_{1,1}|^2}\right) \tag{2.46}$$

Another important parameter of interest is the return loss at the input defined for a matched load as

$$RL_{in}|_{dB} = 20 \log_{10}\left(\frac{1}{|S_{1,1}|}\right) \tag{2.47}$$

This measure specifies the closeness of the input impedance to that of the network. A similar parameter can be defined for the output return loss given as

$$RL_{out}|_{dB} = 20 \log_{10}\left(\frac{1}{|S_{2,2}|}\right) \tag{2.48}$$

Similar to the definition of the input return loss, the output return loss defines the closeness of the output impedance to that of the network. Finally, define the reverse isolation as *absolute value* of the reverse gain or

$$G_{reverse}|_{dB} = 10 \log_{10}\left(|S_{1,2}|^2\right) \tag{2.49}$$

The VSWRs at the input port as well as the output port can also be defined in terms of S-parameters as

$$\begin{aligned} VSWR_{port1} &= \frac{1+S_{1,1}}{1-S_{1,1}} \text{ VSWR at input port (port 1)} \\ VSWR_{port1} &= \frac{1+S_{2,2}}{1-S_{1,1}} \text{ VSWR at output port (port 2)} \end{aligned} \tag{2.50}$$

These design measures are used repeatedly throughout the text.

EXAMPLE 2.5 *S*-PARAMETERS OF A *T*-ATTENUATOR TWO-PORT NETWORK

Consider the two-port network depicted in Figure 2.14. Compute the value of R_3 for which there is no reflected power. The characteristic impedance is 50 ohm

Recall that for a matched load, $S_{1,1} = \Gamma = 0$ implies a return loss of ∞ dB, and hence there is no reflected power, say as opposed to a total reflection of the incident power where $\Gamma = 1$. In order to obtain the reflection coefficient $S_{1,1}$, the network must be terminated by a matched load of 50 ohm. According to Eqn (2.42), we have

$$S_{1,1} = \left.\frac{V_1^-}{V_1^+}\right|_{V_2^+=0} = \left.\frac{Z_{in}-Z_{matched}}{Z_{in}+Z_{matched}}\right|_{Z_{matched}=50 \text{ ohm}} \tag{2.51}$$

Continued

EXAMPLE 2.5 *S*-PARAMETERS OF A *T*-ATTENUATOR TWO-PORT NETWORK—cont'd

where Z_{in} can be computed as

$$
\begin{aligned}
Z_{in} &= R_1 + \frac{R_3(R_2+50)}{R_3+R_2+50} \\
Z_{in} &= 12 + \frac{R_3(R_2+50)}{R_3+12+50}
\end{aligned}
\tag{2.52}
$$

Sweeping the value of R_3 say between 1 and 200 ohm we obtain the return loss in dB as a function of R_3 as shown in Figure 2.15.

It turns out that at approximately 98 ohm, the return loss is $+\infty$. Given that $R_1 = R_2$, the resistive *T*-attenuator depicted in Figure 2.14 is also known as a pad. Pads are circuits that are commonly used when we desire to maintain the input/output matching while increasing isolation between various circuits.

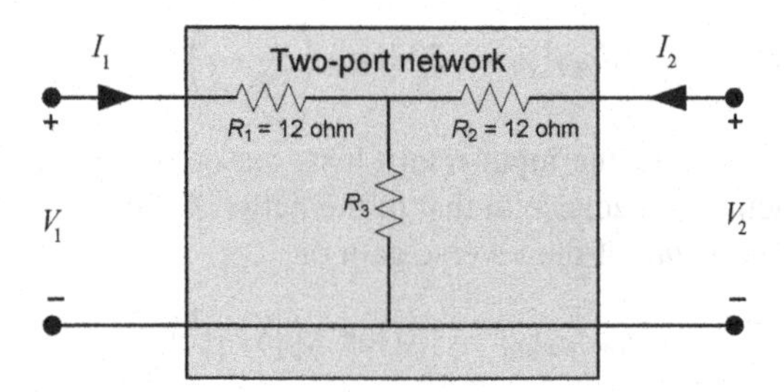

FIGURE 2.14

Simple two-port network.

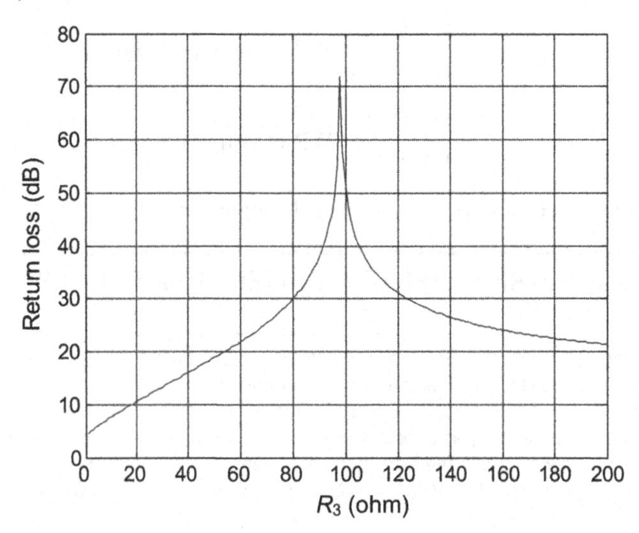

FIGURE 2.15

Return loss in dB versus value of R_3. (For color version of this figure, the reader is referred to the online version of this book.)

EXAMPLE 2.6 *S*-PARAMETERS OF RESISTIVE NETWORK

Consider the two-port resistive network depicted in Figure 2.16. Compute the values $S_{1,1}$ and $S_{2,1}$.

In order to compute the desired parameters, we must first compute the input impedance Z_{in}

$$\begin{aligned} Z_{in} &= R_1 + R_2 \| Z_0 = \frac{R_2 Z_0 + R_1 R_2 + R_1 Z_0}{R_2 + Z_0} \\ &= \frac{100 \times 50 + 30 \times 100 + 30 \times 50}{100 + 50} \approx 63.3 \text{ ohm} \end{aligned} \tag{2.53}$$

$S_{1,1}$ can then be computed as

$$S_{1,1} = \left.\frac{V_1^-}{V_1^+}\right|_{V_2^+=0} = \frac{Z_{in} - Z_0}{Z_{in} + Z_0} = \frac{63.3 - 50}{63.3 + 50} \approx 0.12 \tag{2.54}$$

$S_{2,1}$ on the other hand can be found as a function of $S_{1,1}$ as

$$\begin{aligned} S_{2,1} &= \frac{V_2}{V_1}(1 + S_{1,1}) = \underbrace{\frac{R_2 Z_0}{R_2 Z_0 + R_2 R_1 + R_1 Z_0}}_{=\frac{V_2}{V_1}}(1 + S_{1,1}) \\ &= \frac{100 \times 50}{100 \times 50 + 100 \times 30 + 30 \times 50}(1 + 0.12) \approx 0.59 \end{aligned} \tag{2.55}$$

where obviously according to Eqn (2.55)

$$\frac{V_2}{V_1} = \frac{R_2 \| Z_0}{R_1 + R_2 \| Z_0} = \frac{R_2 Z_0}{R_2 Z_0 + R_2 R_1 + R_1 Z_0} \tag{2.56}$$

Note that for passive networks $S_{2,1} = S_{1,2}$.

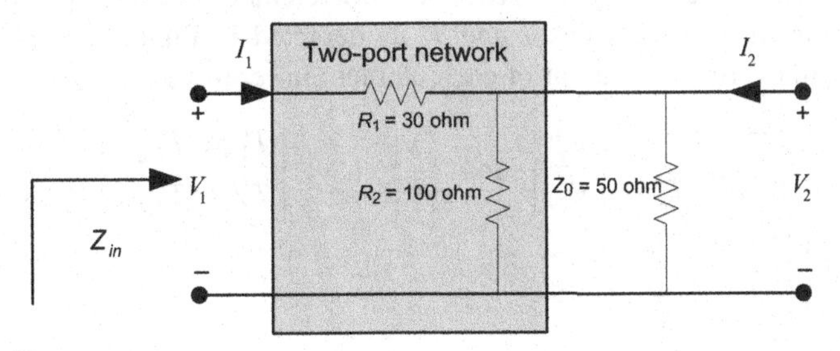

FIGURE 2.16

Two-port resistive network.

2.1.5 Transmission parameters network model

The scattering parameters of a cascade of two-port networks cannot be easily manipulated via matrix multiplication. In order to do so, we resort to the transmission matrix (*T*-matrix) representation. The *T*-matrix is in fact none other than the

ABCD-matrix for high frequency networks. Like the *ABCD*-matrix, the *T*-matrix is simple to manipulate in the sense that a cascade of two or more two-port network circuit elements can be readily obtained via matrix multiplication. To do so, we use the terminology used in Eqn (2.41) and define the two-port *T*-matrix as

$$\begin{bmatrix} V_1^- \\ V_1^+ \end{bmatrix} = \begin{bmatrix} T_{1,1} & T_{1,2} \\ T_{2,1} & T_{2,2} \end{bmatrix} \begin{bmatrix} V_2^+ \\ V_2^- \end{bmatrix} \tag{2.57}$$

However, since most instruments conduct measurements using *S*-parameters, a conversion between *T*-matrix and *S*-matrix elements is necessary. The relation can be stated as

$$\begin{bmatrix} S_{1,1} & S_{1,2} \\ S_{2,1} & S_{2,2} \end{bmatrix} = \begin{bmatrix} \dfrac{T_{1,2}}{T_{2,2}} & \dfrac{T_{1,1}T_{2,2} - T_{1,2}T_{2,1}}{T_{2,2}} \\ \dfrac{1}{T_{2,2}} & -\dfrac{T_{2,1}}{T_{2,2}} \end{bmatrix} \tag{2.58}$$

In a likewise manner, the transmission matrix can be expressed in terms of the *S*-parameters as

$$\begin{bmatrix} T_{1,1} & T_{1,2} \\ T_{2,1} & T_{2,2} \end{bmatrix} = \begin{bmatrix} -\dfrac{S_{1,1}S_{2,2} - S_{1,2}S_{2,1}}{S_{2,1}} & \dfrac{S_{1,1}}{S_{2,1}} \\ -\dfrac{S_{2,2}}{S_{2,1}} & \dfrac{1}{S_{2,1}} \end{bmatrix} \tag{2.59}$$

Note that the definition of the *T*-parameters as presented herein is not unique.

Next, consider the cascade of two network elements characterized by their respective transmission matrices T and T' as depicted in Figure 2.17. From Eqn (2.57), we can express the output of each element in terms of its inputs as

$$\begin{bmatrix} V_1^- \\ V_1^+ \end{bmatrix} = \begin{bmatrix} T_{1,1} & T_{1,2} \\ T_{2,1} & T_{2,2} \end{bmatrix} \begin{bmatrix} V_2^+ \\ V_2^- \end{bmatrix} \text{ and } \begin{bmatrix} V_1'^- \\ V_1'^+ \end{bmatrix} = \begin{bmatrix} T'_{1,1} & T'_{1,2} \\ T'_{2,1} & T'_{2,2} \end{bmatrix} \begin{bmatrix} V_2'^+ \\ V_2'^- \end{bmatrix}$$
$$\text{where } \begin{bmatrix} V_1'^- \\ V_1'^+ \end{bmatrix} = \begin{bmatrix} V_2^+ \\ V_2^- \end{bmatrix} \tag{2.60}$$

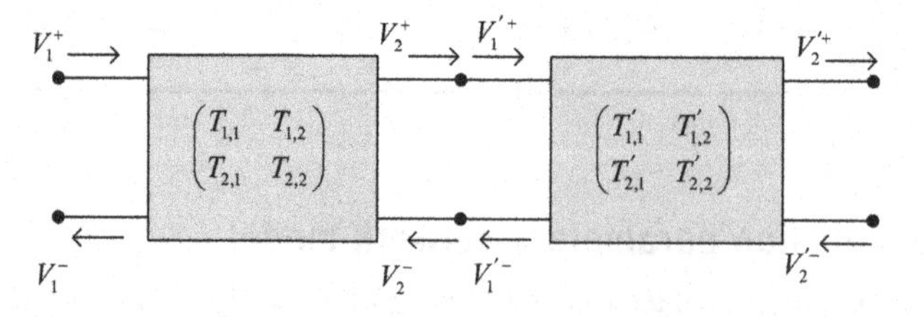

FIGURE 2.17

Cascade of transmission matrices.

Then the cascade matrix is the product of the transmission matrices T and T', and hence the input and output relationships can be expressed as

$$\begin{aligned}\begin{bmatrix} V_1^- \\ V_1^+ \end{bmatrix} &= \begin{bmatrix} T_{1,1} & T_{1,2} \\ T_{2,1} & T_{2,2} \end{bmatrix} \begin{bmatrix} T'_{1,1} & T'_{1,2} \\ T'_{2,1} & T'_{2,2} \end{bmatrix} \begin{bmatrix} V_2'^+ \\ V_2'^- \end{bmatrix} \\ &= \begin{bmatrix} T_{1,1}T'_{1,1} + T_{1,2}T'_{2,1} & T_{1,1}T'_{1,2} + T_{1,2}T'_{2,2} \\ T_{2,1}T'_{1,1} + T_{2,2}T'_{2,1} & T_{2,1}T'_{1,2} + T_{2,2}T'_{2,2} \end{bmatrix}\end{aligned} \tag{2.61}$$

Again as in the case of the *ABCD* matrix multiplication, keep in mind that the matrix multiplication is not commutative.

2.1.6 Power gain based on two-port *S*-parameters model

At this point, it is instructive to consider the power transfer characteristics of a given two-port network as depicted in Figure 2.18. The ultimate goal is to be able to determine vital parameters such as power gain, transducer power gain, and available gain.

To do so, several parameters have to be determined first. Recall from Eqn (2.42) and Eqn (2.43) that the reflection coefficients from the network looking toward the source and from the network looking toward the load can be stated in terms of the characteristic impedance Z_0 as

$$\begin{aligned}\Gamma_1 &= \frac{Z_s - Z_0}{Z_s + Z_0} \quad \text{reflection coefficient from network looking into the source} \\ \Gamma_2 &= \frac{Z_{Load} - Z_0}{Z_{Load} + Z_0} \quad \text{reflection coefficient from network looking into the load}\end{aligned} \tag{2.62}$$

Next let us determine the input reflection coefficient $\Gamma_{in} = V_1^-/V_1^+ = (Z_{in} - Z_{char})/(Z_{in} + Z_{char})$ where Z_{in} is the impedance looking into port 1 when port 2 is terminated. Recall that Γ_2 is the reflection coefficient from the network looking into the load, and hence $\Gamma_2 = V_2^+/V_2^-$, then using Eqn (2.41) we obtain

$$\begin{aligned}V_1^- &= S_{1,1}V_1^+ + S_{1,2}\Gamma_2 V_2^- \\ V_2^- &= S_{2,1}V_1^+ + S_{2,2}\Gamma_2 V_2^-\end{aligned} \tag{2.63}$$

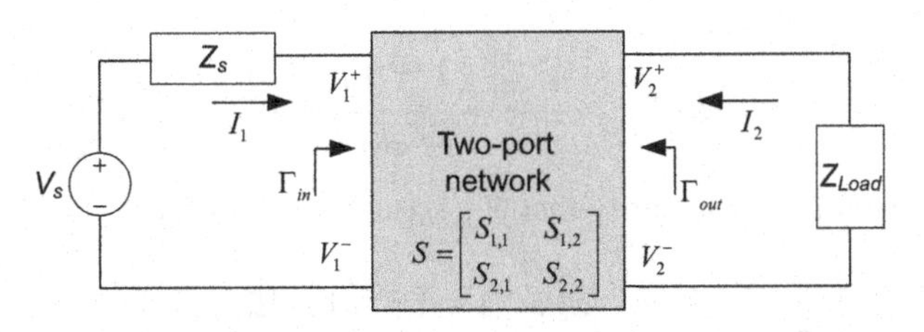

FIGURE 2.18

Two-port network with voltage source, source impedance, and load impedance.

Note that from Eqn (2.63), V_2^- can be expressed as

$$V_2^- = S_{2,1}V_1^+ + S_{2,2}\Gamma_2 V_2^- \Rightarrow V_2^- = \frac{S_{2,1}V_1^+}{1 - S_{2,2}\Gamma_2} \tag{2.64}$$

Furthermore, substituting Eqn (2.64) into Eqn (2.63) reveals that

$$\begin{aligned} V_1^- &= S_{1,1}V_1^+ + S_{1,2}\Gamma_2 V_2^- \Rightarrow \\ \Gamma_{in} &= \frac{V_1^-}{V_1^+} = S_{1,1} + \frac{S_{1,2}\Gamma_2 V_2^-}{V_1^+} = S_{1,1} + \frac{S_{1,2}S_{2,1}\Gamma_2}{1 - S_{2,2}\Gamma_2} \end{aligned} \tag{2.65}$$

In a similar fashion, we can determine the reflection coefficient Γ_{out} as

$$\Gamma_{out} = \frac{V_2^-}{V_2^+} = S_{2,2} + \frac{S_{1,2}S_{2,1}\Gamma_1}{1 - S_{1,1}\Gamma_1} \tag{2.66}$$

Again, recall that Γ_{out} is the reflection coefficient obtained by looking into port 2 given that port 1 is terminated by the source impedance Z_s.

The average power delivered to the network can then be determined as

$$P_{input} = \frac{|V_1^+|^2\left(1 - |\Gamma_{in}|^2\right)}{2Z_{char}} \tag{2.67}$$

where one can obtain V_1^+ via voltage division as

$$V_1 = V_1^+ + V_1^- = V_1^+(1 + \Gamma_{in}) = \left.\frac{V_s Z_{in}}{Z_{in} + Z_s}\right|_{Z_{in} = Z_{char}\frac{1+\Gamma_{in}}{1-\Gamma_{in}}} \tag{2.68}$$

Substituting Eqn (2.65) into Eqn (2.68) and placing the result in Eqn (2.67), we obtain

$$P_{input} = \frac{|V_s|^2|1 - \Gamma_s|^2\left(1 - |\Gamma_{in}|^2\right)}{8Z_{char}|1 - \Gamma_{in}\Gamma_s|^2} \tag{2.69}$$

In a similar analytical fashion, the power delivered to the load can be found as

$$P_{Load} = \frac{|V_2^-|^2\left(1 - |\Gamma_{Load}|^2\right)}{2Z_{char}} \tag{2.70}$$

After manipulation, we can verify that P_{Load} can be written as

$$P_{Load} = \frac{|V_s|^2|S_{2,1}|^2|1 - \Gamma_s|^2\left(1 - |\Gamma_{Load}|^2\right)}{8Z_{char}|1 - \Gamma_{in}\Gamma_s|^2|1 - S_{2,2}\Gamma_{Load}|^2} \tag{2.71}$$

The power gain can then be obtained as the ratio of Eqn (2.71) over Eqn (2.69) as

$$G_{\text{Power Gain}} = \frac{P_{Load}}{P_{input}} = \frac{\left(1 - |\Gamma_{Load}|^2\right)|S_{2,1}|^2}{|1 - S_{2,2}\Gamma_{Load}|^2\left(1 - |\Gamma_{in}|^2\right)} \tag{2.72}$$

There is another power gain parameter that is of interest, namely the ratio of the maximum power that can be delivered to the load that is available from the network to the maximum power delivered to the network that is available from the source. First, it can be shown that the maximum power delivered to the load that is available from the network $P_{network,load}$ is

$$P_{network,load} = P_{input}\big|_{\Gamma_{in}=\Gamma_s^*} = \frac{|V_s|^2|1 - \Gamma_s|^2}{8Z_{char}\left(1 - |\Gamma_s|^2\right)} \tag{2.73}$$

Similarly, the maximum power delivered to the network that is available from the source $P_{source,network}$ is

$$P_{source,network} = P_{Load}\big|_{\Gamma_{Load}=\Gamma_{out}^*} = \frac{|V_s|^2|S_{2,1}|^2|1 - \Gamma_s|^2}{8Z_{char}|1 - S_{1,1}\Gamma_s|^2\left(1 - |\Gamma_{out}|^2\right)} \tag{2.74}$$

The available power gain can then be expressed as the ratio of Eqn (2.73) to Eqn (2.74), that is

$$G_{\text{Available Power Gain}} = \frac{P_{network,load}}{P_{source,network}} = \frac{\left(1 - |\Gamma_s|^2\right)|S_{2,1}|^2}{|1 - S_{1,1}\Gamma_s|^2\left(1 - |\Gamma_{out}|^2\right)} \tag{2.75}$$

Note that both Eqn (2.72) and Eqn (2.75) are independent of V_s and that Eqn (2.73) and Eqn (2.74) are both independent of the input or load impedances.

2.2 Signal flow graphs

Signal flow graph is a design technique that allows the engineer to follow the incident and reflected waves through the network. The technique itself will provide a unique insight into the working of microwave networks [1,6].

2.2.1 Basis operations

Consider the two-port S-parameter relationship given in Eqn (2.41) or $V_1^- = S_{1,1}V_1^+ + S_{1,2}V_2^+$ and $V_2^- = S_{2,1}V_1^+ + S_{2,2}V_2^+$. Designate each variable $\{V_1^-, V_1^+, V_2^-, V_2^+\}$ as a node. Each of the S-parameters will constitute a branch. Each node is the sum of the branches entering it. Consider first the relation

$$V_1^- = S_{1,1}V_1^+ + S_{1,2}V_2^+ \tag{2.76}$$

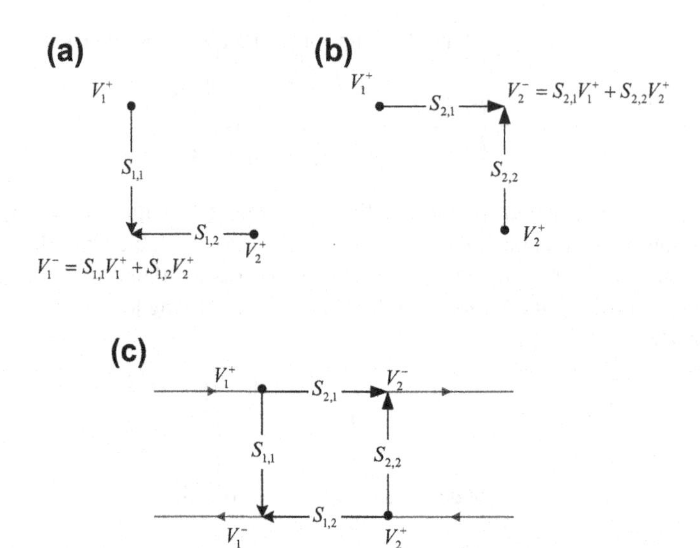

FIGURE 2.19

Signal flow graph for a two-port network.

The signal flow graph for Eqn (2.76) is depicted in Figure 2.19(a). Similarly, V_2^- can be expressed as

$$V_2^- = S_{2,1}V_1^+ + S_{2,2}V_2^+ \tag{2.77}$$

The relationships between the incident and reflected waves can be readily derived. It can be seen that the reflected wave V_1^- is made up of both incident waves V_1^+ and V_2^+ as illustrated in Figure 2.19(a). Similarly, the reflected wave V_2^- is also made up of a combination of the incident waves V_1^+ and V_2^+ as shown in Figure 2.19(b). Combining Figure 2.19(a) and Figure 2.19(b) produces the flow graph for the two-port network as depicted in Figure 2.19(c). Now, it can be seen that the incident wave V_1^+, for example, is split where part of it propagates out of the network and contributes to V_2^- and part of it is reflected back and contributes to V_1^-. The *S*-parameters serve as complex multipliers to reflected waveforms.

Signal flow graphs can be used to effectively model linear systems with combinations of feedforward and feedback paths. The transfer function of such systems can be derived by using Mason's gain rule.[1] Before we delve further into the analysis of signal flow graphs and Mason's gain rule, certain ground rules and terminology

[1]Mason's gain rule or Mason's gain formula is named after Samuel Jefferson Mason. This method can be used to obtain transfer function of a linear continuous time or a discrete time system.

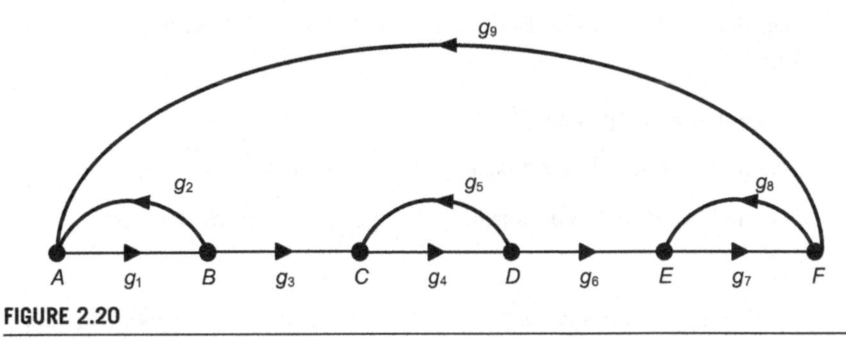

FIGURE 2.20

Signal flow graph comprised of several forward and feedback paths.

need to be established. Consider the signal flow graph depicted in Figure 2.20. Consider the following rules:

1. A loop is a closed signal path that starts and ends at the same node. In a loop, no single node may be encountered more than once. For example, consider the loop $(\widehat{A\,B}A)$ depicted in the signal flow graph shown in Figure 2.20. The signal enters the A node, then proceeds to the B node via the AB path with gain g_1, and then loops back to the A node via the BA path with path gain g_2. Other loops in Figure 2.20 include $(\widehat{C\,D}C)$, $(\widehat{E\,F}E)$, and $(\widehat{A\,B}CDEFA)$.
2. Two loops are said to be non-touching if said loops do not share a common signal node. One example of non-touching loops, according to Figure 2.20 are $(\widehat{A\,B}A)$ and $(\widehat{C\,D}C)$. On the other hand, an example of two touching loops would be $(\widehat{A\,B}A)$ and $(\widehat{A\,B}CDEFA)$.
3. The gain of a loop is the product of all path gains in the loop. For example, the gain of the loop $(\widehat{E\,F}E)$ is $g_7 g_8$.

At this point, we state Mason's gain rule as the output-to-input ratio defined as

$$G = \frac{\sum_i g_i \delta^{(i)}}{\Delta} \tag{2.78}$$

where g_i is the linear gain of the ith-forward path and

$$\begin{aligned} \delta^{(i)} &= 1 - \sum \ell_1^{(i)} + \sum \ell_2^{(i)} - \sum \ell_3^{(i)} + \cdots \\ \Delta &= 1 - \sum \ell_1 + \sum \ell_2 - \sum \ell_3 + \cdots \end{aligned} \tag{2.79}$$

A further explanation of Eqn (2.79) is necessary. The summation $\sum \ell_1$ is comprised of the sum of all loop gains. In Figure 2.20, this implies

$$\sum \ell_1 = g_1 g_2 + g_4 g_5 + g_7 g_8 + g_1 g_3 g_4 g_6 g_7 g_9 \tag{2.80}$$

The summation $\sum \ell_2$ is the sum of *products of non-touching loop gains* taken two at a time, that is

$$\sum \ell_2 = g_1 g_2 g_4 g_5 + g_1 g_2 g_7 g_8 + g_4 g_5 g_7 g_8 \tag{2.81}$$

Hence the determinant Δ in Eqn (2.78) and Eqn (2.79) can be textually expressed as

$$\begin{aligned}\Delta = 1 - &\text{ sum of all loop gains} \\ &+ \text{sum of products of non-touching loop gains taken two at a time} \\ &- \text{sum of products of non-touching loop gains taken three at a time} \\ &+ \cdots\end{aligned} \tag{2.82}$$

Next we examine the determinant $\delta^{(i)}$ *associated* with the *i*th-forward path. The summation $\sum \ell_1^{(i)}$ is comprised of the sum of all loop gains that do not touch the *i*th-forward path at any node. Similarly, the summation $\sum \ell_2^{(i)}$ is the sum of the products of non-touching loop gains taken two at a time and excluding the *i*th-forward path. This can be best illustrated via some examples.

EXAMPLE 2.7 SINGLE-PORT GENERATOR CIRCUIT

Figure 2.21 depicts a single-port network model of an ideal voltage source $V_{s,ideal}$ connected to source impedance Z_S. In accordance with the *S*-parameter network models discussed thus far, the terminal current is conventionally entering the port. Assume the characteristic impedance Z_{char} at the port to be known. Determine the terminal voltage V_S and the source reflection coefficient Γ_S. Draw the signal flow graph of Figure 2.21.

The total terminal voltage V_S can be found as the sum of incident and reflected voltages as

$$\begin{aligned}V_S &= V_{s,ideal} + I_S Z_S \\ &= V_{s,ideal} + \left(I_S^+ + I_S^-\right) Z_S \\ &= V_{s,ideal} + \left(\frac{V_S^+ - V_S^-}{Z_{char}}\right) Z_S\end{aligned} \tag{2.83}$$

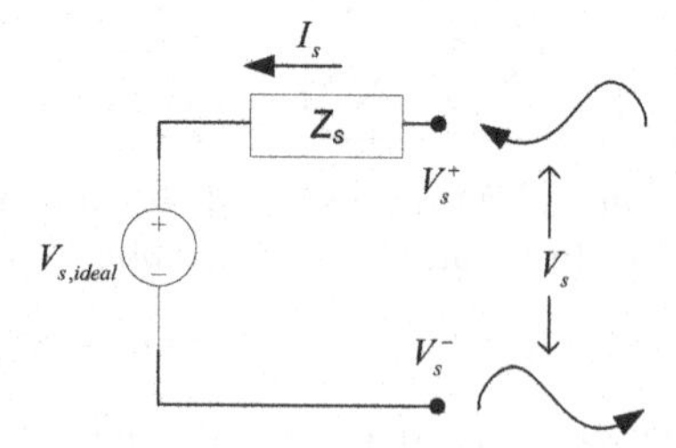

FIGURE 2.21

Single-port voltage source circuit.

EXAMPLE 2.7 SINGLE-PORT GENERATOR CIRCUIT—cont'd

where I_S^+ and I_S^- are the incident and reflected currents. The reflected voltage V_S^- can be computed by manipulating Eqn (2.83)

$$\begin{aligned} V_S &= V_S^+ + V_S^- = V_{S,ideal} + \left(\frac{V_S^+ - V_S^-}{Z_{char}}\right) Z_S \\ \Rightarrow \left(\frac{Z_{char} + Z_S}{Z_{char}}\right) V_S^- &= V_{S,ideal} - \left(\frac{Z_{char} - Z_S}{Z_{char}}\right) V_S^+ \\ \Rightarrow V_S^- &= \left(\frac{Z_{char}}{Z_{char} + Z_S}\right) V_{S,ideal} - \left(\frac{Z_{char} - Z_S}{Z_{char} + Z_S}\right) V_S^+ \\ \Rightarrow V_S^- &= \left(\frac{Z_{char}}{Z_{char} + Z_S}\right) V_{S,ideal} + \Gamma_S V_S^+ \end{aligned} \tag{2.84}$$

where the source reflection coefficient is $\Gamma_S = (Z_S - Z_{char})/(Z_S + Z_{char})$. Note that if the source impedance is matched to the characteristic impedance ($Z_{char} = Z_S$) then the reflected voltage becomes

$$V_S^- = \left(\frac{Z_{char}}{Z_{char} + Z_S}\right) V_{S,ideal} \tag{2.85}$$

The signal flow graph is depicted in Figure 2.22.

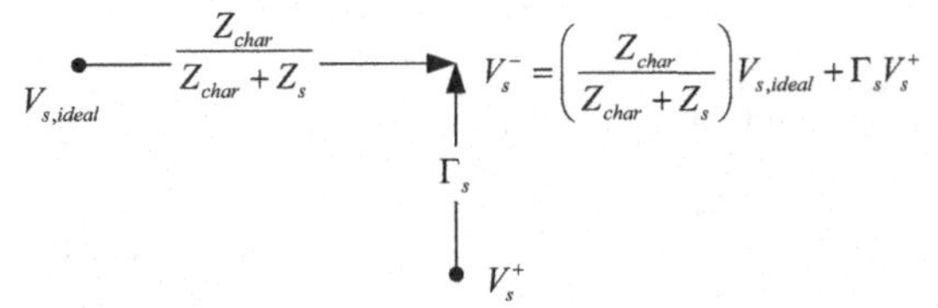

FIGURE 2.22

Signal flow diagram of voltage source circuit of Figure 2.21.

EXAMPLE 2.8 TWO-PORT NETWORK ENCOUNTERED IN AMPLIFIER DESIGN

Consider the typical amplifier circuit represented by a generator connected to a load via a two-port network as shown in Figure 2.23. Using a signal flow graph and Mason's gain rule, determine the input and output reflection coefficients. In order to simplify the analysis, assume that Γ_{Source} is negligible or $\Gamma_{Load} \approx 0$.

Let $b_g = \left(\frac{1}{Z_{char}+Z_S}\right) V_{S,ideal}$ then the signal flow graph corresponding to the circuit depicted in Figure 2.23 is shown in Figure 2.24.

First, the input reflection coefficient can be determined by computing the ratio of V_1^- to V_1^+. In order to apply Mason's rule, a close examination of Figure 2.24 reveals two forward paths that lead from V_1^+ to V_1^- as shown in Figure 2.25.

Continued

EXAMPLE 2.8 TWO-PORT NETWORK ENCOUNTERED IN AMPLIFIER DESIGN—cont'd

Based on Figure 2.25, the first path gain is $gain_{path1} = S_{1,1}$ whereas the second path gain is $gain_{path2} = S_{2,1} \times 1 \times \Gamma_{Load} \times 1 \times S_{1,2}$ or $gain_{path2} = S_{2,1}\Gamma_{Load}S_{1,2}$. Next, according to Figure 2.24, there is only one non-touching loop that starts and ends at the same node as shown in Figure 2.26.

Again, based on Figure 2.26, the loop gain is given as $gain_{loop} = 1 \times \Gamma_{Load} \times 1 \times S_{2,2} = \Gamma_{Load}S_{2,2}$, then according to Mason's rule, the input reflection coefficient is

$$\Gamma_{input} = \frac{V_1^-}{V_1^+} = \frac{gain_{path1}(1 - gain_{loop}) + gain_{path2}}{1 - gain_{loop}}$$
$$= gain_{path1} + \frac{gain_{path2}}{1 - gain_{loop}} = S_{1,1} + \frac{S_{2,1}\Gamma_{Load}S_{1,2}}{1 - \Gamma_{Load}S_{2,2}} \tag{2.86}$$

In a similar manner, the output reflection coefficient can also be computed. First, we must compute the ratio $\Gamma_{output} = V_2^-/V_2^+$.

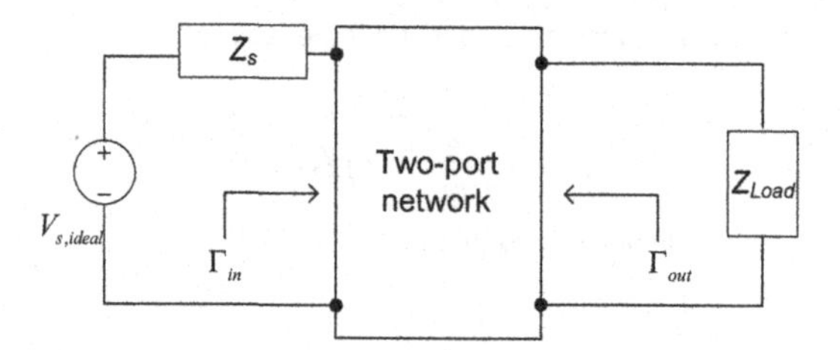

FIGURE 2.23

Voltage source with two-port network and load termination.

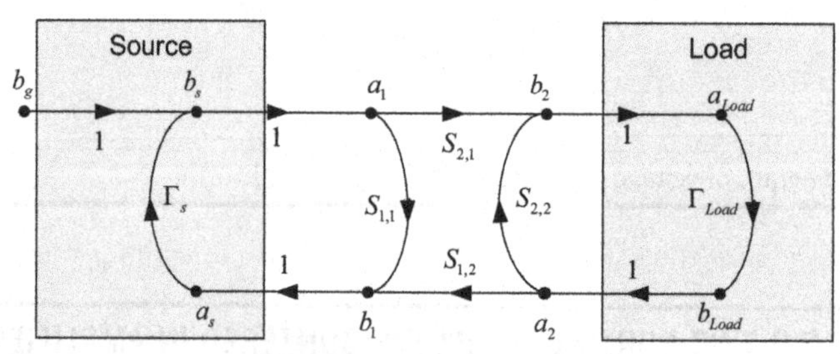

FIGURE 2.24

Signal flow graph representation of the network shown in Figure 2.23.

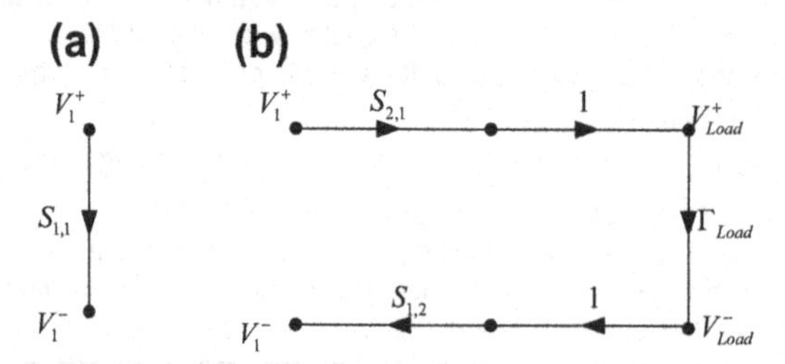

FIGURE 2.25

Two distinct forward signal paths that lead from V_1^+ to V_1^-.

EXAMPLE 2.8 TWO-PORT NETWORK ENCOUNTERED IN AMPLIFIER DESIGN—cont'd

There are two forward paths that lead from V_2^+ to V_2^- as shown in Figure 2.27.

The first gain path is $gain_{path1} = S_{2,2}$ whereas the second gain path is the product $gain_{path2} = S_{1,2}\Gamma_s S_{2,1}$. Furthermore, there is one non-touching loop as shown in Figure 2.28.

The gain of the loop depicted in Figure 2.28 is estimated as $gain_{loop} = \Gamma_s S_{1,1}$. According to Mason's rule, the output reflection coefficient is ratio

$$\begin{aligned}\Gamma_{output} &= \frac{V_2^-}{V_2^+} = \frac{gain_{path1}\left(1 - gain_{loop}\right) + gain_{path2}}{1 - gain_{loop}} \\ &= gain_{path1} + \frac{gain_{path2}}{1 - gain_{loop}} = S_{2,2} + \frac{S_{1,2}\Gamma_s S_{2,1}}{1 - \Gamma_{Load} S_{1,1}}\end{aligned} \tag{2.87}$$

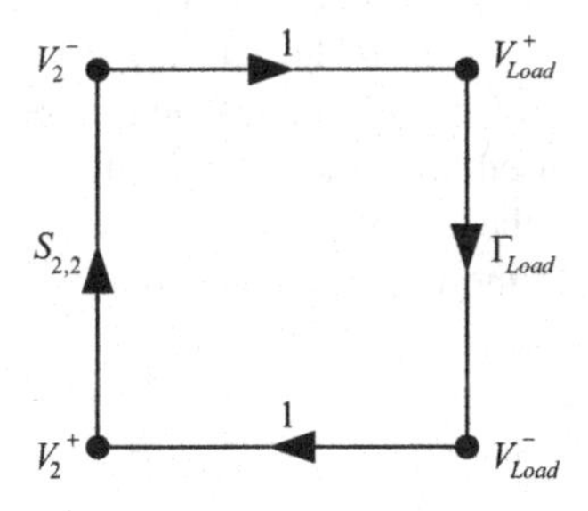

FIGURE 2.26

First non-touching loop of Figure 2.24.

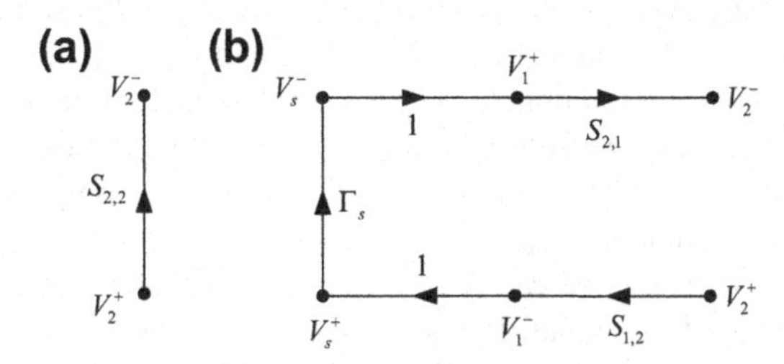

FIGURE 2.27

Two distinct forward signal paths that lead from V_2^+ to V_2^-.

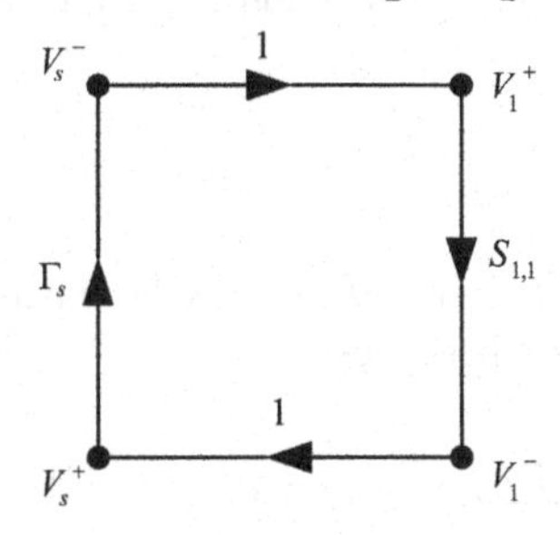

FIGURE 2.28

Second non-touching loop of Figure 2.24.

2.2.2 Power gain equations based on signal flow graphs

Signal flow graph principles can be used to develop the various power gain relationships in a two-port or N-port network in a simple manner. To do so, consider the traveling voltage wave relationships defined in Eqns (2.76) and (2.77), respectively. Define

$$\begin{aligned} a_1 &= \frac{V_1^+}{\sqrt{Z_{char}}} \quad a_2 = \frac{V_2^+}{\sqrt{Z_{char}}} \\ b_1 &= \frac{V_1^-}{\sqrt{Z_{char}}} \quad b_2 = \frac{V_2^-}{\sqrt{Z_{char}}} \end{aligned} \tag{2.88}$$

where again Z_{char} is the characteristic impedance of the transmission line. Note that the square of the magnitude of the new variables a_1, a_2, b_1, and b_2 divided by one-half are traveling power waves. That is, based on the variables a_1 and a_2, we can estimate the incident power waves on ports 1 and 2, respectively. Similarly, the variables b_1 and b_2 can be used to estimate the reflected power waves on ports 1 and 2, respectively. Given the variable definitions in Eqn (2.88), we can state that the S-parameters relate to the new variables as

$$\begin{aligned} b_1 &= S_{1,1}a_1 + S_{1,2}a_2 \\ b_2 &= S_{2,1}a_1 + S_{2,2}a_2 \end{aligned} \tag{2.89}$$

Thus far, it has been shown that in order to allow for maximum power transfer, the load source reflection coefficient Γ_s is equal to the conjugate of the load reflection coefficient Γ_{Load}. However, given an amplifier for example, the source can supply the maximum power to the input if and only if the two-port input reflection coefficient Γ_{input} is the complex conjugate of the source reflection coefficient Γ_s. In a similar vein, the load will receive maximum power from the two-port device if and only if the output reflection coefficient Γ_{output} is the complex conjugate of the load reflection coefficient Γ_{Load}.

Based on this discussion, we can define various useful power gain relationships and analyze them based on signal flow graphs. One such useful relationship is the transducer power gain. The transducer power gain is defined as the ratio of the power delivered to the load to the power delivered from the source. To determine this relationship, consider the two-port signal graph depicted in Figure 2.24. The power delivered to the load can be determined as

$$P_{Load} = \frac{|a_{Load}|^2 - |b_{Load}|^2}{2} \tag{2.90}$$

Recall, however, that from Eqn (2.89), it can be easily derived that $|b_{Load}| = |\Gamma_{Load}||a_{Load}|$, then Eqn (2.90) becomes

$$P_{Load} = \frac{1}{2}|a_{Load}|^2\left(1 - |\Gamma_{Load}|^2\right) = \frac{1}{2}|b_2|^2\left(1 - |\Gamma_{Load}|^2\right)\Big|_{|b_2|=|a_{Load}|} \tag{2.91}$$

The power available from the source can be computed as

$$P_s = \frac{|b_s|^2 - |a_s|^2}{2} \tag{2.92}$$

Relying on Figure 2.24, we can deduce that $b_s = b_g + \Gamma_s a_s$. For maximum power transfer, however, we can claim that

$$b_s = b_g + \Gamma_s a_s = b_g + \Gamma_s b_{Load} = b_g + \Gamma_s \Gamma_{Load} a_{Load} \tag{2.93}$$

where $|a_{Load}|^2$ is the power incident on the load and $|b_{Load}|^2 = |\Gamma_{Load}|^2 |a_{Load}|^2$ is the power reflected from the load. Again, for maximum power transfer, that is $a_{Load} = b_s$, we can rewrite Eqn (2.93) as

$$b_s = b_g + \Gamma_s \Gamma_{Load} \quad b_s \Rightarrow b_s = \frac{b_g}{1 - \Gamma_s \Gamma_{Load}} \tag{2.94}$$

The power delivered by the source can be found by substituting Eqn (2.94) into Eqn (2.92), and allowing for maximum power transfer, we obtain

$$P_s = \frac{1}{2} \frac{|b_g|^2}{|1 - \Gamma_s \Gamma_{Load}|^2} \left(1 - |\Gamma_{Load}|^2\right) \tag{2.95}$$

Finally the transducer power gain can be obtained by taking the ratio of Eqn (2.91) over Eqn (2.95) or

$$G_{transducer} = \frac{P_{Load}}{P_s} = \frac{\frac{1}{2}|b_2|^2\left(1 - |\Gamma_{Load}|^2\right)\Big|_{|b_2| = |a_{Load}|}}{\frac{1}{2}\frac{|b_g|^2}{|1 - \Gamma_s \Gamma_{Load}|^2}\left(1 - |\Gamma_{Load}|^2\right)} = \frac{|b_2|^2}{|b_g|^2}|1 - \Gamma_s \Gamma_{Load}|^2 \tag{2.96}$$

Next, we employ Mason's gain rule to find a closed form solution for Eqn (2.96). More specifically, we need to compute the ratio b_2/b_g as depicted in Figure 2.29. Note there is only one forward path going from b_g to b_2. The path gain can be found upon examination of Figure 2.29 as $gain_{path1} = 1 \times 1 \times S_{21} = S_{21}$. Next we examine

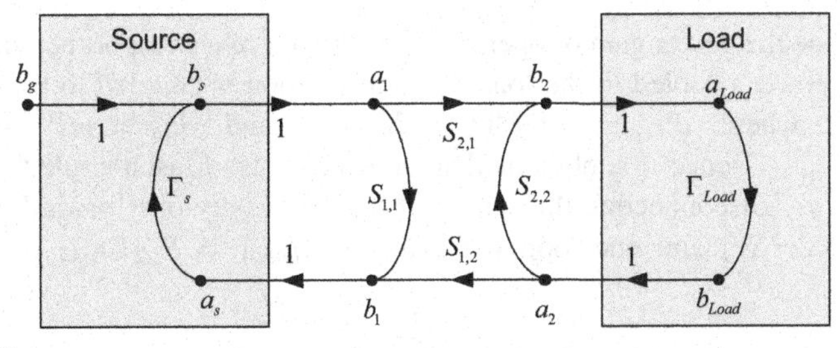

FIGURE 2.29

Signal flow graph of Figure 2.24 using reflected power waves variables.

the three loops $(\overset{\frown}{b_s}\, a_1 b_1 a_s b_s)$, $(\overset{\frown}{b_s}\, a_1 b_2 a_{Load} b_{Load} a_2 a_s b_s)$, and $(\overset{\frown}{b_2}\, a_{Load} b_{Load} a_2 b_2)$. The respective gain of these loops is given as

$$
\begin{aligned}
\left(\overset{\frown}{b_s}\, a_1 b_1 a_s b_s\right) &\Rightarrow gain_{Loop1} = 1 \times S_{1,1} \times 1 \times \Gamma_s = S_{1,1}\Gamma_s \\
\left(\overset{\frown}{b_s}\, a_1 b_2 a_{Load} b_{Load} a_2 a_s b_s\right) &\Rightarrow gain_{Loop2} = 1 \times \Gamma_{Load} \times 1 \times S_{2,2} = \Gamma_{Load} S_{2,2} \\
\left(\overset{\frown}{b_2}\, a_{Load} b_{Load} a_2 b_2\right) &\Rightarrow gain_{Loop3} \\
&= 1 \times S_{2,1} \times 1 \times \Gamma_{Load} \times 1 \times S_{1,2} \times 1 \times \Gamma_s \\
&= S_{2,1}\Gamma_{Load} S_{1,2}\Gamma_s
\end{aligned}
\tag{2.97}
$$

The ratio b_2/b_g can then be computed as

$$
\begin{aligned}
\frac{b_2}{b_g} &= \frac{gain_{path1}}{1 - \left(gain_{Loop1} + gain_{Loop2} + gain_{Loop3}\right) + gain_{Loop1}\, gain_{Loop2}} \\
&= \frac{S_{21}}{1 - \left(S_{1,1}\Gamma_s + \Gamma_{Load}S_{2,2} + S_{2,1}\Gamma_{Load}S_{1,2}\Gamma_s\right) + S_{1,1}\Gamma_s + \Gamma_{Load}S_{2,2}}
\end{aligned}
\tag{2.98}
$$

Substituting Eqns (2.98) and (2.65) into Eqn (2.96), we obtain after simplification the ratio

$$
\begin{aligned}
G_{transducer} &= \frac{P_{Load}}{P_s} = \frac{|b_2|^2}{|b_g|^2}\left(1 - |\Gamma_{Load}|^2\right)\left(1 - |\Gamma_s|^2\right) \\
&= |S_{21}|^2 \frac{\left(1 - |\Gamma_{Load}|^2\right)\left(1 - |\Gamma_s|^2\right)}{\left|1 - S_{1,1}\Gamma_s - S_{2,2}\Gamma_{Load} + S_{1,1}S_{2,2}\Gamma_s\Gamma_{Load} - S_{2,1}S_{1,2}\Gamma_{Load}\Gamma_s\right|} \\
&= |S_{21}|^2 \left(\frac{1 - |\Gamma_{Load}|^2}{|1 - S_{2,2}\Gamma_{Load}|^2}\right)\left(\frac{1 - |\Gamma_s|^2}{|1 - \Gamma_{in}\Gamma_s|^2}\right)
\end{aligned}
\tag{2.99}
$$

Next we define the power gain or operating power gain of a two-port network as the ratio of the power coupled to the load P_{Load} to the power coupled to the two-port circuit P_{in} where $P_{Load} = |b_2|^2(1 - |\Gamma_{Load}|^2)$ and $P_{in} = |a_1|^2 - |b_1|^2 = |a_1|^2(1 - |\Gamma_{in}|^2)$. Hence it is obvious that we need to use Mason's rule to evaluate the ratio b_2/a_1. Disconnecting the source, there exists only one forward path with gain $gain_{path1} = S_{2,1}$ and one loop with gain $gain_{Loop1} = S_{2,2}\Gamma_{Load}$ resulting in the ratio

$$
\frac{b_2}{a_1} = \frac{S_{2,1}}{1 - S_{2,2}\Gamma_{Load}} \tag{2.100}
$$

Hence, using Eqn (2.100), the operating power gain can be expressed as

$$\begin{aligned} G_p &= \frac{|b_2|^2\left(1-|\Gamma_{Load}|^2\right)}{|a_1|^2\left(1-|\Gamma_{in}|^2\right)} \\ &= \frac{|S_{2,1}|^2}{\left(1-|\Gamma_{in}|^2\right)|1-S_{2,2}\Gamma_{Load}|^2} \end{aligned} \tag{2.101}$$

In a similar manner, the available power gain is defined as the available output power of the two-port circuit to the available power from the source. Recall from Eqn (2.91) that the network will deliver maximum power to the load if and only if the load reflection coefficient is the complex conjugate of the output reflection coefficient, and hence available power gain can be expressed as

$$G_{AP} = \frac{|b_2|^2}{|b_g|^2}\left(1-|\Gamma_s|^2\right)\left(1-|\Gamma_{out}|^2\right) \tag{2.102}$$

where the ratio b_2/b_g can be estimated by employing Mason's rule to be

$$\begin{aligned} \frac{b_2}{b_g} &= \frac{S_{2,1}}{1-\left(S_{1,1}\Gamma_s+S_{2,2}\Gamma_{Load}+S_{2,1}S_{1,2}\Gamma_s\Gamma_{Load}\right)+S_{1,1}S_{2,2}\Gamma_s\Gamma_{Load}} \\ &= \left.\frac{S_{2,1}}{(1-S_{1,1}\Gamma_s)+(1-\Gamma_{out}\Gamma_{Load})}\right|_{\Gamma_{Load}=\Gamma^*_{out}} = \frac{S_{2,1}}{(1-S_{1,1}\Gamma_s)+\left(1-|\Gamma_{out}|^2\right)} \end{aligned} \tag{2.103}$$

Finally, substituting Eqn (2.103) into Eqn (2.102) results in the available power gain expression

$$G_{AP} = |S_{2,1}|^2\frac{\left(1-|\Gamma_s|^2\right)}{(1-S_{1,1}\Gamma_s)} \tag{2.104}$$

2.2.3 Stability analysis

When designing amplifiers, for example, an always important question arises. In the process of maximizing the transducer gain of the network, will the amplifier be driven into an unstable state during the matching process? Consequently, is the circuit conditionally or unconditionally stable?

From control theory, we can state that the amplifier circuit depicted in Figure 2.30 is said to be conditionally stable if and only if $\text{Re}\{Z_{in}\}>0$ and $\text{Re}\{Z_{out}\}>0$ for some but not all passive source Z_s and load Z_{Load} impedances. This, however, is

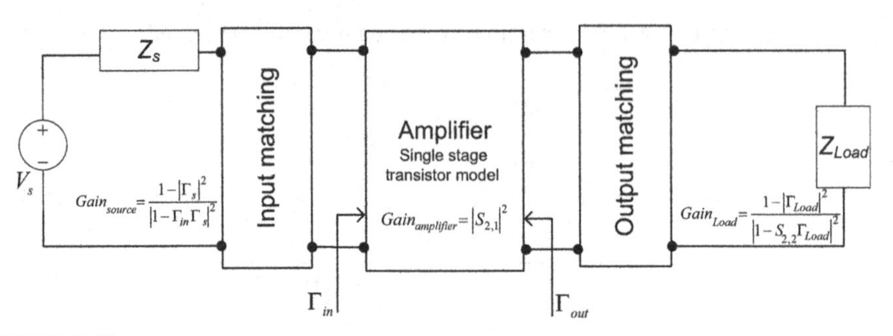

FIGURE 2.30

Amplifier circuit with input and output matching.

only true at a given frequency. Note that $\text{Re}\{Z_{in}\} > 0$ and $\text{Re}\{Z_{out}\} > 0$ implies that $|\Gamma_{in}| < 1$ and $|\Gamma_{out}| < 1$ or

$$|\Gamma_{in}| = \left|S_{1,1} + \frac{S_{1,2}S_{2,1}\Gamma_{Load}}{1 - S_{2,2}\Gamma_{Load}}\right| < 1 \qquad (2.105)$$

and

$$|\Gamma_{out}| = \left|S_{2,2} + \frac{S_{1,2}S_{2,1}\Gamma_s}{1 - S_{1,1}\Gamma_s}\right| < 1 \qquad (2.106)$$

On the other hand, a network is said to be unconditionally stable if and only if $\text{Re}\{Z_{in}\} > 0$ and $\text{Re}\{Z_{out}\} > 0$ ($|\Gamma_{in}| < 1$ and $|\Gamma_{out}| < 1$) for all passive source and load impedances, that is, ($|\Gamma_s| < 1$ and $|\Gamma_{Load}| < 1$). The stability criteria discussed herein is only true at a single frequency. Therefore, in order to ensure the stability of broadband networks, one must repeat the following analysis at a representative set of frequencies representing the band of interest. The relationships in Eqns (2.105) and (2.106) imply that the source and reflection coefficients Γ_s and Γ_{Load}, lie within a certain range of values where the amplifier is said to be stable. The stability of the amplifier can be analyzed with the aid of a Smith chart in conjunction with an input or output *stability circle*. The information provided by the chart and the stability circles can be used to determine the region of stability.

The input stability circle can be determined as the locus of all values of Γ_s for which $|\Gamma_{in}| = 1$. Similarly, the output stability circle can be determined as the locus of all values of Γ_{Load} for which $|\Gamma_{out}| = 1$. Both Γ_s and Γ_{Load} are bounded in magnitude for passive matching networks such that $|\Gamma_s| < 1$ and $|\Gamma_{Load}| < 1$ and both parameters lie on the Smith chart. Note that unconditional stability implies that both input and output stability circles are either completely outside the Smith chart, that is they do not intersect the chart, or the Smith chart is completely enclosed in the circles. Conditional stability implies that the stability circles intersect the Smith chart.

To establish the output stability circle first, consider the input reflection coefficient presented in Eqn (2.105). Setting $|\Gamma_{in}| = 1$ establishes a boundary condition on the input reflection coefficient and implies that in order to ensure stability, Γ_{Load} must lie within a unit circle defined by $|\Gamma_{Load}| = 1$ for a given passive load and matching network. Therefore restricting $|\Gamma_{in}| < 1$ implies that Γ_{Load} will lie on a circle with radius r_{Load} given as

$$r_{Load} = \left|\frac{S_{1,2}S_{2,1}}{|S_{2,2}|^2 - |S_{1,1}S_{2,2} - S_{1,2}S_{2,1}|^2}\right|, r_{Load} \in \mathbb{R}^+ \tag{2.107}$$

where r_{Load} is real. The center of the circle c_{Load} is given as

$$c_{Load} = \frac{\left(S_{2,2} - (S_{1,1}S_{2,2} - S_{1,2}S_{2,1})S^*_{1,1}\right)^*}{|S_{2,2}|^2 - |S_{1,1}S_{2,2} - S_{1,2}S_{2,1}|^2}, c_{Load} \in \mathbb{C} \tag{2.108}$$

where c_{Load} is complex. To obtain the radius and the circle, the S-parameters must be given or measured for a certain frequency. The locus of all values of Γ_{Load} such that $|\Gamma_{in}| < 1$ can be found by plotting the circle defined in Eqns (2.107) and (2.108) over a Smith chart using the upper bound $|\Gamma_{in}| = 1$. The stable region would then be either inside or outside the circle boundary as will be shown later on.

The input stability circle is defined in a like manner to the output stability circle. The circle radius is defined similar to Eqn (2.107) by substituting $S_{1,1}$ for $S_{2,2}$ and vice versa, that is the radius is given as

$$r_s = \left|\frac{S_{1,2}S_{2,1}}{|S_{1,1}|^2 - |S_{1,1}S_{2,2} - S_{1,2}S_{2,1}|^2}\right|, r_s \in \mathbb{R}^+ \tag{2.109}$$

and the circle is centered at

$$c_s = \frac{\left(S_{1,1} - (S_{1,1}S_{2,2} - S_{1,2}S_{2,1})S^*_{2,2}\right)^*}{|S_{1,1}|^2 - |S_{1,1}S_{2,2} - S_{1,2}S_{2,1}|^2}, c_s \in \mathbb{C} \tag{2.110}$$

In order to determine the output stability region, assume that the load is matched to the amplifier, that is $Z_{Load} = Z_0$, then the load reflection coefficient Γ_{Load} is equal to zero, which in turn implies that for $|S_{1,1}| < 1$, $|\Gamma_{in}| < 1$ according to Eqn (2.105). This, in turn, implies that the center of the Smith chart where $\Gamma_{Load} = 0$ is in the stable region, and hence the stable region is the intersection of the exterior of the output stability circle with the Smith chart as shown in Figure 2.31. On the other hand, if $|S_{1,1}| > 1$, this implies $|\Gamma_{in}| > 1$ for $\Gamma_{Load} = 0$, and hence the stability region does not include the center of the Smith chart, and hence it is the intersection of the output stability circle with the Smith chart as shown in Figure 2.32. Similar results can be obtained for input stability circles.

As mentioned earlier, unconditional stability implies that the stability circles are either completely enclosed or completely outside the Smith chart where additional

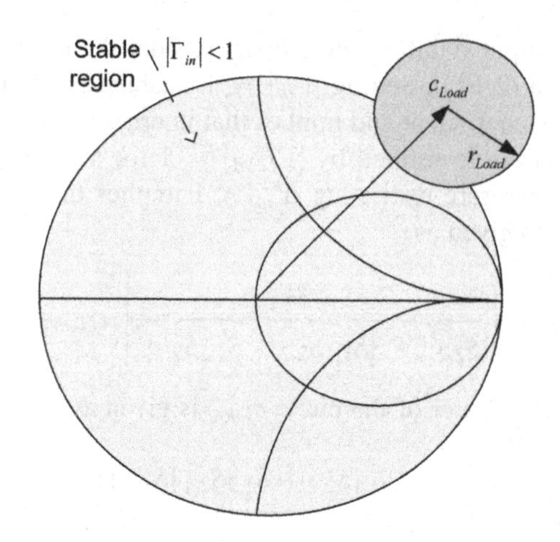

FIGURE 2.31

Stability region as an intersection between the Smith chart and the exterior of the output stability circle for conditionally stable amplifier circuit as indicated by the arrow for $|\Gamma_{in}| < 1$ and $|S_{1,1}| < 1$. (For color version of this figure, the reader is referred to the online version of this book.)

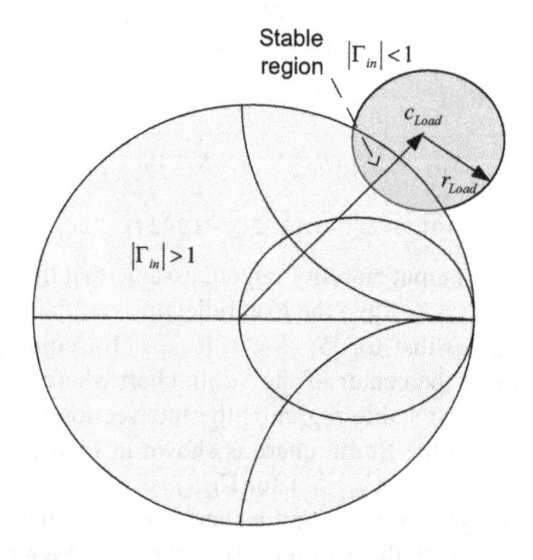

FIGURE 2.32

Stability region as an intersection between the Smith chart and the output stability circle for a conditionally stable amplifier circuit as indicated by the arrow for $|\Gamma_{in}| < 1$ and $|S_{1,1}| > 1$. (For color version of this figure, the reader is referred to the online version of this book.)

constraints are placed on Eqns (2.107)–(2.110). Concerning the output stability circle, unconditional stability implies that for $|S_{1,1}| < 1$

$$||c_{Load}| - r_{Load}| > 1 \tag{2.111}$$

must be met. Similarly, for the input stability circle, unconditional stability implies that for $|S_{2,2}| < 1$

$$||c_s| - r_s| > 1 \tag{2.112}$$

must be met.

Unconditional stability can also be guaranteed as follows. In order to ensure maximum transducer gain, it was shown that the reflection coefficient of the source has to equal the complex conjugate of the input reflection coefficient $\Gamma_s = \Gamma_{in}^*$ and the reflection coefficient of the load has to equal the complex conjugate of the output reflection coefficient $\Gamma_{Load} = \Gamma_{out}^*$. Then the amplifier is said to be unconditionally stable if and only if

$$\begin{aligned} K &= \frac{1 + |S_{1,1}S_{2,2} - S_{1,2}S_{2,1}|^2 - |S_{1,1}|^2 - |S_{2,2}|^2}{2|S_{1,2}||S_{2,1}|} > 1 \\ &|S_{1,1}S_{2,2} - S_{1,2}S_{2,1}| < 1 \end{aligned} \tag{2.113}$$

Note that in reality, the relationship in Eqn (2.113) does not account for temperature variations nor does it take into account any VSWR changes due to drift in the S-parameters. Therefore, care must be taken when applying Eqn (2.113) to amplifier design, for example, due to the reasons stated above.

2.3 Three-port and four-port networks

Thus far, the discussion has centered on two-port networks. In this section, we will extend the concepts discussed herein to multiport networks and, more precisely, to three and four-port networks. In order to characterize a three-port network, for example, nine parameters will be required as opposed to four parameters for two-port networks. Given the three-port network depicted in Figure 2.33, in order to measure the reflection coefficient $S_{1,1}$, we must terminate the second and third ports each with an impedance equivalent to the characteristic impedance of the corresponding transmission line at those ports. Other S-parameters can be computed in a similar manner, given that the other ports are properly terminated. Note then, that the same method of computation or measurement is also applicable to four-port or N-port networks. The difference, however, lies in the number of computations or measurements needed. It can be easily verified that the number of measurements increases by the square of the number of ports available.

Rather than repeating the theory presented thus far for two-port networks, the aim of this section is to present some typical three-port and four-port components used in radio frequency (RF) designs.

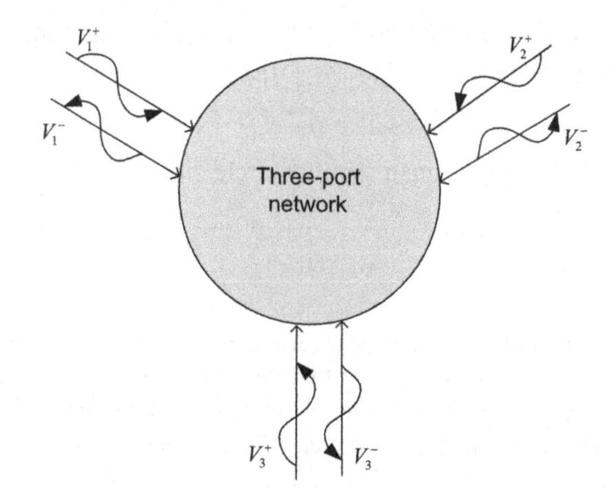

FIGURE 2.33

Three-port network.

2.3.1 Power dividers and power combiners

Power dividers are commonly used in mixers, active circulators, and power amplifiers just to name a few devices. *T*-junction power dividers and combiners are three-port networks, as depicted symbolically in Figure 2.34 [5]. Consider the scattering matrix of a three-port reciprocal network

$$\overline{S} = \begin{bmatrix} 0 & S_{1,2} & S_{1,3} \\ S_{1,2} & 0 & S_{2,3} \\ S_{1,3} & S_{2,3} & 0 \end{bmatrix} \tag{2.114}$$

where $\overline{S}$ is unitary and all ports are assumed to be matched (i.e., $S_{1,1}$, $S_{2,2}$, and $S_{3,3}$ are all equal to zero). If the network is assumed to be lossless, then its characteristics as stated earlier in Eqn (2.38) must hold

$$|S_{1,2}|^2 + |S_{1,3}|^2 = |S_{1,2}|^2 + |S_{2,3}|^2 = |S_{1,3}|^2 + |S_{2,3}|^2 = 1 \tag{2.115}$$

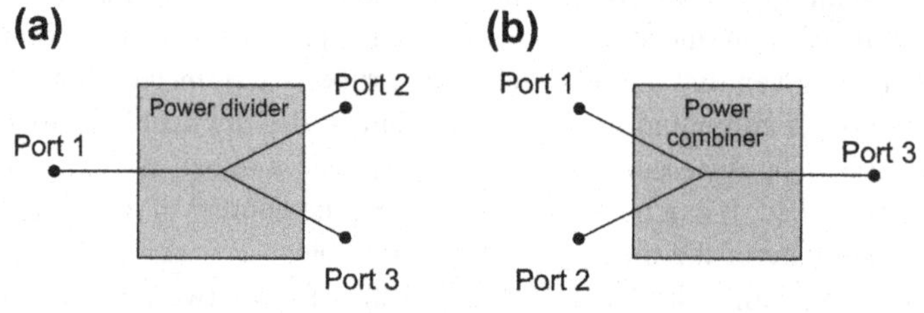

FIGURE 2.34

Three-port network. (a) Power divider, and (b) power combiner.

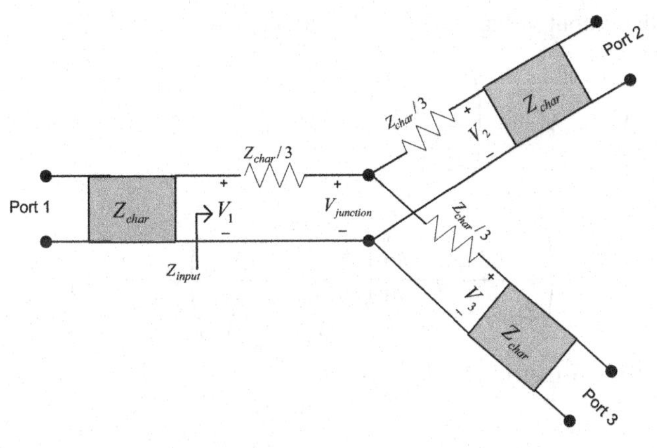

FIGURE 2.35

Three-port *T*-junction resistive power divider.

and

$$S_{1,3}^* S_{2,3} = S_{2,3}^* S_{1,2} = S_{1,2}^* S_{1,3} = 0 \tag{2.116}$$

Note, a simple comparison between Eqns (2.115) and (2.116) implies that all the ports of a three-port *lossless* network cannot be matched.[2] On the other hand, a three-port reciprocal power divider made up of lumped element resistors (passive elements) can be matched at all ports. The network, depicted in Figure 2.35, suffers from poor isolation at the output ports. The reason is that part of the power reflected due to a mismatched output load is coupled to the other output port. The power at the output ports is in theory 3 dB less than the input power.

Given the output impedance Z_{output} looking into the resistors at any of the output ports can be computed as $Z_{output} = \frac{Z_{char}}{3} + Z_{char} = \frac{4Z_{char}}{3}$ provided that all the ports are terminated with Z_{char}. Consequently, the input impedance can then be computed as $Z_{input} = \frac{Z_{char}}{3} + \frac{2Z_{char}}{3} = Z_{char}$ and hence the input is matched to the feedline with impedance equal to the characteristic impedance. Furthermore, due to symmetry, we observe that the parameters $S_{1,1}$, $S_{2,2}$, and $S_{3,3}$ are all equal to zero given that the output ports are also matched.

In order to obtain the voltages V_2 and V_3, we must relate the junction voltage $V_{junction}$ to the voltage of the input port V_1, that is, using voltage division on the network depicted in Figure 2.35, we obtain

$$V_{junction} = \left(\frac{\frac{2Z_{char}}{3}}{\frac{2Z_{char}}{3} + \frac{Z_{char}}{3}} \right) V_1 = \frac{2}{3} V_1 \tag{2.117}$$

[2]From Eqns (2.115) and (2.116), this implies that either two of the parameters, $S_{1,3}$ or $S_{2,3}$, are equal to zero.

And hence, the output voltages V_2 and V_3 can then be obtained:

$$V_2 = V_3 = \left(\frac{Z_{char}}{Z_{char} + \frac{Z_{char}}{3}} \right) V_{junction}$$
$$= \left(\frac{Z_{char}}{Z_{char} + \frac{Z_{char}}{3}} \right) \left(\frac{\frac{2Z_{char}}{3}}{\frac{2Z_{char}}{3} + \frac{Z_{char}}{3}} \right) V_1 = \frac{1}{2} V_1 \tag{2.118}$$

Thus the resulting S-matrix is computed as

$$\overline{S} = \begin{bmatrix} 0 & \frac{1}{2} & \frac{1}{2} \\ \frac{1}{2} & 0 & \frac{1}{2} \\ \frac{1}{2} & \frac{1}{2} & 0 \end{bmatrix} \tag{2.119}$$

This in turn implies that the output power of ports 2 and 3 P_{port2} and P_{port3}, respectively, are one-fourth the input power P_{input} supplied into the network. The other half of the power is dissipated in the resistors.

In order to overcome the matching and isolation problems of a T-junction, designers resort to using a Wilkinson power combiner or divider instead. A Wilkinson divider, for example, is said to be lossless provided that all of its output ports are matched. A Wilkinson divider can be made to split the power evenly or unevenly amongst the output ports. A microstrip Wilkinson power divider is shown in Figure 2.36. Wilkinson power dividers are known to offer broad bandwidths of approximately one octave and equal phase characteristics at the output of each port. Port isolation is obtained by terminating the output ports with a series resistor.

The quarter wave lines depicted in Figure 2.36 and Figure 2.37 have a characteristic impedance of $\sqrt{2}Z_{char}$, while the output ports 2 and 3 are connected with a lumped resistor $2Z_{char}$. This in turn ensures that all the ports are matched to a characteristic impedance of Z_{char}.

2.3.2 Directional couplers

Directional couplers offer an excellent example of four-port devices. These passive devices are used to couple out a certain amount of the power traveling in one transmission line through another connection or port [4,7]. One common application of directional couplers is used to measure the output power of the transmitter, thus enabling the radio to perform closed automated output-level power control. This enables the radio to sense the output power levels without connecting directly to the transmission line carrying the power. The coupler's four ports can be labeled as

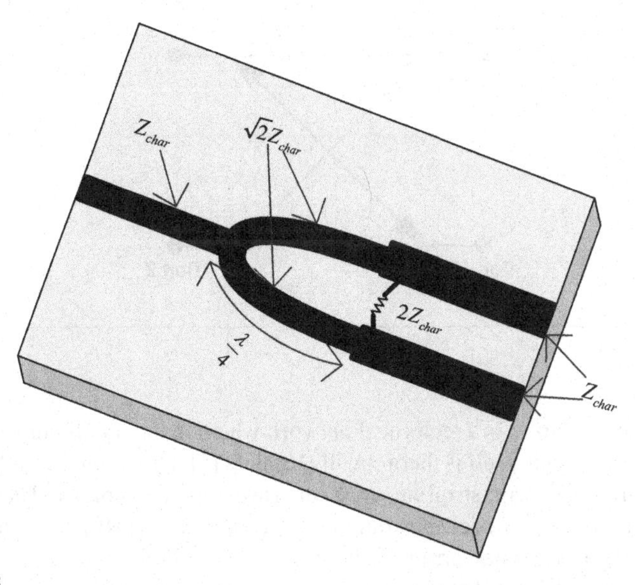

FIGURE 2.36

A Wilkinson equal-power microstrip power divider. (For color version of this figure, the reader is referred to the online version of this book.)

such: the input port; the through port where the incident signal exits the device with minimal degradation to its input power; the coupled port where the input signal appears with a fraction of the incident input signal power; and finally the isolated port. The isolated port is typically terminated. The coupler is called directional because its operation can be reversed, that is, the through port can act as the input port by feeding the input signal through it. The signal then exits through the former input port with minor degradation. The coupled port is now the isolated port and the coupled port is the former coupled port. The coupled port is determined based on which port is used as the incident or input port.

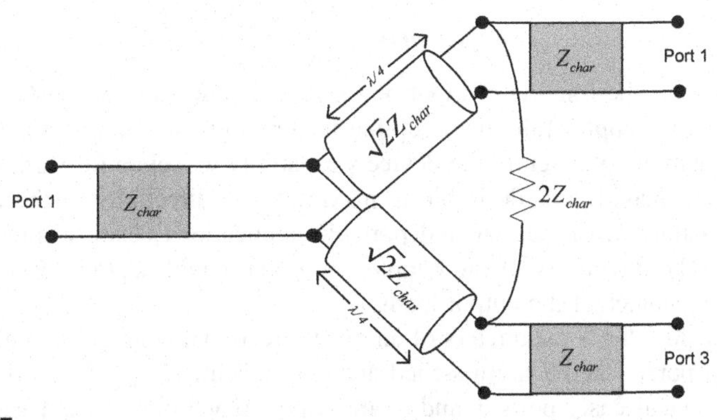

FIGURE 2.37

Transmission line circuit of Wilkinson combiner depicted in Figure 2.36.

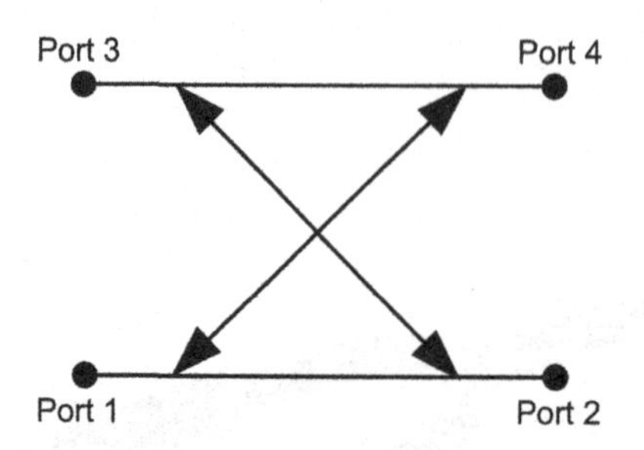

FIGURE 2.38

Directional coupler.

A directional coupler is a reciprocal network where in theory all ports are ideally matched and the circuit itself is theoretically lossless [7]. Directional couplers can be designed using microstrip, stripline, coax, or waveguide technology. The device itself comes in a variety of packages ranging from blocks with RF connectors or solder pins, lie on a substrate carrier, or be designed within a larger radio frequency integrated circuit (RF IC) that performs other functions as well.

Consider the directional coupler depicted symbolically in Figure 2.38. Let the power of the incident waveform on port 1 be P_1. Furthermore, let P_2, P_3, and P_4 be the powers coupled from ports 2, 3, and 4, respectively, to the matched terminations. Define coupling, directivity, and isolations as the following ratios:

$$\begin{aligned} C &= 10\log_{10}\left(\frac{P_1}{P_4}\right) \text{ Coupling} \\ D &= 10\log_{10}\left(\frac{P_4}{P_3}\right) \text{ Directivity} \\ I &= 10\log_{10}\left(\frac{P_1}{P_3}\right) \text{ Isolation} \\ &= (C+D)|_{dB} \end{aligned} \tag{2.120}$$

Coupling denotes the fraction of input power that is coupled to port 4 from port 1. Depending on the application, the coupling factor can be as high as 30 dB. Directivity and isolation both signify the device's capability to isolate forward and backward waves (leakage). It is desirable to maximize the directivity of the device as much as possible. Most cellular and portable applications desire a directivity of 30–40 dBs. The directivity of the coupler, however, could degrade depending on the level of mismatch at the output loads.

Next, consider the S-parameters of an ideal directional coupler. Assume for the moment that ports 1 and 2 are matched and consequently $S_{1,1} = S_{2,2} = 0$. Furthermore, in the ideal case, ports 1 and 3 are completely isolated from each other, and hence $S_{1,3} = 0$. Additionally, assume ports 2 and 4 to also be completely isolated, that is, $S_{2,4}=0$, then by invoking the reciprocity property of the network where $S_{i,j}=S_{j,i}$, the scattering matrix could be expressed as

$$\overline{S} = \begin{bmatrix} 0 & S_{1,2} & 0 & S_{1,4} \\ S_{1,2} & 0 & S_{2,3} & 0 \\ 0 & S_{2,3} & S_{3,3} & S_{3,4} \\ S_{1,4} & 0 & S_{3,4} & S_{4,4} \end{bmatrix} \tag{2.121}$$

Note that for a lossless coupler, the matrix in Eqn (2.121) is unitary, and hence the relations in Eqns (2.38) and (2.39) apply. This in turn implies that ports 3 and 4 are matched, and consequently $S_{3,3}$ and $S_{4,4}$ both equal to zero. This can be established as follows. Consider the complex conjugate product of rows 1 and 4 and 2 and 3:

$$\begin{aligned} 0 \times S_{1,4}^* + S_{1,2} \times 0 + 0 \times S_{3,4}^* + S_{1,4} \times S_{4,4}^* &= S_{1,4} \times S_{4,4}^* = 0 \text{ rows 1 and 4} \\ S_{1,2} \times 0 + 0 \times S_{2,3}^* + S_{2,3} \times S_{3,3}^* + 0 \times S_{3,4}^* &= S_{2,3} \times S_{3,3}^* = 0 \text{ rows 2 and 3} \end{aligned} \tag{2.122}$$

But since $S_{1,4}$ and $S_{2,3}$ cannot be zero, this automatically implies that $S_{3,3}=0$ and $S_{4,4}=0$, and hence Eqn (2.121) becomes

$$\overline{S} = \begin{bmatrix} 0 & S_{1,2} & 0 & S_{1,4} \\ S_{1,2} & 0 & S_{2,3} & 0 \\ 0 & S_{2,3} & 0 & S_{3,4} \\ S_{1,4} & 0 & S_{3,4} & 0 \end{bmatrix} \tag{2.123}$$

Furthermore, consider the following relations. From rows 1 and 3, the dot product can be computed and further manipulated as

$$\begin{aligned} S_{1,2}S_{2,3}^* + S_{1,4}S_{3,4}^* &= 0 \Rightarrow \left|S_{1,2}S_{2,3}^*\right| - \left|S_{1,4}S_{3,4}^*\right| \\ &= \left|S_{1,2}\right|\left|S_{2,3}\right| - \left|S_{1,4}\right|\left|S_{3,4}\right| = 0 \\ &\Rightarrow \left|S_{1,2}\right|\left|S_{2,3}\right| = \left|S_{1,4}\right|\left|S_{3,4}\right| \end{aligned} \tag{2.124}$$

Similarly, the dot product of rows 2 and 4 result in

$$\begin{aligned} S_{1,2}S_{1,4}^* + S_{2,3}S_{3,4}^* &= 0 \Rightarrow \left|S_{1,2}S_{1,4}^*\right| - \left|S_{2,3}S_{3,4}^*\right| \\ &= \left|S_{1,2}\right|\left|S_{1,4}\right| - \left|S_{2,3}\right|\left|S_{3,4}\right| = 0 \\ &\Rightarrow \left|S_{1,2}\right|\left|S_{1,4}\right| = \left|S_{2,3}\right|\left|S_{3,4}\right| \end{aligned} \tag{2.125}$$

Next, consider the ratio of Eqn (2.124) divided by Eqn (2.125), and we obtain

$$\frac{\left|S_{1,2}\right|\left|S_{2,3}\right|}{\left|S_{1,2}\right|\left|S_{1,4}\right|} = \frac{\left|S_{1,4}\right|\left|S_{3,4}\right|}{\left|S_{2,3}\right|\left|S_{3,4}\right|} \Rightarrow \frac{\left|S_{2,3}\right|}{\left|S_{1,4}\right|} = \frac{\left|S_{1,4}\right|}{\left|S_{2,3}\right|} \tag{2.126}$$

which in turn implies that

$$|S_{1,4}| = |S_{2,3}| \tag{2.127}$$

That is the coupling between ports 1 and 4 and is the same as the coupling between ports 2 and 3. Substituting Eqn (2.127) into Eqn (2.124), we obtain

$$|S_{1,2}||S_{2,3}| = |S_{1,4}||S_{3,4}| \Rightarrow |S_{1,2}||S_{1,4}| = |S_{1,4}||S_{3,4}| \Rightarrow |S_{1,2}| = |S_{3,4}| \tag{2.128}$$

Next, choose the reference planes on ports 1 and 3 to be positive, that is, let $S_{1,2}$ and $S_{3,4}$ be equal to a positive real number α and the reference plane of $S_{1,4}$ to be an imaginary number $j\beta$ where β is a positive real number. This in turn implies that $S_{2,3} = j\beta$. The conservation of energy property of the unitary scattering matrix in Eqn (2.123) further implies that

$$\begin{aligned} |S_{1,2}|^2 + |S_{1,4}|^2 &= 1 \\ |S_{1,2}|^2 + |S_{2,3}|^2 &= 1 \\ |S_{2,3}|^2 + |S_{3,4}|^2 &= 1 \\ |S_{1,4}|^2 + |S_{3,4}|^2 &= 1 \end{aligned} \tag{2.129}$$

According to Eqn (2.129), the relationship $|S_{1,2}|^2 + |S_{1,4}|^2 = 1$ also implies that

$$\alpha^2 + \beta^2 = 1 \tag{2.130}$$

and hence the ideal scattering matrix can now be expressed as

$$\bar{S} = \begin{bmatrix} 0 & \alpha & 0 & j\beta \\ \alpha & 0 & j\beta & 0 \\ 0 & j\beta & 0 & \alpha \\ j\beta & 0 & \alpha & 0 \end{bmatrix} \tag{2.131}$$

Note that the linear coupling factor C is the ratio

$$C = \frac{1}{|S_{1,4}|^2} = \frac{1}{\beta^2} \tag{2.132}$$

According to Eqn (2.130), it follows that $\alpha = \sqrt{1 - 1/C}$ since $\beta = \sqrt{1/C}$. A special case is the directional hybrid coupler. A directional hybrid coupler has a 3-dB coupling factor and can be classified as either a 90° directional hybrid coupler or a 180° directional hybrid coupler. A 90° hybrid coupler has a scattering matrix given as

$$\bar{S} = \frac{1}{\sqrt{2}} \begin{bmatrix} 0 & 1 & 0 & j \\ 1 & 0 & j & 0 \\ 0 & j & 0 & 1 \\ j & 0 & 1 & 0 \end{bmatrix} \tag{2.133}$$

A 90° hybrid coupler is also known as a quadrature hybrid (symmetrical coupler) with a 90° phase difference between ports 2 and 4 given that the input is fed through port 1. On the other hand, the 180° directional hybrid coupler (antisymmetrical coupler) has a 180° phase difference between ports 1 and 4, called the Σ ports, given that the input is fed through port 3. Ports 2 and 3 are also known as the Δ ports. The scattering matrix of the 180° hybrid coupler is given as

$$\bar{S} = \frac{1}{\sqrt{2}} \begin{bmatrix} 0 & 1 & 0 & -1 \\ 1 & 0 & 1 & 0 \\ 0 & -1 & 0 & 1 \\ 1 & 0 & 1 & 0 \end{bmatrix} \tag{2.134}$$

A signal incident on a Σ port causes the equal-powered output signals to have the same phase. Similarly, a signal incident on a Δ port causes the equal-powered output signals to have opposite phase.

References

[1] R.E. Collin, Foundations for Microwave Engineering, second ed, McGraw-Hill, New York, NY, 1992.

[2] K. Chang, Microwave Solid State Circuits and Applications, John Wiley & Sons, New York, NY, 1994.

[3] C.G. Montgomery, R.H. Dicke, E.M. Purcell, Principles of Microwave Circuits, in: MIT Radiation Laboratory Series, vol. 8, McGraw-Hill, New York, NY, 1948.

[4] L. Young, Advances in Microwaves, Academic Press, New York, NY, 1966.

[5] D. Pozar, Microwave Engineering, Reading, Addison-Wesley, MA, 1993.

[6] W.A. Davis, Microwave Semiconductor Circuit Design, Van Nostrand Reinhold, New York, NY, 1984.

[7] G. Matthaei, L. Young, E. Jones, Microwave Filters, Impedance Matching Networks, and Coupling Structures, Artech House, Dedham, MA, 1980.

CHAPTER

Noise in Wireless Receiver Systems

3

CHAPTER OUTLINE

The two major limiting factors of a receiver are its nonlinearity and noise thus defining its dynamic range. Nonlinearity dictates the upper limit of the dynamic range and the amount of signal power, desired or undesired, that the receiver can process without distortion. Noise, on the other hand, defines the lower limit of the dynamic range and in turn the smallest signal power from which a desired signal can be received with an acceptable bit error rate performance. Noise comes in different flavors and can be attributed to many phenomena either internal or external to the receiver circuitry. The most common type of noise is thermal noise, also known as Johnson noise or Nyquist noise. Thermal noise, as the name implies, depends

Wireless Receiver Architectures and Design. http://dx.doi.org/10.1016/B978-0-12-378640-1.00003-2

Table 3.1 Noise Colors and their Respective Frequency Content

Noise Color	Frequency f
Purple	f^2
Blue	f
White	$f^0 = 1$
Pink	$1/f$
Red/Brown	$1/f^2$

on the temperature of the various electronic components in the receiver. Phase noise, on the other hand, is attributed mainly to the frequency generation system in the receiver, namely the voltage-controlled oscillator, the VCO, and the components that make up the synthesizer system.

In receivers, there are various types of noise that the designer must be mindful of; all manifest themselves to varying degrees. Besides white and phase noise, there is shot[a] noise, which is independent of frequency and in practical applications tends to taper off at high frequencies. There is transmit-broadband noise in the receive chain due to the power amplifier in the transmitter. This transmit-broadband noise is especially problematic in full duplex radios. There is also flicker or $1/f$ noise, which is present in all active and some passive devices. Quantization noise is attributed to the analog-to-digital conversion process and is mainly dictated by several parameters such as the converter architecture, number of bits, sampling clock, etc. Burst noise, also referred to as popcorn noise, is characterized by high frequency pulses that are due to imperfections in the semiconductor material and heavy ion implants. Avalanche noise is another electronic noise that occurs when a *pn*-junction operates in the reverse breakdown mode. These various types of noise all exist in a receiver, and they are at times hard to separate. Noise can be referred to by its color such as white or thermal noise, pink noise ($1/f$-noise), and red or brown noise. These noise types are listed alongside their noise color in Table 3.1.

This chapter, divided into three major sections, discusses the subject of noise and how it manifests itself in circuits and receiver components. The first section discusses thermal noise. The concept of noise figure as it applies to electronic components is discussed in detail. The section starts out by explaining the basics of thermal noise and the various physical mechanisms that generate it. The concept of noise figure at the component and system level is then discussed. Section 3.2 introduces the concept of phase noise. Various models such as the Leeson's model and the Lee-Hajimiri models are presented. The effect of the phase-locked loop (PLL) as a filter on oscillator and VCO phase noise is discussed in detail. The section

[a]Shot is short for Schottky noise, which is often referred to as quantum noise.

concludes by studying the impact of phase noise on performance. Finally, Section 3.3 discusses external noise sources to the receiver mainly due to coexistence. This subject is receiving significant attention due to the instantiation of multiple radio transceivers and standards on one device or on the same chip. The chapter concludes with a short appendix on thermal noise statistics.

3.1 Thermal noise

Thermal noise is the most common type of noise present in all analog radio frequency (RF) and baseband circuits. Devices ranging from simple resistors to bipolar and MOSFET transistors all exhibit thermal noise. In this section, we discuss the basics of thermal noise theory and the design parameters that relate to wireless receivers.

3.1.1 The basics

Thermal noise present in an electronic component occurs from random currents due to the Brownian motion of electrons. The spectrum of thermal noise is flat over a wide range of frequencies, and hence it is said to be white. The amount of thermal noise present in a component depends on the component's temperature and places a lower bound on the signal strength that can be detected in its presence [1,2].

Consider the resistor depicted in Figure 3.2. The temperature of the resistor is given in kelvin as T. The voltage $v_n(t)$ across the terminals of the resistor for a given measurement bandwidth B_n and a center frequency F_c has zero mean and an *rms* value given as

$$v_{n,rms} = \sqrt{\frac{4hF_cB_nR}{e^{hF_c/KT} - 1}} \tag{3.1}$$

The relationship in Eqn (3.1) is known as Planck's black body radiation law.[b] In Eqn (3.1), h represents Planck's constant where $h = 6.546 \times 10^{-34}$ Js and k represents Boltzman's constant given as $k = 1.38 \times 10^{-23}$ J/K. The voltage fluctuations across the resistor terminals are due to the random motion of electrons whose kinetic energy depends on the temperature T. At microwave frequencies, an approximation of Eqn (3.1) is possible due to the fact that $hF_c \ll KT$ and that $e^{hF_c/KT}$ can be approximated by the first two terms of its Taylor expansion, thus implying

$$e^{hF_c/KT} = 1 + \frac{hF_c}{kT} + \ldots \Rightarrow e^{hF_c/KT} - 1 \approx \frac{hF_c}{kT} \tag{3.2}$$

[b]Named after Nobel laureate German physicist Max Planck who is considered to be the father of quantum theory.

This results in the approximation of Eqn (3.1) as[c]

$$v_{n,rms} \approx \sqrt{4kTB_nR} \tag{3.3}$$

It is interesting to note that Eqn (3.3) is independent of frequency, which implies equal distribution of noise power across the bandwidth B_n. The statistical distribution of thermal noise then appears to be white. Furthermore, thermal noise is *almost* Gaussian distributed and can be treated as such for all practical purposes. Further details concerning the statistics of thermal noise are given in Section 3.4. Recall that Gaussian noise is additive with two-sided noise spectral density

$$S_n(F) = \frac{v_{n,rms}^2}{2B_n} = 2kTR \tag{3.4}$$

EXAMPLE 3.1 THERMAL NOISE VOLTAGE IN RESISTORS

Consider a 1 K-ohm resistor at 72 °C temperature. How often does the instantaneous noise voltage exceed +10 μV in magnitude? Assume the bandwidth of measurements to be 1 GHz.

In order to compute the normalized variable z_n, we first must compute the voltage standard deviation according to Eqn (3.3). To do so, convert the temperature from Celsius to kelvin, that is

$$T^{\circ}_{\text{kelvin}} = T^{\circ}_{\text{kelvin}} + 273 = 72 + 273 = 345 \text{ K} \tag{3.5}$$

Next, compute the *rms* voltage as

$$\begin{aligned}\sigma_n &= v_{n,rms} \approx \sqrt{4kTB_nR} = \sqrt{4 \times 1.38 \times 10^{-23} \times 345 \times 10^9 \times 1000} \\ &= 1.38 \times 10^{-4} V_{rms} = 138 \mu V_{rms}\end{aligned} \tag{3.6}$$

The normalized variable can now be expressed according to Eqn (3.170), found in the appendix, as

$$z_n = \frac{v_n - \mu_n}{\sigma_n} = \frac{v_n - 0}{138} = \left.\frac{v_n}{138}\right|_{v_n = 10\mu V} = \frac{10}{138} = 0.0725 \tag{3.7}$$

where the mean noise voltage $\mu_n = 0$. Therefore, according to Eqn (3.7), at 10 μV, the normalized voltage variable is 0.0725 of the voltage standard deviation. In order to determine the percentage of time that the noise voltage exceeds +10 μV, use the relationship expressed in Eqn (3.172) in the appendix

[c]The relationship defined in Eqn (3.3) is known as the Rayleigh–Jeans approximation named after John William Strut, Baron Rayleigh, an English physicist, and Sir James Jeans also an English physicist.

EXAMPLE 3.1 THERMAL NOISE VOLTAGE IN RESISTORS—cont'd

$$
\begin{aligned}
Q(z_n) &= \left(\frac{1}{(1-\alpha)z_n + \alpha\sqrt{z_n^2+\beta}} \right) \frac{1}{\sqrt{2\pi}} e^{-z_n^2/2} \Bigg|_{z_n=0.0725} \\
&= \left(\frac{1}{(1-0.344)\times 0.0725 + 0.344 \times \sqrt{0.0725^2+5.344}} \right) \frac{1}{\sqrt{2\pi}} e^{-0.0725^2/2} \\
&= 0.4848 \approx 48.5\%
\end{aligned}
\tag{3.8}
$$

According to Figure 3.1, it can be easily shown that as the voltage increases, the normalized cumulative probability density function decreases rapidly until the voltage reaches the standard deviation where the rate of decay starts to slow down. By way of example, it can be easily verified according to Eqn (3.172) that v_n exceeds say +0.2 mV only 7.4% of the time.

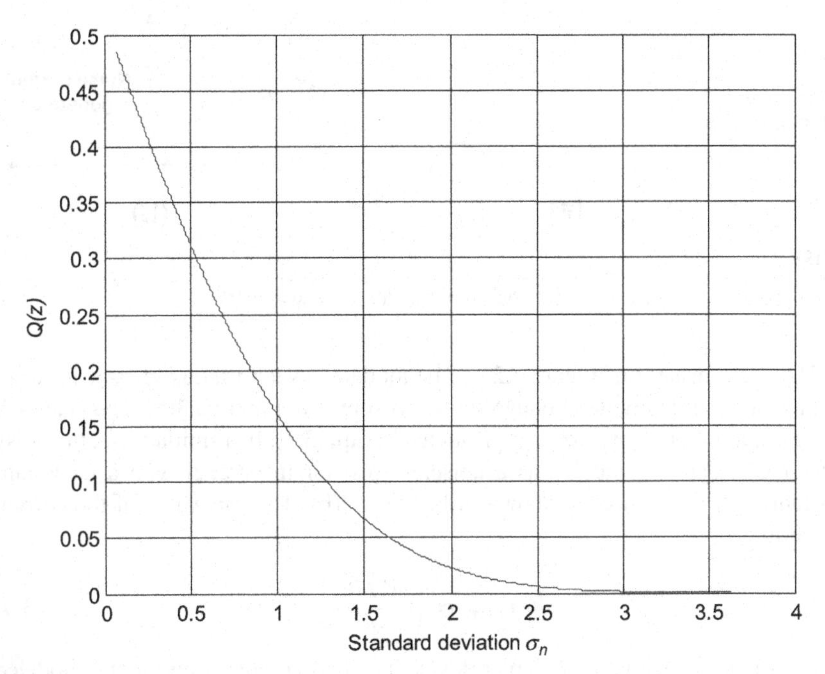

FIGURE 3.1

Cumulative normalized Gaussian probability density function. (For color version of this figure, the reader is referred to the online version of this book.)

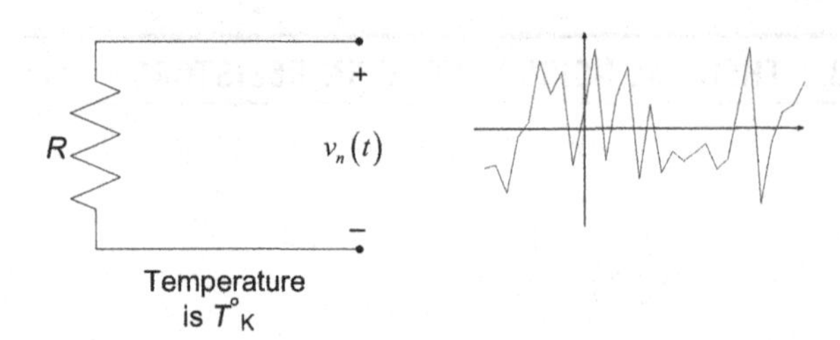

FIGURE 3.2

Thermal noise voltage generated across resistor. (For color version of this figure, the reader is referred to the online version of this book.)

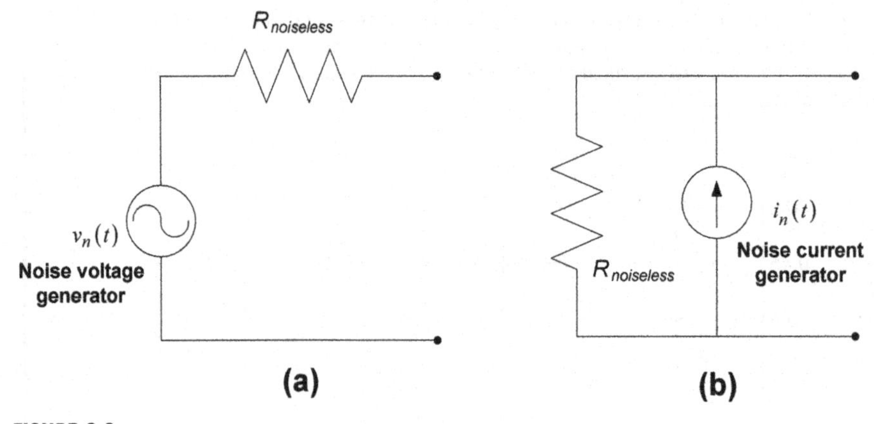

FIGURE 3.3

Voltage source model and current source model for a noisy resistor.

The noisy resistor of Figure 3.2 can be modeled as a *noiseless* resistor in series with a Gaussian-distributed voltage source as shown in Figure 3.3(a). The Gaussian voltage source has an *rms* value as presented in Eqn (3.3). In a similar vein, the noisy resistor can also be modeled as a noiseless resistor in parallel with a Gaussian-distributed current source as shown in Figure 3.3(b). The *rms* value of the current source is

$$i_{n,rms} \approx \sqrt{\frac{4kTB_n}{R}} \tag{3.9}$$

By way of an example, a bipolar device has thermal noise present and modeled by resistor at the base and a resistor at the emitter as shown in Figure 3.4(a). Similarly, the thermal noise present in a MOS device can be represented using a current source model attached between the source and the drain as shown in Figure 3.4(b). The noise current source of the MOS device has an *rms* value given according to Eqn (3.9) [3]

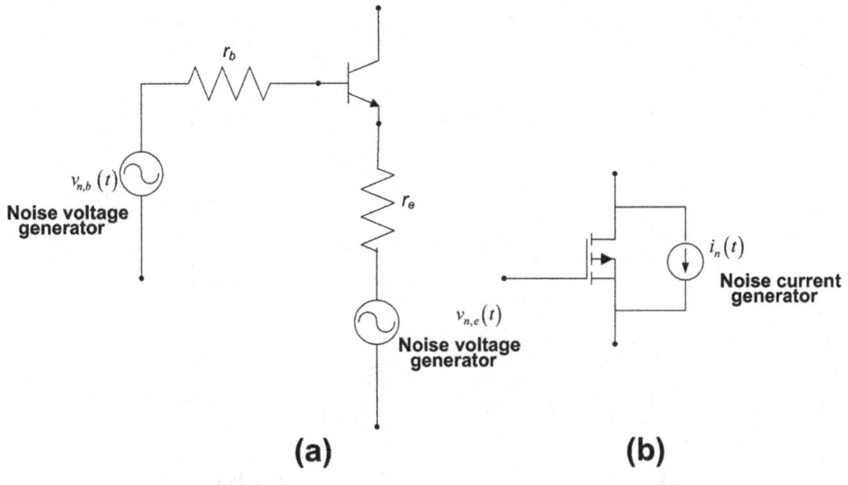

FIGURE 3.4

Thermal noise in transistors (a) Bipolar and (b) MOSFET.

$$i_{n,rms} \approx \sqrt{4kT\left(\frac{2g_m}{3}\right)} \tag{3.10}$$

where g_m is the dynamic transconductance of the transistor.

In a receiver system, it is often the case that multiple noise sources are present and contribute to the degradation of the received signal. For instance, consider the two noise sources shown in Figure 3.5. Due to the statistical nature of thermal noise, one cannot just combine the voltages delivered to the load. However, the power delivered to the noiseless load resistor R_{Load} due to each resistor can be determined separately, and the total noise power can then be determined as the sum of noise powers provided by each resistor.[d] More specifically, let the bandwidth of observation be B_n and that the temperatures of the resistors R_{s1} and R_{s2} are $T_1^\circ K$ and $T_2^\circ K$, respectively. The *rms* noise voltages for every resistor in Figure 3.5(b) can then be determined independently using Eqn (3.3). To determine the noise power dissipated in R_{Load} due to R_{s1} only, we short the voltage due to R_{s2}, that is, we set $V_{n,s2} = 0$. Using Kirchhoff's current law, we can obtain the voltage V_{L1} across R_{Load} as shown in Figure 3.6(a), that is

$$\frac{V_{n,s1} - V_{L1}}{R_{s1}} = \frac{V_{L1}}{R_{s2}} + \frac{V_{L1}}{R_{Load}} \Rightarrow V_{L1} = \frac{R_{s2}R_{Load}}{R_{s1}R_{s2} + R_{s1}R_{Load} + R_{s2}R_{Load}} V_{n,s1} \tag{3.11}$$

[d]Note that the *rms* noise voltages due to the various resistors do not add, however, the respective noise powers do.

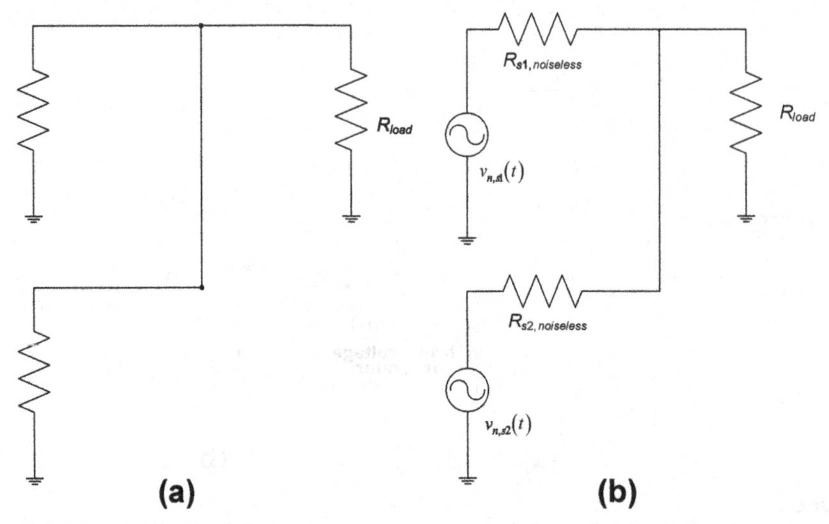

FIGURE 3.5

Two independent noise sources: (a) Two noisy resistors and (b) Two voltage noise sources in series with noiseless resistors.

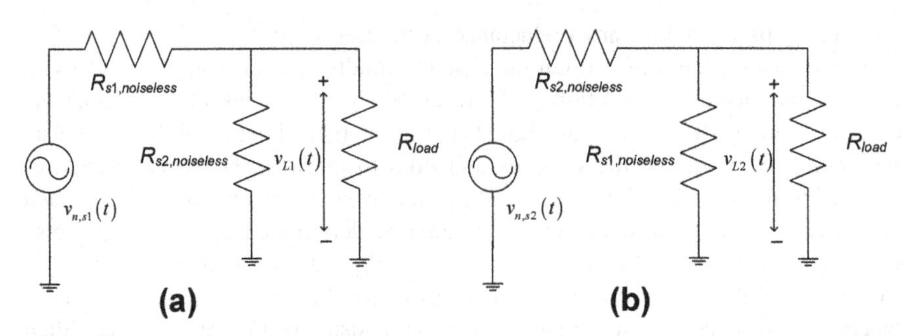

FIGURE 3.6

Total noise power delivered to the load resistance due to two independent noise sources.

The power delivered to the load resistor can now be estimated

$$P_{n,1} = \frac{V_{L1}^2}{R_{Load}} = \left(\frac{R_{s2}R_{Load}}{R_{s1}R_{s2} + R_{s1}R_{Load} + R_{s2}R_{Load}}\right)^2 \frac{V_{n,s1}^2}{R_{Load}} \quad (3.12)$$

But according to Eqn (3.3), $V_{n,s1} \approx \sqrt{4kTB_nR_{Load}}$, then Eqn (3.12) becomes

$$P_{n,1} = \left(\frac{R_{s2}R_{Load}}{R_{s1}R_{s2} + R_{s1}R_{Load} + R_{s2}R_{Load}}\right)^2 4kT_1B_n \quad (3.13)$$

Finding the noise power due to R_{s2} can be found in the same manner. Shorting the noise voltage due to R_{s1} implies that the noise voltage delivered to the load resistor is

$$P_{n,2} = \left(\frac{R_{s1}R_{Load}}{R_{s1}R_{s2} + R_{s1}R_{Load} + R_{s2}R_{Load}}\right)^2 4kT_2B_n \tag{3.14}$$

The total noise power delivered to the load resistor is the sum of the powers given in Eqn (3.22) and Eqn (3.14), that is

$$P_{n,Total} = P_{n,1} + P_{n,2} = \left(\frac{R_{Load}}{R_{s1}R_{s2} + R_{s1}R_{Load} + R_{s2}R_{Load}}\right)^2 (R_{s1}T_2 + R_{s2}T_1)4kB_n \tag{3.15}$$

3.1.2 System noise figure

The purpose of this section is to further discuss thermal noise analysis in the context of individual device or block noise as well the total noise figure of a cascade of devices or blocks.

3.1.2.1 Noise factor and noise figure

The noise factor of a device or RF block[e] is the amount of noise power that gets added to the desired signal, thus degrading its quality. The noise factor can be defined as the ratio of noise power delivered by a noisy component over the noise power delivered by a *noiseless component* whose input noise power is $N_0 = kT_0B_n$ where $T_0 = 290$ K is the absolute reference temperature. A further definition based on the previous statement can be expressed in terms of input and output carrier-to-noise ratios (CNRs) of a given RF block. Consider the CNR of a given input signal to an RF block to be CNR_{input} and the output CNR of the signal exiting the same RF block to beCNR_{output}, then the noise factor can be defined as the ratio

$$F = \frac{CNR_{input}}{CNR_{output}} \tag{3.16}$$

In the coming discussion, we will explore both definitions to clarify the noise factor. The noise figure is related to the noise factor as

$$NF = 10\log_{10}(F) = 10\log_{10}\left(\frac{CNR_{input}}{CNR_{output}}\right) \tag{3.17}$$

The thermal noise properties of a device or block can be described by the noise factor F or the excess input noise temperature T_{exc} [4]. Consequently, assume that

[e]A system block in this context could indicate a single device or an RF block comprised of more than one device.

the gain (or loss) of a given RF block to be G, then the noise factor of a linear two-port system is

$$F = \frac{N_{output}}{GN_{input}} \tag{3.18}$$

where $N_{input} = kT_0B_n$ is the available noise power in a given bandwidth B_n from a matched resistive termination to the characteristic impedance of the line connected to the input device at absolute reference temperature $T_0 = 290$ K. N_{output} is the total noise power in a given bandwidth B_n available at the output of the two-port system for a given input noise power N_{input}, where $N_{output} = Gk(T_0 + T_{exc})B_n$. The parameter G is the available power gain of the two-port system for noncoherent signal input to the two-port system within a *noise equivalent bandwidth* B_n and measured at the output of the two-port system within the same bandwidth. Given N_{input} and N_{output}, the relationship in Eqn (3.18) can be further expressed as

$$F = \frac{N_{output}}{GN_{input}} = \frac{Gk(T_0 + T_{exc})B_n}{GkT_0B_n} \tag{3.19}$$

At this juncture, based on Eqn (3.19), we can define the excess noise temperature T_{exc} in terms of the noise factor F and the absolute reference temperature T_0 as

$$T_{eff} = (F - 1)T_0 \; or \; F = 1 + \frac{T_{exc}}{T_0} \tag{3.20}$$

Given the resistor R at a given temperature T_0, the *rms* noise voltage expressed in Eqn (3.3) implies that the available noise power due to R to another resistor with the same resistance is

$$P_{noise} = \left(\frac{v_{n,rms}}{2R}\right)^2 R = kT_0B_n \tag{3.21}$$

which is equivalent to the input noise power N_{input}. By comparing the results in Eqn (3.19) and Eqn (3.23), one can imply the relationship

$$F = \frac{N_{output}}{GN_{input}} = \frac{N_{output}}{GkT_0B_n} \Rightarrow N_{output} = FGkT_0B_n \tag{3.22}$$

Furthermore, the ratio of the signal powers is the gain, or $S_{output} = GS_{input}$. Using simple mathematics, we can see that the noise factor is the ratio of the input CNR to the output CNR or

$$F = \frac{(C/N)_{input}}{(C/N)_{output}} \tag{3.23}$$

The noise factor is then the amount of noise degradation that has impacted the input CNR at the output of an RF block or device.

At this point, it is important to discuss the noise equivalent bandwidth B_n. In practical applications, in order to measure the noise within the bandwidth B_n, a filter

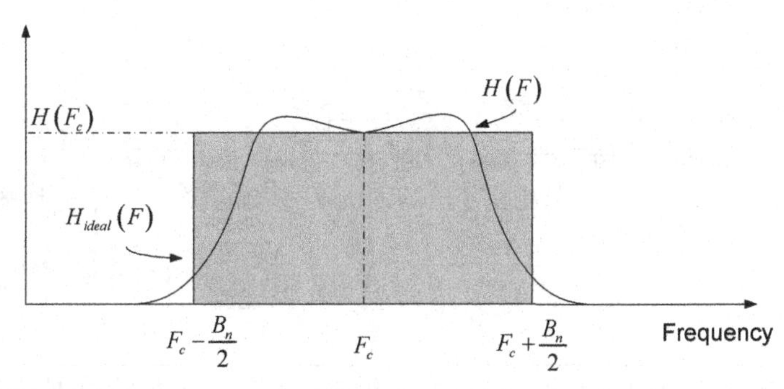

FIGURE 3.7

Noise equivalent bandwidth for an ideal (shaded) and nonideal filter.

is placed at the output of the RF block. The filtering operation is by no means ideal. The noise equivalent bandwidth for a bandpass filter $H(F)$, illustrated in Figure 3.7, is defined as the bandwidth of an ideal filter $H_{ideal}(F)$ centered around a frequency F_c such that the power at the output of this filter, if excited by white Gaussian noise, is equal to that of the real filter given the same input signal. In other words, we can express the noise equivalent bandwidth mathematically as

$$B_n = \frac{1}{|H(F_c)|^2} \int_{F_c - B_n/2}^{F_c + B_n/2} |H(F)|^2 \mathrm{d}F \tag{3.24}$$

The estimation of the noise equivalent bandwidth allows us to compute the amount of in-band noise and its effect on the received signal CNR regardless of the filter's transfer function.

3.1.2.2 Cascaded noise factor and noise figure

In the receiver, the desired signal passes through multiple analog blocks between the antenna and the analog to digital converters (ADCs), as shown in the receive chain depicted in Figure 3.8. These blocks all vary in function. Some are amplifiers such as the low noise amplifier (LNA) and the voltage gain amplifier (VGA), some are filters, some are mixers, and some are signal attenuators or pads. All of these blocks, however, exhibit a certain noise figure and all in turn degrade the signal-to-noise ratio. The question then becomes: What is the total or cumulative noise figure due to various blocks cascaded in a row?

In order to develop a relationship for cascade noise factor and noise figure, consider the cascade of analog blocks depicted in Figure 3.9. The linear model implies that a certain signal plus noise entering a certain analog block will be first either amplified, or attenuated, by a certain gain G_l and that a certain amount of noise N_l will be added to the noisy signal. To illustrate this, consider two analog RF blocks connected in cascade. Let the noise factor of the first block be the ratio of the input CNR, say $(C/N)_{input}$ to the output CNR $(C/N)_{output_1}$. Note, however, that

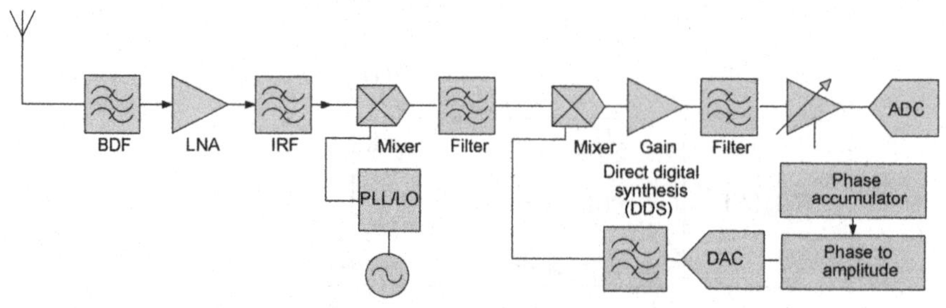

FIGURE 3.8

Conceptual receive chain of an IF sampling transceiver. BDF, band definition filter; LNA, low noise amplifier; PLL/LO, phase-locked loop/local oscillator; ADC, analog to digital converter; DAC, digital to analog converter; IRR, Image reject filter.

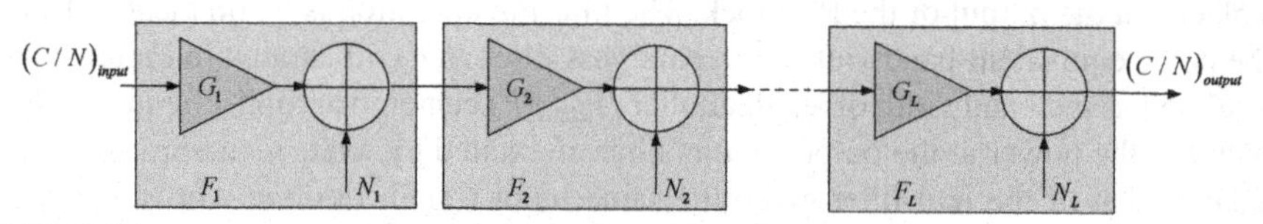

FIGURE 3.9

Cascade of analog radio frequency (RF) building blocks. (For color version of this figure, the reader is referred to the online version of this book.)

$$\left(\frac{C}{N}\right)_{output_1} = \frac{C_{output_1}}{N_{output_1}} = \frac{G_1 C_{input}}{G_1 N_{input} + N_1} \tag{3.25}$$

Then, the noise factor of the first block can be computed as

$$F_1 = \frac{(C/N)_{input}}{(C/N)_{output_1}} = \frac{\frac{C_{input}}{N_{input}}}{\frac{G_1 C_{input}}{G_1 N_{input} + N_1}} = 1 + \frac{N_1}{G_1 N_{input}} \tag{3.26}$$

By the same thought process, the noise figure of the second block is given as

$$F_2 = \frac{(C/N)_{output_1}}{(C/N)_{output}} = 1 + \frac{N_2}{G_2 N_{out_1}} \tag{3.27}$$

where $N_{out_1} = G_1 N_{input} + N_1$ is the noise output of the first block. The cascaded noise factor at the output of the second block

$$F_{\text{cascade}} = \frac{(C/N)_{input}}{(C/N)_{output}} = \frac{\left(\frac{C_{input}}{N_{input}}\right)}{\left(\frac{C_{output}}{N_{output}}\right)} \tag{3.28}$$

The output carrier signal is the input carrier signal multiplied by the various stages of gain, or in this case $C_{output} = G_1 G_2 C_{input}$. The output noise is the output noise of the first stage N_{out_1} multiplied by the gain of the second stage G_2 and added to N_2. The cascaded noise figure can then be computed as the ratio

$$F_{\text{cascade}} = \frac{\left(\frac{C_{input}}{N_{input}}\right)}{\left(\frac{C_{output}}{N_{output}}\right)} = \frac{\frac{C_{input}}{N_{input}}}{\frac{G_1 G_2 C_{input}}{(G_1 N_{input} + N_1) G_2 + N_2}} = \frac{(G_1 N_{input} + N_1) G_2 + N_2}{G_1 G_2 N_{input}}$$

$$= 1 + \frac{N_1}{G_1 N_{input}} + \frac{N_2}{G_1 G_2 N_{input}} = F_1 + \frac{F_2 - 1}{G_1} \tag{3.29}$$

The final result in Eqn (3.29) implies that the total noise factor of a cascade of two RF blocks is comprised of the sum of the noise factor of the first block plus a noise factor of the second block scaled by the gain of the first block. Two important observations come to mind. The first observation is that the noise figure of the first block dictates the total noise figure of the system. The second observation has to do with the second term in Eqn (3.29). The impact of the noise figure of the second block can to a great degree be diminished by increasing the gain of the first block. Therefore, it makes sense that in a receiver system, the first amplifier is a low noise amplifier with large power gain.

In general, for an arbitrary number of blocks, the total noise factor can be found according to the Friis noise formula as

$$F_{\text{cascade}} = F_1 + \frac{F_2 - 1}{G_1} + \frac{F_3 - 1}{G_1 G_2} + \ldots + \frac{F_L - 1}{G_1 G_2 \ldots G_{L-1}} = F_1 + \sum_{l=2}^{L} \left(\frac{F_l - 1}{\prod_{n=1}^{L-1} G_n} \right) \tag{3.30}$$

The total noise figure is given as

$$NF_{\text{cascade}} = 10 \log_{10}(F_{\text{cascade}}) \tag{3.31}$$

The gain values in Eqn (3.30) can be either the power gain or the square of the voltage gain depending on the source impedance gain. This point will be discussed further when discussing the thermal noise of the mixer. For now, however, let us assume that the gain values correspond to power gain of the various stages.

At this point, it is important to note that the output gain of the cascade chain can be implied as the linear product of all the gain stages in the cascade, or the output signal can be related to the input signal in terms of signal strength as

$$\begin{aligned} \frac{C_{output}}{C_{input}} &= G_1 G_2 \ldots G_{L-1} = \prod_{l=1}^{L-1} G_l \\ \left.\frac{C_{output}}{C_{input}}\right|_{dB} &= G_{1,dB} + G_{2,dB} \ldots G_{L-1,dB} = \sum_{l=1}^{L-1} G_{l,dB} \end{aligned} \tag{3.32}$$

where $G_{i,dB} = 10\log_{10}(G_i)$. From Eqn (3.30), we can extend the previous observations to imply that the noise figure of the first block in the cascade is the most dominant noise figure in the system, and that the noise figure of each consecutive block becomes less and less dominant since each block is normalized by the aggregate gain up to that block in the cascade. So the noise figure of the last block is the least dominant. It is important to note that Eqn (3.30) and Eqn (3.32) imply that all input and output impedances are perfectly matched to the system impedance as will be explained later on. At this point, it is instructive to examine the relationship between total noise figure of a multistage amplifier as it is specified in Eqn (3.30) and the overall power efficiency. To do so, let us first define the power-added efficiency of a single-stage amplifier as the ratio of the difference of the output power minus the input power divided by the source power due to the DC bias, or

$$\eta = \frac{P_{output} - P_{input}}{P_s} = \frac{GP_{input} - P_{input}}{P_s} = \frac{P_{input}(G-1)}{P_s} \tag{3.33}$$

In a multistage amplifier, where the single stage amplifiers are connected in tandem, the input power to the second stage is the input power to the first stage multiplied by the gain of the first stage $P_{input,2} = G_1 P_{input,1}$. For an N-stage amplifier, as shown in Figure 3.10, the relationship can be extended as

$$P_{input,N} = \left(\prod_{n=1}^{N-1} G_n\right) P_{input,1} = (G_1 G_2 \ldots G_{N-1}) P_{input,1} \tag{3.34}$$

The output of the Nth-stage amplifier is related to the input power of the Nth-stage amplifier as $P_{out,N} = G_N P_{input,N}$ which according to Eqn (3.33) and Eqn (3.34) implies

$$\begin{aligned}
\eta_N &= \frac{P_{output,N} - P_{input,N}}{P_{s,N}} = \frac{(G_N - 1)P_{input,N}}{P_{s,N}} \\
&= \frac{(G_N - 1)}{P_{s,N}} \left(\prod_{n=1}^{N-1} G_n\right) P_{input,1} \quad \text{or} \quad P_{s,N} = \frac{(G_N - 1)}{\eta_N} \left(\prod_{n=1}^{N-1} G_n\right) P_{input,1}
\end{aligned} \tag{3.35}$$

where η_N designates the efficiency of the Nth stage.

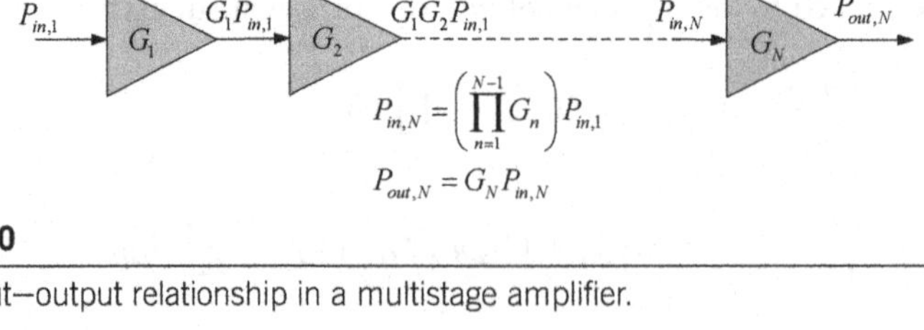

FIGURE 3.10

Power input–output relationship in a multistage amplifier.

Recall that the efficiency of a multistage amplifier is expressed as the ratio

$$\eta_{Total} = \frac{P_{output,N} - P_{input,1}}{P_{s,1} + P_{s,2} + \ldots + P_{s,N}} = \frac{P_{output,N} - P_{output,1}}{\sum_{n=1}^{N} P_{s,n}} \tag{3.36}$$

At this point, it is instructive to see how the total efficiency relates to the various stage gains and efficiencies. To do so, let us examine both the numerator and the denominator of Eqn (3.36)

$$P_{output,N} - P_{input,1} = \prod_{n=1}^{N} G_n P_{input,1} - P_{input,1} = P_{input,1}\left(\prod_{n=1}^{N} G_n - 1\right) \tag{3.37}$$

and

$$\begin{aligned}\sum_{n=1}^{N} P_{s,n} &= \frac{G_1 - 1}{\eta_1} P_{input,1} + \frac{G_2 - 1}{\eta_2} P_{input,2} + \ldots + \frac{G_N - 1}{\eta_N} P_{input,N} \\ &= P_{input,1}\left\{\frac{G_1 - 1}{\eta_1} + \frac{G_2 - 1}{\eta_2} G_1 + \ldots + \frac{G_N - 1}{\eta_N}\left(\prod_{n=1}^{N-1} G_n\right)\right\}\end{aligned} \tag{3.38}$$

Hence the ratio of Eqn (3.37) over Eqn (3.38) gives

$$\eta_{Total} = \frac{\prod_{n=1}^{N} G_n - 1}{\frac{G_1-1}{\eta_1} + \frac{G_2-1}{\eta_2} G_1 + \ldots + \frac{G_N-1}{\eta_N}\left(\prod_{n=1}^{N-1} G_n\right)} \tag{3.39}$$

The result obtained in Eqn (3.39) can be further simplified by assuming that each gain stage has a significant linear gain such that $G_n - 1 \approx G_n$, then Eqn (3.39) becomes

$$\eta_{Total} = \frac{\prod_{n=1}^{N} G_n}{\frac{G_1}{\eta_1} + \frac{G_1 G_2}{\eta_2} + \ldots + \frac{\prod_{n=1}^{N} G_n}{\eta_N}} = \frac{1}{\frac{1}{\prod_{n=2}^{N} G_n \eta_1} + \frac{1}{\prod_{n=3}^{N} G_n \eta_2} + \ldots + \frac{1}{\eta_N}} \tag{3.40}$$

The relationship in Eqn (3.40) implies that the efficiency of the last stage in a multistage amplifier has the most impact to the overall efficiency η_{Total}, and hence the last stage has to be the most efficient.

Finally, it is important to note that in practice, the relationships in Eqn (3.30) and Eqn (3.32) are not entirely accurate due to the fact that not all the terminal impedances are matched to the characteristic impedance of the system. If that were the case, then each stage in the cascade will completely transfer the signal power from the stage preceding it.[f] The relationships in Eqn (3.30) and Eqn (3.32) deviate from actual performance especially in the case of broadband amplifiers and filters, for example. In this case, the mismatch between loads is expressed in the maximum input/output voltage standing wave ratio (VSWR) specified over the entire band of interest.[g] Given the terminal impedance Z_{Terminal} then the load impedance can vary as

$$R_{Load} = \frac{Z_{\text{Terminal}}}{VSWR} \text{ or } R_{Load} = Z_{\text{Terminal}} \times VSWR \tag{3.41}$$

By way of an example, if the VSWR of a certain wideband amplifier is 3:1, then the load impedance could be either $R_{Load} = \frac{1}{3}Z_{\text{Terminal}}$ or $R_{Load} = 3Z_{\text{Terminal}}$. Given that $Z_{\text{Terminal}} = 50$ ohm then $R_{Load} = 16.67$ ohm or $R_{Load} = 150$ ohm. That is over the operational frequency of the amplifier, the input impedance may vary between 16.67 ohm and 150 ohm. Incidentally, the gain of the amplifier may vary over the same operational band. Recall that the transmission loss increases due to VSWR by a factor

$$\rho = \frac{VSWR^2 + 1}{2VSWR} \tag{3.42}$$

Therefore, it is important to take into account when analyzing the cumulative noise figure of a system the variations on the individual components noise figure over temperature and frequency as well as the impedance matching of the system.

3.1.2.3 System sensitivity and the link budget equation

Given a certain modulation scheme and a certain desired *minimum* raw bit error rate (BER) performance, it is often desired to determine the minimum received signal power required to properly demodulate the received signal. This minimum received signal power is known as the system sensitivity. The system sensitivity is defined as

$$P_{\text{sensitivity}} = 10\log_{10}(kT) + 10\log_{10}(B_n) + NF_{\text{cascade}} + CNR \tag{3.43}$$

where k is Boltzmann's constant, and T is the temperature in kelvin. At room temperature, the quantity $10\log_{10}(kT) = -174$ dBm/Hz and hence, based on Eqn (3.43), we obtain the well-known relationship

$$P_{\text{sensitivity}} = -174\,\text{dBm/Hz} + 10\log_{10}(B_n) + NF_{\text{cascade}} + CNR \tag{3.44}$$

Note that in both Eqn (3.43) and Eqn (3.44), the noise equivalent bandwidth was used instead of the channel bandwidth. After digital filtering, a certain processing

[f]Furthermore, note that the output impedance of the previous stage also influences the NF of the next stage.

[g]The band of interest is the operational band over six octaves.

gain is theoretically obtained. This processing gain is the ratio of the noise equivalent bandwidth to the channel bandwidth. This can be best understood by relating the required CNR to the required SNR, or

$$CNR = 10\log_{10}\left(\frac{E_b R_b}{N_0 B_n}\right) = \left(\frac{E_b}{N_0}\right)_{dB} + \underbrace{10\log_{10}\left(\frac{R_b}{B_n}\right)}_{\text{Processing Gain}} \tag{3.45}$$

where R_b is the data rate. In order to determine the necessary transmit power in an interference-free environment, the Friis link budget equation, defined in Chapter 1, can be restated as [5]

$$\begin{aligned} P_{Tx} &= P_{\text{sensitivity}} + P_{loss} - G_{Tx} - G_{Rx} \\ &= 10\log_{10}(kT) + 10\log_{10}(B_n) + NF_{\text{cascade}} + CNR + P_{loss} - G_{Tx} - G_{Rx} \end{aligned} \tag{3.46}$$

where P_{loss} is the path loss of the signal, and G_{Tx} and G_{Rx} are the transmit and receive antenna gains respectively.

EXAMPLE 3.2 CASCADE NOISE FIGURE ANALYSIS OF GPS RECEIVER

Consider the IF GPS receiver depicted in Figure 3.11. The GPS signal is centered at 1575.42 MHz. Assume that the system sensitivity at the output of the antenna connector to be −140 dBm. The noise equivalent bandwidth of the GPS receiver is 9.548 MHz. Determine the CNR given the receive-chain parameters in Table 3.2. In the ensuing analysis, assume that the experiment is conducted at room temperature. Furthermore, consider all input and output impedances to be perfectly matched to the system impedance. All ports are conjugate matched for 50 Ohm impedance.

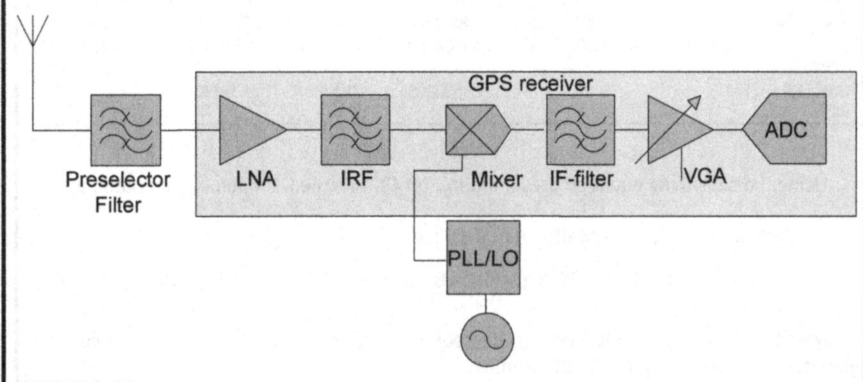

FIGURE 3.11

GPS receiver with external preselection filter. LNA, low noise amplifier; VGA, voltage gain amplifier; IRF, Image reject filter; PLL/LO, phase-locked loop/local oscillator; ADC, analog to digital converter. (For color version of this figure, the reader is referred to the online version of this book.)

Continued

EXAMPLE 3.2 CASCADE NOISE FIGURE ANALYSIS OF GPS RECEIVER—cont'd

What is the impact of lowering the noise figure of the LNA by 1 dB on the cascaded system noise figure? What is the effect of increasing the LNA gain by 4 dB on the cascaded system noise figure?

The first step in determining the operating CNR is to determine the cumulative noise figure of the system. Using the parameters in Table 3.2 and the relationship in Eqn (3.30) and Eqn (3.31), the cascaded or total noise figure of the receiver can be determined. The results are summarized in Table 3.3. The cascaded noise figure is shown to be 4.36 dB.

Table 3.2 GPS Receiver Noise Figure and Gain Parameters

Parameter	Preselector	LNA	Filter	Mixer	IF Filter	VGA
Noise Figure (dB)	1.5	2.3	1.4	9	4	13
Gain (dB)	−1.5	18	−1.4	11.5	−4	50

LNA, low noise amplifier; VGA, voltage gain amplifier.

Table 3.3 Cascade Noise Figure Analysis of Receiver Shown in Figure 3.11

Parameter	Preselector	LNA	Filter	Mixer	IF Filter	VGA
Noise Figure (dB)	1.5	2.3	1.4	9	4	13
Gain (dB)	−1.5	18	−1.4	11.5	−4	50
F (linear)	1.41253754	1.6982437	1.38038426	7.94328235	2.51188643	19.9526231
Gain (linear)	0.70794578	63.095734	0.72443596	14.1253754	0.39810717	100,000
Casc. Gain	1	0.7079458	44.6683592	32.3593657	457.08819	181.970086
Casc. F (linear)	1.41253754	2.3988329	2.40734866	2.6219166	2.62522425	2.72937666
Casc. NF (dB)	1.5	3.8	3.81538995	4.18618873	4.19166407	4.36063473

Using the sensitivity equation presented in Eqn (3.44), the CNR can be computed as

$$\begin{aligned} CNR &= P_{\text{sensitivity}} + 174\ \text{dBm/Hz} - 10\log_{10}(B_n) - NF_{\text{cascade}} \\ &= -140 + 174 - 10\log_{10}\left(9.548 \times 10^6\right) - 4.36 = -40.16\ \text{dB} \end{aligned} \tag{3.47}$$

Note that the above result does not incorporate the gain due to filtering or the processing gain due to de-spreading and multidwelling.

Next, if we decrease the noise figure of the LNA by 1 dB from 2.3 to 1.3 dB and repeat the cascade noise figure analysis, as shown in Table 3.4, we see that the cascade noise figure of the receiver improves by approximately 0.87 dB from 4.36 to 3.49 dB. Since the LNA is not the first block in the receiver, improving the LNA noise figure does not normally improve the overall noise dB for dB.

Next, we increase the LNA gain by 4 dB from 18 to 22 dB. The cascade noise figure performance is summarized in Table 3.5. Note the improvement in the cascade noise

EXAMPLE 3.2 CASCADE NOISE FIGURE ANALYSIS OF GPS RECEIVER—cont'd

figure is 0.33 dB from 4.36 to 4.03 dB. Recall that increasing the gain in the LNA only serves to lower the noise figure of the subsequent gain stages. It can be concluded that increasing the gain of the LNA in this case did not, by itself, significantly improve the performance of the overall receiver. A further negative effect of increasing the gain is increasing the linearity requirements for the next stages due to higher signal level.

Table 3.4 Cascade Noise Figure Analysis of the Receiver Shown in Figure 3.11 with Lowered LNA Noise Figure

Parameter	Preselector	LNA	Filter	Mixer	IF Filter	VGA
Noise Figure (dB)	1.5	1.3	1.4	9	4	13
Gain (dB)	−1.5	18	−1.4	11.5	−4	50
F (linear)	1.41253754	1.3489629	1.38038426	7.94328235	2.51188643	19.9526231
Gain (linear)	0.70794578	63.095734	0.72443596	14.1253754	0.39810717	100,000
Casc. Gain	1	0.7079458	44.6683592	32.3593657	457.08819	181.970086
Casc. F (linear)	1.41253754	1.9054607	1.91397646	2.1285444	2.13185205	2.23600446
Casc. NF (dB)	1.5	2.8	2.81936592	3.28082713	3.28757061	3.49472665

LNA, low noise amplifier; VGA, voltage gain amplifier.

Table 3.5 Cascade Noise Figure Analysis of Receiver Shown in Figure 3.11 with Lowered LNA Noise Figure

Parameter	Preselector	LNA	Filter	Mixer	IF Filter	VGA
Noise Figure (dB)	1.5	2.3	1.4	9	4	13
Gain (dB)	−1.5	22	−1.4	11.5	−4	50
F (linear)	1.41253754	1.6982437	1.38038426	7.94328235	2.51188643	19.9526231
Gain (linear)	0.70794578	158.48932	0.72443596	14.1253754	0.39810717	100,000
Casc. Gain	1	0.7079458	112.201845	81.2830516	1148.15362	457.08819
Casc. F (linear)	1.41253754	2.3988329	2.4022231	2.48764413	2.48896093	2.53042475
Casc. NF (dB)	1.5	3.8	3.80613338	32.95788253	3.96018079	4.03193427

LNA, low noise amplifier; VGA, voltage gain amplifier.

3.1.3 Thermal noise in components

In this subsection, we will discuss noise in two-port systems. The analysis is then applied to various RF components such as LNAs and mixers. The discussion offers the reader a certain insight into the generation and workings of thermal noise in RF components.

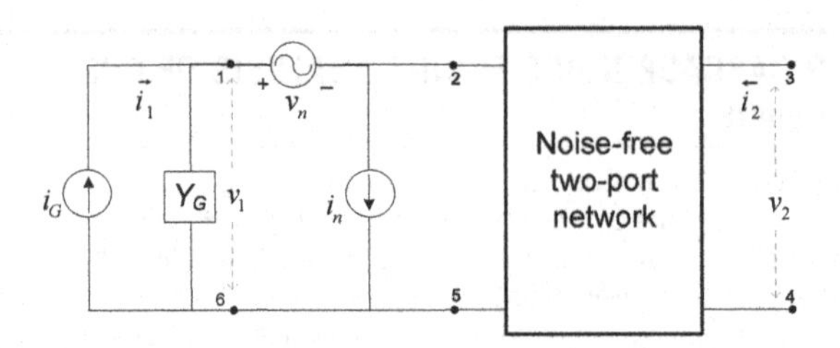

FIGURE 3.12

Two-port noiseless circuit with noise sources referred to the input. (For color version of this figure, the reader is referred to the online version of this book.)

3.1.3.1 Noise in two-port systems

Rohte and Danlke [6] and Haus [7] have shown that a noisy two-port network can be modeled as a noiseless two-port network with a voltage noise source v_n and a current noise source i_n as shown in Figure 3.12. The following analysis follows the excellent development provided in [8] and [9]. Note that due to the random nature of thermal noise, the polarity of the noise sources is irrelevant. The input termination Y_G produces the input noise current i_G.

Using the *ABCD* matrix representation developed in the previous chapter, we can express the voltage v_1 and the current i_1 as

$$\begin{aligned} v_1 &= Av_2 + Bi_2 + v_n \\ i_1 &= Cv_2 + Di_2 + i_n \end{aligned} \tag{3.48}$$

The Thévenin equivalent circuit of Figure 3.12 is depicted in Figure 3.13. The short-circuit average current power at ports 2–5 can be expressed as the sum of the average termination source current power and the noise current power due to i_n and $Y_G v_n$ or

$$\langle i_{sc}^2 \rangle = \langle i_G^2 \rangle + \left\langle |i_n + Y_G v_n|^2 \right\rangle \tag{3.49}$$

where the $\langle .^2 \rangle$ denotes average power. Expanding the notation in Eqn (3.49), we obtain

$$\begin{aligned} \langle i_{sc}^2 \rangle &= \langle i_G^2 \rangle + \langle (i_n + Y_G v_n)(i_n + Y_G v_n)^* \rangle \\ &= \langle i_G^2 \rangle + \langle i_n^2 \rangle + |Y_G|^2 \langle v_n^2 \rangle + Y_G^* \langle i_n v_n^* \rangle + Y_G \langle i_n^* v_n \rangle \end{aligned} \tag{3.50}$$

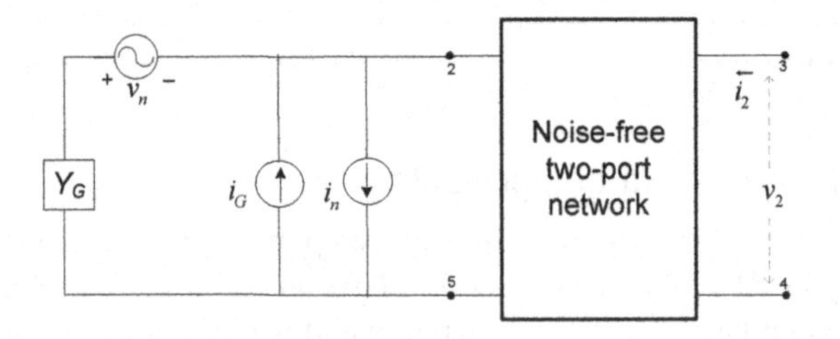

FIGURE 3.13

Equivalent Thévenin circuit of Figure 3.12. (For color version of this figure, the reader is referred to the online version of this book.)

As previously stated, the two-port noiseless network does not contribute any noise to the system.

Next, we examine the cross-correlation terms in Eqn (3.50) between i_n and v_n. The current i_n is made up of the sum of two components $i_n = u_n + Y_C v_n$ where Y_C is the correlation admittance. Hence, a reexamination of the cross-correlation term $Y_G \langle i_n^* v_n \rangle$ reveals that

$$Y_G \langle i_n^* v_n \rangle = Y_G \langle (u_n + Y_C v_n)^* v_n \rangle = Y_G \underbrace{\langle u_n^* v_n \rangle}_{=0} + Y_G Y_C^* \langle v_n^* v_n \rangle = Y_G Y_C^* \langle v_n^2 \rangle \tag{3.51}$$

where $\langle u_n^* v_n \rangle = 0$. In a like manner, it could be shown that

$$Y_G^* \langle i_n v_n^* \rangle = Y_G^* Y_C \langle v_n^2 \rangle \tag{3.52}$$

Next, recall that each noise source can be expressed according to Eqn (3.3) for a given bandwidth B_n as

$$\langle v_n^2 \rangle = 4kTB_n R_n \tag{3.53}$$

where R_n is the equivalent noise resistance for noise-source voltage v_n. By the same token, the uncorrelated current u_n and the input noise current can be expressed in like manner as

$$\begin{aligned} \langle u_n^2 \rangle &= 4kTG_u B_n \\ \langle i_G^2 \rangle &= 4kTG_G B_n \end{aligned} \tag{3.54}$$

where G_u is the equivalent noise conductance for the noise current u_n and G_G is the equivalent noise conductance for the noise current i_G. The excitation noise or input termination admittance is given as $Y_G = G_G + jB_G$.[h] Recall that it is possible to choose a value for Y_G such that one can minimize the noise figure, however, that can only be accomplished at the expense of reducing the gain.

Given the definition of the total noise current above, the current noise power $\langle i_n^2 \rangle$ can be further expressed as

$$\langle i_n^2 \rangle = \left\langle (u_n + Y_C v_n)^2 \right\rangle = \langle u_n^2 \rangle + |Y_C|^2 \langle v_n^2 \rangle \tag{3.55}$$

Substituting Eqn (3.53) and Eqn (3.54) into Eqn (3.55), we obtain

$$\langle i_n^2 \rangle = 4kTB_n \left(G_u + R_n |Y_C|^2 \right) \tag{3.56}$$

Next, substituting the results from Eqn (3.55) and Eqn (3.56) into Eqn (3.50), we obtain

---

[h] B_G in this case indicates the susceptance and not the bandwidth.

$$\begin{aligned}\langle i_{sc}^2\rangle &= \langle i_G^2\rangle + \langle u_n^2\rangle + \left(|Y_C|^2 + |Y_G|^2 + Y_G^* Y_C + Y_G Y_C^*\right)\langle v_n^2\rangle \\ &= \langle i_G^2\rangle + \langle u_n^2\rangle + (|Y_C| + |Y_G|)^2 \langle v_n^2\rangle \end{aligned} \tag{3.57}$$

Recall that in the model shown in Figure 3.12 and Figure 3.13, the noise sources are referred to the input of the two-port network system. The noise factor for this two-port network is simply the ratio

$$F = \frac{\langle i_{sc}^2\rangle}{\langle i_G^2\rangle} = 1 + \frac{\langle u_n^2\rangle + (|Y_C| + |Y_G|)^2 \langle v_n^2\rangle}{\langle i_G^2\rangle} \tag{3.58}$$

Substituting Eqn (3.53) and Eqn (3.54) into Eqn (3.58), we obtain, after simplification, the result for noise factor

$$F = \frac{\langle i_{sc}^2\rangle}{\langle i_G^2\rangle} = 1 + \frac{G_u + R_n\left\{(G_G + G_C)^2 + (B_G + B_C)^2\right\}}{G_G} \tag{3.59}$$

The relationship in Eqn (3.59) shows that the noise factor is a function of $Y_G = G_G + jB_G$. The noise factor can be minimized for a given optimal value of G_G and B_G. The optimum value for B_G is simply

$$B_{G,opt} = B_G = -B_C \tag{3.60}$$

whereas the optimal value for G_G can be obtained by taking the partial derivative of the noise factor and setting it to zero, or $\frac{\partial F}{\partial G_G} = 0$, resulting in an optimal solution for G_G as

$$G_{G,opt} = \sqrt{\frac{G_u + R_n G_c^2}{R_n}} \Rightarrow G_u = R_n\left(G_{G,opt}^2 - G_C^2\right) \tag{3.61}$$

The resulting minimum noise factor can then be found by substituting Eqn (3.61) into Eqn (3.59) and further manipulating the mathematics to arrive at

$$F_{\min} = 1 + 2R_n\left(G_{G,opt} + \sqrt{\frac{R_n G_{G,opt}^2 - G_u}{R_n}}\right) \tag{3.62}$$

Furthermore, substituting the value obtained in Eqn (3.61) and Eqn (3.62) into Eqn (3.59), we obtain the expression for the noise factor in terms of $F_{\min}$ simply as

$$F = F_{\min} + \frac{R_n}{G_G}\left\{\left(G_G - G_{G,opt}\right)^2 + \left(B_G - B_{G,opt}\right)^2\right\} \tag{3.63}$$

Note that in the event where $G_G = G_{G,opt}$ and $B_G = B_{G,opt}$, the noise factor F reverts to the minimum noise factor $F_{\min}$. Therefore, when performing noise analysis, care must be taken to examine the best-case scenario with the minimum noise figure associated with the various blocks in the receive chain as well as typical and worst-case scenarios that arise due to input termination admittance mismatch with the two-port network. Furthermore, the noise factor relationship in Eqn (3.63) relies on parameters such as $F_{\min}$, $G_{G,opt}$, $B_{G,opt}$, G_G, and B_G that are not always readily available to the designer. In many cases, an empirical formula, such as the one provided by Fukui [10] for transistors, may be used.

3.1.3.2 Thermal noise in LNAs

The impact of the LNA noise figure is paramount on the receiver system noise figure. The LNA is typically the first gain stage in the receiver chain. Its gain as well as its noise figure dominates the amount of noise present in the receiver and *dictates* the receiver sensitivity. Let us further express the relationship in Eqn (3.63) in terms of the normalized noise resistance $r_n = R_n/Z_{char}$ and normalized source admittances $y_G = Y_G Z_{cha} = Y_G/Y_{char}$ and $y_{G,opt} = Y_{G,opt} Z_{char} = Y_{G,opt}/Y_{char}$ as

$$F = F_{\min} + \frac{r_n}{\mathrm{Re}\{y_G\}} \left| y_G - y_{G,opt} \right|^2 \tag{3.64}$$

Furthermore, the normalized admittances can be expressed in terms of the reflection coefficients as

$$\begin{aligned} y_G &= \frac{1-\Gamma_G}{1+\Gamma_G} \Leftrightarrow \Gamma_G = \frac{1-y_G}{1+y_G} \\ y_{G,opt} &= \frac{1-\Gamma_{G,opt}}{1+\Gamma_{G,opt}} \Leftrightarrow \Gamma_{G,opt} = \frac{1-y_{G,opt}}{1+y_{G,opt}} \end{aligned} \tag{3.65}$$

The noise factor in Eqn (3.64) can now be expressed in terms of reflection coefficients as

$$F = F_{\min} + 4r_n \frac{\left| \Gamma_G - \Gamma_{G,opt} \right|^2}{\left| 1 + \Gamma_{G,opt} \right|^2 \left(1 - \left| \Gamma_G \right|^2 \right)} \tag{3.66}$$

The center and radius of the noise circle according to Eqn (3.66) can now be found. Let

$$\Theta = \frac{F - F_{\min}}{4r_n} \left| 1 + \Gamma_{G,opt} \right|^2 \tag{3.67}$$

Then the center and radius of the noise circle overlaid on the Smith chart are

$$\begin{aligned} C_F &= \frac{\Gamma_{G,opt}}{1+\Theta} \\ \rho_F &= \frac{1}{1+\Theta} \sqrt{\Theta^2 + \Theta\left(1 - \left| \Gamma_{G,opt} \right|^2 \right)} \end{aligned} \tag{3.68}$$

where C_F and ρ_F are the center and the radius of the circle, respectively.

The analysis concerning noise figure for two-port networks has indicated that for every LNA, itself a two-port network, there exists an optimum noise figure derived from optimum source resistance. As it turns out, the optimum gain occurs when $\Gamma_{G,opt} = S^*_{1,1}$, whereas the minimum noise figure occurs when the admittance is $Y_G = Y_{G,opt}$, which implies that these two conditions are not met at the same time. However, procedures, such as the one described in [11], exist that optimize for both gain and noise figure. As a case in point, consider the popular high-electron-mobility transistor (HEMT) device for designing LNAs. HEMT devices are essentially FET transistors that provide the highest gain for the lowest noise figure. This makes it ideal for the first stage of a high power LNA. In order to optimize for noise figure and gain performance, a small inductor is placed between the transistor source contact and the ground brings $\Gamma_{G,opt}$ and $S^*_{1,1}$ closer together as shown in Figure 3.14. The resulting effect is known as series inductive feedback, which tends to decrease the gain in favor of improving the stability factor [12]. When fabricated in an MMIC (monolithic microwave integrated circuits) process, spiral inductors used in the input matching network add a significant amount of resistive loss to the LNA and contribute directly to the noise figure. Therefore, in applications requiring low system noise figure, it becomes necessary to employ an off-chip inductor to perform the matching. This is only true, however, when the application is operating at low enough frequency bands such that the bond wire connections to the off-chip components are minimally reactive.

Thus far, we have shown that the LNA noise figure, gain, and input and output impedances are of paramount importance to the performance of the receiver at sensitivity. However, from a system and design point of view, it is important to realize that there are other parameters that impact the performance of the LNA and hence the performance of the receiver in general. Parameters such as the second

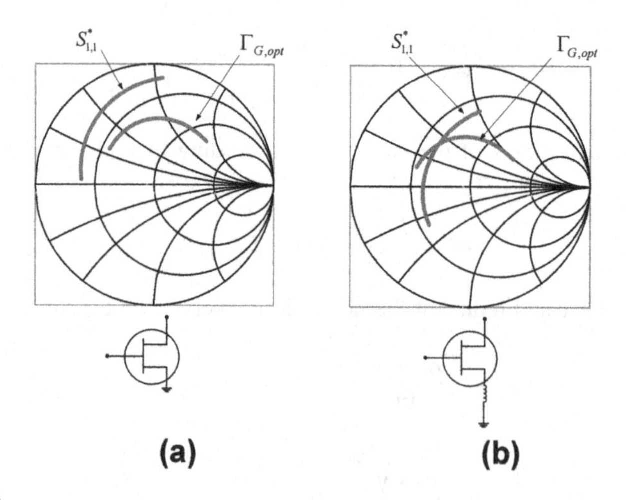

FIGURE 3.14

$S^*_{1,1}$ and $\Gamma_{G,opt}$ for a HEMT device: (a) without inductor, and (b) source to ground inductor. (For color version of this figure, the reader is referred to the online version of this book.)

and third order input-referred intercept points, reverse isolation, and stability are also very important and somehow indirectly influence noise figure. The required noise figure and gain of the LNA are mainly governed by the receiver architecture and the signaling waveform to achieve the required sensitivity. By way of an example, consider a superheterodyne receiver, which unlike a direct conversion receiver, requires an image reject filter (IRF). The IRF introduces insertion loss to the lineup, thus dictating higher LNA gain. Furthermore, the performance of the mixer in terms of its noise figure and third order input-referred intercept point also dictate to some extent the LNA parameters. The input and output impedances of the LNA are also influenced by the architecture and signaling waveform. For example, a full-duplex waveform requires a duplexer. An external duplexer has standard 50 ohm resistive input-impedance, which in turn imposes certain impedance to the input of the LNA. The output impedance of the LNA may also be designed with a standard termination if it is required to interface to an external IRF such as the case of a superheterodyne receiver. On the other hand, this is not a concern for the direct conversion receiver since the architecture does not have an image noise problem.

Another LNA parameter that is largely dictated by the architecture is the reverse isolation $|S_{1,2}|$. This parameter determines the amount of local oscillator (LO) signal leakage from the mixer's RF port to the antenna. The amount of required reverse isolation depends on the architecture. In superheterodyne receivers, the LO leakage at the RF port is somewhat attenuated by the IRF. On the other hand, the effect of LO leakage in full-duplex direct conversion receivers has to be mostly prevented by the reverse isolation of the LNA. In direct conversion receivers that support TDMA signaling waveforms, a transmit-receive (TR) switch typically precedes the LNA and somewhat alleviates the reverse isolation requirement.

3.1.3.3 Thermal noise and gain in mixers

The mixer performs the function of translating the signal's center frequency F_{c1} to another center frequency F_{c2}. The mixer, in its theoretical form, is a linear time-variant device. A mixer can be either passive or active. A *passive* mixer is a device that does not provide any amplification to the input signal. As a matter of fact, if the LO signal is a square wave with 50% duty cycle, as opposed to a pure sinusoid, the resulting LO signal is made up of a series of harmonics of which the fundamental frequency F_{LO} has a magnitude of $1/\pi$. Hence as a result of the frequency convolution of the desired signal with the LO signal, the voltage gain of the mixing operation is $1/\pi$. Active mixers on the other hand provide a certain gain to the RF input signal and hence it is desirable for signal waveforms that operate at low sensitivity. The reason is obviously the noise reduction due to gain in the subsequent stages as dictated by the Friis formula.

Furthermore, the mixing operation can also be classified as single sideband (SSB) or double sideband (DSB) operation. First, consider a theoretical mixing operation with a noiseless mixer whereby a *desired* single tone at RF, say $x(t) = A(t)\cos[(\omega_{LO} + \omega_{IF})t]$ is down-converted by the mixer to an IF frequency via an LO signal $\Upsilon(t) = \cos(\omega_{LO}t)$. The signal's tone is at the RF frequency $\omega_{RF} = \omega_{LO} + \omega_{IF}$. Furthermore, assume another tone signal $y(t)$ to be present at RF such that $y(t) = B(t)\cos[(\omega_{LO} - \omega_{IF})t]$. Assume both $A(t)$ and $B(t)$ are slowly moving signals over time compared to the IF and LO frequencies.

Next, if we perform a simple mixing operation, we obtain

$$\begin{aligned}(x(t) + y(t))\Upsilon(t) &= A(t)\cos[(\omega_{LO} + \omega_{IF})t]\cos(\omega_{LO}t) \\ &\quad + B(t)\cos[(\omega_{LO} - \omega_{IF})t]\cos(\omega_{LO}t) \\ &= \frac{A(t)}{2}\{\cos[(2\omega_{LO} + \omega_{IF})t] + \cos(\omega_{IF}t)\} \\ &\quad + \frac{B(t)}{2}\{\cos[(2\omega_{LO} - \omega_{IF})t] + \cos(\omega_{IF}t)\}\end{aligned} \tag{3.69}$$

After further filtering at the output of the mixer, the IF signal can be obtained from (3.69) simply as

$$(x(t) + y(t))\Upsilon(t) = \frac{A(t)}{2}\cos(\omega_{IF}t) + \frac{B(t)}{2}\cos(\omega_{IF}t) \tag{3.70}$$

The relationship in Eqn (3.70) implies that whatever signal exists at the *image frequency* $\omega_{LO} - \omega_{IF}$ gets *translated or moved* on top of the desired signal. In a super heterodyne architecture, however, the IRF attenuates whatever signal exists at the image frequency to a certain acceptable degradation level. It is then apparent that the higher the IF frequency the better rejection we get from the IRF. Assume, for a moment, that whatever signal existed at the image frequency is completely filtered by the IRF. This leaves the *image noise* in the image band that nonetheless still gets frequency translated on top of the desired signal as shown in Figure 3.15.

The operation depicted in Figure 3.15 implies that the output SNR of the mixer is 3 dB lower than the input SNR thus implying that the noise figure of this theoretical mixer is 3 dB! This is definitely the case in super heterodyne receivers where an SSB signal, that is a signal that exists on either the upper or lower side of the LO frequency, gets mixed on top of the image noise, and hence the resulting theoretical SSB mixing operation's noise figure is 3 dB.

In most cases, we have to be especially weary of port-to-port isolation in the mixer. Note, since the LO exhibits higher power in most cases than the received RF signal, the LO to RF port isolation is of great importance. In the presence of a weak RF signal, we can anticipate the presence of an LO signal at the RF port. In direct conversion receivers, the leaked LO signal mixes with the LO signal to produce a high in-band DC signal.

Next, consider the case of a direct conversion receiver where the mixer downconverts the signal from RF directly to baseband. In this case, the IF frequency in Eqn (3.69) is zero and hence the receiver is also known as a zero IF receiver. In this case, it is obvious that no image of the desired signal exists, or more accurately

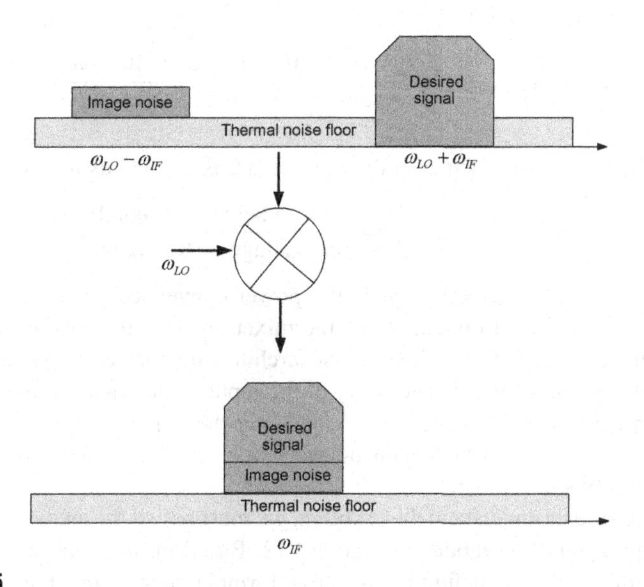

FIGURE 3.15

Image noise folding onto desired signal at IF. (For color version of this figure, the reader is referred to the online version of this book.)

the image exits in the negative frequency, and hence no thermal noise is added on top of the desired signal. In this case, the theoretical output SNR is equal to the input SNR and hence the DSB noise figure is 0 dB and it is 3 dB less than that of the SSB mixer.

A further subject of interest concerning the mixer is its topology. A mixer can be either a single-ended, single-balanced (SB) or a double-balanced mixer, or an image-reject mixer. In this context, we will limit our discussion to balanced mixers. An SB mixer is one in which the input RF signal is single ended whereas the LO signal is differential. On the other hand, the DB mixer implies that both the RF as well as the LO signal are differential. The RF signal incoming into the mixer is either single ended or differential. In the former case, the signal is converted from single ended to differential via a balun. An SB mixer displays less input referred noise over a DB mixer, whereas the latter has a higher second order input-referred intercept point (IIP2). The advantage of DB mixers, as will be discussed in more detail in upcoming chapters, is in reducing the ½ IF distortion in superheterodyne receivers. The output of both mixers can be either single ended or differential with the latter being less susceptible to noise. However, in IF sampling receivers that require external single-ended surface acoustic wave (SAW) filters at IF, the mixer's output is single ended to interface with the filter.

At this point, it is instructive to further discuss the mixer gain. The mixer exhibits two different types of gain, namely voltage conversion gain and power conversion gain. The power conversion gain can be defined as the ratio

$$G_{\text{Power Conversion}} = \frac{\text{Power at IF delivered to the load}}{\text{Available RF power to the source}} \tag{3.71}$$

The voltage conversion gain, on the other hand, is defined as the ratio

$$G_{\text{Voltage Conversion}} = \frac{rms \text{ voltage of IF signal}}{rms \text{ voltage of RF signal}} \tag{3.72}$$

The voltage conversion gain equals the power conversion gain only when the input impedance and load impedance of the mixer equal to the source impedance. Again, in the case of the superheterodyne architecture that employs an external IR filter, conjugate matching is necessary at the input of the mixer to match the filter's standard termination impedance. A question then arises: In a cascade noise figure and gain analysis, which gain do we use given that the impedance levels may vary from block to block?

In order to further understand the problem, we must revisit the analysis employed on the two-port noiseless model of Figure 3.12. Based on the analysis of Section 3.1.3, and following the definition of Friis formula, neglecting the correlation between noise current and voltage current noise, the voltage noise can be written in terms of the noise current and source admittance

$$\left(F_{n+1,G1} - 1\right)\left\langle i_G^2\right\rangle - |Y_{G1}|^2\left\langle v_n^2\right\rangle = \left(F_{n+1,G2} - 1\right)\left\langle i_G^2\right\rangle - |Y_{G2}|^2\left\langle v_n^2\right\rangle \tag{3.73}$$

where the index $n+1$ implies the $(n+1)^{th}$ stage. Alternatively, Eqn (3.73) can be expressed in terms of noise currents and source resistances as

$$\left(F_{n+1,G1} - 1\right)4kT_0R_{G1} - |R_{G1}|^2\left\langle I_n^2\right\rangle = \left(F_{n+1,G2} - 1\right)4kT_0R_{G2} - |R_{G2}|^2\left\langle I_n^2\right\rangle \tag{3.74}$$

The relationships in Eqn (3.73) and Eqn (3.74) cannot be computed unless the noise voltage or noise currents are known. However, in the event that the noise current of the amplifier is negligible, that is

$$\begin{aligned} \left(F_{n+1,G1} - 1\right)4kT_0R_{G1} - |R_{G1}|^2\left\langle I_n^2\right\rangle &\approx \left(F_{n+1,G1} - 1\right)4kT_0R_{G1} \\ \left(F_{n+1,G2} - 1\right)4kT_0R_{G2} - |R_{G2}|^2\left\langle I_n^2\right\rangle &\approx \left(F_{n+1,G2} - 1\right)4kT_0R_{G2} \end{aligned} \tag{3.75}$$

then the relationship in Eqn (3.74) can be approximated as

$$\left(F_{n+1,G1} - 1\right)R_{G1} \approx \left(F_{n+1,G2} - 1\right)R_{G2} \tag{3.76}$$

At this point, it is important to address the gain values associated with the Friis formula given in Eqn (3.30). The question that was raised earlier is whether the gain values used in Eqn (3.30) are the available power gain of the staged or the square of the voltage gain. It turns out that the noise factor $(F_{n+1} - 1)$ is divided by the available power gain $G_{\text{Power Conversion}}$ of the preceding stage if and only if noise factor is calculated with respect to R_{G1}. On the other hand, if the noise factor F_{n+1} is computed with respect to say R_{G2}, then the noise factor $(F_{n+1} - 1)$ is divided by the square of the voltage gain, or the power conversion gain in the case of the mixer, $G_{\text{Voltage Conversion}}$. This can be best illustrated with an example.

EXAMPLE 3.3 CASCADE NOISE FIGURE ANALYSIS WITH DIFFERENT INTERSTAGE IMPEDANCES

Consider the receiver chain where the source and load impedances of the various receiver blocks are as defined in Figure 3.16. What is the noise figure of the PMA with respect to a 50 ohm input impedance? What is the cumulative system noise figure of the mixer and PMA? The impedances indicated next to the noise figures are the source impedances of the various blocks. Assume input noise current in the PMA to be negligible.

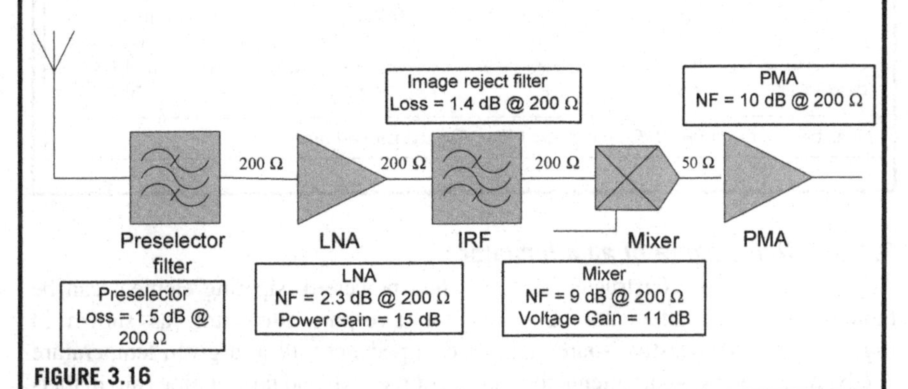

FIGURE 3.16

Receiver with different interstage impedances. IRF, Image reject filter; LNA, low noise amplifier; PMA, postmixer amplifier.

The first observation we need to make in reference to the receiver depicted in Figure 3.16 is that the input and output impedances of the mixer are 200 ohm and 50 ohm, respectively. The postmixer amplifier (PMA) noise figure is given for a source impedance of 200 ohm and therefore it must be computed for the input source impedance of 50 ohm. According to Eqn (3.76), the PMA noise figure at 50 ohm is given as

$$F_{PMA,50\Omega} = \frac{(F_{PMA,200\Omega} - 1)R_{PMA,200\Omega}}{R_{PMA,50\Omega}} + 1 = \frac{(10^{10/10} - 1) \times 200}{50} + 1 = 37 \quad (3.77)$$

$$NF_{PMA,50\Omega} = 10\log_{10}(37) = 15.68 \text{ dB}$$

Next, we need to compute the cumulative system noise figure. Assume that the noise figure of the PMA is available only for 200 ohm, then the combined noise factor of the mixer and PMA is obtained by dividing $(F_{PMA,200\Omega} - 1)$ by the square of the voltage conversion gain of the mixer, that is

$$F_{mixer,PMA} = F_{\text{mixer}} + \frac{F_{PMA,200\Omega} - 1}{\left(10^{G_{\text{Voltage Conversion}}/10}\right)^2} = 10^{9/10} + \frac{10^{10/10} - 1}{\left(10^{11/10}\right)^2} = 8 \quad (3.78)$$

$$NF_{mixer,PMA} = 10\log_{10}(8) = 9.03 \text{ dB}$$

Continued

EXAMPLE 3.3 CASCADE NOISE FIGURE ANALYSIS WITH DIFFERENT INTERSTAGE IMPEDANCES—cont'd

The cascaded system noise figure of the system depicted in Figure 3.16 is summarized in Table 3.6.

Table 3.6 Cascaded Noise Figure Analysis of the System Depicted in Figure 3.16

Parameter	Preselector	LNA	IRF	Mixer/PMA	PMA
Noise Figure (dB)	1.5	2.3	1.4	9.030937061	9.030937061
Gain (dB)	−1.5	15	−1.4	15	10
F (linear)	1.41253754	1.6982437	1.380384265	8.000068508	8.000068508
Gain (linear)	0.70794578	31.622777	0.72443596	31.6227766	10
Casc. Gain	1	0.7079458	22.38721139	16.21810097	512.861384
Casc. F (linear)	1.41253754	2.3988329	2.41582406	2.847444785	2.861093831
Casc. NF (dB)	1.5	3.8	3.830653022	4.544553115	4.56532101

LNA, low noise amplifier; IRF, image reject filter; PMA, postmixer amplifier.

3.1.3.4 Noise figure of an attenuator

Any passive device constructed as a matched power–dissipating element can be thought of as an attenuator. Consider the two-port attenuator, as shown in Figure 3.17, with resistive source and load impedances all at a given temperature T. Given that the two-port attenuator and its source and load terminations are in thermal equilibrium, the noise power delivered by the source impedance and attenuator to the load impedance is statistically equal to the noise power delivered by the load impedance to the attenuator and source. Let the noise temperature generated in the attenuator and delivered to the load be T_{att}, then the noise temperature delivered to the attenuator and source impedance is

$$T = G_{transducer}T + T_{att} \tag{3.79}$$

where $G_{transducer} < 1$ is the transducer gain of the attenuator. Note that, from a physics point of view, we can simply refer T_{att} to the input, and hence the attenuator noise temperature can be expressed as

$$T_{att,input} = \frac{T_{att}}{G_{transducer}} = T\left(\frac{1}{G_{transducer}} - 1\right) \tag{3.80}$$

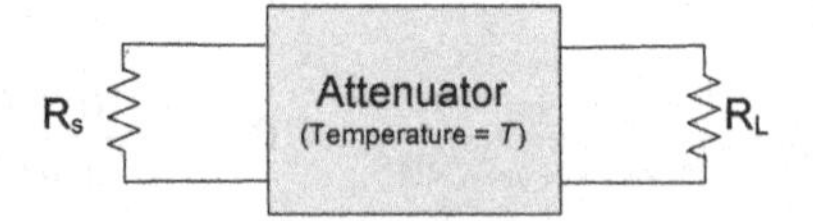

FIGURE 3.17

Attenuator at thermal equilibrium.

Given $T = T_0 = 290$ K, the noise factor of the attenuator is given as

$$F_{att} = \frac{T_{att,input}}{T_0} + 1 \tag{3.81}$$

Substituting the result obtained in Eqn (3.80) into Eqn (3.81), the noise factor of the attenuator becomes

$$F_{att} = \frac{1}{G_{transducer}} \tag{3.82}$$

This is an interesting result that seems to imply that the noise factor of an attenuator is simply equal to the loss factor $1/G_{transducer}$.

3.2 Phase noise

Thus far, in the previous section, we have discussed thermal noise that is additive to the desired signal and can be characterized in terms of noise figure. As it turns out, increasing the signal strength increases the SNR and hence minimizes the impact of thermal noise on the receiver performance. In this section, however, we study phase noise, which is multiplicative in the time domain and cannot be simply mitigated by increasing the received signal power.

Depending on the transceiver architecture, a frequency synthesizer is typically employed to generate a spectrally *pure* signal in order to up-convert (modulate) or down-convert (de-modulate) the desired signal to and from RF or IF to baseband. The output of the frequency synthesizer, or local oscillator, exhibits both amplitude and phase variations with the latter being by far more detrimental to the desired signal quality. These phase variations, or phase noise, contaminate the purity of the modulating signal, as depicted in Figure 3.18 for a pure sinusoid. This phase impurity translates into signal degradation in terms of bit error rate performance and tends to reduce the channel selectivity of the receiver.

There are various synthesis techniques from which frequency can be generated [13]. These can mainly be classified into three categories:

- Incoherent synthesis relies on the use of multiple crystals to generate the various output frequencies needed for modulation and demodulation. An incoherent synthesizer can be constructed using various crystal oscillators, mixers, and bandpass filters.
- Coherent direct synthesis, unlike incoherent synthesis, relies only on one crystal oscillator to generate the various frequency outputs. The stability and accuracy of the various frequency outputs depends on the stability and accuracy of the sole crystal oscillator. Coherent direct synthesis can be accomplished via various frequency generation techniques, such as the brute force technique, the harmonic approach, and double and triple mixing approach. In the brute force approach, for example, a series of frequency multipliers and dividers, mixers,

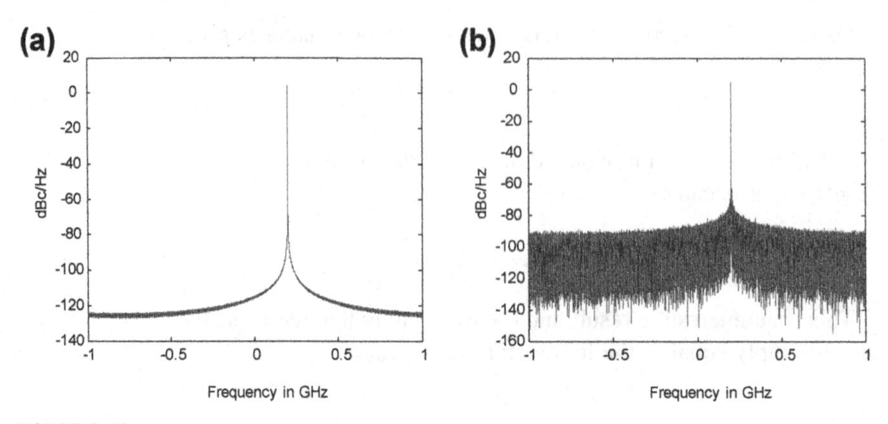

FIGURE 3.18

Effect of phase noise on a single tone: (a) single tone with $1/f$ noise at −120 dBc/Hz and (b) single tone with $1/f$ noise at −70 dBc/Hz. (For color version of this figure, the reader is referred to the online version of this book.)

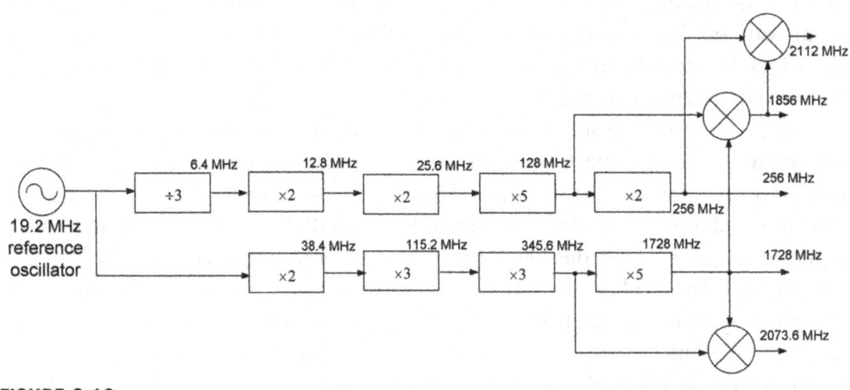

FIGURE 3.19

Brute force frequency synthesis.

and reference frequency source are used. This approach, depicted in Figure 3.19, is sometimes preferred when multiple fixed frequencies need to be generated simultaneously. Another popular approach, often used in software-defined radios, is direct digital synthesis (DDS). In DDS, the output frequency is generated totally in the digital domain as shown in Figure 3.20. The DDS system can exhibit spreading of the spectral lines at its output, thus contributing impurity to the output signal. This phenomenon is mainly due to the jitter in the system clock. The largest source of impurity in a DDS, however, is charge glitches in the DAC.

- Coherent indirect synthesis is also known as phase locking via a feedback system, or simply a PLL. A conceptual PLL is depicted in Figure 3.21. In a PLL,

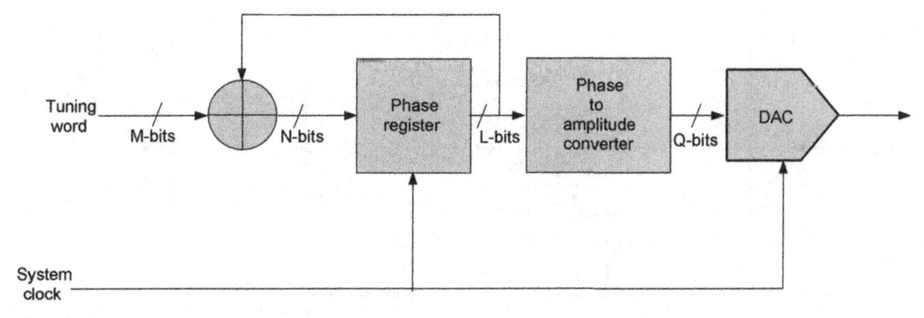

FIGURE 3.20

Frequency tunable direct digital synthesis (DDS). DAC, digital to analog converter.

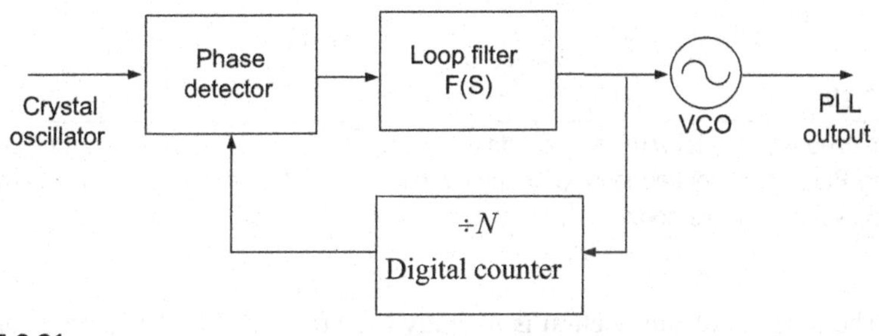

FIGURE 3.21

Frequency synthesis via PLL. VCO, voltage-controlled oscillator; PLL, phase-locked loop.

the crystal reference oscillator provides the reference frequency, which tends to operate at a much lower frequency than the voltage controlled oscillator. The VCO's output frequency is controlled by the output of the loop filter. The input to the loop filter is an error voltage that corresponds to the difference between the reference oscillator and a digital counter output. In the ensuing analysis, we will focus on frequency synthesis only in the context of coherent indirect synthesis. We will analyze the impact of the various PLL blocks on phase noise.

3.2.1 The PLL as a filter

The intent of this section is not to analyze PLL concepts and synthesizer techniques. This subject will be fully explored in Chapter 7. The aim is to examine the impact of phase noise on the desired signal. The phase noise at the output of a PLL or synthesizer is impacted by many sources, including the reference oscillator, the VCO, and the components of the PLL or synthesizer. Before we explore the nature of the noise sources in each component of the PLL, it is important to understand the role the loop plays in shaping the various noise densities. This in turn will clarify why we place certain emphasis on the close-in phase noise of the reference oscillator, for example, versus its far-out noise density, and why to some extent the opposite is true for the

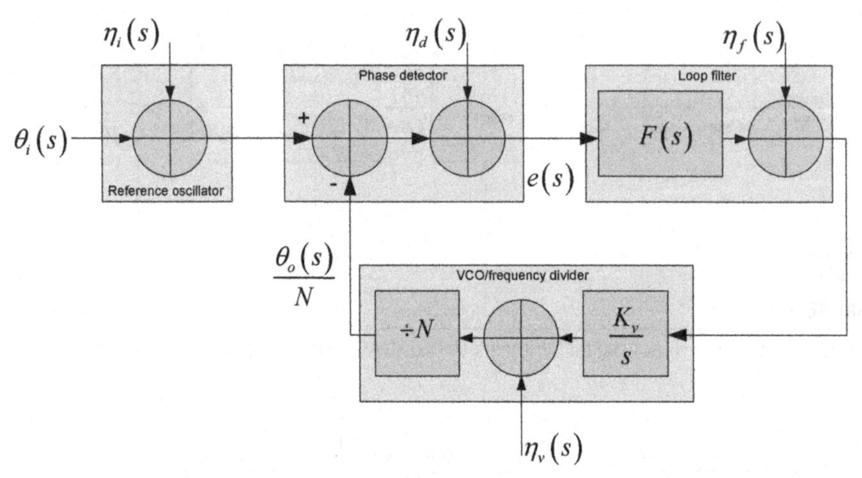

FIGURE 3.22

The linearized PLL model with various additive noise sources. VCO, voltage-controlled oscillator; PLL, phase-locked loop. (For color version of this figure, the reader is referred to the online version of this book.)

VCO. The purpose of this section is to study the effect of the PLL on the various noise sources using a linearized loop model as shown in Figure 3.22. Specific emphasis will be placed on the shaping of the reference oscillator and VCO.

The input/output relationship inferred from Figure 3.22 can be expressed in the Laplace domain as

$$\theta_i(s) + \eta_i(s) - \frac{\theta_o(s)}{N} + \eta_d(s) = e(s) \tag{3.83}$$

where $\theta_i(s)$ is the phase of the reference oscillator, $\theta_o(s)$ is the phase of the LO, N is the PLL up-conversion ratio that is the ratio between the reference frequency and the LO frequency, $\eta_i(s)$ and $\eta_d(s)$ are the additive phase noise of the reference oscillator and phase/frequency or loop detector, and $e(s)$ is the PLL error function. The output phase of the loop can be further expressed in terms of the loop error function and vice versa as

$$\begin{aligned} \theta_o(s) &= \eta_v(s) + \frac{K}{s}\{F(s)e(s) + \eta_F(s)\} \\ e(s) &= \frac{1}{F(s)}\left\{\frac{s}{K}[\theta_o(s) - \eta_v(s)] - \eta_F(s)\right\} \end{aligned} \tag{3.84}$$

where $F(s)$ and $\eta_F(s)$ are the loop filter transfer function and loop filter additive noise, K is the VCO gain, and $\eta_v(s)$ is the VCO phase noise. In the ensuing analysis, assume that $F(s)$ is a lowpass function. Note that all the noise densities are assumed to be additive and somehow contribute to the phase noise of the LO, where the two noise densities of concern are the phase noise of the reference oscillator and that of the VCO.

Substituting the value of the error function found in Eqn (3.84) into Eqn (3.83), we obtain

$$\theta_i(s) + \eta_i(s) - \frac{\theta_o(s)}{N} + \eta_d(s) = \frac{1}{F(s)}\left\{\frac{s}{K}[\theta_o(s) - \eta_v(s)] - \eta_F(s)\right\} \quad (3.85)$$

Consequently, the output of the PLL can be expressed as

$$\begin{aligned}\theta_o(s) &= \frac{NKF(s)}{Ns + KF(s)}\{\theta_i(s) + \eta_i(s) + \eta_d(s)\} + \frac{Ns}{Ns + KF(s)}\eta_v(s) \\ &+ \frac{NK}{Ns + KF(s)}\eta_F(s)\end{aligned} \quad (3.86)$$

The relationship in Eqn (3.86) expresses the output phase of the loop $\theta_o(s)$ in terms of the input reference oscillator and the various additive noise as *shaped* by the loop. According to Eqn (3.86), it is of great interest to see what influences the output phase at or near DC frequencies. That is,

$$\lim_{\Omega \to 0} \theta_o(j\Omega) = N\theta_i(0) + N\{\eta_i(0) + \eta_d(0)\} + \frac{N}{F(0)}\eta_F(0), F(0) \neq 0 \quad (3.87)$$

where again $F(s)$ is a lowpass function. The relationship in Eqn (3.87) *implies* certain important characteristics of the PLL as a filter at *low frequencies* that is at or near DC:

- The phase noise density of the VCO does not contribute, or contributes minimally, to the phase noise density of the LO.
- The phase noise of the LO is dominated by the close-in phase noise of the reference oscillator. The phase noise density of the LO is the same as the phase noise density of the reference multiplied by N^2. The increase in phase noise density is proportional to the synthesizer ratio or simply $10 \log_{10} (N^2)$ dB. Note that, for the same reason mentioned in the case of the crystal oscillator, the phase noise of the LO is further deteriorated by the additive phase noise effects of the detector increased by a factor $10 \log_{10} (N^2)$ dB.
- The phase noise density of the LO is also degraded by the phase noise of the loop filter. The phase noise due to the filter is additive and proportional to the ratio $N^2/|F(0)|^2$ or in dB $10 \log_{10}(N^2/|F(0)|^2)$dB.
- In Eqn (3.87), the dominant phase noise degrading the performance of the LO is by far that of the reference oscillator. The reference oscillator is required to have a high Q at the required RF frequency.

Next, define the loop transfer function as

$$H(s) = \frac{NKF(s)}{Ns + KF(s)} \quad (3.88)$$

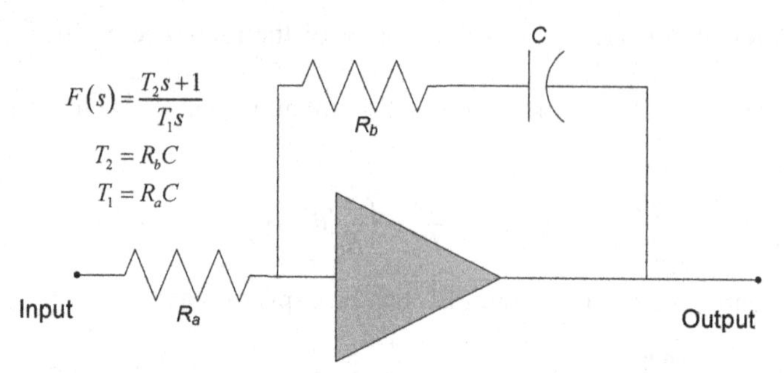

FIGURE 3.23

A perfect integrator active loop filter.

Then according to Eqn (3.88), we have

$$1-\frac{1}{N}H(s) = \frac{Ns}{Ns+KF(s)} \tag{3.89}$$

Substituting the relations in Eqn (3.88) and Eqn (3.89) into Eqn (3.86), we obtain

$$\begin{aligned}\theta_o(s) &= H(s)\{\theta_i(s)+\eta_i(s)+\eta_d(s)\}+\left\{1-\frac{1}{N}H(s)\right\}\eta_v(s)\\ &+\frac{K}{s}\left\{1-\frac{1}{N}H(s)\right\}\eta_F(s)\end{aligned} \tag{3.90}$$

For the sake of simplicity, and in order to explain the effect of the loop on the VCO and loop-filter phase noise, assume that the PLL is a second order loop, and let the loop filter be a perfect integrator of the form

$$F(s) = \frac{T_2s+1}{T_1s} \tag{3.91}$$

as shown in Figure 3.23.

Then the loop transfer function can be expressed by substituting Eqn (3.91) into Eqn (3.88) as

$$H(s) = \frac{NK\frac{T_2s+1}{T_1s}}{Ns+K\frac{T_2s+1}{T_1s}} = \frac{K\frac{T_2}{T_1}s+\frac{K}{T_1}}{s^2+K\frac{T_2}{NT_1}s+\frac{K}{NT_1}} \tag{3.92}$$

The relationship in Eqn (3.92) is of the form

$$H(s) = \frac{1}{N}\frac{2\xi\Omega_ns+\Omega_n^2}{s^2+2\xi\Omega_ns+\Omega_n^2} \tag{3.93}$$

The equivalent loop noise bandwidth of Eqn (3.93) is

$$B_L = \frac{\xi\Omega_n}{2}\left(1+\frac{1}{4\xi^2}\right)\ \text{Hz} \tag{3.94}$$

where Ω_n is the natural frequency and ξ is the loop damping ratio. The equivalent loop noise bandwidth will be examined in more detail in Chapter 7. Then, according to Eqn (3.89) the filter transfer function acting on the VCO can be found by substituting Eqn (3.93) into Eqn (3.89), or

$$1 - \frac{1}{N}H(s) = 1 - \frac{1}{N^2}\frac{2\xi\Omega_n s + \Omega_n^2}{s^2 + 2\xi\Omega_n s + \Omega_n^2} = \frac{s^2 + 2\left(1 - \frac{1}{N^2}\right)\xi\Omega_n s + \left(1 - \frac{1}{N^2}\right)\Omega_n^2}{s^2 + 2\xi\Omega_n s + \Omega_n^2} \tag{3.95}$$

Again, in this *particular* case, it is important to understand the impact of the loop transfer function $H(s)$ and of $1 - \frac{1}{N}H(s)$ at or near DC, that is at very close-in frequencies, as well as at high frequencies, that is as $\Omega \to \infty$. According to Eqn (3.93), we have

$$\begin{aligned} \lim_{\Omega \to 0} H(j\Omega) &= \frac{1}{N}\frac{2j\xi\Omega_n\Omega + \Omega_n^2}{-\Omega^2 + 2j\xi\Omega_n\Omega + \Omega_n^2}\bigg|_{\Omega=0} = \frac{1}{N} \\ \lim_{\Omega \to \infty} H(j\Omega) &= \frac{1}{N}\frac{2j\xi\Omega_n\Omega + \Omega_n^2}{-\Omega^2 + 2j\xi\Omega_n\Omega + \Omega_n^2}\bigg|_{\Omega=\infty} \\ &= \frac{1}{N}\frac{2j\xi\Omega_n/\Omega + \Omega_n^2/\Omega^2}{-1 + 2j\xi\Omega_n/\Omega + \Omega_n^2/\Omega^2}\bigg|_{\Omega=\infty} = 0 \end{aligned} \tag{3.96}$$

The relationship in Eqn (3.96) implies that the loop transfer function $H(s)$ exhibits lowpass characteristic, where in this particular case, the loop transfer function gain at DC is inversely proportional to N, as depicted by way of example in Figure 3.24. Likewise, upon examination of Eqn (3.95), we obtain

$$\begin{aligned} \lim_{\Omega \to 0}\left\{1 - \frac{1}{N}H(j\Omega)\right\} &= \frac{-\Omega^2 + 2j\left(1 - \frac{1}{N^2}\right)\xi\Omega_n\Omega + \left(1 - \frac{1}{N^2}\right)\Omega_n^2}{-\Omega^2 + 2j\xi\Omega_n\Omega + \Omega_n^2}\Bigg|_{\Omega=0} = 1 - \frac{1}{N^2} \\ \lim_{\Omega \to \infty}\left\{1 - \frac{1}{N}H(j\Omega)\right\} &= \frac{-1 + 2j\left(1 - \frac{1}{N^2}\right)\xi\Omega_n/\Omega + \left(1 - \frac{1}{N^2}\right)\Omega_n^2/\Omega^2}{-1 + 2j\xi\Omega_n/\Omega + \Omega_n^2/\Omega^2}\Bigg|_{\Omega=\infty} = 1 \end{aligned} \tag{3.97}$$

For this particular loop filter, in the event where $N = 1$, the function expressed Eqn (3.97) has highpass characteristics. However, as N increases, the function in Eqn (3.97) becomes less of a highpass function, as can be seen in Figure 3.25. Note that this is not a generalization of PLL behavior for any given loop filter but rather for the PLL under discussion with integrator loop filter as expressed in Eqn (3.91). Therefore, in this case, it can be seen that N affects the transfer function characteristic of the loop, and in turn the shaping of the various noise density functions, particularly of the reference oscillator and VCO. The other two factors that affect the

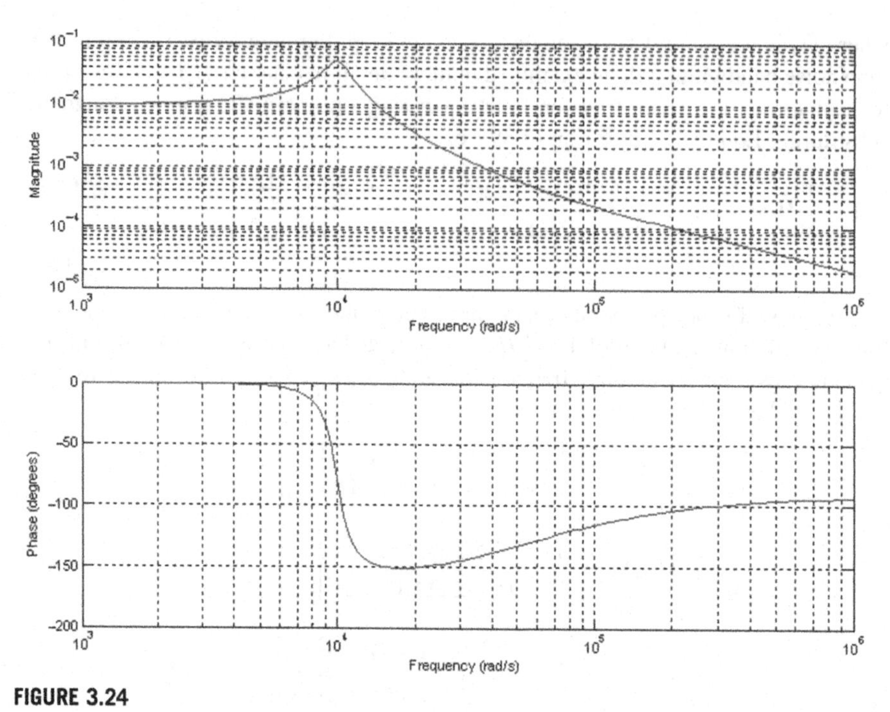

FIGURE 3.24

Second order loop transfer function $H(s)$ for loop bandwidth of 10 KHz damping factor of 0.1, and $N = 100$. (For color version of this figure, the reader is referred to the online version of this book.)

loop transfer function and the shaping of the noise densities are the choice of the natural frequency Ω_n as well as the loop damping ratio ξ. The impact of decreasing Ω_n on the overall phase noise of the LO serves to generally decrease the amount of phase noise degradation due to the reference oscillator and increase the amount of phase degradation attributed to the VCO. Increasing the natural frequency of the PLL tends to accomplish the opposite. Figure 3.26 and Figure 3.27 illustrate the impact of increasing the damping factor on the transfer functions $H(s)$ and $1 - \frac{1}{N}H(s)$. Decreasing ξ serves to boost the response of the filters around the natural frequency, whereas increasing ξ serves to smooth out the response of the filters in the passband.

Next, consider a more realistic third order PLL with a loop filter realized using passive elements as depicted in Figure 3.28. The reference oscillator frequency is 10 MHz and the LO output frequency is 1.6 GHz. The closed loop gain response of the PLL shows a loop bandwidth of 10 KHz as illustrated in Figure 3.29. Figure 3.30 depicts the loop's total phase noise density along with that of the reference oscillator, the VCO, and the loop filter. It is interesting to see that the total phase noise of the PLL is dominated by that of reference oscillator at low frequency up to and around 1 KHz. On the other hand, one can also observe that the total phase noise

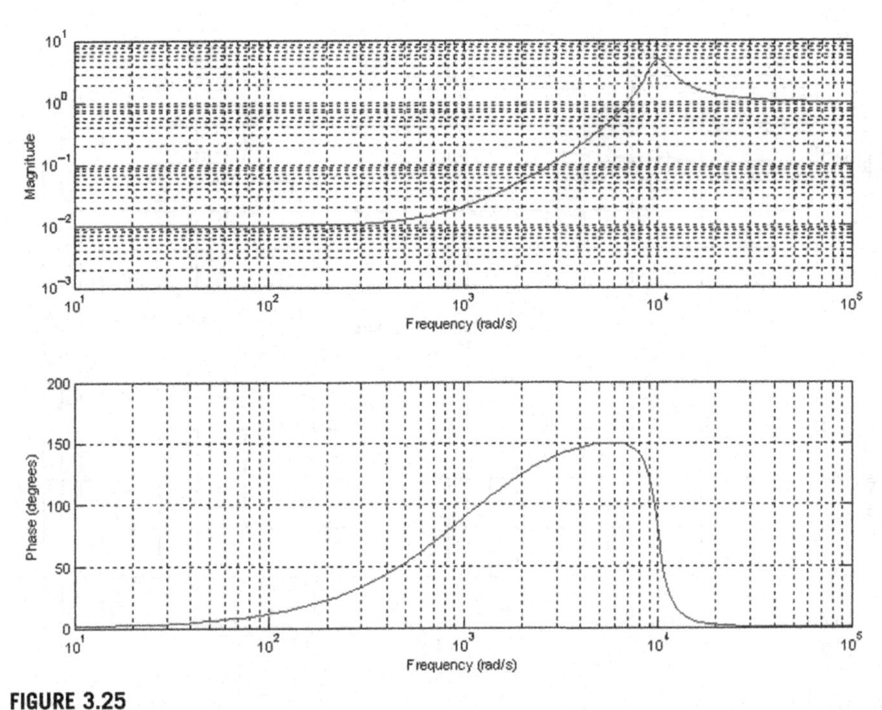

FIGURE 3.25

Transfer function $1 - \frac{1}{N}H(s)$ for loop bandwidth of 10 KHz damping factor of 0.1, and $N = 100$ acting on the voltage-controlled oscillator (VCO) phase noise. (For color version of this figure, the reader is referred to the online version of this book.)

of the PLL is dominated by that of the VCO at high frequencies and as low as 1 KHz. In most cases, the phase noise due to the loop filter is considered significant but not dominant at high frequencies. Furthermore, note how the various phase noise densities are shaped by the PLL. To further illustrate this point, the analysis is repeated for a natural frequency of 50 KHz as shown in Figure 3.31. By comparing Figure 3.30 and Figure 3.31, one can readily observe that increasing the loop bandwidth increases the impact of the reference oscillator phase noise on the total phase noise of the PLL, while it diminishes the impact of the VCO phase noise on the total phase noise of the PLL. The impact of increasing the loop bandwidth on the total phase noise and the loop's filtering effect on the reference oscillator and VCO can be further illustrated in Figure 3.32, Figure 3.33, and Figure 3.34. In conclusion, it is obvious that the amount of phase noise admitted by the reference oscillator and VCO and their respective impact on the LO's total phase noise is mainly dictated by the choice of the PLL's order and type, its loop bandwidth, and its damping factor.

Thus far, we have discussed the impact of the digital divider block in Figure 3.22 and Figure 3.28 on phase noise. Note that any AM noise at the input to the digital divider causes time jitter at the output of the divider and hence the AM noise transforms into PM noise. This process is an AM-to-PM conversion process that can be

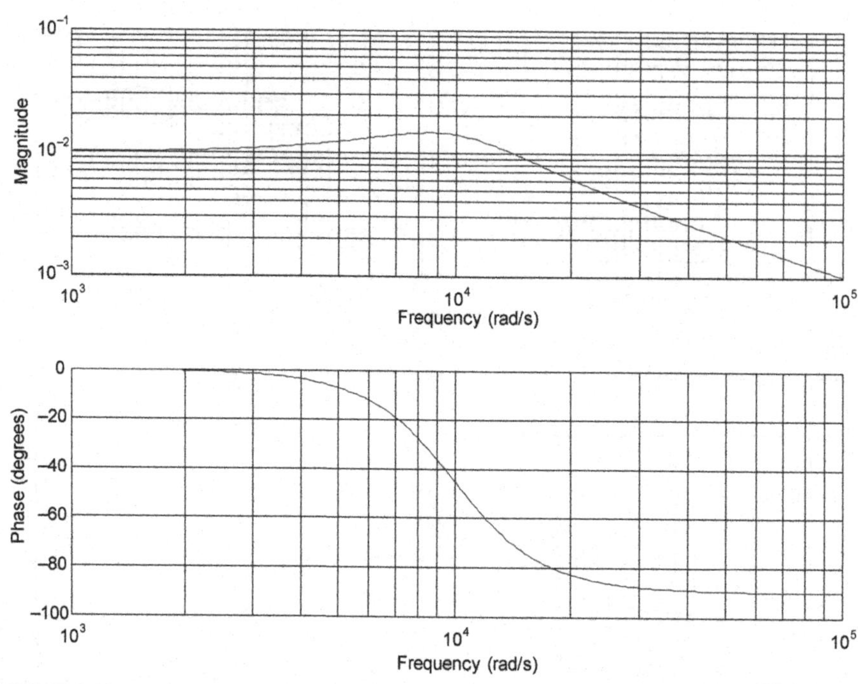

FIGURE 3.26

Second order loop transfer function $H(s)$ for loop bandwidth of 10 KHz damping factor of 0.5, and $N = 100$. (For color version of this figure, the reader is referred to the online version of this book.)

analyzed by differentiating the waveform and computing the power spectral density of the resulting waveform [14]. That is, given a power spectral density

$$S_{\chi}(F) = \left(\frac{1}{2\pi F}\right)^2 S_{\chi'}(F) \tag{3.98}$$

According to [14], the power spectral density of the input signal to the digital divider, given a digital input waveform, can be expressed as

$$S_{\chi}(F) = \frac{1}{4\pi^2F^2T^2}\frac{1-e^{-4\pi^2F^2\sigma_{jitter}^2}}{\left(1-e^{-2\pi^2F^2\sigma_{jitter}^2}\right)^2+4e^{-2\pi^2F^2\sigma_{jitter}^2}\cos^2(\pi FT)} \tag{3.99}$$

where σ_{jitter}^2 is the timing jitter variance and T is the period of the input pulse. The power spectral density at the output of the digital divider is the same as the power spectral density of the input signal as expressed in Eqn (3.99) reduced by division ratio N or

$$S_y(F) = \frac{1}{N^2}S_{\chi}(F) \tag{3.100}$$

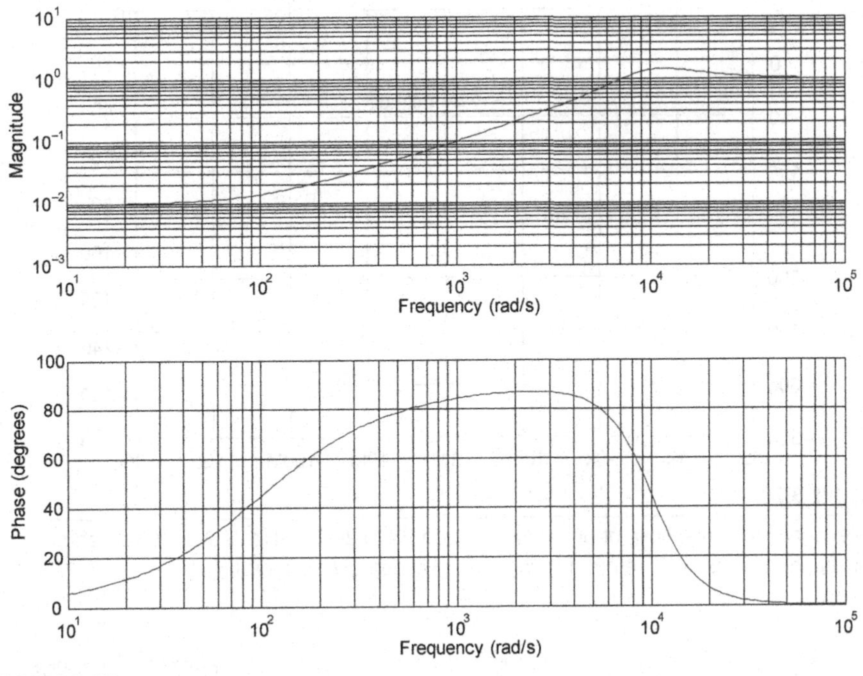

FIGURE 3.27

Transfer function $1-\frac{1}{N}H(s)$ for loop bandwidth of 10 KHz damping factor of 0.5, and $N=100$ acting on the voltage-controlled oscillator (VCO) phase noise. (For color version of this figure, the reader is referred to the online version of this book.)

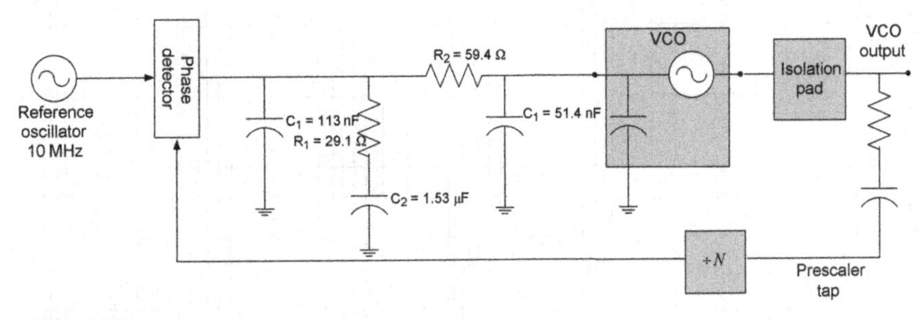

FIGURE 3.28

Third order PLL using passive loop filter resulting in 10 KHz natural frequency. VCO, voltage-controlled oscillator; PLL, phase-locked loop.

One can readily see upon examination of Eqn (3.100) that the phase noise at the input of the loop is reduced as a function of the divider ratio.

The use of a digital divider is not limited to the feedback chain, as illustrated in Figure 3.35. In applications where a frequency synthesizer must cover a wide range

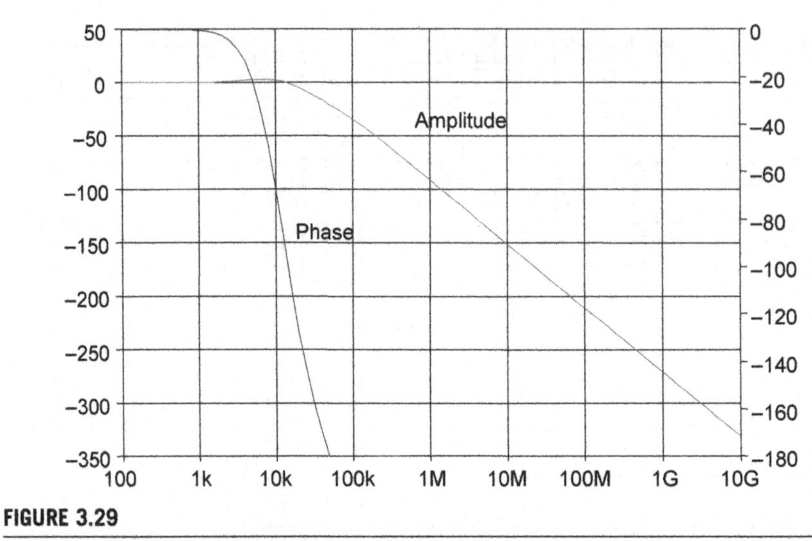

FIGURE 3.29

Closed loop gain response of phase-locked loop (PLL) at 1.6 GHz. (For color version of this figure, the reader is referred to the online version of this book.)

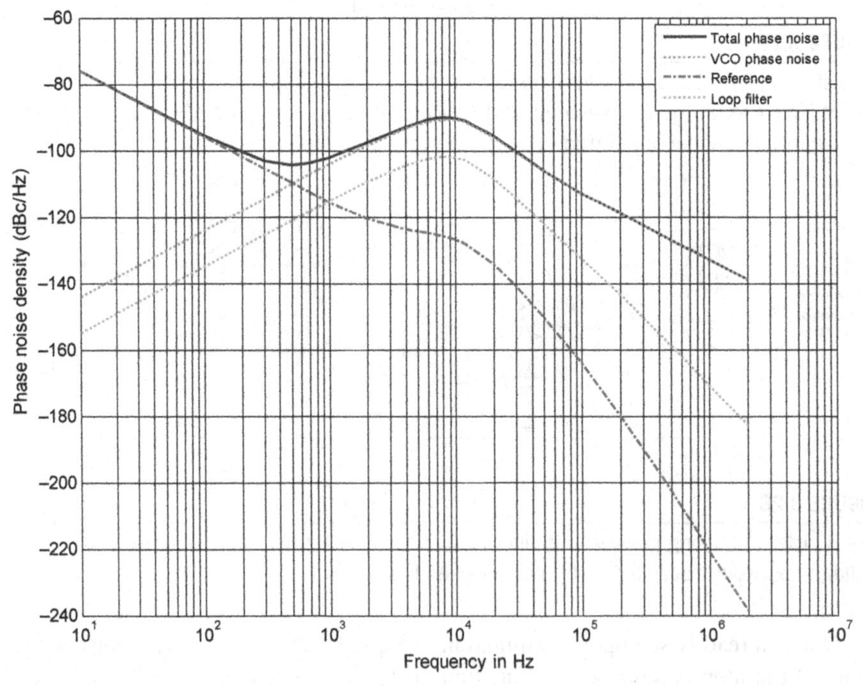

FIGURE 3.30

Total phase noise density at 1.6 GHz with natural frequency set at 10 KHz. VCO, voltage-controlled oscillator. (For color version of this figure, the reader is referred to the online version of this book.)

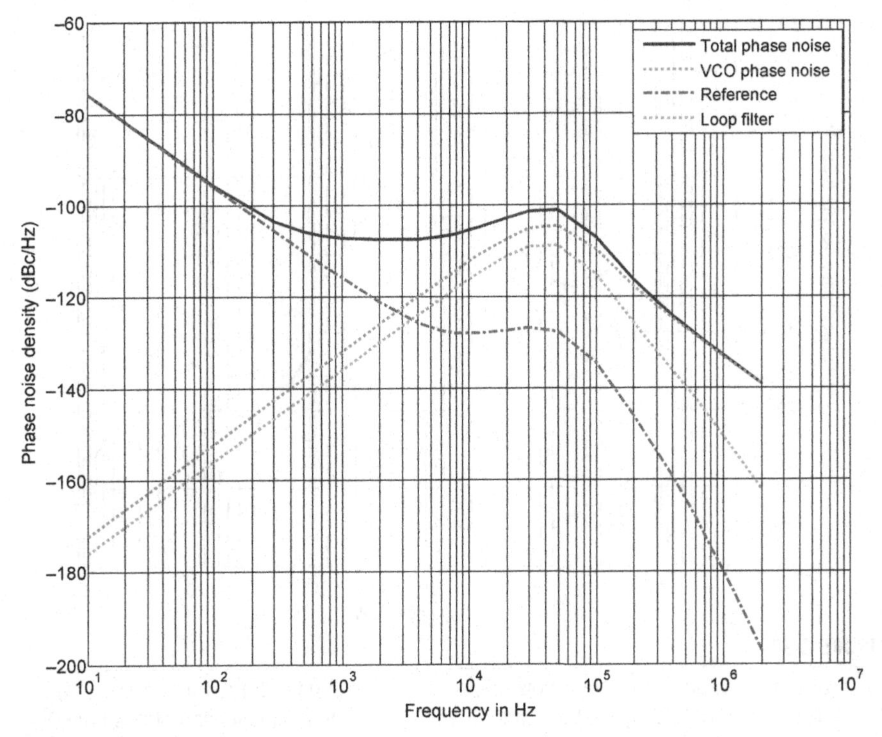

FIGURE 3.31

Total phase noise density at 1.6 GHz with natural frequency set at 50 KHz. VCO, voltage-controlled oscillator. (For color version of this figure, the reader is referred to the online version of this book.)

of frequencies, a programmable divider in the feedback loop is used along with a fixed divider situated after the reference oscillator and before the frequency detector to allow the synthesizer to cover several channels within a given frequency band. The resolution of the synthesizer is dictated by the ratio of the oscillator frequency divided by the fixed digital divider F_{osc}/M. Under steady state, when the PLL is locked, the following relationship between the output frequency of the VCO and that of the crystal oscillator must apply:

$$\frac{F_{osc}}{M} = \frac{F_{VCO}}{N} \Rightarrow F_{VCO} = \frac{N}{M} F_{osc} \tag{3.101}$$

The ratio F_{osc}/M is the frequency resolution of the synthesizer.

In order to extend the range of the synthesizer, multiple VCOs may be used in a multiloop architecture as shown in Figure 3.36 [15]. The frequency output of the first VCO is simply obtained as $F_{VCO_1} = N_1 F_{osc}/M$. Similarly, the frequency output of the second VCO is also given as the ratio $F_{VCO_2} = N_2 F_{osc}/M$. In order to provide a

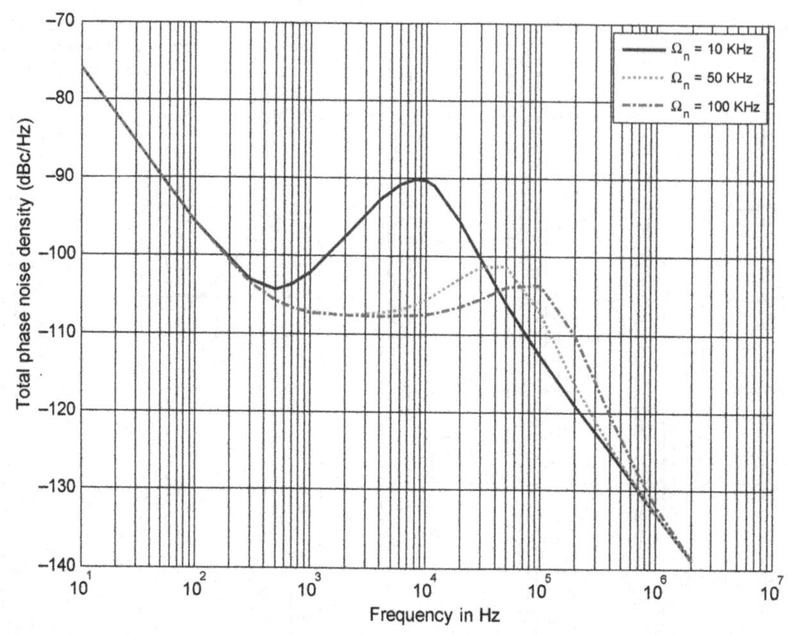

FIGURE 3.32

Comparison of various total phase noise densities for three different natural frequencies. (For color version of this figure, the reader is referred to the online version of this book.)

reference to the third VCO, the frequency output of the first VCO is then divided by N_3. The output frequency of the third VCO is mixed with the output frequency of the second VCO and filtered to obtain the frequency

$$F_{BP} = F_{VCO_3} - \frac{N_2 F_{osc}}{M} \tag{3.102}$$

Finally, the output of the third VCO must be such that it satisfies a zero-error condition at the output of the detector, or

$$F_{VCO_3} - \frac{N_2 F_{osc}}{M} - \frac{N_1}{N_3}\frac{F_{osc}}{M} = 0 \Rightarrow F_{VCO_3} = \frac{F_{osc}}{M}\left(N_2 + \frac{N_1}{N_3}\right) \tag{3.103}$$

And hence, the output of the various dividers, N_1, N_2, and N_3, can be set in order to obtain a wide range of frequency coverage and fast settling time.

In frequency synthesis, multiplying the frequency reference is also common. For example, in the analog domain, if the reference frequency is $V_{osc}(t) = \cos(2\pi F_{osc}t + \varphi(t))$, then multiplying the frequency N times implies that the output of the multiplier becomes $V_{ssc}^{N}(t) = \cos^N(\phi(t)) = \cos^N(2\pi F_{osc}t + \varphi(t))$. Based on De Moivre's theorem of complex numbers, which states that

$$\{\cos(\phi(t)) + j\sin(\phi(t))\}^N = e^{jN\phi(t)} = \cos(N\phi(t)) + j\sin(N\phi(t)) \tag{3.104}$$

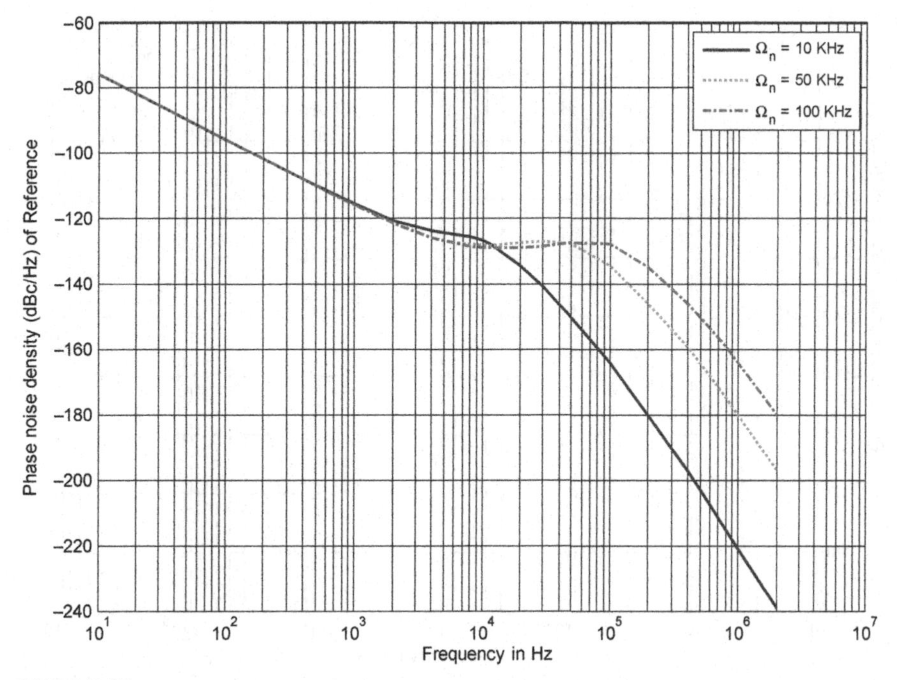

FIGURE 3.33

Comparison of various reference oscillator phase noise densities shaped by the phase-locked loop (PLL) for three different natural frequencies. (For color version of this figure, the reader is referred to the online version of this book.)

the trigonometric expansion of $\cos^N(\phi(t))$ can be expressed as[i]

$$\cos^N(\phi(t)) = \cos(N\phi(t)) + \binom{N}{2}\cos^{N-2}(\phi(t))\sin^2(\phi(t)) - \binom{N}{4}\cos^{N-4}(\phi(t))\sin^4(\phi(t)) + \ldots \quad (3.105)$$

According to Eqn (3.105), there are two cases, namely the case where N is even and the case where N is odd. For even N, the relationship in Eqn (3.105) implies that only even harmonics of the fundamental ranging from DC to the Nth harmonic are produced. Conversely, for odd N, only odd harmonics of the fundamental will be produced ranging from the fundamental up to the Nth harmonic. Therefore, in order to obtain a multiply-by-N version of the reference, a filter must follow the mixing operation in order to filter the harmonics and obtain the resulting up-converted

[i]Note that the combination $C_{(k)}^{(N)}$ of k elements taken from N elements is $C_{(k)}^{(N)} = \binom{N}{k} = \frac{N!}{(N-k)!k!}$

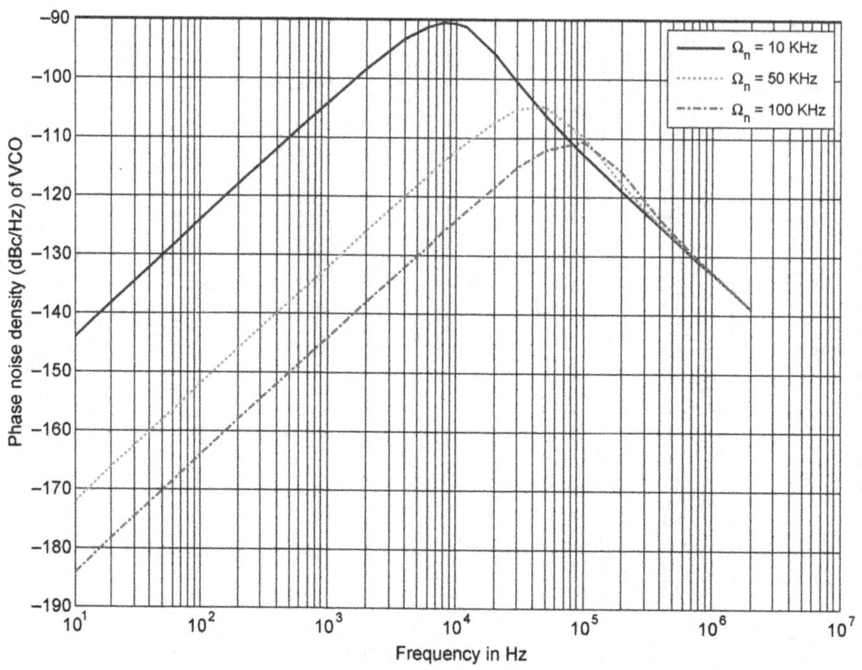

FIGURE 3.34

Comparison of various VCO phase noise densities shaped by the PLL for three different natural frequencies. VCO, voltage-controlled oscillator; PLL, phase-locked loop. (For color version of this figure, the reader is referred to the online version of this book.)

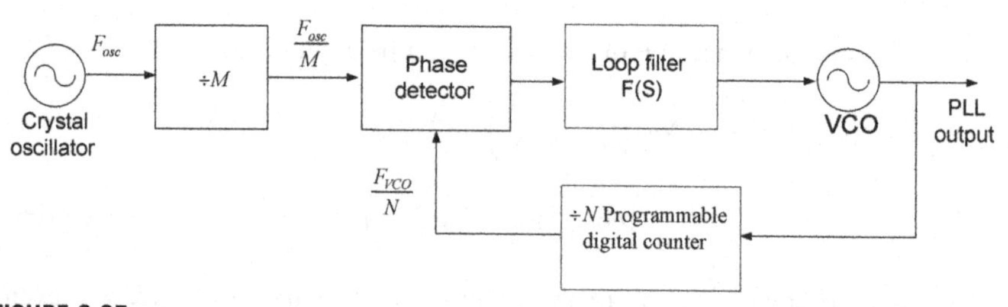

FIGURE 3.35

Frequency synthesizer with programmable digital divider. VCO, voltage-controlled oscillator; PLL, phase-locked loop.

reference that is $V_{osc}^{N}(t)\big|_{filtered} \approx \cos(2\pi N F_{osc} t + N\varphi(t))$. The spectral density of the phase fluctuations is then given as

$$S_{\varphi}\left(F_{offset}\right) = \frac{\varphi_{rms}^{2}\left(F_{offset}\right)}{1\ \text{Hz}}\ (radians)^{2} \tag{3.106}$$

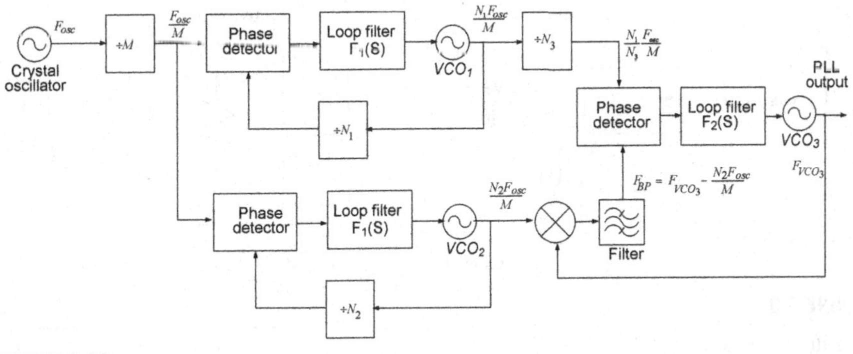

FIGURE 3.36

Frequency synthesizer with multiloop architecture. VCO, voltage-controlled oscillator; PLL, phase-locked loop.

Again, the power spectrum defined as the ratio of the power in 1-Hz bandwidth measured at F_{offset} from F_{osc} divided by the total signal power, or single-sided phase noise, is given as

$$L_{osc}\left(F_{offset}\right) = \frac{\varphi_{rms}^2\left(F_{offset}\right)}{2} \tag{3.107}$$

After up-converting the reference frequency N times, the single-sided phase noise becomes

$$L_{osc,\times N}\left(F_{offset}\right) = N^2\frac{\varphi_{rms}^2\left(F_{offset}\right)}{2} \tag{3.108}$$

The theoretical ratio between the up-converted reference phase noise density to that of the original reference is simply

$$10\log_{10}\left(\frac{L_{osc,\times N}\left(F_{offset}\right)}{L_{osc}\left(F_{offset}\right)}\right) = 20\log_{10}(N) \tag{3.109}$$

The relationship in Eqn (3.109) is applicable at the output of the multiplier and simply means that the phase noise of the reference has increased in proportion to the multiplication factor. However, in the context of a PLL, the loop will act on the multiplier as a filter. If the output of the multiplier is at the input of the loop, as is the case with the reference oscillator, then only the close-in phase noise is increased in proportion to N.

3.2.2 Noise in oscillators

In this section, we study the oscillator in the context of a feedback model as opposed to a negative resistance model depicted in Figure 3.37. In some contexts, it is convenient to apply the feedback circuit model as, for example, is the case with ring

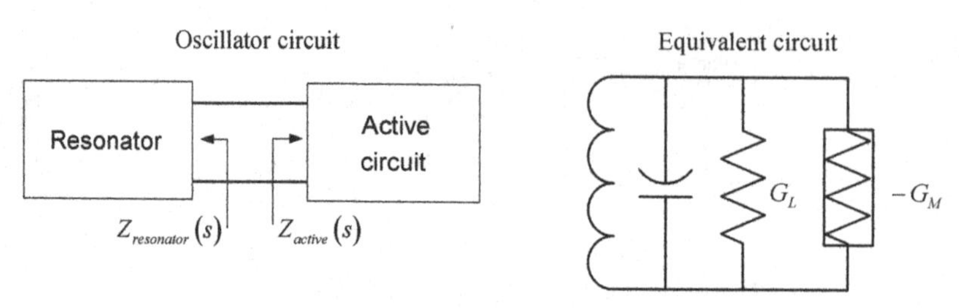

FIGURE 3.37

Negative resistance oscillator model.

oscillators. On the other hand, the negative resistor model provides more insight into the working of resonator-based oscillators. In this book, we use the feedback model to perform high-level system analysis. Then, we study two phase noise models. The first model is Leeson's model, which is by far the most popular model due to its simplicity. Leeson's model is based on frequency-domain analysis of noise in an oscillator loop. The second method is based on a time-varying model, which gives practical insight into the mechanics of phase noise in oscillators.

However, before we continue our study of oscillators, and for the sake of completion, we will very briefly discuss the working of the negative oscillator model. Given a resonator tank modeled as an *RLC* circuit as depicted in Figure 3.38, if we excite the circuit with a current impulse, the output of the circuit will oscillate as it decays until the energy is completely depleted in the resistor R_p. In order to ensure oscillation, suppose we place a *negative resistor* $-R_p$ in parallel with R_p as shown in Figure 3.39. Note that in this case, the output of the circuit due to current impulse will oscillate without decay. This is true since $R_p \| -R_p = \infty$ and hence the tank will indefinitely oscillate at the frequency of oscillation F_{osc}. Hence, if an active one-port circuit is placed in parallel with an *RLC* tank, the combination will oscillate. This model is known as the negative resistance model. One example of such a topology is the negative resistance provided by the cross-coupled transistors in an *LC* circuit as depicted in Figure 3.40.

3.2.2.1 Oscillator as a feedback system

We begin our discussion of the reference oscillator by discussing the sinusoidal model. To begin with, we note with caution[j] that all amplifier-based oscillators are intrinsically nonlinear, thus causing distortion to the output signal. However, one can still use linear-analysis techniques in order to provide first order design parameters and give the reader a basic system understanding of the oscillator itself. At

[j] As we shall see later, according to [17] and [18], nonlinear behavior alone cannot account for the phase noise behavior of the oscillator.

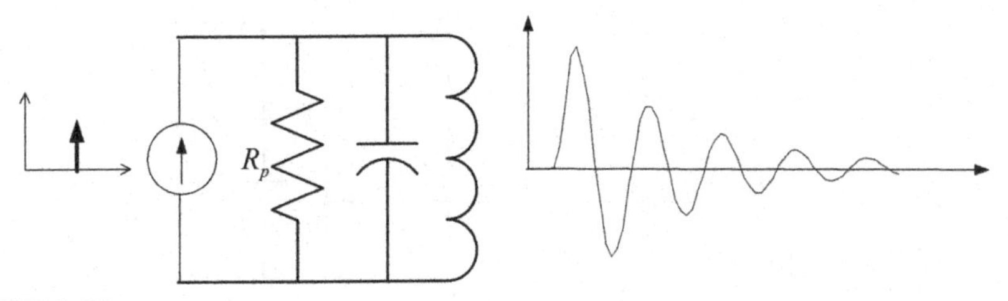

FIGURE 3.38

Oscillation decay in a *RLC* circuit. (For color version of this figure, the reader is referred to the online version of this book.)

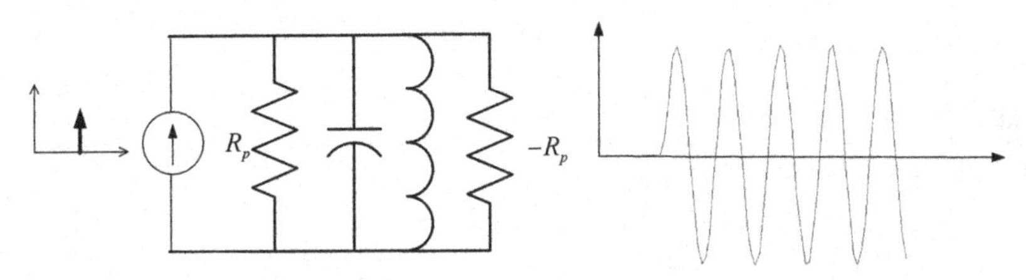

FIGURE 3.39

Negative resistance compensation for oscillation decay in a *RLC* circuit. (For color version of this figure, the reader is referred to the online version of this book.)

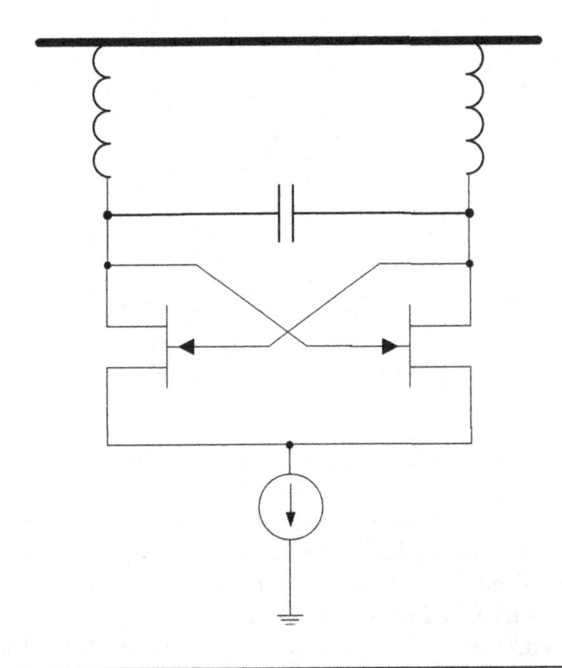

FIGURE 3.40

Cross-coupled transistors in *LC* circuit.

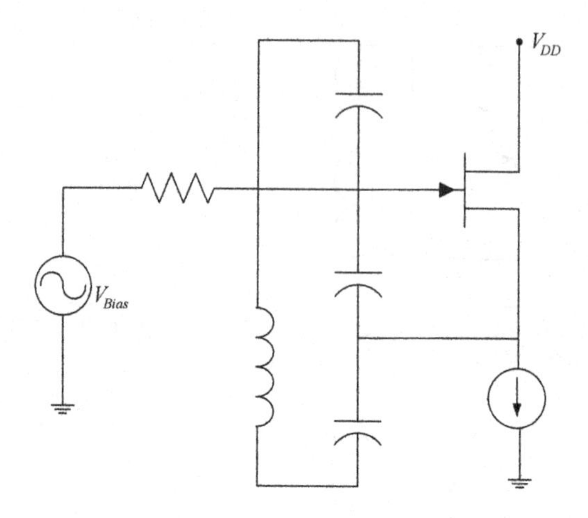

FIGURE 3.41

Simplified Clapp oscillator.

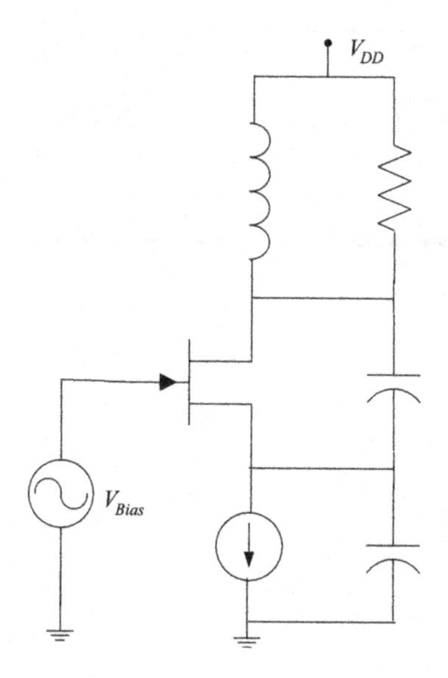

FIGURE 3.42

Simplified Colpitts oscillator where the resonator operates as an inductor.

this juncture it is important to note that there are several established oscillator circuit architectures such as the Clapp and Colpitts oscillators depicted in Figure 3.41 and Figure 3.42, respectively. The intent is not to study these various circuits but to establish a common theory of phase noise behavior.

A linearized model of the oscillator is shown in Figure 3.43. The gain in the forward path $G(s)$, as well as the filter in the feedback path $H(s)$, are frequency

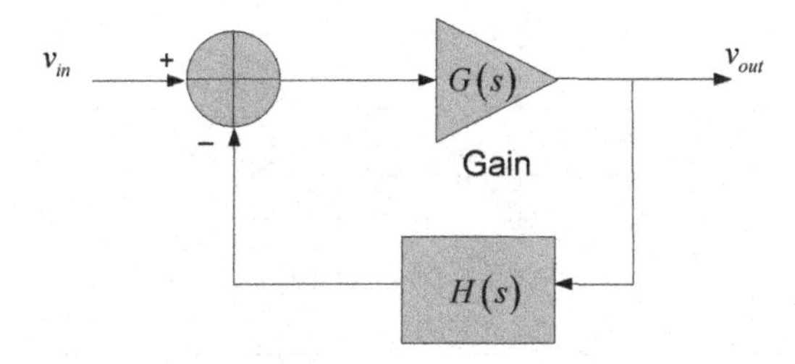

FIGURE 3.43

Linearized model of the oscillator shown as a feedback loop.

dependent. The output voltage $V_o(s)$ can be expressed in terms of the input voltage $V_i(s)$ as

$$V_o(s) = \frac{G(s)}{1 + G(s)H(s)} V_i(s) \tag{3.110}$$

To ensure that the circuit will oscillate, the output of the oscillator must be none-zero even when the input voltage is zero. This is only true if the denominator of Eqn (3.110) is zero, or in the frequency domain, for a given resonant frequency Ω_0, the following relationship must hold:

$$1 + G(j\Omega_0)H(j\Omega_0) = 0 \Rightarrow G(j\Omega_0)H(j\Omega_0) = -1 \tag{3.111}$$

According to the Nyquist criterion, the relationship in (3.111),[k] implies that oscillation will occur at the frequency Ω_0. It further implies that the magnitude of the open loop transfer function is equal to unity, or that the open loop gain satisfies

$$|G(j\Omega_0)H(j\Omega_0)| = 1 \tag{3.112}$$

and the phase shift of the open loop response is π, or the open loop phase

$$\arg\{G(j\Omega_0)H(j\Omega_0)\} = \pi \tag{3.113}$$

The conditions stated in Eqn (3.112) and Eqn (3.113) imply that if the open loop gain response of a feedback system is unity at a certain frequency Ω_0 and its corresponding phase shift is 180 at the same frequency, then the loop will oscillate at Ω_0. If the open loop gain is less than unity at 180, then the system is said to be stable. In contrast, if the open loop gain is greater than unity at 180, then the system is said to be unstable.[l]

[k]The Barkhausen criterion states the condition for stability in a manner equivalent to the Nyquist criterion, that is, given the closed loop transfer function $\frac{V_o}{V_i} = \frac{\alpha}{1-\alpha\beta}$, the system will oscillate if the product of the forward voltage gain α with feedback voltage gain β is such that $\alpha\beta = 1$.

[l]This is true of systems representing oscillators as stated in the simple linearized system model and is not necessarily true of more complex systems.

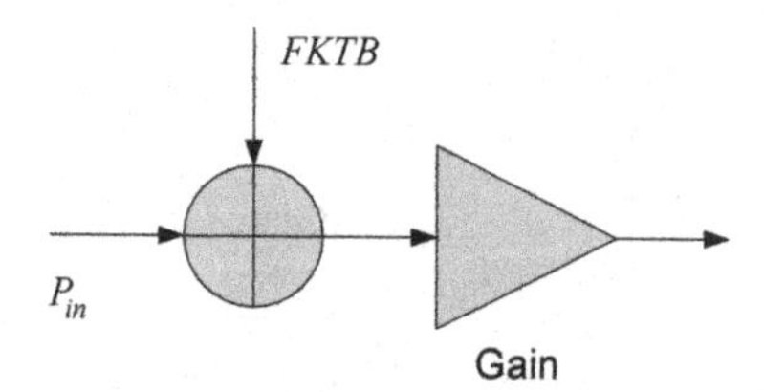

FIGURE 3.44

Additive thermal noise model of oscillator amplifier. K is Boltzmann's constant, T is the temperature in Kelvin, and B is the bandwidth.

Next, assume that the amplifier in Figure 3.43 has a noise factor F, then the output noise power of the amplifier N_{out} can be expressed in terms of its input noise power N_{in} as

$$N_{out} = GF_{noise,\text{amplifier}}N_{in} = GF_{noise,\text{amplifier}}KTB \tag{3.114}$$

where $F_{noise,\text{amplifier}}$ is the noise factor of the amplifier. Let P_{in} be the input power[m] of the signal driving the gain amplifier of Figure 3.43, and given the amplifier model of Figure 3.44, let $V_{noise,rms}$ be the *rms* noise density in Volts/$\sqrt{\text{Hz}}$ and $V_{in,rms}$ as the *rms* voltage signal at the input of the limiting amplifier then the total *rms* single side-band (SSB) phase deviation at a given frequency offset F_{offset} from the carrier is

$$\Delta\theta_{rms} = \frac{V_{noise,rms}}{V_{in,rms}} = \sqrt{\frac{F_{noise,\text{amplifier}}KT}{P_{in}}} \tag{3.115}$$

And consequently, the power spectral density of the phase noise can be expressed as

$$S_\theta\left(F_{offset}\right) = \Delta\theta_{rms}^2 = \frac{F_{noise,\text{amplifier}}KT}{P_{in}} \tag{3.116}$$

where the frequency offset F_{offset} is *significantly* far away from the carrier frequency. That is, for a given input power to the amplifier, the spectral density of the phase noise in 1 Hz is the noise floor of the amplifier or

$$S_{\theta,dB}\left(F_{offset} > F_c\right) = -174 + \underbrace{NF}_{10\log_{10}(F)} - P_{in} \tag{3.117}$$

According to Eqn (3.117), the noise floor decreases as the amplitude of oscillation and hence the input power increases.

3.2.2.2 The Leeson's model

In this section, we continue the analysis started in the previous section and further examine the nature of phase noise as we get closer and closer to the carrier

[m] P_{in} is often referred to as the oscillator power at the limiting port.

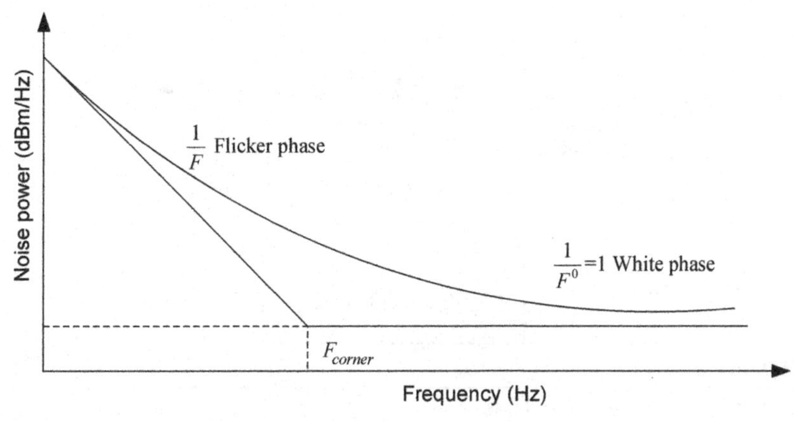

FIGURE 3.45

Active device flicker noise.

frequency. This discussion should lead to the development of Leeson's model and Leeson's phase noise density equation.

Our first consideration in understanding oscillator phase noise starts with the examination of the gain amplifier. As with all active devices, we note that a flicker noise component starts to exhibit itself clearly near the flicker corner F_{corner} closer to the carrier, as shown in Figure 3.45, and hence degrades the noise figure. This degradation can be accounted for by replacing the noise factor in Eqn (3.116) by

$$F_{noise,\text{amplifier}} = F_{\text{noise,amplifier}}\left(1+\frac{F_{corner}}{F_{offset}}\right) \tag{3.118}$$

resulting in the phase noise density

$$S_{\theta}\left(F_{offset}\right) = \frac{F_{noise,\text{amplifier}}KTB}{P_{in}}\left(1+\frac{F_{corner}}{F_{offset}}\right)\Bigg|_{B=1\text{Hz}} \tag{3.119}$$

The noise below the flicker corner is referred to as $1/f$ noise. The flicker corner depends on the device technology used. In bipolar transistors, the flicker corner is near 1 KHz and thus its effect is almost negligible, whereas in CMOS transistors the flicker corner can be up to several 100 KHz. Flicker noise is present in all active devices and many passive devices as well. It is important to note that flicker noise increases with a slope of $1/F$ as we get closer to the carrier. The equivalent phase noise model of Figure 3.43 can now be modified to include flicker noise as shown in Figure 3.46.

Next, we consider the effect of the resonator in the context of a feedback loop. The resonator circuit is a bandpass filter whose primary function is to set the oscillation frequency of the feedback system. In the ensuing analysis, we will show that the resonator will transform the output of the gain amplifier with its flat and $1/f$ noise

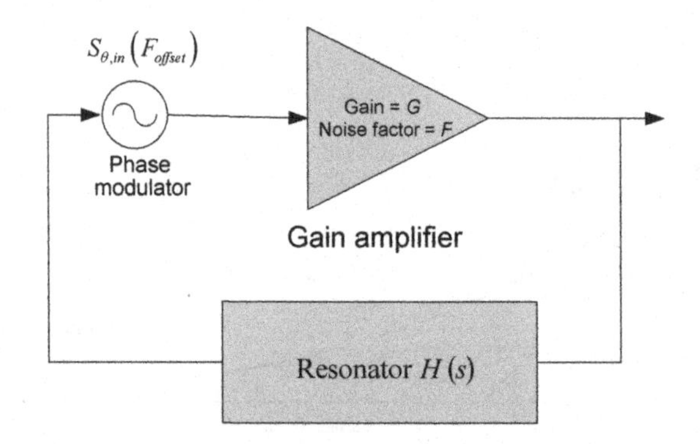

FIGURE 3.46

Oscillator feedback phase noise model.

components into $1/f^2$ and $1/f^3$ components. And consequently, the output of the loop will contain noise with various slope characteristics.

The transfer function of the bandpass resonator $H(F)$ can be expressed in terms of its lowpass equivalent $H_L(F)$ as

$$H(F) = H_L(F - F_{osc}) + H_L(F + F_{osc}) \tag{3.120}$$

where F_{osc} is the frequency of oscillation, and $H_L(F)$ has the form

$$H_L(F) = \frac{1}{1 + j\frac{2Q_{Loaded}(F - F_{osc})}{F_{osc}}} \tag{3.121}$$

In Eqn (3.121), Q_{Loaded} refers to the loaded Q of the feedback network. Later on in this section we will study in more detail the nature of the oscillator. However, for now, in order to uphold the Barkhausen criterion for oscillation,[n] assume that

$$H(F = F_{osc}) = 1, \text{ and } G = 1 \tag{3.122}$$

Furthermore, assume the input signal to the loop $v_{in}(t)$, as depicted in Figure 3.43, to be statistically additive and zero-mean white Gaussian distributed, that is, thermal noise in the practical sense. The noise power is

$$P_n = \frac{N_0}{2} = \frac{KTF_{noise}}{2} \text{ Watt/Hz} \tag{3.123}$$

where F is the noise factor of the amplifier. Let $v_h(t)$ be the output of the bandpass resonant tank circuit, then the output of the oscillator can be expressed as the sum

$$v_{out}(t) = v_{in}(t) + v_h(t) \tag{3.124}$$

[n]Note that the Barkhausen's criteria is necessary but not sufficient for oscillation. For example, if the phase shift around the loop is 2π at DC with sufficient gain, the oscillator will tend to latch rather than oscillate.

At this point let us examine the autocorrelation of the output signal

$$\begin{aligned} r_{v_{out},v_{out}}(\tau) &= E\{(v_{in}(t)+v_h(t))(v_{in}(t+\tau)+v_h(t+\tau))\} \\ &= E\{v_{in}(t)v_{in}(t+\tau)\}+E\{v_h(t)v_h(t+\tau)\}+E\{v_{\mathrm{in}}(t)v_h(t+\tau)\} \\ &\quad + E\{v_h(t)v_{in}(t+\tau)\} \\ &= r_{v_{in},v_{in}}(\tau)+r_{v_h,v_h}(\tau)+r_{v_{in},v_h}(\tau)+r_{v_h,v_{in}}(\tau) \end{aligned} \tag{3.125}$$

Since $v_{in}(t)$ is additive white Gaussian, then $r_{v_{in},v_{in}}(\tau)$ can be expressed as

$$r_{v_{in},v_{in}}(\tau) = \frac{N_o}{2}\delta(\tau) \tag{3.126}$$

and hence

$$R_{v_{in},v_{in}}(F) = \frac{N_o}{2}, \text{ for all } F \tag{3.127}$$

Next, consider the autocorrelation of $v_h(t)$. The output of the resonator can be expressed as the convolution between the output signal $v_{out}(t)$ and the resonator impulse response $h(t)$ or

$$v_h(t) = h(t) * v_{out}(t) = \int_{-\infty}^{\infty} h(\alpha)v_{out}(t-\alpha)d\alpha \tag{3.128}$$

The mean of $v_h(t)$ is given as

$$E\{v_h(t)\} = \int_{-\infty}^{\infty} h(\alpha)E\{v_{out}(t-\alpha)\}d\alpha = \mu_{v_{out}}\int_{-\infty}^{\infty} h(\alpha)d\alpha = \mu_{v_{out}}H(0) \tag{3.129}$$

where $\mu_{v_{out}} = E\{v_{out}(t-\alpha)\}$ is the mean of the output signal $v_{out}(t)$ and $H(0)$ is the frequency response of the resonator at DC. The autocorrelation function of $v_h(t)$ is given according to the relationship

$$\begin{aligned} r_{v_h,v_h}(\tau) &= E\{v_h(t)v_h(t+\tau)\} \\ &= \int_{-\infty}^{\infty}\int_{-\infty}^{\infty} h(\alpha)h(\beta)E\{v_{out}(t)v_{out}(t+\tau)\}d\alpha\, d\beta \\ &= \int_{-\infty}^{\infty}\int_{-\infty}^{\infty} h(\alpha)h(\beta)r_{v_{out},v_{out}}d\alpha\, d\beta \end{aligned} \tag{3.130}$$

The Fourier transform of Eqn (3.130) can be expressed as

$$\begin{aligned} R_{v_h,v_h}(F) &= \int_{-\infty}^{\infty} r_{v_h,v_h}(\tau)e^{-j2\pi F\tau}d\tau \\ &= \int_{-\infty}^{\infty}\left[\int_{-\infty}^{\infty}\int_{-\infty}^{\infty} h(\alpha)h(\beta)r_{v_{out},v_{out}}(\tau+\alpha-\beta)d\alpha\, d\beta\right]e^{-j2\pi F\tau}d\tau \\ &= |H(F)|^2 R_{v_{out},v_{out}}(F) \end{aligned} \tag{3.131}$$

The relationship in Eqn (3.131) is the power spectral density of the output signal multiplied by the magnitude squared of the frequency response of the resonator filter.[o]

Next, let us examine the cross-correlation functions $r_{v_{in},v_h}(\tau)$ and $r_{v_h,v_{in}}(\tau)$. According to [16], the cross-correlation functions are related in the time domain as

$$r_{v_{in},v_h}(\tau) = r_{v_h,v_{in}}(-\tau) \tag{3.132}$$

Given that $v_{in}(t)$ and $v_h(t)$ are real valued, then

$$R_{v_{in},v_h}(F) = R^*_{v_h,v_{in}}(F) \tag{3.133}$$

Furthermore, given the nature of the resonator tank transfer function as expressed in Eqn (3.121), it turns out that both $R_{v_{in},v_h}(F)$ and $R_{v_h,v_{in}}(F)$ are imaginary [17] and

$$\begin{aligned} \mathrm{Re}\{R_{v_{in},v_h}(F)\} &= 0, F \neq F_{osc} \\ R_{v_{in},v_h}(F) + R_{v_h,v_{in}}(F) &= 0, F \neq F_{osc} \end{aligned} \tag{3.134}$$

And, consequently,

$$R_{v_{in},v_h}(F) = \frac{N_0}{2}\frac{1}{j\frac{2Q_{Loaded}(F-F_{osc})}{F_{osc}}} \tag{3.135}$$

Then, according to Eqn (3.134), the relationship in Eqn (3.131) implies that

$$R_{v_{out},v_{out}}(F) = R_{v_h,v_h}(F) + \frac{N_0}{2} \Rightarrow R_{v_{out},v_{out}}(F) = \frac{N_0}{2}\frac{1}{1-|H(F)|^2} \tag{3.136}$$

For $F > 0$ and $H(F) \approx H_L(F - F_{osc})$[p] then

$$H(F) \approx \frac{1}{1 + j\frac{2Q_{Loaded}(F-F_{osc})}{F_{osc}}}, F > 0 \tag{3.137}$$

[o]The reader needs to be aware of the notational difference between F denoting frequency and F_{noise} denoting the noise factor.

[p]This is true if $H(F)$ is extremely narrowband such that the influence of $H_L(F + F_{osc})$ for $F > 0$.

Next, substitute Eqn (3.137) and Eqn (3.123) into Eqn (3.136), then substituting the resulting relation into Eqn (3.119), we obtain the Leeson's equation.

$$S_\theta(F_{offset}) = \left.\frac{F_{\text{noise,amplifier}}KTB}{P_{in}} \frac{1}{1+\left(\frac{2Q_{Loaded}(F-F_{osc})}{F_{osc}}\right)^2}\right|_{B=1Hz} \tag{3.138}$$

And, consequently, the phase noise density in dBc/Hz, where $|F - F_{osc}| > F_{corner}$ and $0 < F \neq F_{osc}$, can then be expressed based on Eqn (3.138) as [16]

$$L(F - F_{offset}) = 10\log_{10}\left\{\frac{1}{2}\frac{F_{noise,\text{amplifier}}KTB}{P_{in}} \left\|\frac{1}{1+\left(\frac{2Q_{loaded}(F-F_{osc})}{F_{osc}}\right)^2}\right\|\right\} \tag{3.139}$$

On the other hand, when $|F - F_{osc}| < F_{\text{corner}}$ and $0 < F \neq F_{osc}$, then the Leeson's equation must be modified according to (3.118) as

$$L(F - F_{offset}) = 10\log_{10}\left\{\frac{1}{2}\frac{F_{\text{noise,amplifier}}KTB}{P_{in}}\left(1+\frac{F_{\text{corner}}}{F_{offset}}\right) \left\|\frac{1}{1+\left(\frac{2Q_{Loaded}(F-F_{osc})}{F_{osc}}\right)^2}\right\|\right\} \tag{3.140}$$

According to Eqn (3.140), the phase noise density versus frequency is depicted in Figure 3.47. The challenges in building an oscillator that covers a certain RF frequency range is to ensure low close-in phase noise for improved receiver selectivity, low noise floor that is deemed necessary for blocking immunity, and, of course, transmitter spectral purity.

Furthermore, according to Eqn (3.140), it is also important to observe that the phase noise density improves as Q_{Loaded} increases or as the signal power P_{in} increases. Increasing Q_{Loaded} decreases the ratio quadratically since the tank's impedance falls off as shown in the denominator of Eqn (3.140). On the other hand, increasing the signal power while the noise power remains the same increases the signal-to-noise ratio and hence decreases the amount of phase noise at the output of the oscillator.

3.2.2.3 Linear time-varying phase noise model

The limiting property of the amplifier in Figure 3.46 implies that nonlinearity is a fundamental property of oscillators. In many phase noise models, this nonlinear property is invoked to explain phase noise behavior. Hajimiri and Lee [17,18] challenge this assumption and proceed to show that oscillators are intrinsically time-varying systems. In the ensuing analysis, we follow the development of the time-varying model presented in [17,18].

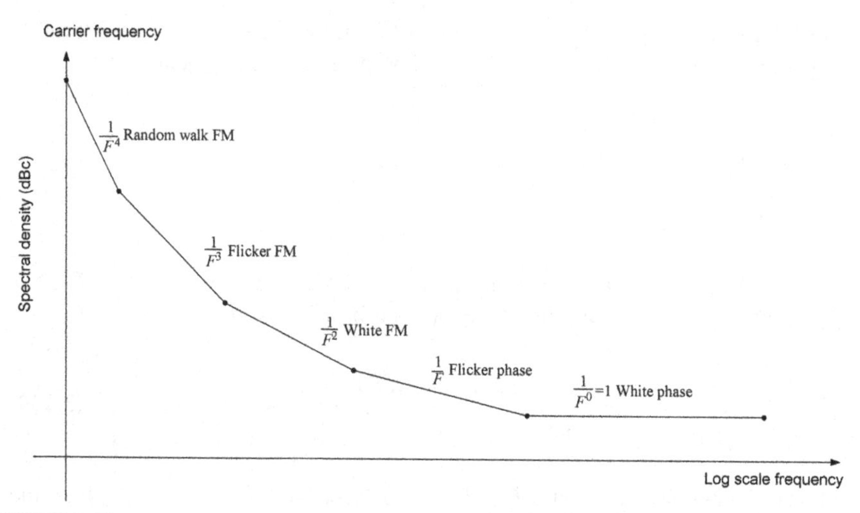

FIGURE 3.47

Power spectral density of phase noise based on Leeson's model.

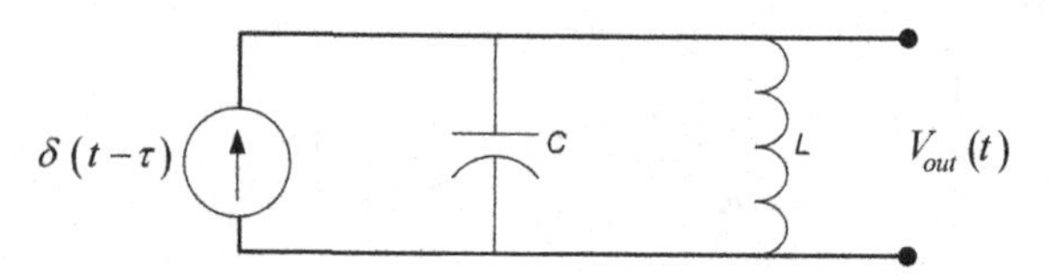

FIGURE 3.48

Oscillator with *LC*-tank circuit excited by a current pulse.

In order to prove the time-variance property of oscillators, consider a system oscillating at constant amplitude until a certain disturbance (current impulse) occurs and hence examine the system's response to this disturbance. To do so, let's examine how this current impulse affects the sinusoidal output of a resonator with a lossless *LC* tank, as shown in Figure 3.48.

At time $t < \tau$, the output of the oscillator in Figure 3.48 is a sinusoidal signal with constant amplitude. At time $t = \tau$, an impulse current occurs thus affecting the output in three possible manners:

- First, assume that the impulse occurs at the maximum voltage of the sinusoid as shown in Figure 3.49(a). In this case, we observe an abrupt increase in voltage amplitude. However, the zero crossing point of the waveform remains unaffected, thus indicating that the impulse simply served to alter the magnitude of the oscillator output but not its zero crossing point, and the system remains time invariant.
- Second, assume that the impulse occurs exactly at the zero crossing point of the oscillator's output waveform. In this case, the zero crossing point gets affected but not the amplitude of the waveform, thus causing the system to be time variant as shown in Figure 3.49(b).

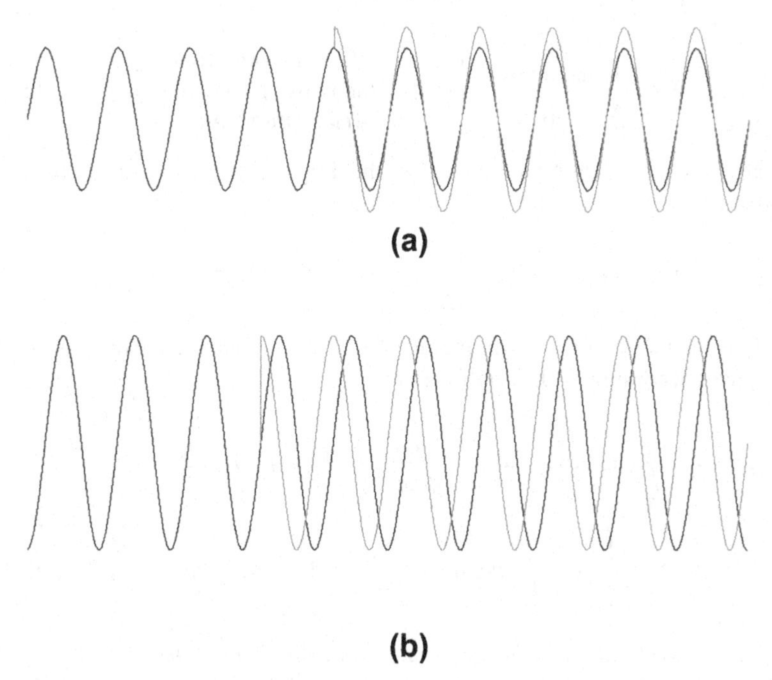

FIGURE 3.49

Impact of impulse oscillator response of a lossless *LC* tank: (a) impulse occurs at the output's maximum voltage and, (b) impulse occurs at the zero crossing point. (For color version of this figure, the reader is referred to the online version of this book.)

- Third, in the event where the impulse occurs anywhere between the maximum amplitude and the zero crossing point, the waveforms undergo both amplitude and time variance.

Based on our discussion thus far, we conclude that in the presence of random impulse noise, with some exceptions, the oscillator tends to react in a time-variant manner. In this case, the current impulse at the input produces a phase impulse response given as

$$h_{\varphi}(t,\tau) = \frac{1}{q_{\max}}\Gamma(\omega_{osc}\tau)u(t-\tau) \tag{3.141}$$

where $u(.)$ is the unit step function, $q_{\max}$ is the maximum charge displacement across the capacitor C, and $\Gamma(.)$ is defined in [17] and [18] as the impulse sensitivity function, which is dimensionless, periodic in 2π, and frequency and amplitude independent. The impulse sensitive function can be determined via simulation or analytically as is the case with the lossless tank oscillator under discussion. Under the circumstance, it can be shown that $\Gamma(\omega_{osc}\tau)$ is none other than the derivative of the sinusoidal oscillating function. In this case, we note that

$$\Gamma(\omega_{osc}\tau) = \begin{cases} \text{maximum} & v_{out}(t) = 0 \text{ or the zero-crossing point of the noise free oscillation} \\ 0 & v_{out}(t) = \text{maximum} \end{cases} \tag{3.142}$$

In any case, periodicity implies that the impulse sensitive function can be expressed as the Fourier series

$$\Gamma(\omega_{osc}\tau) = \frac{1}{2}c_0 + \sum_{n=1}^{\infty} c_n \cos(n\omega_{osc}\tau + \theta_n), \quad c_n \in \mathbb{R} \tag{3.143}$$

where θ_n is the phase of the nth harmonic of $\Gamma(\omega_{osc}\tau)$. Finally, the excess noise may be computed according to the relationship:

$$\begin{aligned} \varphi(t) &= \int_{-\infty}^{\infty} h_{\varphi}(t,\tau)i(\tau)d\tau = \frac{1}{q_{max}} \int_{-\infty}^{t} \Gamma(\omega_{osc}\tau)i(\tau)d\tau \\ &= \frac{1}{q_{max}}\left\{\frac{1}{2}c_0 \int_{-\infty}^{t} i(\tau)d\tau + \sum_{n=1}^{\infty} c_n \int_{-\infty}^{t} i(\tau)\cos(n\omega_{osc}\tau)\right\} \end{aligned} \tag{3.144}$$

where we assumed that the noise components are all uncorrelated, making the relative phase immaterial, and hence θ_n can be ignored. As is apparent from Eqn (3.144), the computation of the impulse sensitivity function is not easy and very much depends on the topology of the oscillator. In this case, the phase noise equation can be expressed as [19]

$$L(F - F_{offset}) = \begin{cases} 10\log_{10}\left\{\dfrac{c_0^2}{q_{max}^2}\dfrac{i_n^2/B_{noise}}{8F_{offset}^2}\dfrac{F_{corner}}{F_{offset}}\right\}, \dfrac{1}{f^3}\text{region} \\ 10\log_{10}\left\{10\log_{10}\left(\dfrac{\Gamma_{rms}^2}{q_{max}^2}\dfrac{i_n^2/B_{noise}}{4F_{offset}^2}\right)\right\}, \dfrac{1}{f^2}\text{region} \end{cases} \tag{3.145}$$

where B_{noise} is the noise bandwidth, F_{corner} is the $1/f$ flicker corner frequency, i_n^2/B_{noise} is the noise power spectral density and Γ_{rms}^2 is defined in terms of $\Gamma(\omega_{osc}\tau)$ as

$$\Gamma_{rms}^2 = \frac{1}{\pi}\int_0^{2\pi} |\Gamma(\omega_{osc}\tau)|^2 d\tau = \sum_{n=0}^{\infty} c_n^2 \tag{3.146}$$

The relationship Eqn (3.145) does not constitute an exact design rule, but nonetheless provided with accurate design data it could lead to fairly accurate results. Furthermore, the result in Eqn (3.145) implies that the noise near DC gets up-converted and weighted by the coefficient c_0 and the $1/f$ device noise gets up-converted to $1/f^3$ noise near the carrier. Noise near the carrier, on the other hand, remains near the carrier and all noise near integer multiples of the carrier undergo down-conversion to noise near the carrier in the $1/f^2$ region.

3.2.2.4 Types of oscillators

A VCO uses a voltage signal in order to control the output frequency of the oscillator. VCOs come in two different flavors [20], namely resonator-based or waveform-based, as illustrated in Figure 3.50. Resonator-based oscillators output a sinusoid whereas waveform-based oscillators output a square or triangular waveform. Each type of oscillator relies on a certain frequency tuning technique peculiar to it. For example, *LC* oscillators employ varactors to tune their frequencies, whereas ring oscillators utilize current steering techniques for that purpose. Oscillator types vary widely in terms of performance and form factor. For example, a ring oscillator can be integrated on chip whereas an *LC* oscillator provides better phase noise performance. There are various types of *LC*-based oscillators, namely, SAW oscillators,

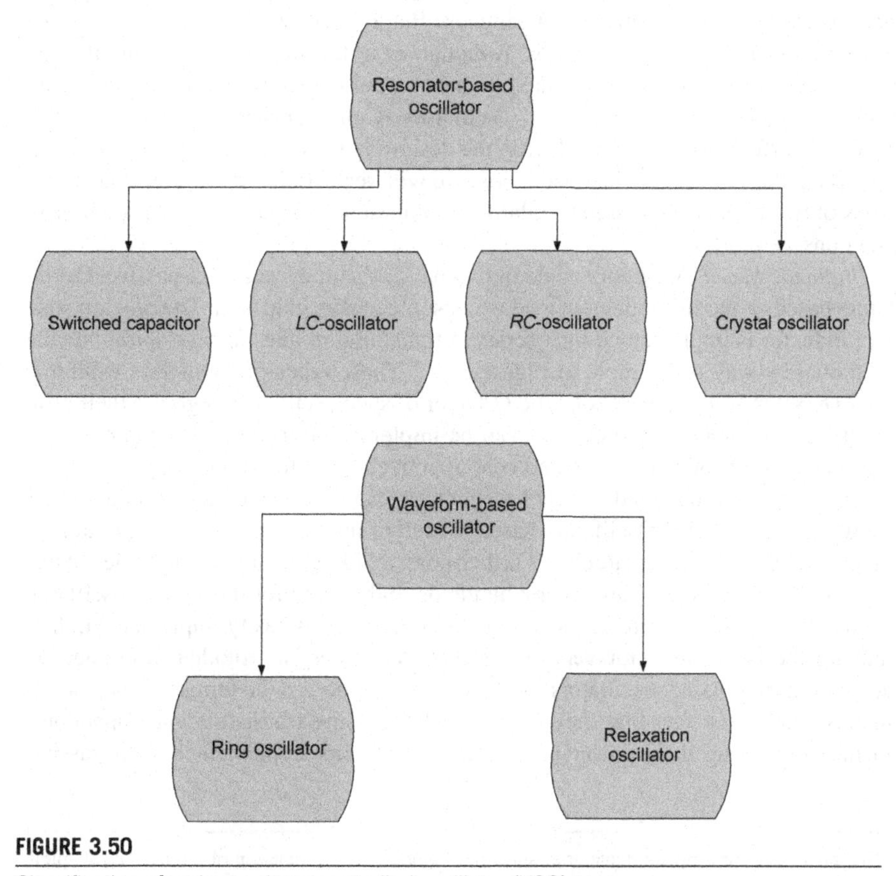

FIGURE 3.50

Classification of various voltage-controlled oscillator (VCO) types.

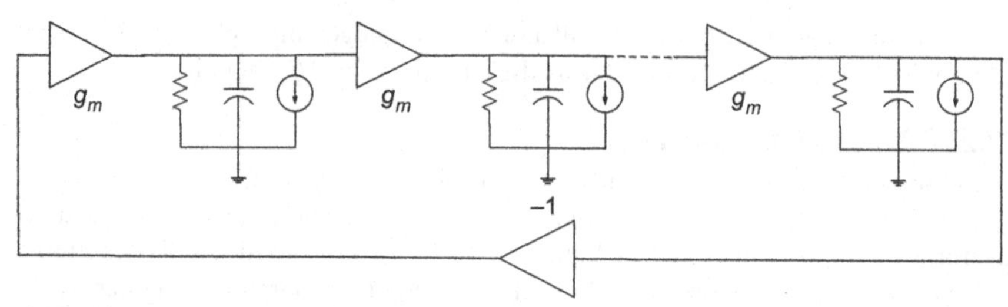

FIGURE 3.51

Multiphase ring oscillator.

stripline oscillators, dielectric-based oscillators, crystal oscillators, and *LC*-tank oscillators. By their very nature, *LC*-based oscillators tend to preserve energy thus resulting in high quality *Q*-factor. RC-based oscillators do not contain any inductors, only capacitors and resistors. This results in a *Q*-factor close to unity and poor phase noise performance. A ring oscillator, depicted in Figure 3.51 is an example of an *RC*-based oscillator. Although easy to integrate, they exhibit very poor phase noise characteristics making them a poor choice for RF-design. Relaxation oscillators are also *RC*-based oscillators. Consequently, depending on the required performance and form factor, the design may dictate one type of oscillator versus another. In the ensuing discussion, we will very briefly discuss the characteristics of two popular *LC*-based oscillators used in radio design and discuss their pros and cons.

Inductor-based oscillators come in two flavors, namely, active or passive. Oscillators based on passive inductors tend to consume a large chip area. The on-chip passive inductor is implemented by a series of transmission lines using a spiral layout, as shown by way of example in Figure 3.52.[q] These types of oscillators exhibit a poor *Q*-factor as well. Increasing the *Q*-factor of such oscillators degrades their reliability. Passive inductors, however, can be implemented using active components, thus achieving better reliability and cost effectiveness in the design.

A gyrator implemented in MOS technology can emulate an active inductor, as shown in Figure 3.53. Oscillators based on active inductors possess a wide tuning range with simple tuning circuitry. Furthermore, a high *Q*-factor can be achieved using active inductors reaching several hundreds. Inductance for this type of oscillator is controlled via bias current. The tuning circuitry itself is easily implemented, thus making the active inductor very suitable for VCO design. Another advantage of active inductor-based oscillators is their compact design in terms of chip area, mainly due to the fact that they can be realized using transistors and capacitors, in turn increasing their power dissipation. Oscillators implemented with passive

[q]The inductance for a square inductor as shown in Figure 3.52 is a function of the number of spiral turns N and the radius of the spirals in meters R and given as $L = 4\pi \times 10^{-7} N^2 R$.

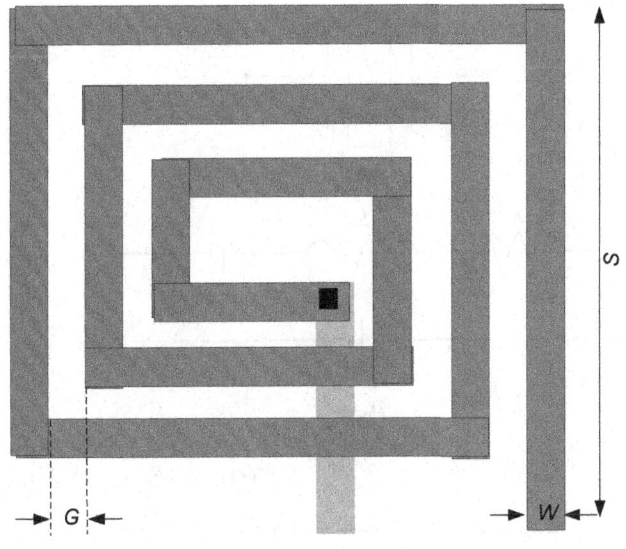

FIGURE 3.52

Square spiral inductor.

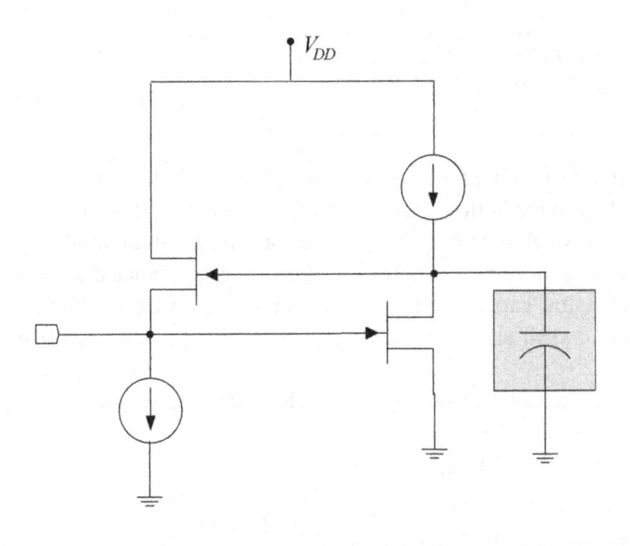

FIGURE 3.53

A gyrator circuit implemented in MOS.

inductors on the other hand are likely to be very large. The phase noise performance of active inductor-based oscillators tends to be poor.

Crystal oscillators are popular oscillators used to generate clock signals for digital circuits and synthesizers [21]. The unloaded Q-factor of crystal resonators is

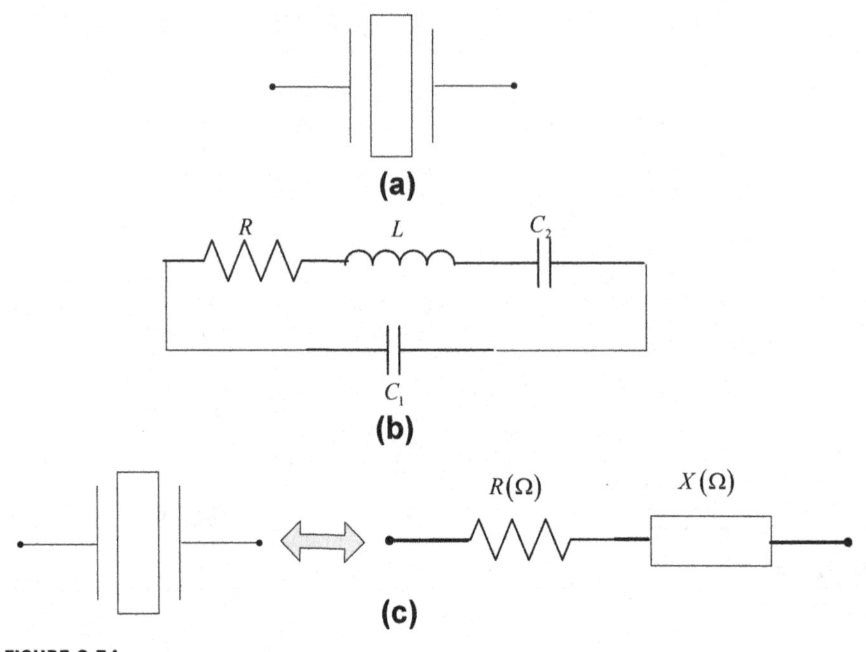

FIGURE 3.54

Piezoelectric oscillator: (a) symbol (b) equivalent circuit of fundamental mode, and (c) equivalent crystal resonator at $\Omega = \Omega_{osc}$.

extremely high thus making them perfect candidates for reference oscillators. Piezoelectric crystals provide both serial and parallel resonance.[r] The circuit equivalent of a piezoelectric crystal operating in parallel or fundamental mode is depicted in Figure 3.54. The capacitor C_1 denotes the parasitic capacitance due to crystal casing and holder. Parasitic capacitance plays a major role in determining the resonant behavior of the crystal simply because the series capacitor C_2 is extremely small, that is, $C_2 \ll C_1$.[s]

The serial and parallel resonant frequencies can be expressed as

$$\begin{aligned} \Omega_{series} &= \frac{1}{\sqrt{LC_2}} \\ \Omega_{parallel} &= \frac{1}{\sqrt{LC_2C_1/(C_2+C_1)}} \end{aligned} \tag{3.147}$$

[r]Parallel resonant circuits have their highest impedance at the resonant frequency while series resonant circuits have their lowest impedance at the resonant frequency.

[s]Usually on the order of a hundred.

The Q-factor can then defined as the ratio

$$Q = \frac{\Omega_{osc} L}{R} \tag{3.148}$$

where Ω_{osc} is the frequency of oscillation. The Q-factor referred to in Eqn (3.148) is the unloaded Q since it is relevant to the losses due to the resonant network only. The relationships given in Eqn (3.147) imply that the operating frequency is bounded by the parallel and series resonant frequencies, that is,

$$\Omega_{series} < \Omega_{osc} < \Omega_{parallel} \tag{3.149}$$

It turns out that at the resonant frequency Ω_{osc}, the crystal itself can be viewed as a resistance $R(\Omega)$ in series with an inductive reactance $X(\Omega)$ as shown in Figure 3.54(c). Both $R(\Omega)$ and $X(\Omega)$ are frequency dependent and can be obtained analytically as

$$R(\Omega) = \frac{\Omega_{parallel} - \Omega_{series}}{2QC_1} \frac{1}{\left(\Omega - \Omega_{parallel}\right)^2} \tag{3.150}$$

and

$$X(\Omega) = j\frac{1}{\Omega C_1}\left(\frac{\Omega - \Omega_{series}}{\Omega_{parallel} - \Omega}\right) \tag{3.151}$$

The values for the capacitance and inductance can be easily measured in the laboratory using a network analyzer.

The high Q-factor enables the oscillator design to be extremely stable despite possible variations of values of external components. External varactor diodes, for example, can be used to tune the resonant frequency of crystal oscillators, however, the tunable range is extremely limited. Piezoelectric crystals are plagued with high power dissipation and typically operate at low frequencies (e.g., 19.2, 20, 33.3 MHz, etc.).

3.2.2.5 Reciprocal mixing

The phase noise at the output of the LO is mainly due to the reference oscillator and the VCO shaped by the PLL as a filter. After mixing, the phase noise impacts any incoming signal both desired and interference in turn degrading the desired signal's SNR. In the event where only the desired signal is present, the phase noise in the LO directly affects the desired signal's purity and increases the noise floor. Additionally, if a strong interfering signal is present outside the desired signal's occupied bandwidth, the effect of phase noise on the interferer after mixing also increases the noise floor. This degradation is known as reciprocal mixing. To further illustrate this phenomenon, consider a desired GMSK signal received at an RF frequency F_c and a strong interfering tone situated at a frequency $F_c + \Delta F_c$ as illustrated in Figure 3.55(a) [13]. After mixing with the LO signal, both the desired signal and the interferer suffer from phase noise translated from the LO signal itself as shown

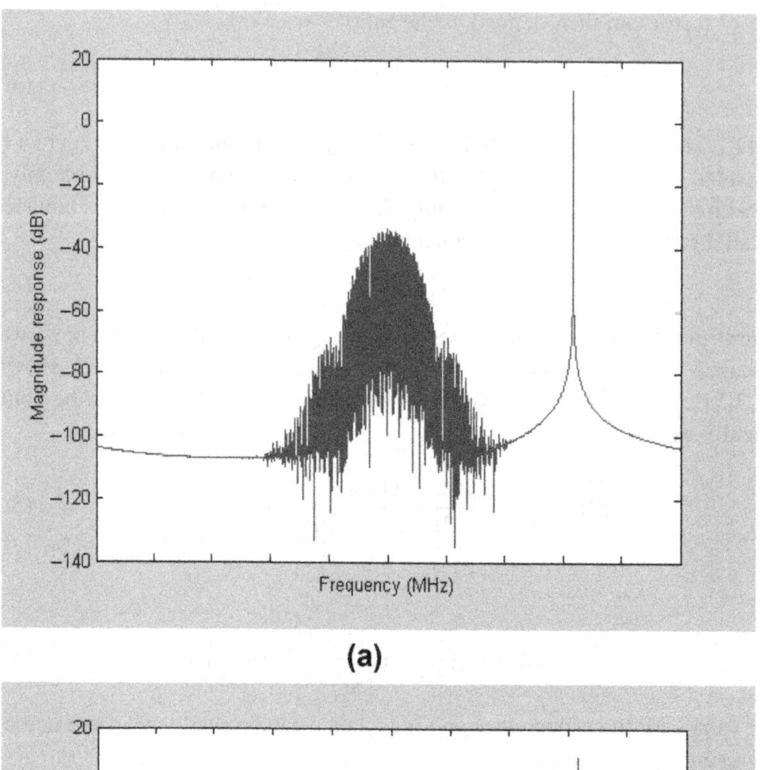

(a)

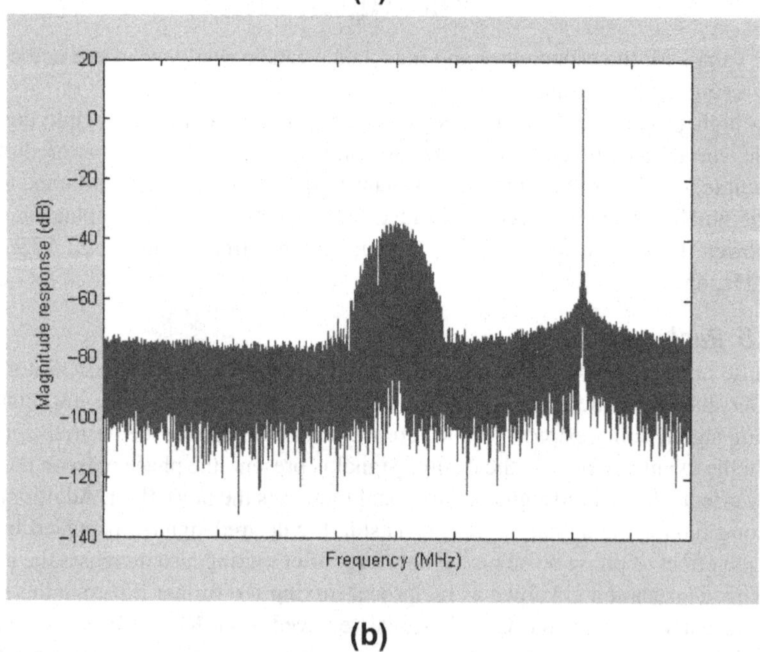

(b)

FIGURE 3.55

A desired GMSK signal in the presence of a strong narrowband interferer: (a) before mixing with the low noise amplifier (LO) signal and (b) after the mixing with the LO signal. (For color version of this figure, the reader is referred to the online version of this book.)

in Figure 3.55(b). It is important to note that the phase noise due to the narrowband tone *spills over* to the desired signal thus degrading its SNR and causing what is known as reciprocal mixing. The amount of noise power P_{pn} caused by the interfering tone alone can be estimated as

$$P_{pn} = P_{\text{interferer}} + 10\log_{10}\left[\int_{F_L}^{F_H} L(F)dF\right] \quad (3.152)$$

where $P_{\text{interferer}}$ is the interferer signal power in dBm and $L(f)$ is the SSB phase noise power spectral density in dBc/Hz. The upper and lower integration limits and F_H and F_L are the integration limits equal to the channel filter's noise equivalent bandwidth. Given that the interferer is situated far enough from the desired signal such that the phase noise can be considered flat in the desired signal band, then the relationship in Eqn (3.152) can be further simplified as

$$P_{pn} = P_{\text{interferer}} + L_{\text{dBc/Hz}}(F) + 10\log_{10}(B) \quad (3.153)$$

where $B = F_H - F_L$.

EXAMPLE 3.4 RECEIVER DESENSITIZATION

Given a narrowband jammer situated sufficiently farther away from the desired signal such that the phase noise density is flat within the desired band. Let $P_{jammer} = 10$ dBm be the jamming signal power before the band definition filter. The band definition filter is the filter placed before the LNA. Let the desired signal bandwidth be 10 MHz with a system noise figure of 4 dB. The phase noise density in the desired band due to the reciprocal mixing of the LO with the jammer with no filtering is $L(F_c + \Delta F_c) = -155$ dBc/Hz. It is desired that the impact of the desensitization due to reciprocal mixing on the system sensitivity be less than 1 dB. Compute the necessary rejection due to the band definition filter in the receiver's front end.

The first step is to compute the signal noise floor at room temperature given the system noise figure

$$\begin{aligned} P_{noisefloor} &= -174\text{ dBm/Hz} + 10\log_{10}(B) + NF \\ &= -174\text{ dBm/Hz} + 10\log_{10}\left(10\times 10^6\text{Hz}\right) + 4 = -100\text{ dBm} \end{aligned} \quad (3.154)$$

The total noise floor can be obtained by adding the noise degradation due to reciprocal mixing to the noise floor due to thermal noise found in Eqn (3.154), that is

$$P_{Total} = P_{noisefloor} + 1\text{ dB} = -100 + 1 = -99\text{ dBm} \quad (3.155)$$

The noise due to reciprocal mixing can then be computed as

$$\begin{aligned} P_{pn} &= 10\log_{10}\left(10^{P_{total}/10} - 10^{P_{noisefloor}/10}\right) = 10\log_{10}\left(10^{-99/10} - 10^{-100/10}\right) \\ &= -105.87\text{ dBm} \end{aligned} \quad (3.156)$$

According to Eqn (3.153), the jammer power after filtering can be obtained as

$$P_{interferer} = P_{pn} - L_{\text{dBc/Hz}}(F_c + \Delta F_c) - 10\log_{10}(B) = -105.87 + 155 - 70 = -21\text{ dBm} \quad (3.157)$$

Continued

EXAMPLE 3.4 RECEIVER DESENSITIZATION—cont'd

The required rejection by the band definition filter is simply the difference between the interference power obtained Eqn (3.157) and the jammer power before the filter or

$$P_{\text{Rejection}} = P_{jammer} - P_{\text{interferer}} = -10\text{ dBm} + 21\text{ dBm} = 11\text{ dB} \quad (3.158)$$

The result in Eqn (3.158) implies that the band definition filter must have an average power rejection of 11 dB across the 10 MHz signal band.

3.3 Coexistence

In this section, we are concerned with noise[t] or interference generated by one or more transmitters that somehow end up in the desired frequency band of the receiver. Today's personal communication devices, be it a smartphone or a laptop, are hosts to a multitude of radios and radio protocols that operate simultaneously. The radios themselves are designed either into a single chipset that supports multiple standards and protocols, or multiple chipsets each dedicated to a single standard (e.g., GPS, WCDMA and LTE, Bluetooth, etc.). A commonplace scenario, for example, involves three standards operating simultaneously as depicted in Figure 3.56: the use of a cellular standard such as WCDMA to communicate via a cell tower with another WCDMA user, a Bluetooth headset for hands-free communication, and GPS for navigation.

The source of interference or noise in the receive band, due to coexistence, is mainly due to:

- PA broadband noise that could extend into the receive band. This type of noise is especially pronounced in full duplex systems, for example.
- Modulated harmonics and intermodulation products due to various transmitters generated at the output of the LNA in the desired band.
- Intermodulation products due to a transmit signal and an incoming interfering signal.

In this section, we will mainly focus on modulated harmonics and intermodulation products. In order to do so, let us first discuss certain characteristics of polynomial nonlinearity. Given a set of variables, say $\{y_1(x), y_2(x), \ldots, y_k(x)\}$, then

$$(y_1(x), y_2(x), \ldots, y_k(x))^n = \sum_{\substack{n_1, n_2, \ldots, n_k \geq 0 \\ n_1+n_2+\ldots+n_k=n}} \frac{n!}{n_1!n_2!\ldots n_k!} y_1^{n_1}(x) y_2^{n_2}(x) \ldots y_k^{n_k}(x) \quad (3.159)$$

[t]Noise, in coexistence, can refer to degradations due to various sources such as phase noise, PA thermal noise, noise due to intermodulation products and harmonics, etc.

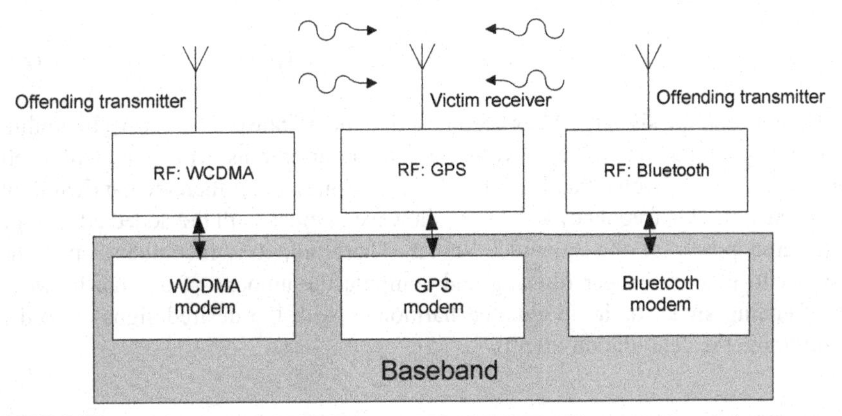

FIGURE 3.56

A conceptual view of a handheld device that can simultaneously support multiple standards at various frequency bands. RF, radio frequency.

Furthermore, if we let $y(x) = \cos(x)$ or $y(x) = \sin(x)$, then for n odd, we have

$$\cos^n(x) = \frac{1}{2^{n-1}} \sum_{k=0}^{\frac{n-1}{2}} \binom{n}{k} \cos((n-2k)x) \text{ or}$$
$$\sin^n(x) = \frac{1}{2^{n-1}} \sum_{k=0}^{\frac{n-1}{2}} (-1)^{\left(\frac{n-1}{2}-k\right)} \binom{n}{k} \sin((n-2k)x) \tag{3.160}$$

And for n even, we have

$$\cos^n(x) = \frac{1}{2^n}\binom{n}{\frac{n}{2}} + \frac{1}{2^{n-1}} \sum_{k=0}^{\frac{n}{2}-1} \binom{n}{k} \cos((n-2k)x) \text{ or}$$
$$\sin^n(x) = \frac{1}{2^n}\binom{n}{\frac{n}{2}} + \frac{1}{2^{n-1}} \sum_{k=0}^{\frac{n}{2}-1} (-1)^{\left(\frac{n}{2}-k\right)} \binom{n}{k} \cos((n-2k)x) \tag{3.161}$$

The relationship in Eqn (3.160) implies that a tone raised to the power n where n is odd produces odd order harmonics ranging from the fundamental to the nth harmonic. On the other hand, the relationship in Eqn (3.161) implies that a tone raised to the power n where n is even produces harmonics ranging from DC to the nth harmonic. Tones, however, have a frequency bandwidth of 1 Hz. If a certain modulated signal has a bandwidth greater than 1 Hz, then raising the signal to the power n due to a certain nonlinearity increases its bandwidth by n-fold. This frequency expansion results due to the multiplication in the time domain or convolution in the frequency domain, that is

$$y(t)g(t) \Leftrightarrow \frac{1}{2\pi}Y(\Omega) * G(\Omega) \tag{3.162}$$

Therefore, if the signal $y(t)$ has been raised to the nth power due to certain nonlinearity, the resulting signal, the convolution of the Fourier transform of $y(t)$ with itself n times, results in a signal that has a bandwidth n-times larger than the bandwidth of the signal $y(t)$. This frequency expansion, however, comes with the added advantage of lowered power of the expanded signal. Therefore, it is incumbent upon the designer to employ proper filtering and compute the amount of overlap between the offending signal or its modulated harmonics with the desired signal in order to minimize the degradation effects.

EXAMPLE 3.5 COEXISTENCE OF COMMERCIAL GPS WITH WCDMA AND BLUETOOTH

Given a handset that hosts three radios, namely WCDMA, Bluetooth, and GPS. WCDMA operates at 825 MHz with maximum average transmit power of +24 dBm and signal bandwidth of 3.84 MHz. WCDMA operates in full duplex. The WCDMA to GPS antenna-to-antenna isolation is 15 dB. Bluetooth, on the other hand, operates in the ISM band of 2.4 GHz with an average maximum power of 10 dBm and instantaneous signal bandwidth of 1 MHz. The transmitter has a duty cycle of 1/6. The Bluetooth to GPS antenna-to-antenna isolation is 12 dB. The GPS LNA is preceded with a band definition filter (BDF). The BDF has an average rejection of 67 dB at 824 MHz. The GPS receiver has an instantaneous channel bandwidth of 8 MHz. The GPS RF center frequency is at 1575.42 MHz. The GPS receiver noise figure referred to the LNA is 2 dB. The GPS has a system IIP2 of −37 dBm. Furthermore, assume that the Bluetooth transmitter is in non-hopping mode. What is the required average BDF rejection in the ISM band in order not to degrade the GPS receiver sensitivity by more than 0.5 dB? The GPS antenna has a gain of 0 dBi. The GPS BDF filter has an insertion loss of 1 dB. It may be helpful for the reader to review Chapter 4 before delving into the details of this example.

First, let us address the mechanism that generates the degradation to the GPS noise floor. Both WCDMA and Bluetooth signals undergo path loss, due to antenna-to-antenna isolation, before both signals impinge on the GPS antenna. Both signals get further attenuation due to the GPS band definition filter as shown in Figure 3.57. At this point, at least theoretically, the nonlinearity of the filter has negligible effect on the incoming interfering signals. However, that is not the case in the LNA.

At the output of the LNA, the second order nonlinearity results in two intermodulation products, one at the sum of 2.4 GHz and 825 MHz and one at the difference of 2.4 GHz and 825 MHz, which incidentally happens to be at center of the GPS band of 1575 MHz. Since this is a second order effect—the interfering signal bandwidth is the sum of the Bluetooth signal bandwidth of 1 MHz and that of the WCDMA signal of 3.84 MHz, resulting in an interferer whose bandwidth is 4.484 MHz totally contained within the GPS instantaneous channel bandwidth. This implies that there is no partial overlap of bandwidths between the interferer and that of the GPS receiver.

The signal due to the WCDMA transmitter is attenuated by 15 dB due to path loss and then further attenuated by 67 dB due to the GPS BDF. The resulting signal impacting the GPS LNA is P_{WCDMA} = 24 dBm − 15 dB − 67 dB = −58 dBm.

EXAMPLE 3.5 COEXISTENCE OF COMMERCIAL GPS WITH WCDMA AND BLUETOOTH—cont'd

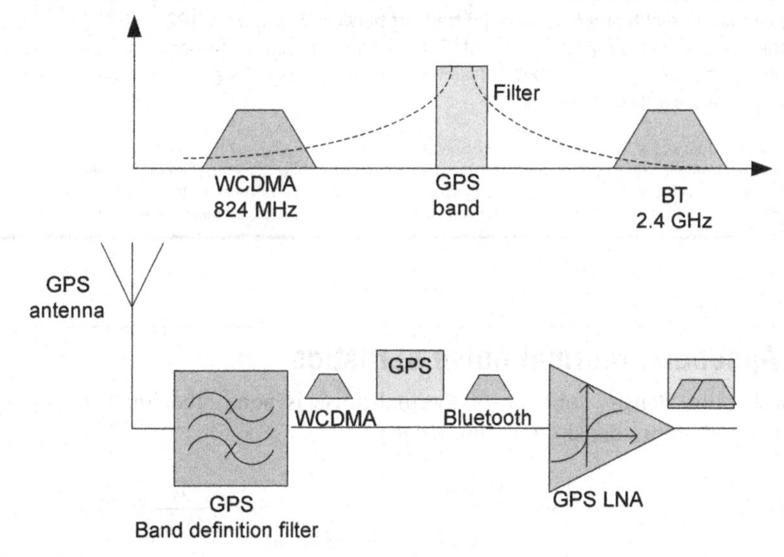

FIGURE 3.57

Coexistence of the GPS radio with Bluetooth and WCDMA. (For color version of this figure, the reader is referred to the online version of this book.)

Next, let us determine the noise floor of the GPS receiver as referred to the LNA. The noise power of the GPS receiver can be computed as

$$P_{noisefloor\ (\text{dBm})} = P_{sensitivity} - CNR_{dB} = -174\ \text{dBc/Hz} + NF + 10\log_{10}(\text{Bandwidth}) = -174 + 2 + 10\log_{10}\left(8\times 10^{6}\right) = -102.97\ \text{dBm} \tag{3.163}$$

A degradation of 0.5 dB to the noise floor implies that the intermodulation product due to second order nonlinearity at the output of the LNA can be computed as

$$P_{IM} = 10\log_{10}\left(10^{\frac{(P_{noisefloor}+\delta)}{10}} - 10^{\frac{P_{noisefloor}}{10}}\right) = 10\log_{10}\left(10^{\frac{(-102.97+0.5)}{10}} - 10^{\frac{-102.97}{10}}\right) = -112.10\ \text{dBm} \tag{3.164}$$

The average power due to the Bluetooth transmitter after the BDF at the input of the LNA can be estimated according to the relationship

$$\begin{aligned} P_{IM} &= P_{WCDMA} + P_{Bluetooth} - IIP_2 \Rightarrow P_{Bluetooth} = P_{IM} - P_{WCDMA} + IIP_2 \\ P_{Bluetooth} &= -112.10\ \text{dBm} + 58\ \text{dBm} - 37\ \text{dBm} = -91.1\ \text{dBm} \end{aligned} \tag{3.165}$$

Keeping in mind that the Bluetooth transmitter has a duty cycle of 1/6,[u] the required BDF rejection in the ISM band before the LNA becomes

$$R = P_{TX,Bluetooth} - P_{\text{path loss}} + 10\log_{10}(\text{Duty Cycle}) - P_{Bluetooth} = 10 - 12 + 10\log_{10}\left(\frac{1}{6}\right) + 91.1 = 81.3\ \text{dB} \tag{3.166}$$

Continued

EXAMPLE 3.5 COEXISTENCE OF COMMERCIAL GPS WITH WCDMA AND BLUETOOTH—cont'd

Note that, in reality, there is a design tradeoff between the system IIP2 and BDF rejection that the designer has to take into account. Furthermore, the Bluetooth signaling waveform is a frequency-hopping waveform that occasionally overlaps the GPS signal, thus lessening the impact on the GPS desired signal.

[u]In practice, the impact of the duty cycle depends heavily on the signal structure of the victim waveform. In simple calculations, taking the duty cycle of the transmitter into account may show little impact on the offended signal, whereas in a real scenario the victim's received signal may be grossly affected. Therefore, in order to more accurately assess the impact of the offending transmit signal on the victim's received signal, a detailed simulation is needed.

3.4 Appendix: thermal noise statistics

The probability density function of thermal noise is none other than the general Gaussian probability density function (PDF)

$$PDF_{general}(v_n) = \frac{1}{\sqrt{2\pi\sigma_n^2}} e^{-\frac{(v_n-\mu_n)^2}{2\sigma_n^2}} \Bigg|_{\mu_n=0 \text{ at equilibrium}} = \frac{1}{\sqrt{2\pi\sigma_n^2}} e^{-\frac{v_n^2}{2\sigma_n^2}} \tag{3.167}$$

where σ_n is the standard deviation and $\mu_n = E\{v_n\}$ is the voltage mean equal to zero at equilibrium. The variance is defined as the square of standard deviation and is statistically expressed as

$$\sigma_n^2 = E\left\{(v_n - \mu_n)^2\right\} = E\left\{v_n^2\right\} - \mu_n^2 \tag{3.168}$$

It is often desirable to use the normalized Gaussian PDF of thermal noise given as

$$PDF_{general}(z_n) = \frac{1}{\sqrt{2\pi}} e^{-\frac{z_n^2}{2}} \tag{3.169}$$

where the normalized variable z_n is

$$z_n = \frac{v_n - \mu_n}{\sigma_n} \tag{3.170}$$

Next define the complementary error function, also known as the cumulative normalized Gaussian PDF, as[v] [13]

[v]Note that the complementary error function is given as $erfc(x) = 1 - erf(x) = 2Q(\sqrt{2}x) \Rightarrow Q(x) = 0.5\, erfc(x/\sqrt{2})$ and $erf(x) = \frac{2}{\sqrt{\pi}} \int_0^x e^{-\alpha^2} d\alpha \approx 1 - \frac{e^{-x^2}}{\sqrt{\pi}x}$ for $x \gg 1$ resulting in. $Q(x) = \frac{1}{\sqrt{2\pi}} \int_x^\infty e^{-\alpha^2/2} d\alpha \approx 1 - \frac{e^{-x^2}}{\sqrt{\pi}x}$ for $x \gg 1$.

$$Q(z_n) = \frac{1}{\sqrt{2\pi}} \int_{z_n}^{\infty} e^{-z_n^2/2} dz_n \quad (3.171)$$

Mathematically, the relationship presented in Eqn (3.171) is the integral of the normalized Gaussian PDF and can also approximated numerically as

$$Q(\text{valid for } z_n \geq 0) = \left(\frac{1}{(1-\alpha)z_n + \alpha\sqrt{z_n^2 + \beta}} \right) \frac{1}{\sqrt{2\pi}} e^{-z_n^2/2} \quad (3.172)$$

where $\alpha = 0.344$ and $\beta = 5.344$.

References

[1] J.B. Johnson, Thermal agitation of electricity in conductors, Phys. Rev. 32 (July 1928) 97.

[2] H. Nyquist, Thermal agitation of electric charge in conductors, Phys. Rev. 32 (July 1928) 110.

[3] B. Rezavi, RF Microelectronics, Prentice Hall, Upper Saddle River, NJ, 1998.

[4] W. Mumford, E. Scheibe, Noise Performance Factors in Communication Systems, Horizon House-Microwave, Dedham, MA, 1968.

[5] T.-S. Chu, L.J. Greenstein, A quantification of link budget differences between cellular and PCS bands, IEEE Trans. Veh. Technol. 48 (1) (January 1999) 60–65.

[6] H. Rohte, W. Danlke, Theory of noisy fourpoles, Proc. IRE 44 (1956) 811–818.

[7] H.A. Haus, Representation of noise in linear twopoles, Proc. IRE 48 (1960) 69–74.

[8] W.A. Davis, K. Agarwal, Radio Frequency Circuit Design, Wiley, New York, NY, 2001.

[9] T.H. Lee, Planar Microwave Engineering, Cambridge University Press, Cambridge, UK, 2004.

[10] H. Fukui, Design of microwave GaAs MESFETs for broad-band low noise amplifiers, IEEE Trans. Electron Devices 26 (1979) 1032–1037.

[11] R.E. Lehman, D.D. Heston, X-band monolithic series feedback LNA, IEEE Trans. Microw. Theory Tech. 33 (1985) 1560–1566.

[12] D. Henkes, LNA design uses series feedback to achieve simultaneous low input VSWR and low noise, Appl. Microwave Wireless 10 (October 1998) 26–32.

[13] T. Rouphael, RF and Digital Signal Processing for Software Defined Radio, Elsevier, Boston, MA, 2009.

[14] N.M. Blachman, S. Mayerhofer, An astonishing reduction in the bandwidth of noise, Proc. IEEE 63 (7) (July 1975) 1077–1078.

[15] D.C. Green, Radio Systems Technology, Longman Scientific & Technical, Essex, UK, 1990.

[16] A. Papoulis, Probability, Random Variables, and Stochastic Processes, McGraw-Hill, New York, NY, 1991.

[17] T. Lee, Oscillator phase noise: a tutorial, IEEE Custom Integr. Conf. 16 (1999) 373–380.

[18] A. Hajimiri, T. Lee, A general theory of phase noise is electrical oscillators, IEEE J. Solid State Circuits 33 (2) (February 1998) 179–194.
[19] M. Odyniec, RF and Microwave Oscillator Design, Artech House, Boston, MA, 2002.
[20] D.A. Jones, K. Martin, Analog Integrated Circuit Design, John Wiley & Sons, Hoboken, NJ, 1997.
[21] A. Luzzato, G. Shirazi, Wireless Transceiver Design, John Wiley and Sons, Hoboken, NJ, 2007.

CHAPTER

System Nonlinearity

4

CHAPTER OUTLINE

Thus far, it can be established that analysis developed using linear algebraic tools, such as scattering parameters, is essentially used to perform *linear analysis* of radio frequency (*RF*) networks. Circuits that exhibit nonlinearity with or without memory *cannot* be simply analyzed using linear techniques to sufficiently characterize their performance. In practice, both linear and nonlinear techniques are used for analysis, simulation, and design. For example, consider the frequency multiplier circuit shown in Figure 4.1(a). The circuit is comprised of input and output matching networks, and a nonlinear multiplier. The circuit can be modeled as a cascade of three blocks: a two-port input matching network, a nonlinear block, followed by a two-port output matching network as shown in Figure 4.1(b). The matching networks can be analyzed using the linear techniques developed in

Wireless Receiver Architectures and Design. http://dx.doi.org/10.1016/B978-0-12-378640-1.00004-4

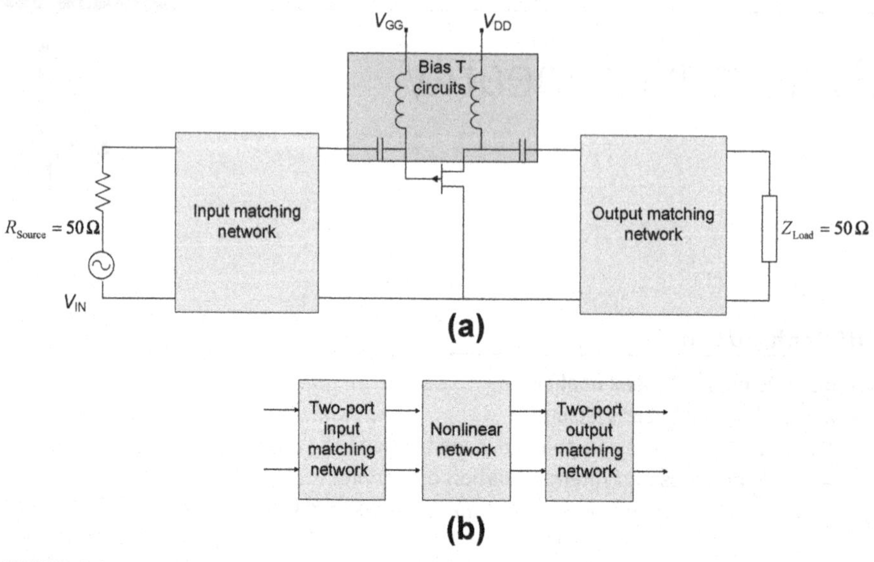

FIGURE 4.1

Basic frequency multiplier circuit. (For color version of this figure, the reader is referred to the online version of this book.)

Chapter 2. The nonlinear network, however, depending on its instantiation within the receiver, has to be treated with the appropriate nonlinear techniques. These nonlinear techniques are the subject of this chapter. In a similar manner, if several nonlinear systems or circuits are presented in cascade, for example, the circuit itself as a whole can be modeled as a cascade of linear and nonlinear elements as shown in Figure 4.2.

Nonlinear systems can be classified as either weakly nonlinear or strongly nonlinear.

Nonlinear systems can also be modeled as memoryless nonlinear systems or nonlinear systems with memory. In this chapter we mainly focus on weakly nonlinear systems with brief mention of strongly nonlinear systems.

The chapter is divided into three sections. Section 4.1 discusses and classifies the various nonlinearities encountered in *RF* design. Section 4.2 delves into the subject of memoryless nonlinearities using power series methods. And finally, systems with

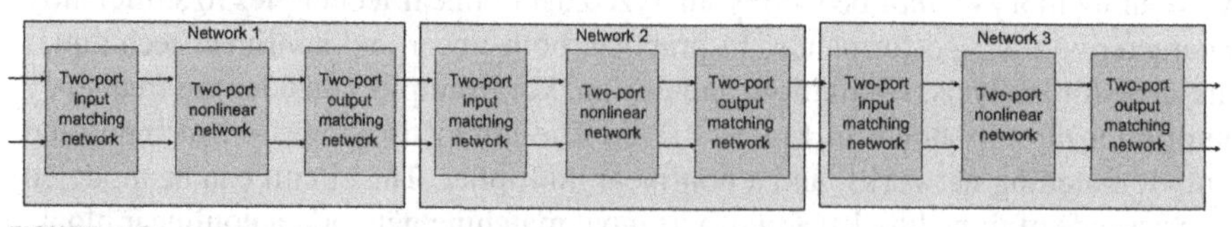

FIGURE 4.2

Cascade of two-port linear and nonlinear networks.

memory are treated in Section 4.3 with focus on techniques that employ the Volterra series. The chapter also contains several appendices.

4.1 Classification of system nonlinearity

Before we proceed, let us recall the definition of a linear system. A system is said to be linear if the principle of superposition holds. That is, given an input signal $x(t)$ and an output signal $y(t)$ such that

$$y(t) = h[x(t)] \tag{4.1}$$

where t is an independent variable representing time. Given two inputs $x_1(t)$ and $x_2(t)$ the system produces two outputs $y_1(t) = h[x_1(t)]$ and $y_2(t) = h[x_2(t)]$. If the system transfer function is linear, then an input $ax_1(t) + bx_2(t)$ produces the output

$$h[ax_1(t) + bx_2(t)] = ah[x_1(t)] + bh[x_2(t)] = y_1(t) + y_2(t) \tag{4.2}$$

where a and b are scalars. A system is said to be nonlinear if and only if it is *not linear*, that is, it does not satisfy Eqn (4.2). Furthermore, and for the sake of completion, a system is said to be time invariant if and only if for a given time constant τ we have

$$y(t-\tau) = h[x(t-\tau)], \forall\tau \tag{4.3}$$

Then, a system is said to be time-varying if it is *not time invariant*, that is it does not satisfy Eqn (4.3). A system that is both linear and time invariant is referred to as a linear time-invariant (LTI) system. An important implication of an LTI system is that the spectrum of the output $y(t)$ does not contain any spectral components that are not in the spectrum of $x(t)$. More specifically, an LTI system does not generate any new frequencies[1]. The same cannot be said about nonlinear systems. Consider, for example, the nonlinear but time-invariant polynomial

$$y(t) = \underbrace{\beta_1 x(t)}_{\text{Linear term}} + \underbrace{\beta_2 x^2(t)}_{\text{Nonlinear term}} \tag{4.4}$$

where β_1 and β_2 are scalar coefficients associated with the nonlinearity of the system. Let the input signal $x(t)$ be the sum of two tones situated at two distinct frequencies F_1 and F_2, that is $x(t) = V_1\sin(2\pi F_1 t) + V_2\sin(2\pi F_2 t)$, where V_1 and V_2 are scalar coefficients associated with the input waveform. Then, the output signal according to Eqn (4.4) is given as

$$y(t) = \underbrace{\beta_1\{V_1\sin(2\pi F_1 t) + V_2\sin(2\pi F_2 t)\}}_{\text{Linear term}} + \underbrace{\beta_2\{V_1\sin(2\pi F_1 t) + V_2\sin(2\pi F_2 t)\}^2}_{\text{Nonlinear term}} \tag{4.5}$$

[1]Note that linear time-varying systems can generate additional mixing products.

The nonlinear term in Eqn (4.5) can be further expanded as

$$\{V_1 \sin(2\pi F_1 t) + V_2 \sin(2\pi F_2 t)\}^2 = V_1^2 \sin^2(2\pi F_1 t) + V_2^2 \sin^2(2\pi F_2 t) + 2V_1 V_2 \sin(2\pi F_1 t)\sin(2\pi F_2 t) \tag{4.6}$$

Using the identities provided in Appendix A, we note that for any frequency F_n, we have

$$\sin^2(2\pi F_n t) = \frac{1}{2} - \frac{1}{2}\cos(4\pi F_n t) \tag{4.7}$$

and

$$\sin(2\pi F_1 t)\sin(2\pi F_2 t) = \frac{1}{2}\cos(2\pi(F_1 - F_2)t) - \frac{1}{2}\cos(2\pi(F_1 + F_2)t) \tag{4.8}$$

Substituting Eqns (4.7) and (4.8) into Eqn (4.6) and subsequently into Eqn (4.5), we obtain

$$\begin{aligned} y(t) = {} & \underbrace{\beta_1\{V_1 \sin(2\pi F_1 t) + V_2 \sin(2\pi F_2 t)\}}_{\text{Linear term}} \\ & + \underbrace{\frac{1}{2}\beta_2\left(V_1^2 + V_1^2\right)}_{\text{DC term}} - \underbrace{\frac{1}{2}\beta_2\left(V_1^2 \sin(4\pi F_1 t) + V_1^2 \sin(4\pi F_2 t)\right)}_{\text{Harmonics term}} \\ & + \underbrace{\beta_2 \frac{V_1 V_2}{2}\cos(2\pi(F_1 - F_2)t) - \beta_2 \frac{V_1 V_2}{2}\cos(2\pi(F_1 + F_2)t)}_{\text{Intermodulation term}} \end{aligned} \tag{4.9}$$

Note that the content of the input signal is comprised of two distinct tones at the frequencies F_1 and F_2. The output signal, on the other hand, is comprised of a linear term that is manifested in a scaled version of the input signal and three other terms with frequencies other than F_1 and F_2. The first term is a direct current (DC) term at the frequency $F = 0$ Hz. The second term, known as the harmonic term, is comprised of two tones or *harmonics* situated at $2F_1$ and $2F_2$ Hz. Lastly, the intermodulation term is comprised of two tones located at the frequencies $F_1 - F_2$ and $F_1 + F_2$ Hz. Therefore, one can observe that the output of a nonlinear system with second order nonlinearity contains *new* tones at new frequencies that are not present in the input signal. In addition to being nonlinear, a device operating at high frequency contains frequency and phase dependency in its response. Consider by way of example the two-port network depicted in Figure 4.3. Recall from Chapter 2 that

$$S_{2,1} = \left.\frac{V_2^-}{V_1^+}\right|_{V_2^+ = 0} \tag{4.10}$$

Given that the incident wave is $V_1^+(t) = A\cos(2\pi F_0 t)$, then the resulting reflected wave is

$$V_2^-(t) = A\left|S_{2,1}(F_0)\right|\cos\left(2\pi F_0 t + \arg\{S_{2,1}(F_0)\}\right) \tag{4.11}$$

According to the relationship in Eqn (4.11), there is a definite dependence on frequency in both magnitude and phase. Most analysis or synthesis techniques of *RF* systems make certain assumptions about the components under study. Some concentrate on the nonlinear aspect of the device and maintain independence of frequency in the input–output relationship concerning phase and amplitude. Other techniques concentrate on the input–output relationship and its impact on the system performance with little regard to the nonlinearity in the device.

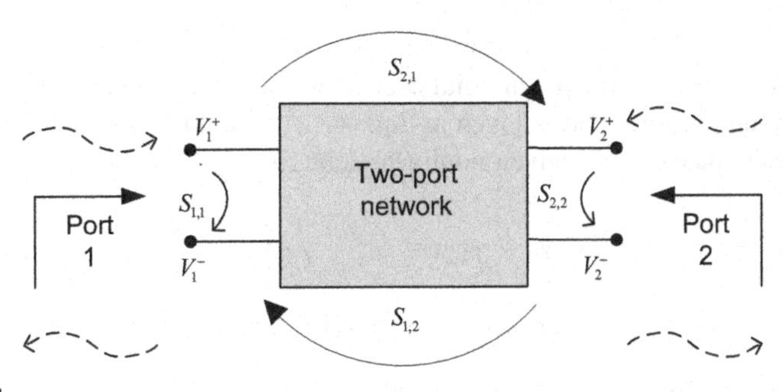

FIGURE 4.3

Two-port network.

Nonlinear systems can be further classified as either weakly nonlinear or strongly nonlinear. Weakly nonlinear systems can be modeled and analyzed using power series if the system is memoryless or Volterra series if the system has memory. A system is said to be weakly nonlinear if the nonlinearity itself along with the *RF* signal drive are *weak* such that the DC operating point in a transistor, for example, is not perturbed [1]. Hence, weak nonlinearity applies when the excitation voltage of such devices as transistors and passive components is typically within the component's normal operating range. A strong nonlinearity, on the other hand, can be sufficiently modeled using harmonic balance or time domain techniques. Strong nonlinearity occurs when a component is operating well beyond its saturation point.

The degree of nonlinearity in *RF* circuits depends on the component type. For instance, purely passive components display nonlinear behavior only when the applied power to the component approaches its maximum allowable limit. By contrast, active and semiconductor devices exhibit considerable nonlinear effects at lower input power than the maximum allowable input power. Furthermore, it is important to note that certain components, such as frequency multipliers, mixers[2], and detectors, for example, rely on nonlinear behavior to realize their purpose.

[2]Although the circuit behavior of a mixer may be nonlinear, the desired effect of mixing the incoming signal with the *LO* is a linear operation.

4.2 Memoryless nonlinear systems

A nonlinear system is said to be memoryless if the output at a given time t depends only on the instantaneous input values at time t and not on any past values of the input signal. Hence, the output $y(t)$ can be simply expressed as a power series of the form

$$y(t) = \sum_{n=0}^{\infty} \beta_n x^n(t) = \beta_0 + \beta_1 x(t) + \beta_2 x^2(t) + \beta_3 x^3(t) + \cdots \tag{4.12}$$

where $\{\beta_0, \beta_1, \beta_2, \cdots\}$ are the polynomial coefficients and $x(t)$ is the input signal. The nonlinear system representation given in Eqn (4.12) is simply an instantiation of the Taylor series expansion of a given nonlinear function $\Lambda(x)$ such that

$$\Lambda(x) \cong \sum_{k=0}^{N} \gamma_k (x - x_0)^k = \gamma_0 + \gamma_1 (x - x_0) + \gamma_2 (x - x_0)^2 + \gamma_3 (x - x_0)^3 + \cdots \gamma_N (x - x_0)^N \tag{4.13}$$

where the coefficient γ_k can be represented as

$$\gamma_k = \frac{1}{k!}\left[\frac{d^k}{dx^k}\Psi(x)\right]_{x=x_0} \tag{4.14}$$

The input and output of Eqn (4.12) could represent either currents, voltages, or incident and reflected waves in one or many-port networks. In the ensuing analysis of this section, we will assume that the system nonlinearity is weak and memoryless, and we will define the parameters that characterize such a system.

4.2.1 1-dB compression point, desensitization, and blocking

The 1-dB compression point (1-dB CP) is an important design parameter. The 1-dB CP is defined as the point at which the linear output power and the output power of the amplifier or nonlinear device differ by 1 dB as depicted in Figure 4.4. Increasing the input power beyond the 1-dB CP serves to further compress the desired signal and degrade its quality. To quantify that, consider the single-tone input signal

$$x(t) = \alpha_1 \cos(\Omega_1 t) \tag{4.15}$$

The output of the nonlinear device, say an amplifier with memoryless nonlinear characteristics, can be expressed according to Eqn (4.12) as

$$y(t) = \beta_1 x(t) + \beta_2 x^2(t) + \beta_3 x^3(t) \tag{4.16}$$

In this case, we have limited the nonlinearity to third order for practical reasons. Substituting Eqn (4.15) into Eqn (4.16), we obtain

$$y(t) = \beta_1 \alpha_1 \cos(\Omega_1 t) + \beta_2 \alpha_1^2 \cos^2(\Omega_1 t) + \beta_3 \alpha_1^3 \cos^3(\Omega_1 t) \tag{4.17}$$

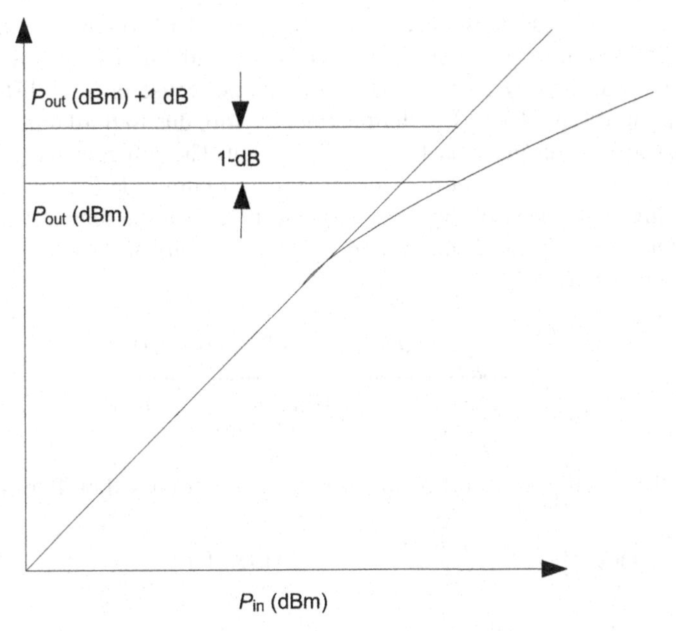

FIGURE 4.4

The 1-dB compression point as a function of input (*x* axis) versus output power (*y* axis). DAC, Digital to analog converter; ADC, Analog to digital converter; Rx PLL, Receive Phase Locked Loop; Tx PLL, Transmit Phase Locked Loop; LNA, Low Noise Amplifier; LPF, Lowpass Filter; IRF, Image reject filter; PA, Power Amplifier; PM, Phase modulation; LO, Local oscillator; RC, Resistor-capacitor; CR, capacitor resistor; HF, no reference; BER, Bit error rate; CNR, carrier to noise ratio.

Furthermore, using the trigonometric relations expressed in Eqns (4.232) and (4.233), the output signal in Eqn (4.17) can be further manipulated as

$$\begin{aligned} y(t) &= \beta_1\alpha_1 \cos(\Omega_1 t) + \frac{1}{2}\beta_2\alpha_1^2[\cos(2\Omega_1 t) + 1] \\ &\quad + \frac{1}{4}\beta_3\alpha_1^3[\cos(3\Omega_1 t) + 3\cos(\Omega_1 t)] \end{aligned} \tag{4.18}$$

Rearranging Eqn (4.18) to further explain the implication of the various nonlinearities, we obtain

$$\begin{aligned} y(t) &= \underbrace{\frac{1}{2}\beta_2\alpha_1^2}_{\text{DC term}} + \underbrace{\beta_1\alpha_1\cos(\Omega_1 t)}_{\substack{\text{Linear term}\\ \text{Desired signal}}} + \underbrace{\frac{3}{4}\beta_3\alpha_1^3\cos(\Omega_1 t)}_{\substack{\text{Degradation due to third}\\ \text{order nonlinearity}}} \\ &\quad + \underbrace{\frac{1}{2}\beta_2\alpha_1^2\cos(2\Omega_1 t)}_{\text{Second harmonic term}} + \underbrace{\frac{1}{4}\beta_3\alpha_1^3\cos(3\Omega_1 t)}_{\text{Third harmonic term}} \end{aligned} \tag{4.19}$$

In Eqn (4.19), the DC term as well as the second order harmonic are due to the second order nonlinearity. Both terms can be filtered out and do not impact the desired signal directly. That is, both terms do not cause in-band distortion to the linear term of Eqn (4.19). The third harmonic term, due to third order nonlinearity, is also out of band and as such can be filtered out. The only remaining term that can cause degradation to the desired signal is due to third order nonlinearity. The term falls directly in-band of the desired signal, in this case a tone at Ω_1 radians, namely $\frac{3}{4}\beta_3\alpha_1^3\cos(\Omega_1 t)$. Hence the desired signal plus the in-band degradation signal are according to Eqn (4.19)

$$\widehat{y}(t) = \underbrace{\beta_1\alpha_1\cos(\Omega_1 t)}_{\substack{\text{Linear term} \\ \text{Desired signal}}} + \underbrace{\frac{3}{4}\beta_3\alpha_1^3\cos(\Omega_1 t)}_{\substack{\text{Degradation term due to third} \\ \text{order nonlinearity}}} \tag{4.20}$$

Define the signal plus distortion to desired signal ratio based on Eqn (4.20) as

$$\rho = 10\log_{10}\left(\frac{\frac{1}{2}\left(\beta_1\alpha_1 + \frac{3}{4}\beta_3\alpha_1^3\right)^2}{\frac{1}{2}(\beta_1\alpha_1)^2}\right) = 20\log_{10}\left(1 - \frac{3}{4}\left|\frac{\beta_3}{\beta_1}\right|\alpha_1^2\right) \tag{4.21}$$

where in Eqn (4.21), we assume a compressive nonlinearity and hence $\beta_1 = |\beta_1|$ and $\beta_3 = -|\beta_3|$. The 1-dB CP occurs at $\rho = -1$ dB, that is

$$\begin{aligned}\rho &= 20\log_{10}\left(1 - \frac{3}{4}\left|\frac{\beta_3}{\beta_1}\right|\alpha_1^2\right) = -1 \Rightarrow \alpha_{1\text{ dB CP,Single tone}} \cong \alpha_1 \\ &= \sqrt{\left|\frac{4}{3}\frac{\beta_1}{\beta_3}\left(1 - 10^{-1/20}\right)\right|} = \sqrt{0.145\left|\frac{\beta_1}{\beta_3}\right|}\end{aligned} \tag{4.22}$$

Recall that the root mean square (*rms*) input power with respect to given source resistance R_s is

$$P_{ICP,\text{dB}} = 10\log_{10}\left(\frac{1}{2}\frac{\alpha^2}{R_s}\right) \tag{4.23}$$

The input power at which the 1-dB input CP occurs can be found by substituting Eqn (4.23) into Eqn (4.24)

$$P_{ICP,\text{dB}} = 10\log_{10}\left(\frac{2}{3}\left|\frac{\beta_1}{\beta_3}\right|\frac{1 - 10^{-1/20}}{R_s}\right) \tag{4.24}$$

The analysis thus far applies only to the single tone case. Next, let us extend the analysis presented above into the two-tone case. The two-tone scenario is relevant when addressing desensitization and blocking.

To proceed, consider the input signal

$$x(t) = \alpha_1\cos(\Omega_1 t) + \alpha_2\cos(\Omega_2 t) \tag{4.25}$$

Table 4.1 Components of Output Signal due to Second and Third Order Memoryless Nonlinearity and Two-Tone Input Signal

Frequency	Amplitude
DC	$\frac{1}{2}\beta_2(\alpha_1^2+\alpha_2^2)$
Ω_1	$\beta_1\alpha_1+\frac{3}{2}\beta_3\left(\frac{1}{2}\alpha_1^3+\alpha_1\alpha_2^2\right)$
Ω_2	$\beta_1\alpha_2+\frac{3}{2}\beta_3\left(\frac{1}{2}\alpha_2^3+\alpha_1^2\alpha_2\right)$
$2\Omega_{i=1,2}$	$\frac{1}{2}\beta_2\alpha_i^2$ for $i=1,2$
$3\Omega_{i=1,2}$	$\frac{1}{4}\beta_3\alpha_i^3$ for $i=1,2$
$\Omega_1\pm\Omega_2$	$\beta_2\alpha_1\alpha_2$
$2\Omega_1\pm\Omega_2$	$\frac{3}{4}\beta_3\alpha_1^2\alpha_2$
$2\Omega_2\pm\Omega_1$	$\frac{3}{4}\beta_3\alpha_1\alpha_2^2$

where the desired signal is situated at $\Omega=\Omega_1$ radians and an interfering signal at $\Omega=\Omega_2$, then the output signal can be obtained by substituting Eqn (4.22) into Eqn (4.16). The resulting output signal has components that are summarized in Table 4.1. The output signal components are made up of the desired tone along with the interferer, DC signal, harmonics, and intermodulation products. After filtering, only the degraded tone at Ω_1 remains, that is

$$\widetilde{y}(t) = \underbrace{\beta_1\alpha_1\cos(\Omega_1 t)}_{\text{Linear term}} + \underbrace{\frac{3}{2}\beta_3\left(\frac{1}{2}\alpha_1^3+\alpha_1\alpha_2^2\right)\cos(\Omega_1 t)}_{\text{Degradation term due to third order nonlinearity}} \tag{4.26}$$

As in the single tone case, the desired signal plus degradation to desired signal ratio can be found as

$$\begin{aligned}\rho &= 20\log_{10}\left(\frac{\beta_1\alpha_1+\frac{3}{2}\beta_3\left(\frac{1}{2}\alpha_1^3+\alpha_1\alpha_2^2\right)}{\beta_1\alpha_1}\right)\\ &= 20\log_{10}\left(1-\frac{3}{2}\left|\frac{\beta_3}{\beta_1}\right|\left(\frac{1}{2}\alpha_1^2+\alpha_2^2\right)\right)\end{aligned} \tag{4.27}$$

Again, the distortion to the desired signal occurs in-band due to the third order nonlinearity. Furthermore, we note that the distortion term is affected by the interferer term, as seen clearly in Eqn (4.27).

To compute the 1-dB CP, assume for the sake of simplicity that $\alpha_1=\alpha_2=\alpha$, then the relationship in Eqn (4.27) becomes:

$$\rho = 20\log_{10}\left(1-\frac{3}{2}\left|\frac{\beta_3}{\beta_1}\right|\alpha^2\right) = -1\text{ dB} \tag{4.28}$$

Then, we can compute the point at which the 1-dB CP occurs as

$$\alpha^2 = \frac{2}{3}\left|\frac{\beta_1}{\beta_3}\left(1 - 10^{-1/20}\right)\right| \Rightarrow \alpha_{1\text{ dB CP,Two tone}} = \sqrt{\frac{2}{3}\left|\frac{\beta_1}{\beta_3}\left(1 - 10^{-1/20}\right)\right|} \quad (4.29)$$

The *rms* input power at the 1-dB CP can then be estimated with respect to the source resistor R_s as

$$P_{ICP,\text{dB}} = 10\log_{10}\left(\frac{2}{3}\left|\frac{\beta_1}{\beta_3}\right|\frac{\left(1 - 10^{-1/20}\right)}{R_s}\right)\text{dBm} \quad (4.30)$$

In the receiver chain, it is desired, in most cases, that all amplifiers operate well below the input 1-dB CP to avoid causing distortion to the desired signal. The amount of backoff[3] in the absence of an interferer is typically driven by the peak to average power ratio (PAPR) of the desired signal as defined in Chapter 5. For a single tone, the PAPR is 3 dB or 0.5 on the linear scale.

At this point, it is interesting to compare the 1-dB CP of the desired signal without an interfering tone present with the 1-dB CP of desired signal tone in the presence of an interferer. Comparing Eqn (4.27) with Eqn (4.23), we obtain

$$\begin{aligned} &20\log_{10}\left(1 - \frac{3}{4}\left|\frac{\beta_3}{\beta_1}\right|\alpha^2_{1\text{ dB CP,Single tone}}\right) \\ &= 20\log_{10}\left(1 - \frac{3}{2}\left|\frac{\beta_3}{\beta_1}\right|\left(\frac{1}{2}\alpha^2_{1\text{ dB,Two tone}} + \alpha_2^2\right)\right) = -1 \end{aligned} \quad (4.31)$$

The relationship in Eqn (4.31) implies

$$\frac{3}{4}\left|\frac{\beta_3}{\beta_1}\right|\alpha^2_{1\text{ dB CP,Single tone}} = \frac{3}{2}\left|\frac{\beta_3}{\beta_1}\right|\left(\frac{1}{2}\alpha^2_{1\text{ dB CP,Two tone}} + \alpha_2^2\right) \Rightarrow$$

$$\alpha^2_{1\text{ dB CP,Two tone}} = \alpha^2_{1\text{ dB CP,Single tone}} - 2\alpha_2^2 \quad (4.32)$$

Define the *rms* power associated with the various tones:

$$\begin{aligned} P_{1\text{ dB CP,Single tone}} &= \frac{1}{2}\frac{\alpha^2_{1\text{ dB CP,Single tone}}}{R_s} \\ P_{1\text{ dB CP,Two tone}} &= \frac{1}{2}\frac{\alpha^2_{1\text{ dB CP,Two tone}}}{R_s} \\ P_2 &= \frac{1}{2}\frac{\alpha_2^2}{R_s} \end{aligned} \quad (4.33)$$

[3]Given the input–output power curve of an amplifier, backoff can be defined as the difference between the 1-dB CP or as given in some literature between the saturation point of the amplifier and the desired *rms* operating point of the output signal.

Then, according to Eqns (4.32) and (4.33), we claim that

$$P_{1\ \text{dB CP,Single tone}} = P_{1\ \text{dB CP,Two tone}} + 2P_2 \tag{4.34}$$

It is obvious from Eqn (4.34), that the 1-dB CP due to single tone occurs at a higher level than that of a two tone, which is what one would expect.[4] In the event where the two tones are equal in power, that is $P_{1\ \text{dB CP,Two tone}} = P_2$, then we conclude that in the linear scale $P_{1\ \text{dB CP,Single tone}} = 3P_{1\ \text{dB CP,Two tone}}\big|_{P_{1\ \text{dB CP,Two tone}}=P_2}$ or the 1-dB CP due to single tone occurs at approximately 4.77 dB higher than that of a two tone. We further observe that as the interfering tone increases in power at twice the rate as that of the desired signal, it eventually overwhelms the desired signal by reducing its overall average gain and causing a phenomenon known as desensitization. As the desired signal gain approaches 0 dB, the signal is said to be blocked, and the interferer is often referred to as a *blocker*. Given the desired signal power P_{desired} and the blocker signal power P_{blocker}, then given $P_{\text{sensitivity}}$ as the sensitivity of the desired signal, define blocking as the ratio

$$\varsigma_{\text{dB}} = 10\log_{10}\left(\frac{P_{\text{blocker}}}{P_{\text{sensitivity}}}\right) \tag{4.35}$$

such that P_{blocker} is the smallest blocker signal power that degrades the desired signal performance to a level equivalent to the sensitivity of the receiver.

Thus far, the 1-dB CP that we have been discussing relates to the input power at which the devices gain drops by 1 dB from linear gain. Similarly, the 1-dB output compression point $P_{1\ \text{dB OCP,Single tone}}$ can be defined in terms of the 1-dB input compression point simply as

$$P_{1\ \text{dB OCP,Single tone}} = P_{1\ \text{dB CP,Single tone}} + (G_{\text{dB}} - 1) \tag{4.36}$$

where G_{dB} is the device gain in dB. Note that the relationship in Eqn (4.36) differs from that of an input–output power gain relationship where the output signal power is equal to that of the input signal power plus the gain, in this case G_{dB}. The difference is in the −1 dB subtracted from the gain in Eqn (4.36).

Finally, the 1-dB compression dynamic range is defined as the difference between the 1-dB compression point and the receiver noise floor [2,3], that is

$$\begin{aligned}\Delta_{1-\text{dB Dynamic range}} &= P_{1\ \text{dB CP,Single tone}} - P_{\text{noise floor}} \\ &= P_{1\ \text{dB CP,Single tone}} \\ &\quad + 174\ \text{dBm/Hz} - NF - 10\log_{10}(B)\end{aligned} \tag{4.37}$$

[4]It is natural to conclude that the 1-dB compression point of a modulated signal occurs at a lower point than the single-tone 1-dB CP.

where $P_{\text{noise floor}}$ is the receiver noise floor, NF is the receiver's noise figure, and B is the channel bandwidth.

In a similar manner, the desensitization dynamic range is defined as the receiver's ability to process a weak desired signal in the presence of a strong narrowband out-band interferer. The desensitization dynamic range is defined as the difference between the blocking signal power P_{blocker} and the receiver noise floor, that is

$$\begin{aligned}\Delta_{\text{Desensitization dynamic range}} &= P_{\text{blocker}} - P_{\text{noise floor}} \\ &= P_{\text{blocker}} + 174\ \text{dBm/Hz} - NF - 10\log_{10}(B) \end{aligned} \quad (4.38)$$

where NF is the receiver system noise figure and B is the noise equivalent bandwidth. Recall that the sensitivity of a receiver at room temperature is defined as

$$\begin{aligned} P_{\text{sensitivity}} &= -174\ \text{dBm/Hz} + 10\log_{10}(B_n) + NF_{\text{cascade}} + CNR \\ P_{\text{sensitivity}} - CNR &= -174\ \text{dBm/Hz} + 10\log_{10}(B_n) + NF \end{aligned} \quad (4.39)$$

Then substituting Eqn (4.39) into Eqn (4.38), we can define the desensitization dynamic range in terms of blocker signal power, receiver sensitivity, and CNR:

$$\Delta_{\text{Desensitization dynamic range}} = P_{\text{blocker}} - P_{\text{sensitivity}} + CNR \quad (4.40)$$

The desensitization dynamic range is then a theoretical measure of determining the receiver's ability to process a desired signal in the presence of a blocker signal only.

4.2.2 Harmonics and intermodulation distortion

Intermodulation (IM) distortion is caused by external strong nondesired signals or interferers passing through a nonlinear device thus creating an in-band nonharmonic distortion in the desired signal band. In most practical RF receivers, an IM product is mainly due to second or third order nonlinearity. Figure 4.5 depicts

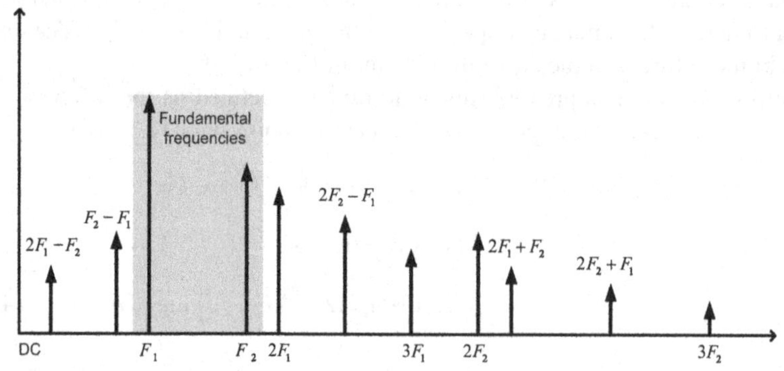

FIGURE 4.5

Ideal spectrum of two-tone signals and associated second and third order harmonics and intermodulation products. Note that the relative magnitudes between the tones have not been maintained whereas their relative locations have. (For color version of this figure, the reader is referred to the online version of this book.)

second and third order harmonics and *IM* products along with the fundamental tones. One method of lessening the negative impact of *IM* products on SNR can be accomplished by preselection or filtering prior to the LNA. An alternate method would be to increase the linearity of the system. In practical receivers, a combination of preselection as well as proper circuit design employed to increase the system linearity are used to minimize the impact of degradation on the desired signal.

4.2.2.1 Degradation effects due to second order nonlinearity

Second order nonlinearity is a major concern in the design of direct conversion receivers. In this type of architecture, depicted in Figure 4.6, the desired signal is modulated down directly from *RF* to baseband without the intermediary intermediate frequency (*IF*) step common in a heterodyne architecture. To further illustrate this point, consider a nonlinear system as shown in Figure 4.7. A degradation due to second order nonlinearity in response to a two-tone input could manifest itself as an *IM* product at the output. Then, depending on the frequencies of the two tones and the bandwidth and center frequency of the desired signal, the resulting *IM* product could fall into the desired signal band. In single-balanced mixers, for example, transistor mismatches coupled with the deviation of the *LO* duty cycle from 50% produces asymmetry in the circuitry allowing for certain signals before the mixer to feedthrough without mixing into the baseband as shown in Figure 4.8.

To further illustrate this phenomenon, assume that the received signal exhibits a certain amount of amplitude modulation (*AM*) variations due to propagation distortion or transmit and receive filtering, then the received signal can be expressed in the simple form:

$$x(t) = [\Gamma + \gamma \cos(\omega_m t)][A \cos(\omega_c t) - B \sin(\omega_c t)] \tag{4.41}$$

where $\Gamma(t) \approx \Gamma$ is a slow varying signal assumed to be constant over an extended period of time, and $\gamma\cos(\omega_m t)$ represents a low frequency *AM* signal as stated

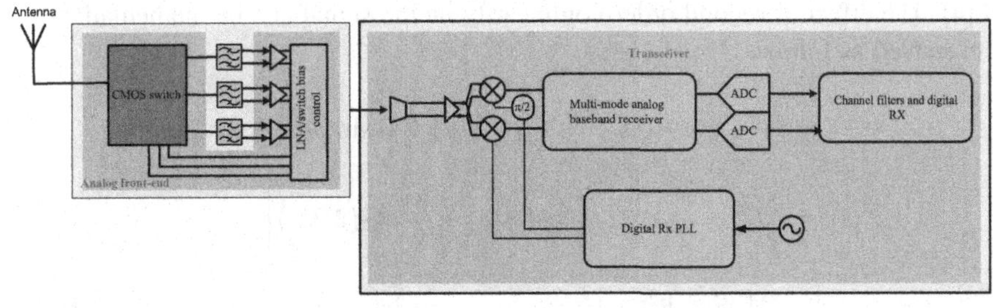

FIGURE 4.6

Conceptual direct conversion receiver. (For color version of this figure, the reader is referred to the online version of this book.)

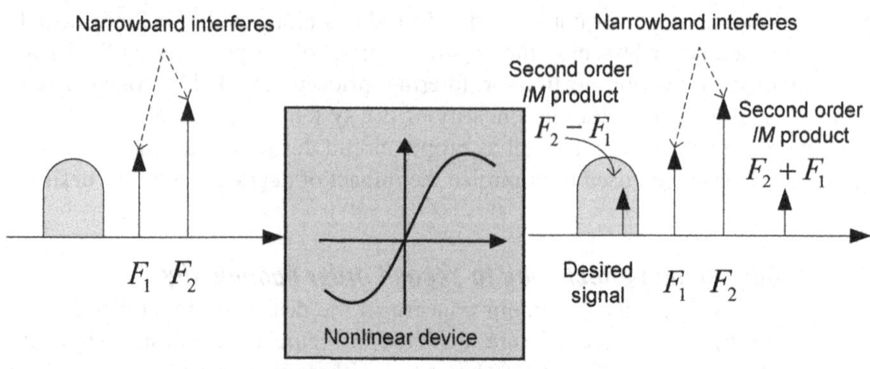

FIGURE 4.7

IM and harmonic distortion due to second order nonlinearity. *IM*, Intermodulation. (For color version of this figure, the reader is referred to the online version of this book.)

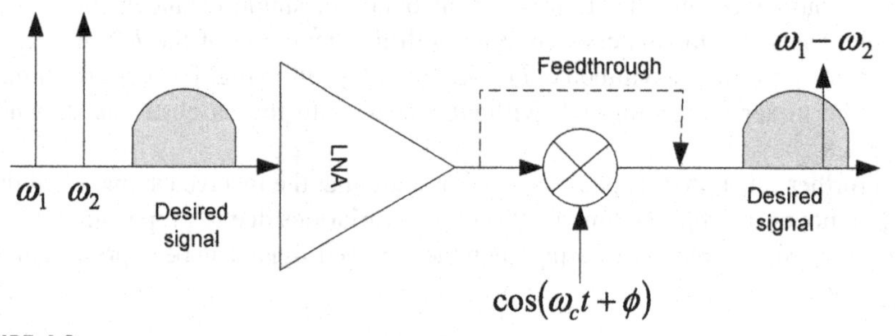

FIGURE 4.8

Effect of mixer feedthrough on the received desired baseband signal due to strong narrowband interference.

in [4]. The effect of second order nonlinearity on the signal can be mathematically represented as follows:

$$x^2(t) = [\Gamma + \gamma \cos(\omega_m t)]^2 [A \cos(\omega_c t) - B \sin(\omega_c t)]^2$$

$$x^2(t) = \left[\Gamma^2 + \frac{\gamma^2}{2} + 2\Gamma\gamma \cos(\omega_m t) + \frac{\gamma^2}{2}\cos(2\omega_m t)\right]$$
$$\times \left[\frac{A^2 + B^2}{2} + \frac{A^2}{2}\cos(2\omega_c t) - \frac{B^2}{2}\cos(2\omega_c t) - AB \sin(2\omega_c t)\right] \tag{4.42}$$

In Eqn (4.42), the product $(A^2 + B^2)\Gamma\gamma\cos(\omega_m t)$ is of great interest.[5] When fed through the mixer, as shown in Figure 4.8, the term appears at baseband in the desired signal band as distortion. This demodulation of *AM* components serves to further corrupt the received signal.

The second order nonlinearity of a device can be described using the second order intercept point. To start, let the input *rms* power of a single tone of a two-tone signal fed to a nonlinear system or device be

$$P_i = \frac{1}{2}\frac{\alpha_i^2}{R_s}, \quad i = 1,2 \tag{4.43}$$

where the tones are centered at $\Omega_i, i = 1,2$ and R_s is the source resistor. The *rms* output power of the nonlinear system or device at the fundamental frequency is

$$P_o = \frac{1}{2}\frac{\beta_1^2\alpha_i^2}{R_L}, \quad i = 1,2 \tag{4.44}$$

according to the polynomial presented in Eqn (4.12), where R_L is the load resistor. Let the distortion power of the *IM* product delivered to a load resistor R_L according to Table 4.1 be

$$P_d = \frac{(\beta_2\alpha_1\alpha_2)^2}{2R_L} \tag{4.45}$$

The *IM* distortion under consideration is due to the frequency $|\Omega_1 - \Omega_2|$, which is assumed to fall in-band of the desired signal as shown previously in Figure 4.8. Next, in order to simplify the math, assume the input and output powers to be normalized to their respective source and load resistors and that the two tones have equal power at the input of the nonlinear device, that is

$$P = P_1 = P_2 \text{ or } \alpha = \alpha_1 = \alpha_2 \tag{4.46}$$

Define the signal amplitude due to the fundamental frequency as

$$\Psi_{\Omega_{1,2}} = |\beta_1|\alpha \tag{4.47}$$

And define the signal amplitude due to the *IM* product as

$$\Psi_{\Omega_1-\Omega_2} = |\beta_2|\alpha_1\alpha_2 = |\beta_2|\alpha^2 \tag{4.48}$$

where absolute values are used to simplify the mathematical assumptions. Define the second order *input-referred second order intercept point* (*IIP*2) as the point for which the *IM* product magnitude equals that of the fundamental signal magnitude as shown in Figure 4.9 depicting the intercept point curve along the *IIP*2

---

[5] In general, given an expression of the form: $[A\cos(\Omega_1 t) + B\cos(\Omega_2 t)]^N$, then the polynomial expansion includes terms of the type $A^k B^l \cos(k\Omega_1 t \pm l\Omega_2 t)$ for $k, l = 0, 1, \cdots, N$, $k + l \leq N$.

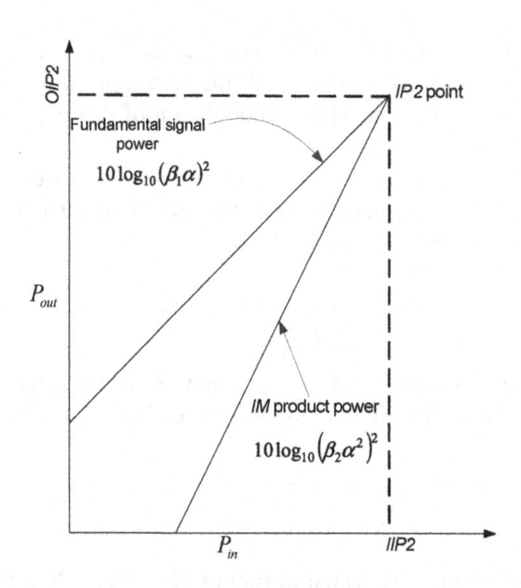

FIGURE 4.9

IIP2 and *OIP2* are the projections to the *x* and *y* axis of the second order intercept point for which the IM product amplitude equals that of the fundamental output signal amplitude. (For color version of this figure, the reader is referred to the online version of this book.)

axis and the *output-referred second order intercept point* (*OIP*2). This simply implies that

$$\Psi_{\Omega_{1,2}} = \Psi_{\Omega_1-\Omega_2} \Rightarrow |\beta_1|\alpha = |\beta_2|\alpha^2 \Rightarrow IIP2 \equiv \alpha = \frac{|\beta_1|}{|\beta_2|} \tag{4.49}$$

Next, consider the ratio of Eqn (4.47) to Eqn (4.48), we obtain

$$\frac{\Psi_{\Omega_{1,2}}}{\Psi_{\Omega_1-\Omega_2}} = \frac{|\beta_1|\alpha}{|\beta_2|\alpha^2} = \frac{|\beta_1|}{|\beta_2|}\frac{1}{\alpha} = \frac{IIP2}{\alpha} \tag{4.50}$$

Then the input referred *IM* product, which is the value of the *IM* product referred to the input of the system or the device is

$$IM_2 = \frac{\Psi_{\Omega_1-\Omega_2}}{|\beta_1|} = \frac{\alpha^2}{IIP2} \tag{4.51}$$

The power of the second order *IM* product in Eqn (4.51), which is itself a tone, is

$$P_{IM_2} = \frac{1}{2}(IM_2)^2 = \frac{1}{2}\frac{\alpha^4}{(IIP2)^2} = \frac{4P_i^2}{2(IIP2)^2} = \frac{P_i^2}{P_{IIP2}} \tag{4.52}$$

Expressing the linear relationship in Eqn (4.52), in dB we obtain

$$P_{IM_2,\text{dB}} = 2P_{i,\text{dBm}} - P_{IIP2,\text{dBm}} \tag{4.53}$$

Note that the $P_{IIP2,\text{dBm}}$ is a made-up quantity used to specify second order nonlinearity. The output power of the system will compress long before the input power reaches $P_{IIP2,\text{dBm}}$. Then if the input power is Δ_{dB} below$P_{IIP2,\text{dBm}}$, we obtain

$$P_{i,\text{dBm}} = P_{IIP2,\text{dBm}} - \Delta_{\text{dB}} \tag{4.54}$$

Substituting Eqn (4.54) into Eqn (4.53), the following relationship emerges:

$$\begin{aligned} P_{IM_2,\text{dB}} &= 2P_{i,\text{dBm}} - P_{IIP2,\text{dBm}} \\ &= 2\left(P_{IIP2,\text{dBm}} - \Delta_{\text{dB}}\right) - P_{IIP2,\text{dBm}} = P_{IIP2,\text{dBm}} - 2\Delta_{\text{dB}} \end{aligned} \tag{4.55}$$

The relationship in Eqn (4.55) implies that the input referred *IM* power is $2\Delta_{\text{dB}}$ below$P_{IIP2,\text{dBm}}$.

In most practical cases, however, the two tones impinging on the receiver have different powers. Hence, if we let $P_{i1,\text{dBm}}$ be the input power of the first tone and $P_{i2,\text{dBm}}$ be the input power of the second tone, then Eqn (4.53) can be reexpressed as

$$P_{IM_2,\text{dB}} = P_{i_1,\text{dBm}} + P_{i_2,\text{dBm}} - P_{IIP2,\text{dBm}} \tag{4.56}$$

As will be seen later, second order distortion is particularly problematic in direct conversion receivers. The key contributors to second order distortion in this architecture are the *RF*-to-baseband mixer as well as the baseband gain stages. Hence, the design of a direct conversion receiver requires a high system *IIP*2.

Associated with the system *IIP*2 is the spur-free dynamic range due to second order nonlinearity. The dynamic range is defined as the difference between the receiver's *IIP*2 and its respective noise floor, that is

$$\Delta_{SFDR-IIP2} = \frac{1}{2}\left(IIP2 + 174 - NF - 10\log_{10}(B)\right) \tag{4.57}$$

where *NF* is the system noise figure, and *B* is the bandwidth. In order to obtain the system *IIP*2, an expression similar to the Friis cascaded noise figure expression can be found. Consider the cascade of analog *RF* blocks depicted in Figure 4.10. The system *IIP*2 of a cascade of *N-RF* blocks can be expressed as

$$IIP2_{\text{system}} = \left(\frac{1}{\sqrt{\frac{1}{IIP2_1}} + \sqrt{\frac{G_1}{IIP2_2}} + \ldots + \sqrt{\frac{G_1 G_2 \ldots G_{N-1}}{IIP2_N}}}\right)^2 \tag{4.58}$$

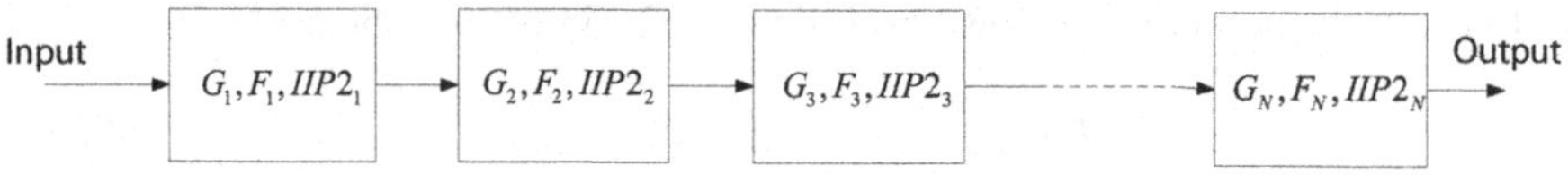

Input → $G_1, F_1, IIP2_1$ → $G_2, F_2, IIP2_2$ → $G_3, F_3, IIP2_3$ - - - → $G_N, F_N, IIP2_N$ → Output

FIGURE 4.10

Cascade of *N*-radio frequency blocks with their respective linear gains, noise factors, and second order nonlinearities.

where $\{G_1, G_2, \ldots G_N\}$ and $\{IIP2_1, IIP2_2, \cdots IIP2_N\}$ are the respective linear gains and *IIP*2s of the various *RF* blocks. In general, the cascaded input-referred intercept point for any nonlinearity can be expressed as

$$\left(\frac{1}{IIPx_{\text{system}}}\right)^{\frac{1}{2}(x-1)} = \left(\frac{1}{IIPx_1}\right)^{\frac{1}{2}(x-1)} + \sum_{n=2}^{N}\left(\frac{\prod_{i=1}^{n-1} G_i}{IIPx_n}\right)^{\frac{1}{2}(x-1)} \tag{4.59}$$

where x in Eqn (4.59) is the order of the nonlinearity.

The relationship in Eqn (4.58) is true when all the narrowband interfering signals happen to be in the passband of each *RF* block. In all other cases, filtering and selectivity have to be accounted for.

EXAMPLE 4.1 DERIVATION OF SYSTEM *IIP2*

Given the occupied signal bandwidth of a certain modulation scheme to be 9 MHz with receiver noise figure of 4.5 dB. The required *CNR* needed to decode the desired signal with acceptable BER is 5 dB. Assume that two interfering narrowband signals are impinging on the receiver's band definition filter at −25 dBm signal power each. The two signals produce an *IM* product in the desired signal band. What is the required system *IIP2* in order not to degrade the receiver's noise floor by more than 1 dB at room temperature?

What is the impact on *IIP2* if you allow the degradation of the noise floor to vary between 0.5 and 3.5 dB? Finally, what is the impact on *IIP2* as you vary the system noise figure between 2 and 5 dB? In each case, keep all parameters at their given original values.

The noise floor of a receiver can be computed as

$$\begin{aligned} P_{\text{noise floor,dBm}} &= -174\ \text{dBm/Hz} + NF + 10\log_{10}(B) \\ &= -174\ \text{dBm/Hz} + 4.5\ \text{dB} + 10\log_{10}\left(9\times 10^{6}\right) \\ &= -99.96\ \text{dBm} \end{aligned} \tag{4.60}$$

The problem states that the degradation due to the *IM* product is $P_{\text{degradation}} = 1$ dB. That is, the degraded noise floor becomes −98.96 dBm. The *IM* product signal power can then be obtained as

$$\begin{aligned} P_{IM_2,\text{dB}} &= 10\log_{10}\left(10^{(P_{\text{noise floor,dBm}}+P_{\text{degradation}})/10} - 10^{P_{\text{noise floor,dBm}}/10}\right) \\ P_{IM_2,\text{dB}} &= 10\log_{10}\left(10^{-98.96/10} - 10^{-99.96/10}\right) = -105.826\ \text{dBm} \end{aligned} \tag{4.61}$$

EXAMPLE 4.1 DERIVATION OF SYSTEM *IIP2*—cont'd

Next, we compute the *IIP2* according to Eqn (4.53) as

$$P_{IM_2,dB} = 2P_{i,dBm} - P_{IIP2,dBm} \Rightarrow P_{IIP2,dBm} = 2P_{i,dBm} - P_{IM_2,dB}$$

$$P_{IIP2,dBm} = 2 \times (-25\ \text{dBm}) + 105.826 = 55.83\ \text{dBm} \tag{4.62}$$

Using the methodology introduced above, we allow the degradation to the noise floor to vary between 0.5 and 3.5 dB. Changing the degradation from 3 to 3.5 dB decreases the input-referred third order intercept point (*IIP3*) requirement by approximately 1 dB. On the other hand, increasing the noise floor degradation from 0.5 to 1 dB decreases the *IIP3* requirement by more than 3 dB as shown in Figure 4.11. Therefore, we conclude that the impact of increasing the noise degradation on the *IIP2* requirement diminishes with increasing the amount of degradation. On the other hand, allowing the noise figure of the receiver to degrade by 0.5 dB decreases the *IIP2* requirement also by the same proportion as seen in Figure 4.12.

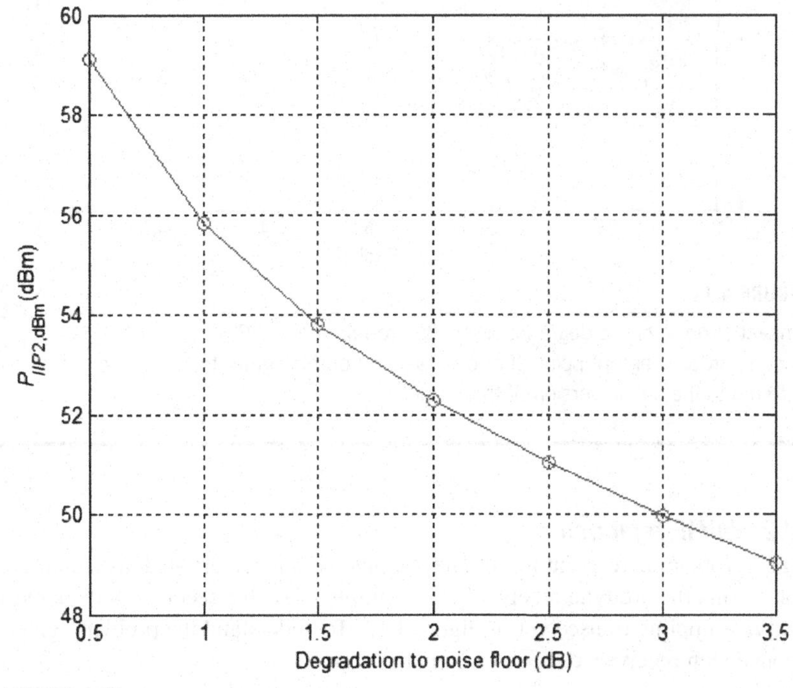

FIGURE 4.11

Impact of noise floor degradation on *IIP2* requirement. *IIP2*, input-referred second order intercept point. (For color version of this figure, the reader is referred to the online version of this book.)

Continued

EXAMPLE 4.1 DERIVATION OF SYSTEM *IIP2*—cont'd

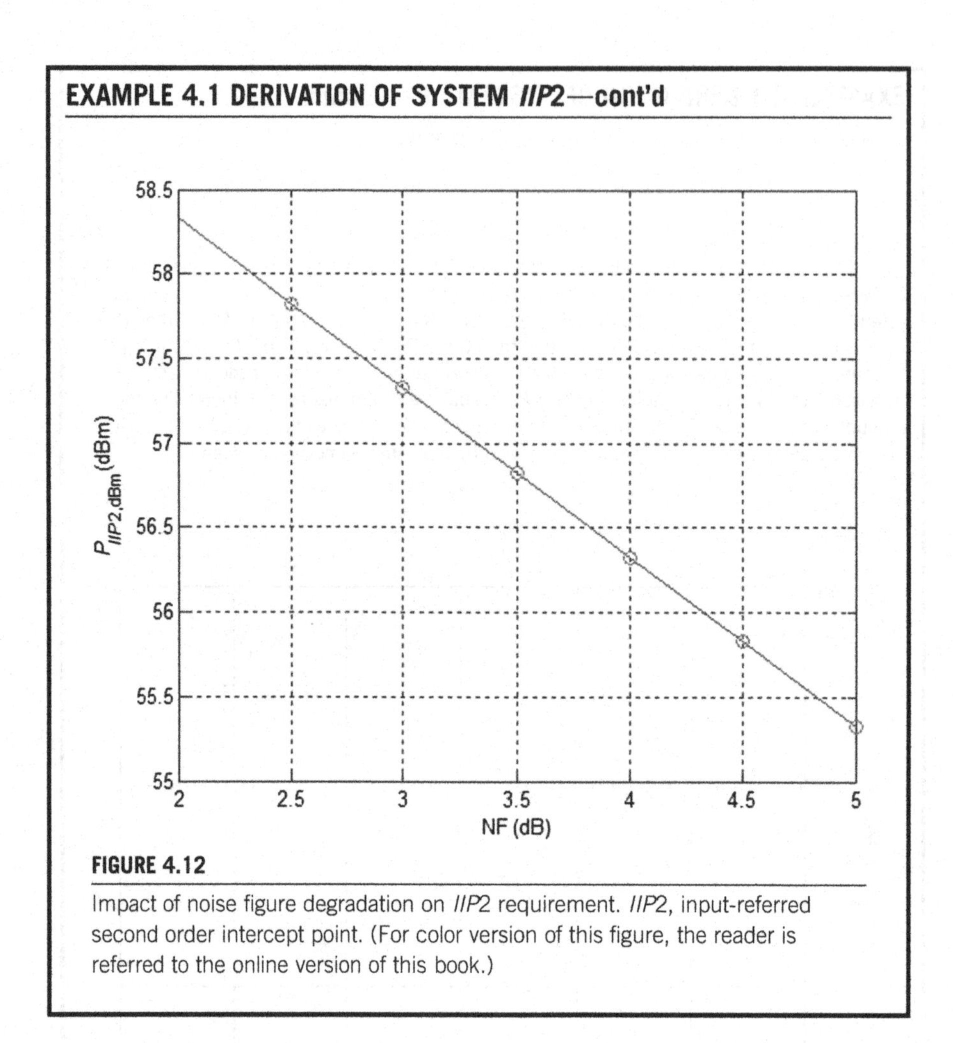

FIGURE 4.12

Impact of noise figure degradation on *IIP2* requirement. *IIP2*, input-referred second order intercept point. (For color version of this figure, the reader is referred to the online version of this book.)

4.2.2.2 *Half* IF *rejection*

The ½ *IF* problem is relevant to nondirect conversion receivers such as a dual conversion or superheterodyne receiver. One example of such a receiver is depicted as an ideal *IF*-sampling transceiver in Figure 4.13. To understand the problem, given a dual conversion receiver, consider a signal $x(t)$ given as

$$x(t) = A\cos\left[\left(\Omega_c + \Omega_{IF/2}\right)t\right] + B\cos[(\Omega_c + \Omega_{IF})t] \tag{4.63}$$

where Ω_c is the center frequency at *RF* and Ω_{IF} is the *IF*. Define $\Omega_{IF/2}$ as the *RF* frequency related to the center frequency such that $\Omega_{IF/2} = \Omega_c + \frac{1}{2}\Omega_{IF}$. Furthermore, define the first *LO* frequency as a function of the center and *IF* frequencies as $\Omega_{LO} = \Omega_c \pm \Omega_{IF}$.

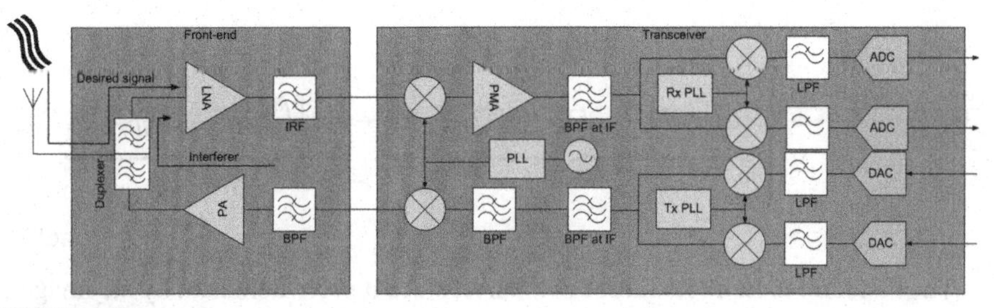

FIGURE 4.13

IF-sampling dual conversion transceiver. *IF*, intermediate frequency; BPF, bandpass filter; PMA, post-mixer filter. (For color version of this figure, the reader is referred to the online version of this book.)

As mentioned earlier, the ½ *IF* degradation is caused by a fourth order nonlinearity manifesting itself as a second order distortion. To expound on this statement, consider the nonlinear system depicted in Figure 4.14 and particularly the term $x^4(t)$. Recall that a fourth order polynomial can be expanded according to Eqn (4.241) as

$$(a+b)^4 = a^4 + 4a^3b + 6a^2b^2 + 4ab^3 + b^4 \qquad (4.64)$$

Examining the term $6a^2b^2$ given the input $x(t)$, we obtain

$$\widehat{y}(t) = 6\beta_4 A^2 B^2 \cos^2\left[\left(\Omega_c + \Omega_{IF/2}\right)t\right]\cos^2[(\Omega_c + \Omega_{IF})t] \qquad (4.65)$$

Using the trigonometric relation $\cos^2 \alpha = (\cos 2\alpha + 1)/2$, the relationship in Eqn (4.65) becomes

$$\widehat{y}(t) = \frac{3\beta_4 A^2 B^2}{2}\left\{\cos\left[2\left(\Omega_c + \Omega_{IF/2}\right)t\right] + 1\right\}\{\cos[2(\Omega_c + \Omega_{IF})t] + 1\} \qquad (4.66)$$

In Eqn (4.66), we are interested in the product

$$\tilde{y}(t) = \frac{3\beta_4 A^2 B^2}{2}\cos\left[2\left(\Omega_c + \Omega_{IF/2}\right)t\right]\cos[2(\Omega_c + \Omega_{IF})t] \qquad (4.67)$$

Using the trigonometric relations provided in Eqns (4.227) and (4.228), the relationship in Eqn (4.67) can be further manipulated as

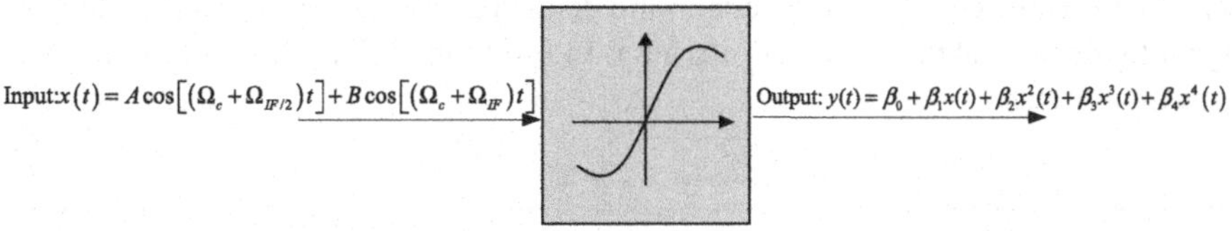

Input: $x(t) = A\cos\left[\left(\Omega_c + \Omega_{IF/2}\right)t\right] + B\cos\left[\left(\Omega_c + \Omega_{IF}\right)t\right]$

Output: $y(t) = \beta_0 + \beta_1 x(t) + \beta_2 x^2(t) + \beta_3 x^3(t) + \beta_4 x^4(t)$

FIGURE 4.14

Impact of fourth order nonlinearity on ½ *IF* distortion. *IF*, intermediate frequency. (For color version of this figure, the reader is referred to the online version of this book.)

$$\tilde{y}(t) = \frac{3\beta_4 A^2 B^2}{4}\left\{\underbrace{\cos\left[\left(4\Omega_c + 2\Omega_{IF} + 2\Omega_{IF/2}\right)t\right]}_{\text{First term}} + \underbrace{\cos\left[\left(2\Omega_{IF} - 2\Omega_{IF/2}\right)t\right]}_{\text{Second term}}\right\} \tag{4.68}$$

The first term in Eqn (4.68) is of no interest since it gets filtered out by the image rejection filter. The second term, however, deserves closer examination.

Note that the frequency of the second term is simply:

$$2\Omega_{IF} - 2\Omega_{IF/2} = 2\Omega_{IF} - \Omega_{IF} = \Omega_{IF} \tag{4.69}$$

since $\Omega_{IF/2} = \frac{1}{2}\Omega_{IF}$.

The analysis thus far implies that if an interferer occurs at $\Omega_c \pm \frac{1}{2}\Omega_{IF}$, it will overlap with the desired signal after mixing. Hence, an interesting design trade emerges. Depending on the band of operation, it would seem reasonable that increasing the *IF* allows the front-end filters such as the BDF to further attenuate the ½ *IF* interferer. However, this is not without implications on the *IF* filter and ADC performance in an *IF* sampling receiver for example. Therefore, the designer must perform a trade-off between a higher *IF*, *IIP2*, co-channel interference, and other design implications at *IF* or baseband.

Let the desired signal power P_{signal} be twice as large as the sensitivity $P_{\text{sensitivity}}$. Then let the smallest interfering signal $P_{IF/2}$ impinge on the antenna at half the *IF* away from the center frequency of the desired signal. Assume that the interferer degrades the power of the desired signal to that of $P_{\text{sensitivity}}$, then the ½ *IF* rejection may be defined in dB as the ratio

$$\Upsilon = 10\log_{10}\left(\frac{P_{IF/2,\text{linear}}}{P_{\text{sensitivity,linear}}}\right) \tag{4.70}$$

Thus, increasing the *IF* interferer by 1 dB serves to increase the degradation by 2 dB since this is a second order phenomenon. Therefore, given the system *IIP2*, the ½ *IF* rejection can be obtained as

$$\Upsilon = \frac{1}{2}\left(IIP2 - P_{\text{sensitivity}} + CCR\right) \tag{4.71}$$

where *CCR* is the co-channel rejection ratio defined as the ratio between the wanted signal to the unwanted blocker or interferer. In FM terms *CCR* is also referred to as the capture ratio.[6]

[6]The capture ratio is associated with the capture effect in FM. If two simultaneous signals, one strong and one weak, are present at the input of a limiting amplifier, the "limiting operation" further strengthen the strong signal at the expense of weakening the weak signal. Thus, the strong signal "captures" the weak signal.

Finally, recall that since the sensitivity can be expressed in terms of *CCR* as

$$\begin{aligned} P_{\text{sensitivity}} &= -174 \text{ dBm/Hz} + NF + 10 \log_{10}(B) + CNR \\ &= P_{\text{noise floor}} + CNR = P_{\text{noise floor}} - CCR\big|_{CCR=-CNR} \end{aligned} \tag{4.72}$$

Hence, according to Eqn (4.72), we can express Eqn (4.71) simply as

$$\Upsilon = \frac{1}{2}\left(IIP2 - P_{\text{noise floor}}\right) \tag{4.73}$$

The ½-*IF* problem will be encountered again when discussing the dual conversion architecture.

4.2.2.3 Degradation effects due to third order nonlinearity

In a similar manner to the analysis presented above, the third order intercept point is an important performance parameter of the receiver that specifies the amount of distortion that affects the desired signal due to two narrowband interferers. This phenomenon is illustrated in Figure 4.15 where the input to a nonlinear device is a desired bandlimited signal along with two-tone interferers. The output of the nonlinear device displays the desired signal with the two-tone interferers along with two *IM* products situated at $2F_1 - F_2$ and $2F_1 + F_2$, respectively. One of these *IM* products, say, for example, $2F_1 - F_2$, could very well fall in band of the desired signal as shown in Figure 4.15.

Consider the output of a third order nonlinear system $y(t)$, as specified in Eqn (4.16), in response to a two-tone input signal as defined in Eqn (4.22):

$$y(t) = \left\{\beta_1\alpha_1 + \frac{3}{2}\beta_3\left(\frac{1}{2}\alpha_1^3 + \alpha_1\alpha_2^2\right)\right\}\cos(\Omega_1 t) + \frac{3}{4}\beta_3\alpha_1^2\alpha_2\cos[(2\Omega_1 - \Omega_2)t] + \cdots \tag{4.74}$$

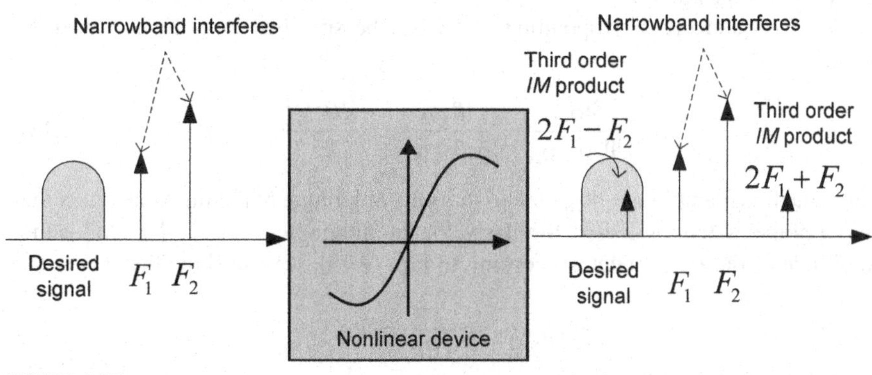

FIGURE 4.15

IM and harmonic distortion due to third order nonlinearity. *IM*, Intermodulation. (For color version of this figure, the reader is referred to the online version of this book.)

Let the two tones have equal input powers, that is let $\acute{\alpha}=\acute{\alpha}_1=\acute{\alpha}_2$ and further assume that the polynomial coefficients β_1 and β_3 are such $|\beta_1| >> |\beta_3|$, then the relationship in Eqn (4.74) can simply be reexpressed as

$$y(t) = \underbrace{\beta_1 \alpha \cos(\Omega_1 t)}_{\text{Linear term}} + \frac{9}{4}\beta_3 \alpha^3 \cos(\Omega_1 t) + \frac{3}{4}\beta_3 \alpha^3 \cos[(2\Omega_1 - \Omega_2)t] + \cdots \tag{4.75}$$

Similar to the definition of *IIP*2, define *IIP*3 as the point where the magnitude of the linear component of Eqn (4.75), that is, of $\cos(\Omega_1 t)$, is equal to the *IM* product term relevant to $\cos[(2\Omega_1 - \Omega_2)t]$. According to Eqn (4.75), this implies that

$$|\beta_1|\alpha = \frac{3}{4}|\beta_3|\alpha^3 \tag{4.76}$$

The *IIP*3 point can then be found according to Eqn (4.76) as

$$IIP3 \equiv \alpha|_{\alpha=IP3} = \sqrt{\frac{4}{3}\frac{|\beta_1|}{|\beta_3|}} \tag{4.77}$$

The projection of the intercept point onto the x axis is known as the *IIP*3 point as illustrated in Figure 4.16. The projection of the same intercept point onto the y axis is the output-referred third order intercept point (*OIP*3). Upon close examination of Figure 4.16, we observe that the amplitude of the *IM* product increases three times faster on the logarithmic scale due to cubic nonlinearity compared to the amplitude of the fundamental input tone α. Note that, like the *IIP*2 and *OIP*2 points, the *IIP*3 and *OIP*3 points are also fictitious points and that the system will compress long before it reaches the power level indicated by these points. Having said that, however, *IIP*3 and *OIP*3 are performance measures by which a system or device nonlinearity can be specified.

The *IIP*3 and the corresponding *OIP*3 can be specified as follows. Consider the ratio

$$\frac{\Psi_{\Omega_{1,2}}}{\Psi_{2\Omega_1-\Omega_2}} = \frac{|\beta_1|\alpha}{\frac{3}{4}|\beta_3|\,\alpha^3} = \frac{(IIP3)^2}{\alpha^2} \tag{4.78}$$

where the linear amplitude $\Psi_{\Omega_{1,2}} = |\beta_1|\alpha$ is the amplitude of the fundamental at the output of the system or device. Similarly, the amplitude $\Psi_{2\Omega_1-\Omega_2} = \frac{3}{4}|\beta_3|\,\alpha^3$ is the amplitude of the *IM* product. According to Eqn (4.78), we can show that

$$\begin{aligned}\Psi_{2\Omega_1-\Omega_2} &= \frac{\alpha^2}{(IIP3)^2}\Psi_{\Omega_{1,2}} \\ &= \frac{\alpha^3}{(IIP3)^2}|\beta_1|\end{aligned} \tag{4.79}$$

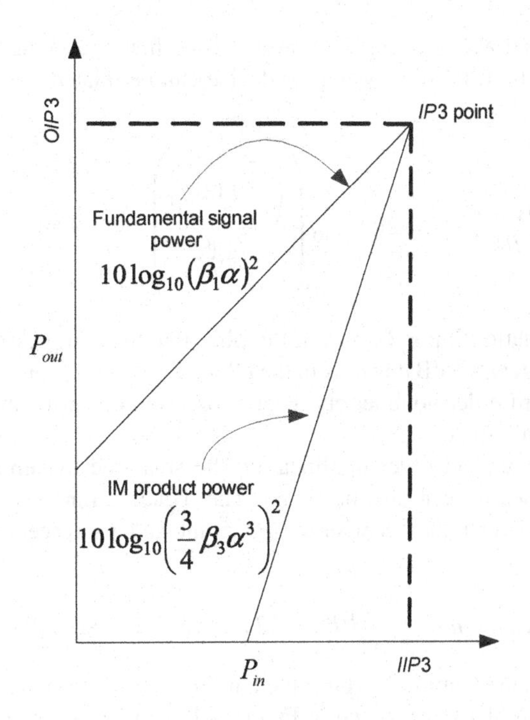

FIGURE 4.16

*IIP*3 and *OIP*3 are the projections to the *x* and *y* axis of the third order intercept point for which the *IM* product amplitude equals that of the fundamental output signal amplitude. *IM*, intermodulation; *IIP*3, input-referred third order intercept point; *OIP*3, output-referred third order intercept point. (For color version of this figure, the reader is referred to the online version of this book.)

Similar to the second order case, the input referred *IM* product for third order nonlinearity can be found by taking Eqn (4.79) and referring to the input of the system or device, that is

$$IM_3 = \frac{\Psi_{2\Omega_1-\Omega_2}}{|\beta_1|} = \frac{\alpha^3}{(IIP3)^2} \tag{4.80}$$

The power of the *IM* product can be expressed in dB as

$$P_{IM_3,\text{dB}} = 3P_{in,\text{dBm}} - 2P_{IIP3,\text{dBm}} \tag{4.81}$$

Thus far, we have assumed that both tones have equal power. In the case where the two tones are not equal in power, the *IM*-product power at the frequency $F_{IM_3} = \pm 2F_1 \pm F_2$ where F_1 and F_2 are the frequencies of the first and second tone, respectively, is specified as

$$P_{IM_3,\text{dB}} = 2P_{in,1,\text{dBm}} + P_{in,2,\text{dBm}} - 2P_{IIP3,\text{dBm}} \tag{4.82}$$

where $P_{in,1,\text{dBm}}$ and $P_{in,2,\text{dBm}}$ are the powers of the first and second tones, respectively. Note that the IPP3 of a system or device can be related *theoretically* to the 1-dB CP as:

$$20\log_{10}\left(\frac{\alpha_{1\text{ dB CP}}}{IIP3}\right) = 20\log_{10}\left(\frac{\sqrt{0.145\left|\frac{\beta_1}{\beta_3}\right|}}{\sqrt{\frac{4}{3}\left|\frac{\beta_1}{\beta_3}\right|}}\right) = -9.6357\text{ dB} \tag{4.83}$$

That is, the relationship in Eqn (4.83) implies that the compression by 1 dB occurs at approximately 9.6 dB down from the $IIP3$ point. The compression in this case is only due to third order nonlinearity. Higher order nonlinearity, higher than third order, is not taken into account.

Similar to the second order nonlinearity, the spur-free dynamic range due to third order nonlinearity can also be found. The dynamic range can be defined as the difference between the receiver's $IIP3$ point and its respective noise floor, that is

$$\Delta_{SFDR-IIP3} = \frac{2}{3}(IIP3 + 174 - NF - 10\log_{10}(B)) \tag{4.84}$$

Next, based on the relationship presented in Eqn (4.59), the *coherent* system $IIP3$ of a cascade of RF blocks as shown in Figure 4.17 can be expressed as

$$IIP3_{\text{system}} = \frac{1}{\frac{1}{IIP3_1} + \frac{G_1}{IIP3_2} + \ldots + \frac{G_1G_2\ldots G_{N-1}}{IIP3_N}} \tag{4.85}$$

where $\{G_1, G_2, \ldots G_N\}$ and $\{IIP3_1, IIP3_2, \cdots IIP3_N\}$ are the respective linear gains and $IIP3$s of the various RF blocks. The relationship in Eqn (4.85) implies that all the intermodulation products in the cascade add up coherently, that is they all have the same phase that is not always a practical assumption. Similar to the second order case in Eqn (4.58), the relationship in Eqn (4.85) is true when all the narrowband interfering signals happen to be in the passband of each RF block and the blocks themselves do not impact the phase coherence of the signals as they propagate in the chain and that the intermodulation signals add up in phase as mentioned earlier. In all other cases, filtering and selectivity have to be accounted for.

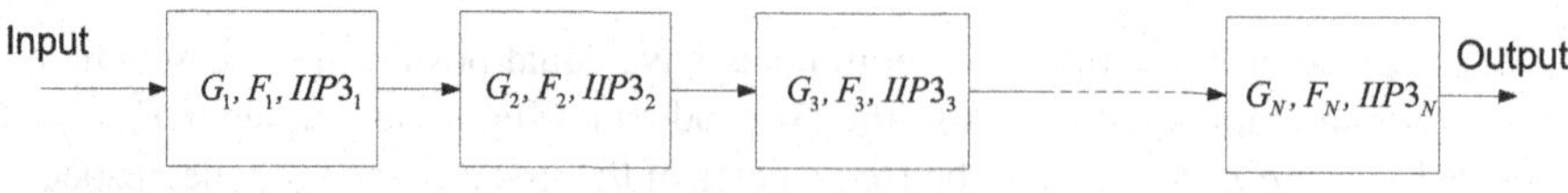

FIGURE 4.17

Cascade of N-radio frequency blocks with their respective linear gains, noise factors, and third order nonlinearities.

Thus far, we have discussed *IM* products relevant to second and third order nonlinearities. In reality, we can extend these relationships to an *nth*-order nonlinearity of the polynomial form $y(t) = \sum_{n=0}^{N} \beta_n x^n(t)$. Let

$$G_{n,\mathrm{dB}} = 10 \log_{10}(\beta_n) \tag{4.86}$$

Then it can shown that the generalized relationship for *nth*-order *IM* power product is

$$P_{IM_n} = nP_i - (n-1)P_{IIPn} \tag{4.87}$$

Similarly, the generalized *nth*-order output *IM* power product can be expressed as

$$P_{OIPn} = G_{1,\mathrm{dB}} + P_{IIPn} = G_{n,\mathrm{dB}} + nP_{IIPn} \tag{4.88}$$

Note that the relationships expressed above are applicable only for weak signal nonlinearities under memoryless conditions.

EXAMPLE 4.2 DERIVATION OF SYSTEM *IIP*3

Given a broadband signal operating in the ISM band with specifications as presented in Table 4.2:

Assume that two-tone interferers with equal signal power impinge at the BDF with signal power of 5 dBm each as shown in Figure 4.18. Given that the antenna gain is 0 dBi, let the first interferer be situated at 2.75 GHz and the second interferer at 3.1 GHz. Determine the signal *IIP*3 of the receiver after the BDF such that the *IM* signal power does not degrade the receiver noise floor by more than 1 dB. Assume that the BDF contributes negligible degradation due to nonlinearity and insignificant insertion loss.

Assume that the BDF can be modeled as a fifth order elliptic filter centered at 2.4 GHz as shown in Figure 4.19. The filter bandwidth is 200 MHz. The filter rejection is approximately 31 and 35 dBs at 2.75 and 3.1 GHz, respectively. Hence, the interferer signal power at $F_1 = 2.75$ GHz becomes 5 dBm − 31 dB = −26 dBm after the BDF. Similarly, the interferer signal power at $F_2 = 3.1$ GHz after the BDF becomes 5 dBm − 35 dB = −30 dBm.

Furthermore, note that the third order *IM* falls in band. The *IM* frequency is:

$$F_{IM} = 2F_1 - F_2 = 2 \times 2.75\ \mathrm{GHz} - 3.1\ \mathrm{GHz} = 2.4\ \mathrm{GHz} \tag{4.89}$$

Table 4.2 Specifications of Desired Signal

Parameter	Value
Receiver signal sensitivity	−110 dBm
Noise equivalent bandwidth	10 MHz
Center frequency of received signal	2.4 GHz
Bandwidth of BDF	200 MHz
CNR for acceptable BER	4 dB
Signaling bandwidth	9 MHz

BDF, bandpass filter.

Continued

EXAMPLE 4.2 DERIVATION OF SYSTEM *IIP*3—cont'd

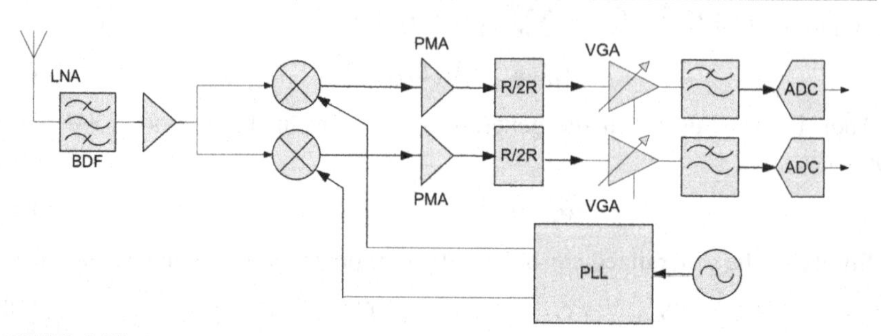

FIGURE 4.18

Simplified receiver block diagram. BPF, bandpass filter; VGA, voltage gain amplifier; PMA, post-mixer filter. (For color version of this figure, the reader is referred to the online version of this book.)

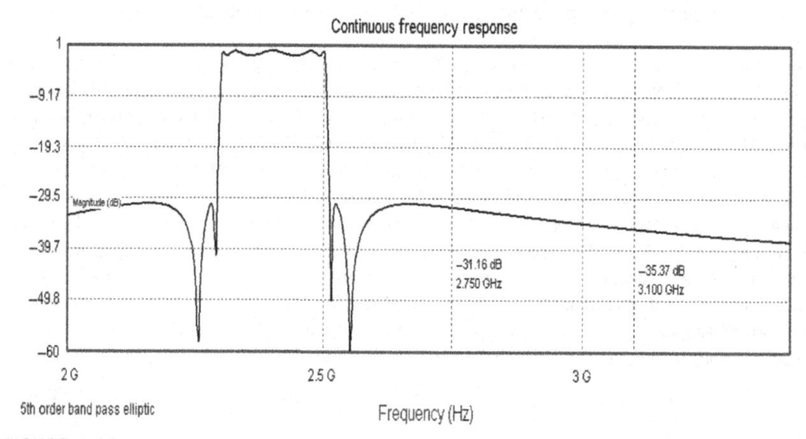

FIGURE 4.19

Elliptic fifth order bandpass filter centered at 2.4 GHz with 200 MHz bandwidth. (For color version of this figure, the reader is referred to the online version of this book.)

Of course, in this rhetorical example, the *IM* falls at the center of the band, which is obviously not the norm in pragmatic examples.

Next, we compute the desired *IM* power. Recall that the noise floor of the receiver is related to the sensitivity of the receiver and the required *CNR*, that is

$$P_{\text{noise floor,dBm}} = P_{\text{sensitivity,dBm}} - CNR$$
$$= -110\text{ dBm} - 4\text{ dB} = -114\text{ dBm} \quad (4.90)$$

EXAMPLE 4.2 DERIVATION OF SYSTEM *IIP3*—cont'd

Next, we determine the power of the allowable *IM* power. Allowing 1 dB of degradation to the noise floor implies that

$$\begin{aligned} P_{IM_3,\text{dB}} &= 10\log_{10}\left(10^{(P_{\text{noise floor,dBm}}+P_{\text{degradation}})/10} - 10^{P_{\text{noise floor,dBm}}/10}\right) \\ P_{IM_3,\text{dB}} &= 10\log_{10}\left(10^{-113/10} - 10^{-114/10}\right) = -119.87\text{ dBm} \end{aligned} \quad (4.91)$$

Finally, using the relationship in Eqn (4.82)

$$\begin{aligned} P_{IIP3,\text{dBm}} &= \frac{2P_{in,1,\text{dBm}} + P_{in,2,\text{dBm}} - P_{IM_3,\text{dB}}}{2} \\ &= \frac{2\times(-26) - 30 + 119.8}{2} = 18.9\text{ dBm} \end{aligned} \quad (4.92)$$

There is a trade to be made when designing the receiver. In order to lower the system *IIP3*, the BDF rejection must increase at the expense of the BDF complexity, size, and insertion loss. Another trade is the allowable degradation to the noise floor. Allowing the degradation to vary between 0.2 and 2 dB in 0.2 dB step varies the *IIP3* roughly between 22.6 dBm and 17 dBm, a difference of 5.6 dB. Allowing the degradation to the noise floor to increase from 1 to 2 dB lowers the *IIP3* by roughly 1.77 dB as shown in Figure 4.20.

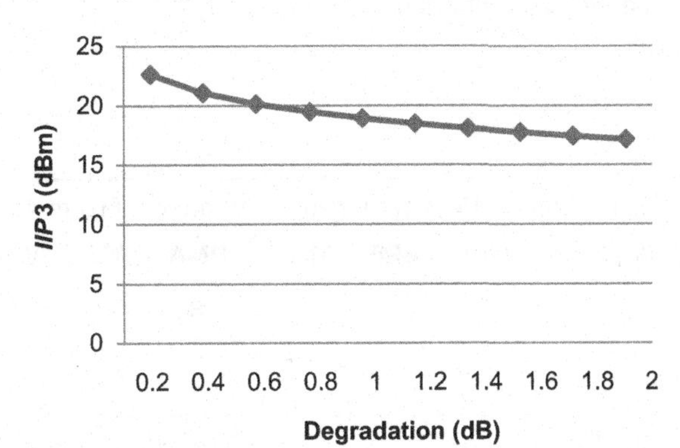

FIGURE 4.20

Allowable degradation to the noise floor and its impact on system *IIP3*. *IIP3*, input-referred third order intercept point.

EXAMPLE 4.3 LONG-TERM EVOLUTION (LTE) LINEUP ANALYSIS, A RHETORICAL EXAMPLE

Consider the LTE receiver depicted in Figure 4.21. LTE as a standard supports multiple bandwidths ranging from 1.4 to 20 MHz as well as various symbol constellations imposed on an orthogonal frequency multiple access (OFDM) signaling scheme. The required system sensitivities for 1.4 and 20 MHz are −102.5 and −90.2 dBm, respectively. Based on the architecture of Figure 4.21, assume that the noise figure, gain, and IPP3 for the various

Continued

EXAMPLE 4.3 LONG-TERM EVOLUTION (LTE) LINEUP ANALYSIS, A RHETORICAL EXAMPLE—cont'd

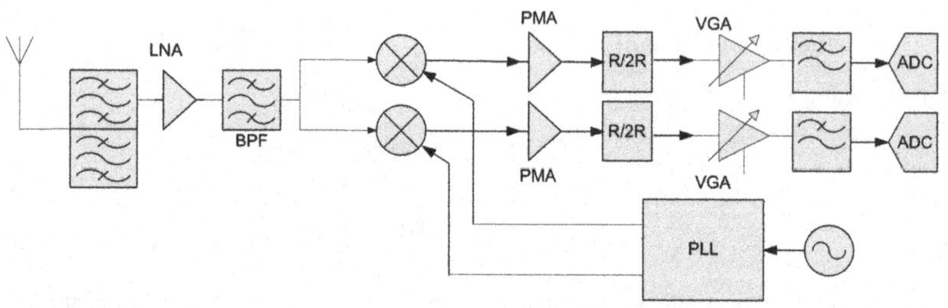

FIGURE 4.21

Simplified direct conversion LTE receiver. LTE, Long-term evolution; *IIP*3, input-referred third order intercept point; BPF, bandpass filter; VGA, voltage gain amplifier; PMA, post-mixer filter. (For color version of this figure, the reader is referred to the online version of this book.)

Table 4.3 Specifications of Receiver Lineup Presented in Figure 4.21

Parameter	Duplexer	LNA	BPF	Mixer	PMA	R2R	VGA	LPF
Gain (dB)	−1.8	15	−2	4	8	−0.1	40	−2
NF (dB)	1.8	1.5	2	7.8	5	4	20	5
*IIP*3 (dBm)	60	−4.5	10	7	2	20	10	40

BDF, bandpass filter; PMA, post-mixer filter; VGA, voltage gain amplifier; NF, *noise figure;* IIP3; *input-referred third order intercept point.*

analog blocks are as presented in Table 4.3. Determine the system noise figure, system *IIP*3, and spur-free dynamic range for 1.4 and 20 MHz bandwidths. Suppose it is feasible to vary the LNA gain between 8 and 22 dB without altering any other *RF* parameter including the LNA's linearity, determine the impact on system noise figure, *IIP*3, and spur-free dynamic range. Perform all analyses at room temperature.

In Figure 4.21, BPF refers to bandpass filter, PMA refers to post-mixer filter, R2R is a resistor-to-resistor attenuator network, and VGA stands for voltage gain amplifier. The VGA has variable gain settings controlled algorithmically by the automatic gain control (AGC) loop.

EXAMPLE 4.3 LONG-TERM EVOLUTION (LTE) LINEUP ANALYSIS, A RHETORICAL EXAMPLE—cont'd

Recall from Chapter 3 that the cumulative system noise figure can be found according to Friis formula as:

$$F_{\text{system}} = F_1 + \sum_{l=2}^{L}\left(\frac{F_l - 1}{\prod_{n=1}^{L-1} G_n}\right)$$

$$= F_{\text{Duplexer}} + \frac{F_{\text{LNA}} - 1}{G_{\text{Duplexer}}} + \frac{F_{\text{IRF}} - 1}{G_{\text{Duplexer}} G_{\text{LNA}}} + \cdots + \frac{F_{\text{LPF}} - 1}{G_{\text{Duplexer}} G_{\text{LNA}} \cdots G_{\text{VGA}}} =$$

$$F_{\text{system}} = 10^{1.8/10} + \frac{10^{1.5/10} - 1}{10^{-1.8/10}} + \frac{10^{2/10} - 1}{10^{-1.8/10} 10^{15/10}}$$

$$+ \cdots + \frac{10^{5/10}}{10^{-1.8/10} 10^{15/10} \ldots 10^{40/10}} = 4.92 \text{ dB} \tag{4.93}$$

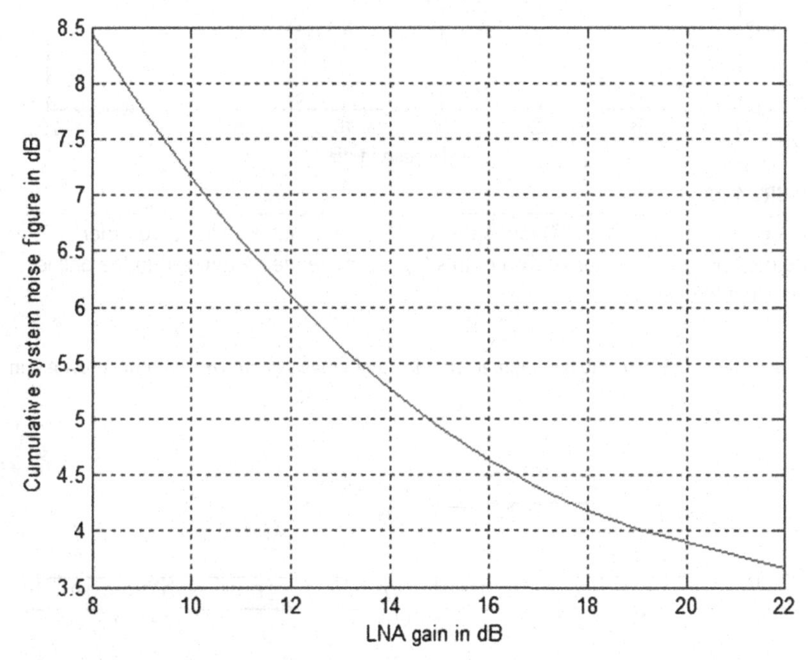

FIGURE 4.22

LNA gain versus system noise figure for Example 4.3. (For color version of this figure, the reader is referred to the online version of this book.)

Continued

EXAMPLE 4.3 LONG-TERM EVOLUTION (LTE) LINEUP ANALYSIS, A RHETORICAL EXAMPLE—cont'd

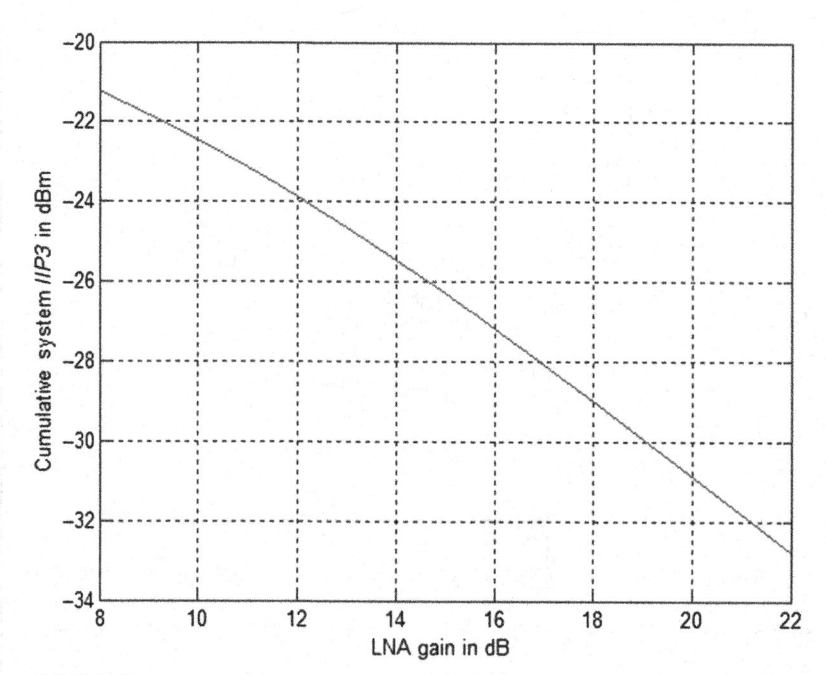

FIGURE 4.23

LNA gain versus system *IIP*3 for Example 4.3. *IIP*3, input-referred third order intercept point. (For color version of this figure, the reader is referred to the online version of this book.)

The system *IIP*3 can also be obtained via the cumulative system *IIP*3 formula provided in Eqn (4.85) as:

$$IIP3_{system} = \frac{1}{\frac{1}{IIP3_1} + \frac{G_1}{IIP3_2} + \frac{G_1 G_2}{IIP3_3} \cdots + \frac{G_1 G_2 \ldots \infty G_{N-1}}{IIP3_N}}$$

$$IIP3_{system} = \frac{1}{\frac{1}{10^{60/10}} + \frac{10^{-1.8/10}}{10^{-4.5/10}} + \frac{10^{-1.8/10}10^{15/10}}{10^{10/10}} \cdots + \frac{10^{-1.8/10}10^{15/10} \ldots 10^{40/10}}{10^{40/10}}}$$

$$= -26.3 \text{ dBm} \tag{4.94}$$

EXAMPLE 4.3 LONG-TERM EVOLUTION (LTE) LINEUP ANALYSIS, A RHETORICAL EXAMPLE—cont'd

Finally, recall that the spur-free dynamic range (SFDR) due to third order nonlinearity, as defined in Eqn (4.84) for 1.4 and 20 MHz is:

$$\begin{aligned}
\Delta_{SFDR-IIP3} &= \frac{2}{3}(IIP3_{\text{system}} + 174 - NF_{\text{system}} - 10\log_{10}(B)) \\
\Delta_{SFDR-IIP3}\Big|_{1.4\text{ MHz case}} &= \frac{2}{3}\left(-26.3 + 174 - 4.92 - 10\log_{10}\left(1.4 \times 10^6\right)\right) = 54.21\text{ dB} \\
\Delta_{SFDR-IIP3}\Big|_{20\text{ MHz case}} &= \frac{2}{3}\left(-26.3 + 174 - 4.92 - 10\log_{10}\left(20 \times 10^6\right)\right) = 46.51\text{ dB}
\end{aligned} \tag{4.95}$$

Next, we vary the LNA gain between 8 and 20 dB keeping all other *RF* parameters the same. Note that in reality there are always interdependencies amongst parameters that relate to the LNA circuit design under consideration. However, increasing the LNA gain decreases the overall system noise figure. The LNA, being at the front end of the receiver, has a major impact on the overall system noise figure as shown in Figure 4.22. The adverse effect of increasing the LNA gain, however, is on the system *IIP3* as shown in Figure 4.23. Increasing the LNA gain serves to decrease the system *IIP3*, thus making the receiver more vulnerable to blockers. One common solution is to employ a high-gain mode and low gain for the LNA controlled via the AGC.

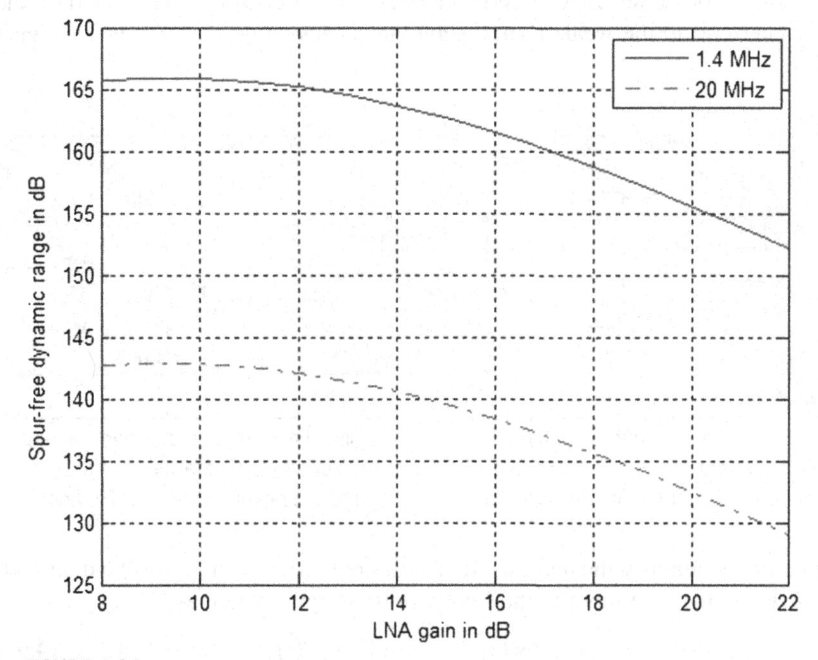

FIGURE 4.24

LNA gain versus spur-free dynamic range for Example 4.3. (For color version of this figure, the reader is referred to the online version of this book.)

Continued

EXAMPLE 4.3 LONG-TERM EVOLUTION (LTE) LINEUP ANALYSIS, A RHETORICAL EXAMPLE—cont'd

Next, we study the impact of varying the LNA gain on the spur-free dynamic range due to third order nonlinearity. First, it is noted that as the bandwidth increases from 1.4 to 20 MHz, the overall dynamic range of the system decreases. Furthermore, increasing the LNA gain serves to decrease the SFDR, as can be seen in Figure 4.24.

4.2.2.4 Cross-modulation distortion

Cross-modulation distortion is degradation due to third order nonlinearity. By definition, it is the amount of *AM* that transfers from a strong undesired signal to a weak desired signal after passing through a nonlinear system or device [5]. An in-band blocker received at the antenna can cross-modulate onto a weak desired signal at the output of a nonlinear device such as an LNA. A similar scenario can develop in a full-duplex system as will be shown shortly.

Consider the full-duplex transceiver depicted in Figure 4.25. By way of an example, suppose that the transmitted signal is an OFDM signal. OFDM fundamentally is *AM*. The strong transmit signal passing through the duplexer leaks to the receiver and gets amplified by the LNA. On the other hand, the weak received signal also passes through the LNA. A certain amount of *AM* distortion transfers from the strong interferer to the weak desired signal thus causing degradation to the signal quality.

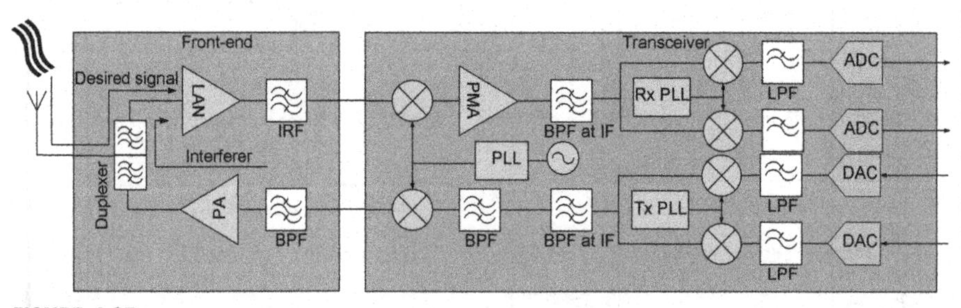

FIGURE 4.25

Cross-modulation between a weak desired–received signal and the transmit signal acting as a strong interferer. BPF, bandpass filter; PMA, post-mixer filter; *IF*, intermediate frequency. (For color version of this figure, the reader is referred to the online version of this book.)

In order to simplify the analysis, let the desired signal be a narrowband unmodulated signal, then the input to the LNA can be simply written as

$$x(t) = \underbrace{\alpha_1 \cos(\Omega_1 t)}_{\text{unmodulated desired signal}} + \underbrace{\alpha_2\{1 + m(t)\}\cos(\Omega_2 t)}_{\text{modulated interferer signal}} \tag{4.96}$$

where Ω_1 and Ω_2 are two distinct frequencies. At the output of the LNA, and due to nonlinearity, the output signal $y(t)$ can be expressed as:

$$
\begin{aligned}
y(t) &= \beta_1 x(t) + \beta_3 x^3(t) \\
&= \left\{ \underbrace{\beta_1\alpha_1}_{\text{linear gain}} + \underbrace{\frac{3}{4}\beta_3\alpha_1^3 + \frac{3}{2}\beta_3\alpha_1\alpha_2^2}_{\text{compression and desensitization}} \underbrace{+\frac{3}{2}\beta_3\alpha_1\alpha_2^2\left(2m(t)+m^2(t)\right)}_{\text{cross-modulation}} \right\} \\
&\quad \times \cos(\Omega_1 t) + \cdots
\end{aligned}
\tag{4.97}
$$

A close examination of Eqn (4.97) reveals that due to third order nonlinearity the amplitude modulation due to the undesired transmit signal or interferer has transferred onto the desired unmodulated signal. In reality, any odd nonlinearity would serve to cross-modulate the undesired signal on top of the desired signal thus causing degradation.

In order to further express the relationship in Eqn (4.97) in terms of $IIP3$, let

$$
\begin{aligned}
y(t) &= \beta_1\alpha_1\left\{1 + \frac{3}{4}\frac{\beta_3}{\beta_1}\alpha_1^2 + \frac{3}{2}\frac{\beta_3}{\beta_1}\alpha_2^2 + \frac{3}{2}\frac{\beta_3}{\beta_1}\alpha_2^2\left(2m(t)+m^2(t)\right)\right\}\cos(\Omega_1 t) + \cdots \\
&= \beta_1\alpha_1\left\{1 + \frac{\alpha_1^2}{(IIP3)^2} + \frac{2\alpha_2^2}{(IIP3)^2} + \frac{2\alpha_2^2}{(IIP3)^2}\left(2m(t)+m^2(t)\right)\right\}\cos(\Omega_1 t) + \cdots \Bigg|_{IIP3=\sqrt{\frac{4}{3}\frac{|\beta_1|}{|\beta_3|}}} \\
&= \beta_1\alpha_1\cos(\Omega_1 t) + \underbrace{\beta_1\alpha_1\frac{1}{(IIP3)^2}\left\{\alpha_1^2 + 2\alpha_2^2 + 2\alpha_2^2\left(2m(t)+m^2(t)\right)\right\}\cos(\Omega_1 t) + \cdots}_{\text{cross-modulation term}}
\end{aligned}
\tag{4.98}
$$

The relationship in Eqn (4.98) points to the following conclusion: the higher the $IIP3$ of the LNA the smaller the degradation due to cross-modulation. As a matter of fact, the cross-modulation term diminishes in inverse proportion to the square value of $IIP3$.

To further simplify the analysis, assume that the output signal in Eqn (4.98) is comprised only of the linear signal $y_{\text{linear}}(t)$ and cross-modulation signal causing the degradation $y_{\text{cross-modulation}}(t)$, then let

$$
\begin{aligned}
\tilde{y}(t) &= y_{\text{linear}}(t) + y_{\text{cross-modulation}}(t) \\
&= \underbrace{\beta_1\alpha_1}_{\substack{\text{linear gain} \\ \text{term}}} \cos(\Omega_1 t) + \underbrace{\frac{3}{2}\beta_3\alpha_1\alpha_2^2 m^2(t)}_{\substack{\text{cross modulation} \\ \text{term}}} \cos(\Omega_1 t)
\end{aligned}
\tag{4.99}
$$

Substituting the value of $IIP3$ found in Eqn (4.77) into Eqn (4.99), we obtain

$$\begin{aligned}\tilde{y}(t) &= \beta_1 \alpha_1 \cos(\Omega_1 t) \\ &+ \beta_1 \alpha_1 \left\{ \frac{2}{(IIP3_{\text{LNA}})^2} \alpha_2^2 m^2(t) \right\} \cos(\Omega_1 t)\end{aligned} \tag{4.100}$$

Next, consider the power ratio of the single tone term to the cross-modulation term in Eqn (4.100), we obtain

$$\begin{aligned}\Delta|_{\text{Power ratio}} &= \frac{P_{\text{Single tone}}}{P_{\text{cross-modulation}}} = \frac{\frac{1}{2}(\beta_1 \alpha_1)^2 \, (IIP3_{\text{LNA}})^4}{\frac{1}{2}(\beta_1 \alpha_1)^2 \, 4\alpha_2^4 P_{\hat{m}^2}(t)} \\ &= \frac{(IIP3_{\text{LNA}})^4}{4\alpha_2^4 P_{\hat{m}^2}(t)}\end{aligned} \tag{4.101}$$

where $P_{\hat{m}^2}(t)$ is the average power of $m^2(t)$. Note that it is not an unreasonable assumption to suppose that the mean of $m(t)$ is zero and that its variance is simply $\sigma_m^2 = E\{m^2(t)\}$. In this case, the voltage ratio of the single tone term to the cross-modulation term can then be expressed as

$$\Delta|_{\text{Voltage ratio}} = \frac{(IIP3_{\text{LNA}})^4}{4\frac{\alpha_2^2}{2}\sigma_m^2} = \left.\frac{(IIP3_{\text{LNA}})^2}{4P_{\text{Single tone}}\sigma_m^2}\right|_{\alpha_2=\alpha_1} \tag{4.102}$$

At this point, it is imperative to draw certain important conclusions from the analysis conducted thus far. First, according to Eqn (4.101), the transmitted signal (or any blocker) passing through the LNA has been squared and cross-modulated onto the desired signal. Squaring a signal in the time domain implies convolution in the frequency domain:

$$m^2(t) \Leftrightarrow \frac{1}{2\pi} M(\Omega) * M(\Omega) \tag{4.103}$$

The frequency convolution in Eqn (4.103) implies that the bandwidth of the interferer $m(t)$ has doubled at the output of the LNA. Therefore, overlap of the desired signal with the interfering signal has to be considered when analyzing the degradation impacting the desired signal. Furthermore, the amplitude of the desired signal is amplified by the same proportion as that of the interferer. As a matter of fact, increasing the amplitude of the desired signal α_1 automatically amplifies the amplitude of the interferer.

4.2.2.5 Harmonic distortion

Harmonics, like *IM* products, appear at the output of a nonlinear device or system as signals with center frequencies situated at integer multiples of the fundamental frequency. By way of an example, assume the input signal to be comprised of a single tone centered at Ω_1, or $x(t) = \alpha_1 \cos(\Omega_1 t)$, then according to Table 4.1, we can express the output of a nonlinear device in terms of its harmonics as

$$y(t) = \underbrace{\frac{1}{2}\beta_2\alpha_1^2}_{\text{DC term}} + \underbrace{\left(\beta_1\alpha_1 + \frac{3}{4}\beta_3\alpha_1^3\right)}_{\text{First harmonic}} \cos(\Omega_1 t) + \underbrace{\frac{1}{2}\beta_2\alpha_1^2}_{\text{Second harmonic}} \cos(2\Omega_1 t)$$
$$+ \underbrace{\frac{1}{4}\beta_3\alpha_1^3}_{\text{Third harmonic}} \cos(3\Omega_1 t) + \cdots \tag{4.104}$$

According to Eqn (4.104), the harmonic signals are centered at multiples of Ω_1 with the second harmonic centered at $2\Omega_1$, and the third harmonic centered at $3\Omega_1$, and so on. The DC term also exists representing the 0th harmonic. If we extend the analysis to include two tones centered at Ω_1 and Ω_2, then harmonic signals will occur at $\pm m\Omega_1 \pm n\Omega_2$ where m and n are integers. Note that even order harmonics are of no consequence in a differential system.

At this point, let us further examine the second order harmonic due to two-tone input. Substitute Eqn (4.22) into the polynomial in Eqn (4.12) and rearrange in terms of second order nonlinearity while ignoring β_0 for practical reasons, we obtain

$$y(t) = \beta_1\{\alpha_1 \cos(\Omega_1 t) + \alpha_2 \cos(\Omega_2 t)\} + \beta_2\left\{\frac{\alpha_1^2 + \alpha_2^2}{2} + \frac{\alpha_1^2}{2}\cos(2\Omega_1 t)\right.$$
$$\left. + + \frac{\alpha_2^2}{2}\cos(2\Omega_2 t) + \alpha_1\alpha_2 \cos[(\Omega_1 - \Omega_2)t] + \alpha_1\alpha_2 \cos[(\Omega_1 + \Omega_2)t]\right\} + \cdots \tag{4.105}$$

For $\alpha = \alpha_1 = \alpha_2$, that is if the two tones have equal magnitudes, then the relationship in Eqn (4.105) becomes

$$y(t) = \beta_1\{\alpha \cos(\Omega_1 t) + \alpha \cos(\Omega_2 t)\} + \beta_2\left\{\alpha^2 + \frac{\alpha^2}{2}\cos(2\Omega_1 t) + \frac{\alpha^2}{2}\cos(2\Omega_2 t)\right.$$
$$\left. + + \alpha^2 \cos[(\Omega_1 - \Omega_2)t] + \alpha^2 \cos[(\Omega_1 + \Omega_2)t]\right\} + \cdots \tag{4.106}$$

The power of the second order harmonic due to $\beta_2\frac{\alpha^2}{2}\cos(2\Omega_1 t)$ is simply:

$$P_{H_2} = \frac{1}{2}\left[\beta_2\frac{\alpha^2}{2}\right]^2 = \frac{1}{8}\beta_2^2\alpha^4 \tag{4.107}$$

whereas the power due to the fundamental term $\beta_1\alpha\cos(\Omega_1 t)$ is

$$P_{\text{fundamental}} = \frac{1}{2}\beta_1^2\alpha^2 \tag{4.108}$$

Then the ratio of Eqn (4.107) divided by Eqn (4.108) and then substituting Eqn (4.49) into the result, we obtain

$$\begin{aligned}\frac{P_{H_{2,Output}}}{P_{\text{fundamental,output}}} &= \frac{\frac{1}{8}\beta_2^2\alpha^4}{\frac{1}{2}\beta_1^2\alpha^2} = \frac{1}{4}\frac{\beta_2^2\alpha^2}{\beta_1^2} \\ &= \frac{1}{4}\frac{\alpha^2}{IIP2^2} = \frac{1}{2}\frac{1}{IIP2^2}P_i \\ &= \frac{1}{4}\frac{P_i}{P_{IIP2}}\end{aligned} \tag{4.109}$$

Recall that it was previously defined that

$$P_i = \frac{1}{2}\alpha_i^2\bigg|_{\alpha_1=\alpha_2=\alpha} = \frac{1}{2}\alpha^2 \text{ and } P_{IIP2} = \frac{1}{2}(IIP2)^2 \tag{4.110}$$

Referring the output power of the harmonic to the input can be obtained by examining the ratio of the power of the second order output harmonic to the output power of the fundamental, that is

$$\frac{P_{H_{2,Output}}}{P_{\text{fundamental,output}}} = \frac{\beta_1^2 P_{H_{2,Input}}}{\beta_1^2 P_i} = \frac{P_{H_{2,Input}}}{P_i} = \frac{P_i}{4P_{IIP2}} \Rightarrow P_{H_{2,Input}} = \frac{P_i^2}{4P_{IIP2}} \tag{4.111}$$

The relationship in Eqn (4.111) can be expressed in dBm as

$$P_{H_{2,Input,\text{dBm}}} = 2P_{i,\text{dBm}} - P_{IIP2,\text{dBm}} - 6 \tag{4.112}$$

where Eqn (4.112) implies that the input referred harmonic power is 6 dB below that of the input referred second order *IM* product.

Next, we turn our attention to harmonics due to third order nonlinearities. Again, given an input two-tone signal where the tones have equal power, then the third order harmonic signals with their respective amplitudes are:

$$\frac{1}{4}\beta_3\alpha^3\cos(3\Omega_1 t) \text{ and } \frac{1}{4}\beta_3\alpha^3\cos(3\Omega_2 t) \tag{4.113}$$

The average harmonic output power for each harmonic is $P_{H_{3,Output}} = \frac{1}{32}\beta_3^2\alpha^6$. In a similar manner to the analysis done for the second order harmonic, let us examine the ratio of linear output power of the third harmonic to that of the fundamental, and substituting Eqn (4.77) into the result, we obtain:

$$\frac{P_{H_{3,Output}}}{P_{\text{fundamental,output}}} = \frac{\frac{1}{32}\beta_3^2\alpha^6}{\frac{1}{2}\beta_1^2\alpha^2} = \frac{1}{16}\frac{\beta_3^2}{\beta_1^2}\alpha^4 = \frac{1}{9}\frac{P_i^2}{P_{IIP3}^2} \tag{4.114}$$

Manipulating the linear power ratios, in a similar manner to the analysis presented above, we find that the power of the third harmonic referred to the input is:

$$P_{H_{3,Input}} = \frac{1}{9}\frac{P_i^3}{P_{IIP3}^2} \tag{4.115}$$

or in the logarithmic scale as

$$P_{H_{3,Input,\text{dBm}}} = 3P_{i,\text{dBm}} - 2P_{IIP3,\text{dBm}} - 9.54 \tag{4.116}$$

Analysis due to higher order harmonics, although more cumbersome to obtain, can be developed in a similar manner.

4.2.3 *AM* to *AM* and *AM* to *PM* distortion

In this section, we discuss two common types of degradations, namely *AM* to *AM* and *AM* to *PM* conversion. *AM* to *AM* conversion is a process that mainly degrades the desired signal by amplitude modulating it with an undesired signal, thus affecting its amplitude. Similarly, *AM* to *PM* conversion is a process that degrades the desired signal by phase modulating the desired signal with an undesired signal, thus affecting its phase. Both impairments can have adverse effects on the performance of the signal.

Assume that an *IM* product, situated for example at $2\Omega_1 \pm \Omega_2$, resulting from a third order nonlinearity of a given device or system due to a two-tone signal, falls $\Delta\Omega$ Hz away from the carrier frequency, or $\Omega_c + \Delta\Omega$. Given an input signal $\alpha\sin(\Omega_c t)$, then at the output of the nonlinear system or device, the desired signal plus interference is given as

$$y(t) = \alpha\sin(\Omega_c t) + \Gamma\cos(\Omega_c t + \Delta\Omega t) \tag{4.117}$$

where A is the magnitude of the desired signal and Γ is the amplitude of the *IM* product. Both amplitudes of the desired signal and *IM* product are given to be very slowly varying and hence, for the sake of our analysis, A and Γ are assumed to be constant.

The *IM* term in Eqn (4.117) can be expanded as

$$\Gamma\cos(\Omega_c t + \Delta\Omega t) = \Gamma[\cos\Omega_c t\cos\Delta\Omega t - \sin\Omega_c t\sin\Delta\Omega t] \tag{4.118}$$

Substituting Eqn (4.118) into Eqn (4.117), we obtain

$$\begin{aligned} y(t) &= [\alpha - \Gamma\sin(\Delta\Omega t)]\sin(\Omega_c t) + \Gamma\cos(\Delta\Omega t)\cos(\Omega_c t) \\ &= [\alpha - \Gamma\sin(\Delta\Omega t)]\left[\sin(\Omega_c t) + \frac{\Gamma\cos(\Delta\Omega t)}{(\alpha - \Gamma\sin(\Delta\Omega t))}\cos(\Omega_c t)\right] \end{aligned} \tag{4.119}$$

Furthermore, the relationship in Eqn (4.119) can be further manipulated by using the trigonometric identity:

$$\tan(\phi(t)) = \frac{\Gamma\cos(\Delta\Omega t)}{(\alpha - \Gamma\sin(\Delta\Omega t))} \tag{4.120}$$

Substituting Eqn (4.120) into Eqn (4.119), we obtain

$$\begin{aligned} y(t) &= [\alpha - \Gamma \sin(\Delta\Omega t)][\sin(\Omega_c t) + \tan(\phi(t))\cos(\Omega_c t)] \\ &= \left[\frac{\alpha - \Gamma \sin(\Delta\Omega t)}{\cos(\phi(t))}\right][\sin(\Omega_c t) + \tan(\phi(t))\cos(\Omega_c t)]\cos(\phi(t)) \\ &= \underbrace{\left[\frac{1 - \frac{\Gamma}{\alpha}\sin(\Delta\Omega t)}{\cos(\phi(t))}\right]}_{\text{Distortion}} \underbrace{\alpha \sin(\Omega_c t + \phi(t))}_{\text{Desired signal}} \end{aligned} \tag{4.121}$$

According to Eqn (4.121), the distortion term can be broken in terms of *AM/AM* distortion and *AM/PM* distortion. That is, the *AM/AM* term can be defined as

$$\delta_{AM/AM}(t) = \left|\frac{1 - \frac{\Gamma}{\alpha}\sin(\Delta\Omega t)}{\cos(\phi(t))}\right|_{\lim\limits_{\delta x \to 0}} \tag{4.122}$$

The *AM/PM* term can also be expressed as the phase distortion

$$\delta_{AM/PM}(t) = \tan^{-1}\left[\frac{1 - \frac{\Gamma}{\alpha}\sin(\Delta\Omega t)}{\cos(\phi(t))}\right] \tag{4.123}$$

The *AM/AM* and *AM/PM* analysis presented thus far can be extended easily to nonlinear memoryless linearity. Following the development presented in Ref. [6], and given the memoryless channel model embodied in the polynomial presented in Eqn (4.12), let the input signal be a single tone $x(t) = \alpha \cos(\Omega t)$. Using the trigonometric expansion of $\cos^n(x)$ given in Eqn (4.237), we can express the output signal for odd order nonlinearity as

$$y(t) = \sum_{\substack{k=1 \\ k \text{ odd}}}^{n} \left(\frac{1}{2}\right)^{k-1} \begin{pmatrix} k \\ \frac{k-1}{2} \end{pmatrix} \beta_k \alpha^{k-1}, \quad n \text{ is odd}$$

$$y(t) = \beta_1 \alpha \left\{1 + \frac{3}{4}\frac{\beta_3}{\beta_1}\alpha^2 + \frac{5}{8}\frac{\beta_5}{\beta_1}\alpha^4 + \frac{35}{64}\frac{\beta_7}{\beta_1}\alpha^6 + \cdots + \left(\frac{1}{2}\right)^{n-1}\begin{pmatrix} n \\ \frac{n-1}{2} \end{pmatrix}\frac{\beta_n}{\beta_1}\alpha^{n-1}\right\}\cos(\Omega t) \tag{4.124}$$

For the sake of simplicity, if we limit the polynomial in Eqn (4.124) to third order nonlinearity, then we can simplify Eqn (4.124) to

$$y(t) = \left\{\beta_1 \alpha + \frac{3}{4}\beta_3 \alpha^3\right\}\cos(\Omega t) = \beta_1 \alpha \left\{1 + \frac{3}{4}\frac{\beta_3}{\beta_1}\alpha^2\right\}\cos(\Omega t) \tag{4.125}$$

Thus far, in our treatment of the memoryless channel, we have implicitly assumed the coefficients $\{\beta_0, \beta_1, \beta_2, \cdots\}$ to be real. This assumption typically results in worse case degradations due to *IM* products and harmonics. In reality, it is reasonable to assume that the channel coefficients are complex valued and not necessarily real. That is, in the case of Eqn (4.125), assume that $\beta_1 = |\beta_1|e^{j\phi_1}$ and $\beta_3 = |\beta_3|e^{j\phi_3}$. Under this assumption, the relationship in Eqn (4.125) becomes

$$y(t) = |\beta_1|e^{j\phi_1}\alpha\left\{1+\frac{3}{4}\frac{|\beta_3|}{|\beta_1|}e^{j(\phi_3-\phi_1)}\alpha^2\right\}\cos(\Omega t) \tag{4.126}$$

A further expansion of Eqn (4.126) results in the output signal

$$y(t) = |\beta_1|e^{j\phi_1}\alpha\left\{1+\frac{3}{4}\frac{|\beta_3|}{|\beta_1|}\alpha^2\cos(\theta)+j\frac{3}{4}\frac{|\beta_3|}{|\beta_1|}\alpha^2\sin(\theta)\right\}\cos(\Omega t)\Bigg|_{\theta=\phi_3-\phi_1} \tag{4.127}$$

Define the *AM/PM* conversion with respect to ϕ_1 as

$$\phi(t) = \tan^{-1}\left(\frac{\frac{3}{4}\frac{|\beta_3|}{|\beta_1|}\alpha^2\sin(\theta)}{1+\frac{3}{4}\frac{|\beta_3|}{|\beta_1|}\alpha^2\cos(\theta)}\right)_{\theta=\phi_3-\phi_1} \tag{4.128}$$

For relatively small values of $\phi(t)$, that is we can use the small angle approximation to assume that $\tan(\phi(t)) \approx \phi(t)$, then Eqn (4.128) becomes

$$\phi(t) \approx \frac{\frac{3}{4}\frac{|\beta_3|}{|\beta_1|}\alpha^2\sin(\theta)}{1+\frac{3}{4}\frac{|\beta_3|}{|\beta_1|}\alpha^2\cos(\theta)} \tag{4.129}$$

Substituting Eqn (4.77) into Eqn (4.129), we obtain the general form for *AM/PM* based on third order nonlinearity as

$$\phi(t) \approx \frac{\alpha^2\sin(\theta)}{IIP3^2+\alpha^2\cos(\theta)} \tag{4.130}$$

In the made-up family of curves shown in Figure 4.26 and generated according to Eqn (4.130), it is obvious that if you increase the *IIP*3 value of the device or system, you decrease the impact of *AM/PM* distortion as the magnitude of the signal increases to near compression. Therefore, we conclude, as it is intuitively expected, that increasing the *IIP*3 or linearity of the system decreases the degradation of the desired signal due to *AM/PM*. The same can be said about *AM/AM* distortion since both distortions go hand in hand.

The *AM/PM* relationship in Eqn (4.129) is similar but not equivalent to Saleh's model, which is generally expressed as [7]

$$\begin{aligned}\phi(\alpha) &= \frac{\Omega\alpha^2}{1+\Delta\alpha^2}\\ \lim_{\alpha\to\infty}\phi(\alpha) &= \frac{\Omega\alpha^2}{1+\Delta\alpha^2}\bigg|_{\alpha\to\infty} = \frac{\Omega}{\Delta}\end{aligned} \tag{4.131}$$

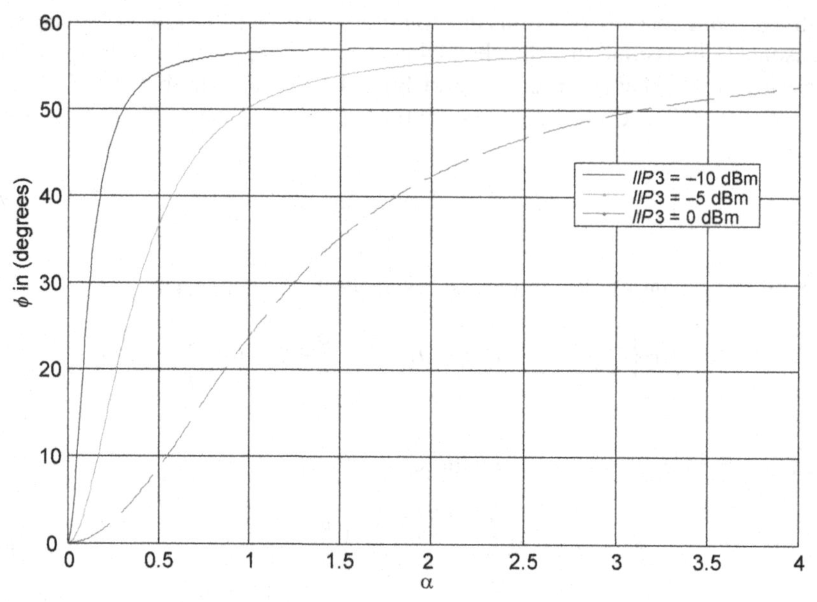

FIGURE 4.26

AM/PM curves for various *IIP3* values and $\theta = \pi/4$. *AM*, amplitude modulation; *IIP3*, input-referred third order intercept point. (For color version of this figure, the reader is referred to the online version of this book.)

In the analysis leading up to Eqn (4.129), we employed only a third order nonlinearity to obtain the *AM/PM* model, whereas in reality, an accurate theoretical model would encompass all higher order odd terms. The coefficients in Saleh's model are derived from curve fitting the data rather than the theoretical treatment presented above and based on the memoryless model.

Next, we derive the *AM/AM* model based on the memoryless channel polynomial. Again, for the sake of simplicity, assume a third order nonlinearity, we can express the magnitude $|A|$ of the output signal as

$$y(t) = (|A|\angle A)\cos(\Omega t) \tag{4.132}$$

where $\angle$ is the phase operator and

$$|A| = |\beta_1|\alpha\sqrt{\left(1+\frac{3}{4}\frac{|\beta_3|}{|\beta_1|}\alpha^2\cos(\theta)\right)^2+\left(\frac{3}{4}\frac{|\beta_3|}{|\beta_1|}\alpha^2\sin(\theta)\right)^2} \tag{4.133}$$

Again, substituting the linear value for *IIP3* expressed in Eqn (4.77) into Eqn (4.133), we obtain the relation

$$|A| = |\beta_1|\alpha\sqrt{\left(1+\left(\frac{\alpha}{IIP3}\right)^2\cos(\theta)\right)^2+\left(\left(\frac{\alpha}{IIP3}\right)^2\sin(\theta)\right)^2} \tag{4.134}$$

The degradation due *AM/AM* conversion then becomes:

$$\delta_{AM/AM}(\alpha,\theta) = \sqrt{\left(1+\left(\frac{\alpha}{IIP3}\right)^2\cos(\theta)\right)^2 + \left(\left(\frac{\alpha}{IIP3}\right)^2\sin(\theta)\right)^2} \quad (4.135)$$

Note that for large values of $\alpha/IIP3$, the relationship in Eqn (4.135) can be simplified as

$$\delta_{AM/AM}(\alpha,\theta) \approx \sqrt{\left(\left(\frac{\alpha}{IIP3}\right)^2\cos(\theta)\right)^2 + \left(\left(\frac{\alpha}{IIP3}\right)^2\sin(\theta)\right)^2} = \frac{\alpha}{IIP3} \quad (4.136)$$

The Saleh's model for *AM/AM* on the other hand is given according to Ref. [7] as

$$\begin{aligned} A(\alpha) &= \frac{\Upsilon\alpha}{1+\Psi\alpha^2} \\ \lim_{\alpha\to\infty} A(\alpha) &= \frac{\Upsilon\alpha}{1+\Psi\alpha^2} = \left.\frac{\Upsilon/\alpha}{1/\alpha^2+\Psi}\right|_{\alpha\to\infty} = 0 \end{aligned} \quad (4.137)$$

Note that as the magnitude of the signal increases, the *AM/AM* model in Eqn (4.137) converges to zero. Although not obvious, the same is true for Eqn (4.136) Again, as is the case for the *AM/PM* relationship, the parameters Υ and Ψ are determined empirically.

At this point, a further examination of Eqn (4.137) is warranted. The ratio itself can be expressed in terms of an infinite series as

$$A(\alpha) = \frac{\Upsilon\alpha}{1+\Psi\alpha^2} = \Upsilon\alpha\{1-\Psi\alpha^2+\Psi^2\alpha^4-\Psi^3\alpha^6+\Psi^4\alpha^8+\cdots\} \quad (4.138)$$

In the event where the memoryless coefficients are real or the imaginary part is negligible, the degradation is then purely due to *AM/AM*. In this case, we can relate the channel coefficients to Υ and Ψ

$$\begin{aligned} A(\alpha) = \beta_1\alpha\Bigg\{ & 1+\frac{3}{4}\frac{\beta_3}{\beta_1}\alpha^2+\frac{5}{8}\frac{\beta_5}{\beta_1}\alpha^4+\frac{35}{64}\frac{\beta_7}{\beta_1}\alpha^6+\cdots \\ & +\left(\frac{1}{2}\right)^{n-1}\begin{pmatrix} n \\ \frac{n-1}{2}\end{pmatrix}\frac{\beta_n}{\beta_1}\alpha^{n-1}\Bigg\} \end{aligned} \quad (4.139)$$

A direct comparison between Eqns (4.138) and (4.139) implies:

$$\begin{aligned} \beta_1 &= \Upsilon, \beta_3 = -\frac{4}{3}\Upsilon\Psi, \beta_5 = -\frac{8}{5}\Upsilon\Psi^2, \beta_7 = -\frac{64}{35}\Upsilon\Psi^3, \cdots\beta_n \\ &= (-1)^{\frac{n-1}{2}}\frac{2^{n-1}}{\begin{pmatrix} n \\ \frac{n-1}{2}\end{pmatrix}}\Upsilon\Psi^{\frac{n-1}{2}} \end{aligned} \quad (4.140)$$

According to Eqn (4.140), the output of the device due to third order nonlinearity only can be expressed as

$$y(t) = \Upsilon\alpha\{1 - \Psi\alpha^2\}\cos(\Omega t) \tag{4.141}$$

Thus far the implication of this analysis is that *AM/AM* and *AM/PM* are independent processes. In reality that is absolutely not true. Both *AM/AM* and *AM/PM* are due mainly to the same nonlinear behavior in the system. The analysis above is presented for the sake of understanding both phenomena and their implications on the system performance.

4.2.4 Receiver selectivity

The selectivity of a receiver is concerned with its ability to reject unwanted interferers in the adjacent and alternate channels, in turn allowing for successful demodulation of the desired signal at an acceptable performance [6]. Receiver selectivity is determined by various parameters in the receive chain, namely filtering, linearity, and noise. The appropriate level of selectivity needed can be quantified in terms of co-channel rejection for example. Co-channel rejection is the ratio between the desired signal to the unwanted interferer or blocker. In FM receivers, co-channel rejection, also known as capture ratio, is concerned with the processing of two signals, one weak and one strong, both present at the input of a limiting amplifier. The limiting operation tends to reinforce the strong signal while at the same time deteriorate the weak signal, thus causing the strong signal to capture the weak signal.

In order to determine the desired selectivity of a receiver, we begin by estimating the maximum allowable degradation, defined as:

$$P_{\text{Degradation}} = 10\log_{10}\left(10^{\frac{P_{\text{Desired,dBm}}-CNR}{10}} - 10^{\frac{P_{\text{noise floor,dBm}}}{10}}\right) \tag{4.142}$$

where $P_{\text{Desired,dBm}}$ is the desired linear signal power in the presence of a blocker or interferer and $P_{\text{noise floor,dBm}}$ is as defined in the relationship given in Eqn (4.60).

As discussed in a previous chapter concerning reciprocal mixing, illustrated in Figure 4.27, consider a narrowband interferer or blocker that is considerably stronger in magnitude than the desired signal at the input of the mixer. The mixer demodulates both the blocker and the desired signal with the *LO*. The demodulation process in the time domain results in the convolution of the phase noise of the *LO* with the desired and blocker signals in the frequency domain as shown in Figure 4.27. Any in-band spurs present in the *LO* signal will also impact the desired signal quality. If the blocker signal is situated such that the phase noise degradation is flat in the desired band, then the noise floor due to phase noise can be expressed in the form

$$P_{\text{Phase noise,dBm}} = P_{\text{blocker,dBm}} + L(F = \Delta F)\text{dBc/Hz} + 10\log_{10}(B) - R \tag{4.143}$$

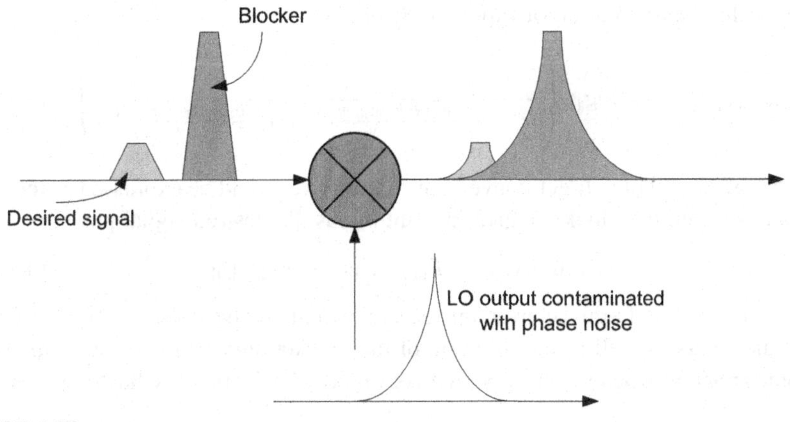

FIGURE 4.27

Reciprocal mixing of blocker and desired signal. (For color version of this figure, the reader is referred to the online version of this book.)

where $P_{\text{Phase noise,dBm}}$ is the amount of degradation due to phase noise in the desired signal bandwidth B, $L(F=\Delta F)$ is the phase noise density in the desired signal considered for the sake of this analysis to be flat at frequency offset ΔF from the *LO* tone. R is any possible rejection due to filtering of the blocker signal. In the linear domain, the relationship in Eqn (4.143) can be expressed as

$$P_{\text{Phase noise, Linear}} = 10^{\frac{P_{\text{Phase noise,dBm}}}{10}} = 10^{\frac{P_{\text{blocker,dBm}}+L(F=\Delta F)\text{dBc/Hz}+10\log_{10}(B)-R}{10}} \tag{4.144}$$

Next, we consider the intermodulation product due to the incoming blocker at the input of the mixer with an *LO* spur expressed in dBc/Hz. The spur power at the output of the mixer is given accordingly as

$$P_{\text{Spur,dBm}} = P_{IM_3,\text{dBm}} + P_{\text{Spur,dBc}} \tag{4.145}$$

Given a certain mixer *IIP*3, then according to Eqn (4.82) the power due to the *IM* product is given as

$$P_{IM_3,\text{dBm}} = 2P_{\text{LO Spur,dBm}} + P_{\text{blocker,dBm}} - 2P_{IIP3,\text{dBm}} \tag{4.146}$$

where we have assumed that the in-band *IM* frequency is $2F_{\text{LO Spur}} - F_{Blocke}$. In the linear domain, the power of the *IM* signal is

$$P_{IM_3,\text{linear}} = 10^{P_{IM_3,\text{dBm}}} = 10^{2P_{\text{LO Spur,dBm}}+P_{\text{blocker,dBm}}-2P_{IIP3,\text{dBm}}} \tag{4.147}$$

The blocker signal power for a certain degradation can then be determined as

$$10^{\frac{P_{\text{blocker,dBm}}}{10}}\left(10^{\frac{L(F=\Delta F)\text{dBc/Hz}+10\log_{10}(B)-R}{10}} + 10^{\frac{P_{\text{LO Spur,dBm}}-2P_{IIP3,\text{dBm}}}{10}}\right) = 10^{\frac{P_{\text{Degradation}}}{10}} \tag{4.148}$$

A further simplification of Eqn (4.148) implies

$$P_{\text{blocker,dBm}} = 10\log_{10}\left(\frac{10^{\frac{P_{\text{Degradation}}}{10}}}{10^{\frac{L(F=\Delta F)\text{dBc/Hz}+10\ \log_{10}(B)-R}{10}}+10^{\frac{P_{\text{LO Spur,dBm}}-2P_{IIP3,\text{dBm}}}{10}}}\right) \tag{4.149}$$

The selectivity of a direct-conversion architecture would be expressed as the difference between the blocker signal in dBm minus the desired signal power or

$$\text{Selectivity}_{\text{dB}} = P_{\text{blocker,dBm}} - P_{\text{signal,dBm}} \tag{4.150}$$

The selectivity of a dual conversion receiver would involve the second mixer phase noise and spurs as well as any filtering of the blocker signal in between. This will become apparent when we discuss dual conversion receivers in a Chapter 8.

4.2.5 Image noise, image rejection, and mixers

Consider a nondirect conversion receiver where the mixer is not designed to reject the image frequency, as depicted in Figure 4.28. Let us first present the problem of image noise and then discuss the remedies. Assume for the moment that the mixer in Figure 4.28 is intended to down-convert a signal $x_{signal}(t)=A(t)\cos(\Omega_c t)$ from a certain *RF* band to an intermediate *IF* Ω_{IF} where $A(t)$ is the desired signal centered at Ω_c. Likewise, assume that there exists an interfering signal or simply noise at the *RF* band $x_{image}(t)=B(t)\cos([\Omega_c+2\Omega_{IF}]t)$ where $B(t)$ is the interferer or noise and is centered at $\Omega_c+2\Omega_{IF}$. So far we have not given a reason as to why it is relevant for the interferer or noise to exist at this particular frequency.

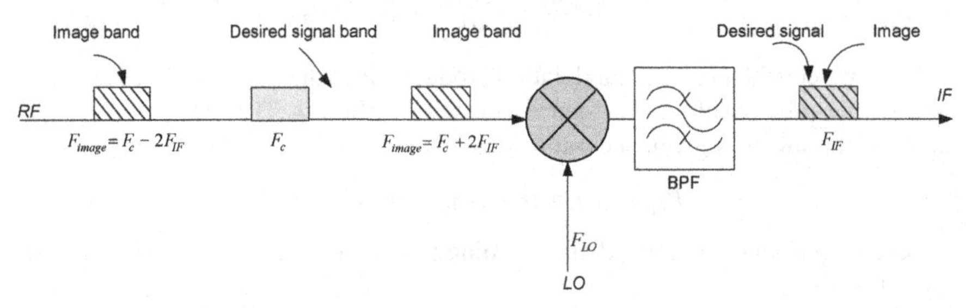

FIGURE 4.28

Image band folding onto the desired signal band after mixing. *IF*, intermediate frequency; BPF, bandpass filter; *RF*, radio frequency. (For color version of this figure, the reader is referred to the online version of this book.)

First, let us consider mixing the signal in the desired *RF* band by the *LO* signal $x_{LO}(t)=\Gamma\cos(\Omega_{LO}t)$, where Γ is the magnitude of the *LO* signal, we obtain

$$\begin{aligned} x_{signal}(t)x_{LO}(t) &= A(t)\Gamma\cos(\Omega_c t)\cos(\Omega_{LO}t) \\ &= \frac{A(t)\Gamma}{2}\{\cos([\Omega_c+\Omega_{LO}]t)+\cos([\Omega_c-\Omega_{LO}]t)\} \end{aligned} \tag{4.151}$$

Define the *IF* simply to be $\Omega_{IF} = \Omega_c - \Omega_{LO}$. Of course, depending on the frequency plan, we could've chosen Ω_{IF} to be $\Omega_c - \Omega_{LO}$, but for the sake of this analysis it does not make any difference which *IF* we pick and hence we picked the former. In this case, the *LO* frequency can be expressed in terms of the *IF* simply as $\Omega_{LO} = \Omega_c - \Omega_{IF}$ and hence Eqn (4.151) becomes:

$$x_{signal}(t)x_{LO}(t) = \frac{A(t)\Gamma}{2}\{\cos([2\Omega_c - \Omega_{IF}]t) + \cos(\Omega_{IF}t)\} \quad (4.152)$$

Assume that the bandpass filter shown in Figure 4.28 to be centered at the *IF* and hence for the sake of this discussion perfectly rejects the signal at $2\Omega_c - \Omega_{IF}$, then the remaining signal at its output is simply:

$$\widehat{x}_{signal}(t) = x_{signal}(t)x_{LO}(t) * h(t) = \frac{A(t)\Gamma}{2}\cos(\Omega_{IF}t) \quad (4.153)$$

Next we apply the same down-conversion analysis presented above to the image signal $x_{image}(t)$:

$$\begin{aligned} x_{image}(t)x_{LO}(t) &= B(t)\Gamma\cos([\Omega_c + 2\Omega_{IF}]t)\cos(\Omega_{LO}t) \\ &= \frac{B(t)\Gamma}{2}\{\cos([\Omega_c + 2\Omega_{IF} + \Omega_{LO}]t) + \cos([\Omega_c + 2\Omega_{IF} - \Omega_{LO}]t)\} \end{aligned} \quad (4.154)$$

Substitute $\Omega_{LO} = \Omega_c - \Omega_{IF}$ into Eqn (4.154) we obtain:

$$\begin{aligned} x_{image}(t)x_{LO}(t) &= \frac{B(t)\Gamma}{2}\{\cos([\Omega_c + 2\Omega_{IF} + \Omega_c - \Omega_{IF}]t) \\ &\quad + \cos([\Omega_c + 2\Omega_{IF} - \Omega_c + \Omega_{IF}]t)\} \\ &= \frac{B(t)\Gamma}{2}\{\cos([2\Omega_c + \Omega_{IF}]t) + \cos(\Omega_{IF}t)\} \end{aligned} \quad (4.155)$$

At the output of the bandpass filter of in Figure 4.28, the down-converted image signal simply becomes:

$$\widehat{x}_{image}(t) = x_{image}(t)x_{LO}(t) * h(t) = \frac{B(t)\Gamma}{2}\cos(\Omega_{IF}t) \quad (4.156)$$

The filter image noise or interferer as presented in Eqn (4.156) implies that at the output of the bandpass filter both filtered signal and interferer coexist in the same band, that is:

$$\widehat{x}_{signal}(t) + \widehat{x}_{image}(t) = \frac{[A(t) + B(t)]\Gamma}{2}\cos(\Omega_{IF}t) \quad (4.157)$$

Note that similar analysis can be performed for the $F_{image} = F_c - 2F_{IF}$ case.

Thus far, it has been shown that the unintended consequence of nonimage reject mixers is to down-convert two frequency bands onto the same *IF* band thus adding noise or interference on top of the desired signal. Of course, this discussion has excluded any noise due to the *LO*. In the event where a properly matched image-reject filter precedes the mixer, there will, in theory, be only thermal noise at the image frequency provided that the filter has sufficient rejection in the image band, as shown in Figure 4.29. Assume for the moment that the mixer does not contribute any noise of its own, then the noise present at *IF* due to mixing is the noise present in the signal band plus the noise present in the image band as illustrated in Figure 4.29. Furthermore, assuming no gain variations across the signal and image noise bands at *RF*, the SNR at the output of the mixer is 3 dB lower than the SNR at the input of the mixer. In this case, the *theoretical mixer* has a 3-dB noise figure! This case is known as the single sideband (SSB) noise figure. This is commonly the case in nondirect conversion or heterodyne receivers.

In direct conversion receiver the *IF* is zero. That is

$$F_{image} = F_c \pm 2F_{IF}|_{F_{IF}=0} \cong F_c \tag{4.158}$$

The relationship in Eqn (4.158) implies that the image noise is the same noise as the noise that exists in the signal band, and hence for the theoretical mixer of Figure 4.29, the SNR at the input mixer is the same as the SNR at the output of the mixer. In this case that is relevant to direct conversion receivers, the noise figure of the mixer is 0 dB. This noise figure is termed as the double sideband (DSB) noise figure of the mixer. The DSB noise figure is 3 dB lower than the SSB noise figure provided that both image and signal bands experience the same gain at the *RF* port of the mixer.

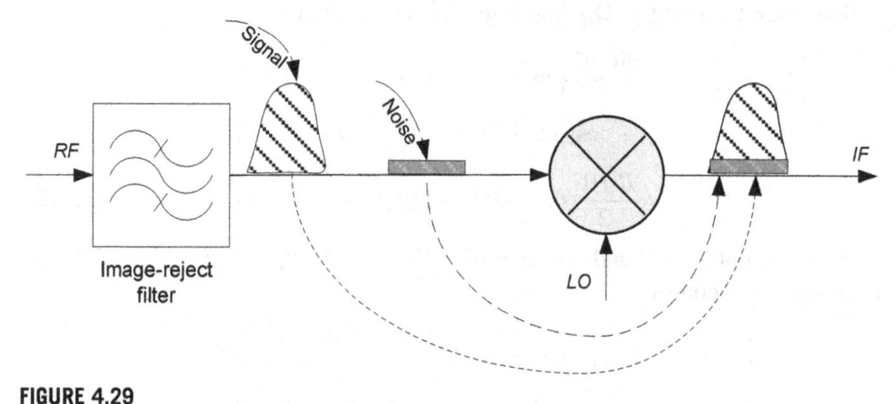

FIGURE 4.29

Signal and image noise folding into *IF*. *IF*, intermediate frequency; *RF*, radio frequency. (For color version of this figure, the reader is referred to the online version of this book.)

At this juncture, it is seems appropriate to discuss some of the key parameters of the mixer. We define the conversion loss of a mixer[7] as the ratio of *IF* power to *RF* power, or

[7]Also known as single sideband (SSB) conversion loss since the measurement is done at the chosen *IF*, which is either the sum of the *LO* and *RF* frequencies or their difference.

$$P_{\text{conversion loss}} = 10 \log_{10}\left(\frac{P_{IF}}{P_{RF}}\right) \tag{4.159}$$

where P_{IF} is the *IF* output power and P_{RF} is the available *RF* input power. Often times, a mixer is specified by its conversion gain, which is none other than the negative of the conversion loss. In its simplest forms, the conversion loss of a mixer is related to the *LO* power level as can be deduced from the analysis above. Given the *theoretical* mixer model used thus far, we see that the conversion loss is directly related to the *LO* power. For this model, decreasing the *LO* signal power serves to increase the conversion loss of the mixer. The *LO* drive level varies between 6 and 22 dBm for a double-balanced mixer. Single-balanced and double-balanced diode mixer circuits are shown in Figure 4.30. Mixer design, however, is not just

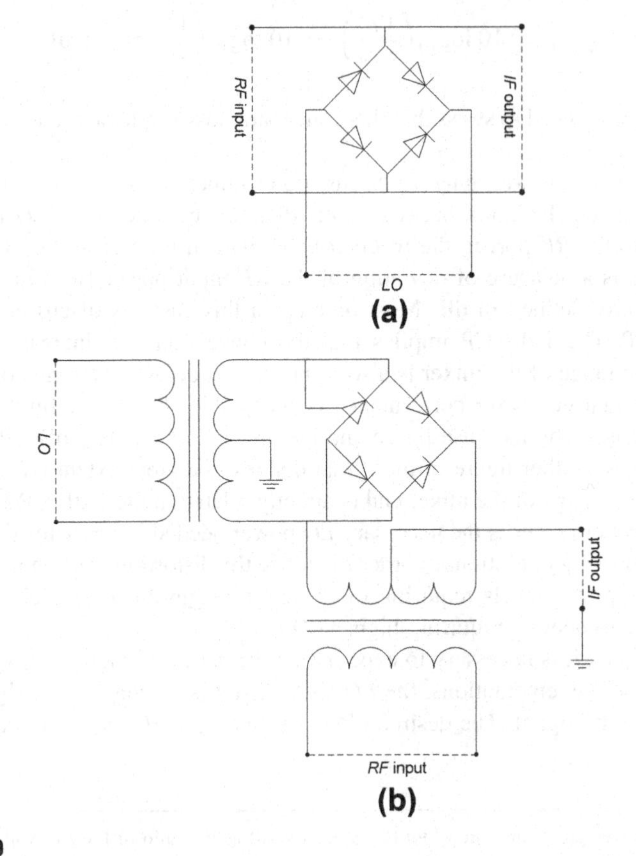

FIGURE 4.30

Single- and double-balanced mixers: (a) single-balanced and (b) double-balanced. *IF*, intermediate frequency; *RF*, radio frequency.

constrained to diodes. Schottky diode mixers are used when high performance mixers are desired. On the other hand, GaAs FETs and CMOS transistors are used in mixer design where cost, not performance, is the main driver.

In both balanced mixers, the *LO* signal is differential. If the *RF*-input connection in the mixer is single ended, then the mixer is said to be single balanced. If the *RF* signal on the other hand is differential, the mixer is said to be double balanced [8]. For a given power dissipation, single-balanced mixers tend to be less prone to input referred noise than double-balanced mixers, while the latter has less even-order distortion, thus reducing the impact of the half *IF* distortion.

The noise figure of the mixer is usually equal to or greater than the conversion loss. In practice, the conversion loss depends on the load of the input *RF* circuit as well as the impedance at the output-*IF* port. For a typical diode mixer, it is desired that $A(t) \times \Gamma \approx 1$, then according to Eqn (4.153), the conversion loss or more precisely the power conversion loss[8] is

$$P_{\text{conversion loss}} = 10\log_{10}\left(\frac{P_{IF}}{P_{RF}}\right) = 10\log_{10}\left(\frac{1}{2}\right)^2 \approx -6\text{ dB} \tag{4.160}$$

Of course, we need to stress that this conversion loss is relevant to a typical diode mixer.

At this point, it is warranted to discuss other important system parameters concerning the mixer. Isolation between ports dictates the amount of *LO* power that leaks back to the *RF* port in the receiver or *IF* port in the transmitter. Conversion compression is a measure of deviation of the *RF* input power from linearity by a certain quantity defined in dB. Most notably in this case, as discussed earlier, is the 1-dB CP. The 1-dB CP implies that the conversion loss increases by 1 dB. The dynamic range of the mixer is also of interest. It defines the range of *RF* input power for which the mixer performance is acceptable. The lower limit of the dynamic range is set by its noise figure and the upper limit by its 1-dB CP. The *IIP*3 of the mixer is another figure of merit that defines to a great extent the upper limit of the dynamic range of the mixer and is directly related to the 1-dB CP. Finally, the *LO* drive level or power is the necessary *LO* power needed in order for the mixer to perform the mixing operation. In order to reduce the distortion, certain mixers operate at higher power levels requiring more than a single diode in each arm of the mixer and consequently requiring higher *LO* power.

At this point, it is necessary to expound on the nature of the *LO* signal itself. In most practical implementations, the *LO* signal itself is a square wave signal rather than a sinusoidal signal. The desired effect of mixing the *RF* down to baseband or

[8]There is also a voltage conversion loss (or gain) defined as the ratio of the *rms* voltage signal at *IF* to the *rms* voltage signal at *RF*. The voltage conversion loss is equal to the power conversion loss if the input impedance and the load impedance of the mixer are both equal to the source impedance [8].

IF is more or less the same. In a single-balanced mixer, the *LO* signal can be modeled as a switch that toggles between a certain positive signal V_{LO} and DC, as shown in Figure 4.31(a). For a 50% duty cycle signal, and $V_{LO} = 1$ V, then the Fourier transform of the *LO* signal is given as

$$x_{LO}(t) = \frac{1}{2} + \sum_{n=1}^{\infty} \frac{\sin\left(\frac{n\pi}{2}\right)}{\frac{n\pi}{2}} \cos(2\pi nt/T) \tag{4.161}$$

where T is the period of the square wave. The relationship in Eqn (4.161) contains a DC term, the fundamental term at $1/T$, and all the odd harmonics scaled by $1/n$ from the fundamental. It does not contain any even harmonics. Modulating the desired signal $x_{signal}(t) = A(t)\cos(\Omega_c t)$ in a manner similar to Eqn (4.152) results in the output signal

$$x_{signal}(t)x_{LO}(t)\big|_{\Omega_{LO}=2\pi/T} = A(t)\cos(\Omega_c t)\left[\frac{1}{2} + \sum_{n=1}^{\infty} \frac{\sin\left(\frac{n\pi}{2}\right)}{\frac{n\pi}{2}} \cos(\Omega_{LO} nt)\right]$$

$$x_{signal}(t)x_{LO}(t)\big|_{\Omega_{LO}=2\pi/T} = \frac{1}{2}A(t)\cos(\Omega_c t) + A(t)\sum_{n=1}^{\infty} \frac{\sin\left(\frac{n\pi}{2}\right)}{\frac{n\pi}{2}} \cos(\Omega_{LO} nt)\cos(\Omega_c t) \tag{4.162}$$

The first term in Eqn (4.162) is the *RF* term at the desired signal frequency. The subsequent terms occur at sums and differences of the *LO* frequency and its odd harmonics and the desired signal's center frequency, that is $\Omega_c \pm n\Omega_{LO}$ for n odd.

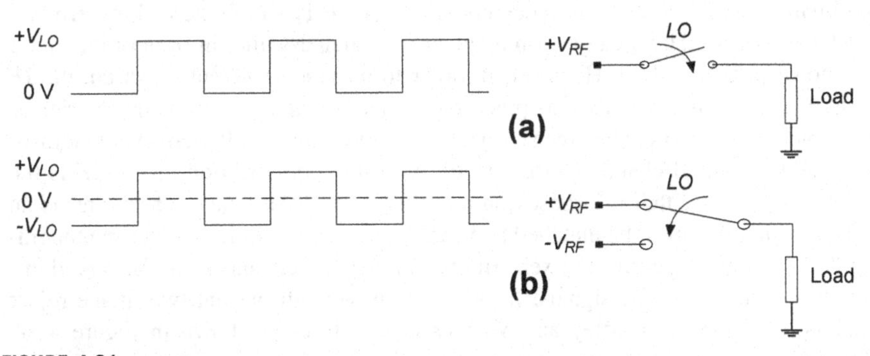

FIGURE 4.31

Single-balanced versus double-balanced mixer models: (a) single-balanced, and (b) double-balanced. *RF*, radio frequency.

Mathematically, there is no *LO*-only component in the resulting output signal of the mixer. However, *LO* leak through typically results due to poor isolation between the *LO* and *IF* or DC port.

In order to suppress the *RF* signal in Eqn (4.162), the DC component must be removed from the *LO* spectrum in Eqn (4.161). In this case, a double-balanced mixer is employed. In theory, a double-balanced mixer allows the *RF* signal to pass through for half a cycle and the negative of the *RF* signal for the other half as shown in Figure 4.31(b). The Fourier transform of the *LO* signal can simply be expressed via the relationship:

$$x_{LO}(t) = 2\sum_{n=1}^{\infty}\frac{\sin\left(\frac{n\pi}{2}\right)}{\frac{n\pi}{2}}\cos(2\pi nt/T) \qquad (4.163)$$

The added complexity of the double-balanced mixer is a small price to pay in comparison to the improvement in intermodulation suppression and low conversion loss. The high linearity and low noise result in an improved dynamic range. Unlike a single-balanced mixer, where the diodes are either on or off, in a double-balanced mixer half the diodes are usually turned on while the other half are turned off during half a cycle and vice versa during the other half depending on the polarity of the *LO* signal. The *LO* to *IF* port isolation is usually poorer in a double-balanced mixer than it is for a single-balanced mixer.

Furthermore, mixers can be classified as either passive or active. Passive mixers do not offer any gain while maintaining high nonlinearity. Active mixers, on the other hand, offer some gain thus reducing the overall noise contribution in the circuit. Therefore, in its simplest term, an active mixer is a mixer that can amplify the signal to be up- or down-converted using DC power. Therefore, Schottky diode mixers are said to be passive mixers, while transistor-based mixers such as MESFET or BJT-based designs are said to be active.

Finally, we address the image reject mixer. Recall that image rejection is not a problem in direct conversion receivers since the *IF* is simply zero. Furthermore, we have addressed image rejection by placing a bandpass filter in front of the mixer as shown in Figure 4.29. However, in order to provide sufficient rejection, the *IF* must be really high in order to place the image as far away from the carrier as possible. Furthermore, the image reject filter must have sufficient bandwidth to pass all the desired channels in the band. A further complication of low *IF* is the high quality factor that it imposes on the filter design. An image reject filter is an off-chip filter that must be matched to 50 Ω. An alternate solution to filtering the image is to use an image reject mixer. An image reject mixer cancels the unwanted image by manipulating the signals. In this section, we will present two image reject mixers based on the Hartley and Weaver architectures as shown in Figure 4.32. The basic premise of an image reject mixer is to split the signal into two nodes whereby the desired signal is combined with the same polarity whereas the image

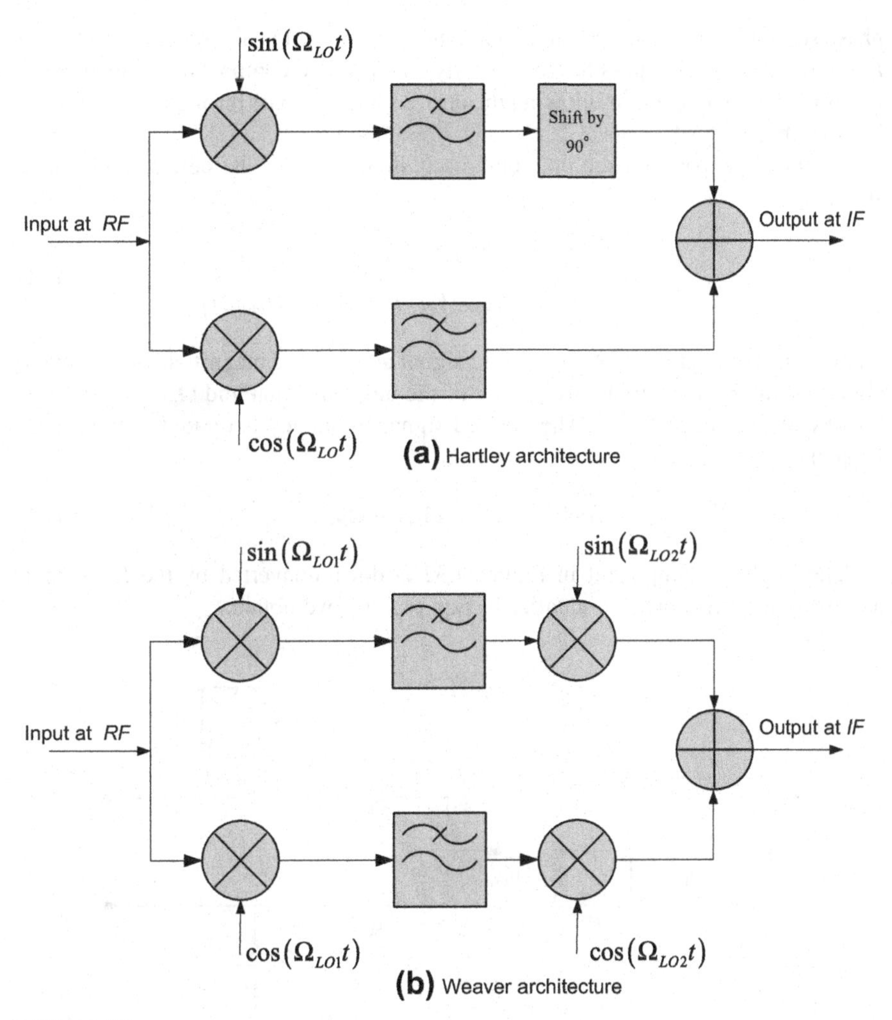

FIGURE 4.32

Image reject architecture, (a) Hartley architecture, and (b) Weaver architecture. *IF*, intermediate frequency; *RF*, radio frequency. (For color version of this figure, the reader is referred to the online version of this book.)

signal with opposite polarity such that when the signals are combined at *IF* the image cancels out.

First, let us consider the operation of the Hartley architecture depicted in Figure 4.32(a). The incoming *RF* signal $x_{RF}(t) = x_{signal}(t) + x_{image}(t)$ is comprised of the desired signal as well as the image signal split into two components. The in-phase component is down-converted by the *LO*, lowpass filtered, and then phase shifted by 90°. Similarly, the quadrature component is down-converted by the *LO* signal and then lowpass filtered. Finally, the in-phase component is added to the quadrature component thus allowing the desired signal to pass through and suppressing the image signal.

To further expand on the previous discussion, consider the desired and image signal $x(t)$:

$$\begin{aligned} x(t) &= x_{signal}(t) + x_{image}(t) \\ &= A\cos(\Omega_{RF}t) + B\cos(\Omega_{image}t) \end{aligned} \tag{4.164}$$

where A is the magnitude of the desired signal and B is the magnitude of the image signal. Similarly, Ω_{RF} is the frequency of the desired signal and Ω_{image} is the frequency of the image signal. The desired signal frequency is related to the image frequency as

$$\Omega_{RF} - \Omega_{LO} = \Omega_{LO} - \Omega_{image} \tag{4.165}$$

The in-phase component in Figure 4.32 is down-converted by the *LO* signal, and using the trigonometric identity in Eqn (4.230), we obtain:

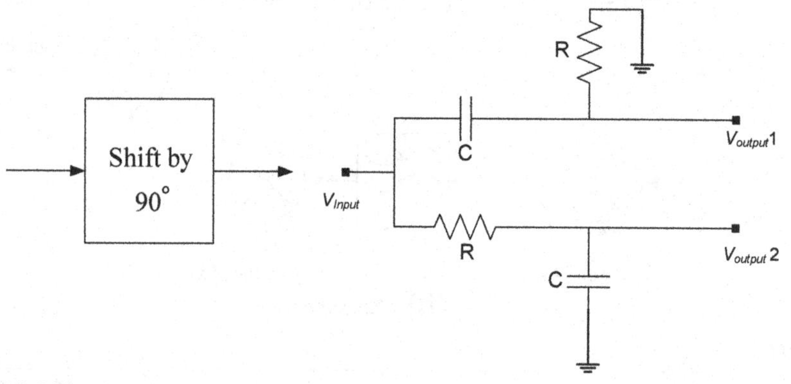

FIGURE 4.33

Shift by 90° circuit as shown in Ref. [8].

$$\begin{aligned} I(t) &= x(t)\sin(\Omega_{LO}t) = \left[A\cos(\Omega_{RF}t) + B\cos(\Omega_{image}t)\right]\sin(\Omega_{LO}t) \\ &= \frac{A}{2}\{\sin[(\Omega_{LO}+\Omega_{RF})t] + \sin[(\Omega_{LO}-\Omega_{RF})t]\} \\ &+ \frac{B}{2}\{\sin[(\Omega_{LO}+\Omega_{image})t] + \sin[(\Omega_{LO}-\Omega_{image})t]\} \end{aligned} \tag{4.166}$$

The signal *I(t)* in Eqn (4.166) is then lowpass filtered resulting in the components $\sin[(\Omega_{LO} + \Omega_{RF})t]$ and $\sin[(\Omega_{LO} + \Omega_{image})t]$ to be *theoretically* suppressed, then Eqn (4.166) becomes:

$$\widehat{I}(t) = \frac{A}{2}\sin[(\Omega_{LO} - \Omega_{RF})t] + \frac{B}{2}\sin\left[\left(\Omega_{LO} - \Omega_{image}\right)t\right] \tag{4.167}$$

Using the identity expressed in Eqn (4.239), the relation in Eqn (4.167) can be rewritten as:

$$\widehat{I}(t) = -\frac{A}{2}\sin[(\Omega_{RF} - \Omega_{LO})t] + \frac{B}{2}\sin\left[\left(\Omega_{LO} - \Omega_{image}\right)t\right] \tag{4.168}$$

The resulting signal $\widehat{I}(t)$ is then phase shifted by 90° via the *RC* circuit depicted in Figure 4.33 to obtain according to Eqn (4.240)

$$\widehat{I}_{\angle 90^\circ}(t) = +\frac{A}{2}\cos[(\Omega_{RF} - \Omega_{LO})t] - \frac{B}{2}\cos\left[\left(\Omega_{LO} - \Omega_{image}\right)t\right] \tag{4.169}$$

A more practical approach to obtain a phase shift by 90° is to implement a +45° phase shift in the in-phase path and a −45° phase shift in the quadrature path as explained in [8]. Hence, a simple *RC* circuit is placed in the in-phase path and a simple *CR* circuit in the quadrature path as shown in Figure 4.34.

Next, we turn our attention to the quadrature component. The signal is down-converted by the *LO* to obtain:

$$\begin{aligned} Q(t) &= x(t)\cos(\Omega_{LO}t) = \left[A\cos(\Omega_{RF}t) + B\cos\left(\Omega_{image}t\right)\right]\cos(\Omega_{LO}t) \\ &= \frac{A}{2}\{\cos[(\Omega_{LO} + \Omega_{RF})t] + \cos[(\Omega_{RF} - \Omega_{LO})t]\} \\ &\quad + \frac{B}{2}\left\{\cos\left[\left(\Omega_{LO} + \Omega_{image}\right)t\right] + \cos\left[\left(\Omega_{image} - \Omega_{LO}\right)t\right]\right\} \end{aligned} \tag{4.170}$$

The quadrature signal in Eqn (4.170) undergoes ideal lowpass filtering to obtain:

$$\widehat{Q}(t) = \frac{A}{2}\cos[(\Omega_{RF} - \Omega_{LO})t] + \frac{B}{2}\cos\left[\left(\Omega_{image} - \Omega_{LO}\right)t\right] \tag{4.171}$$

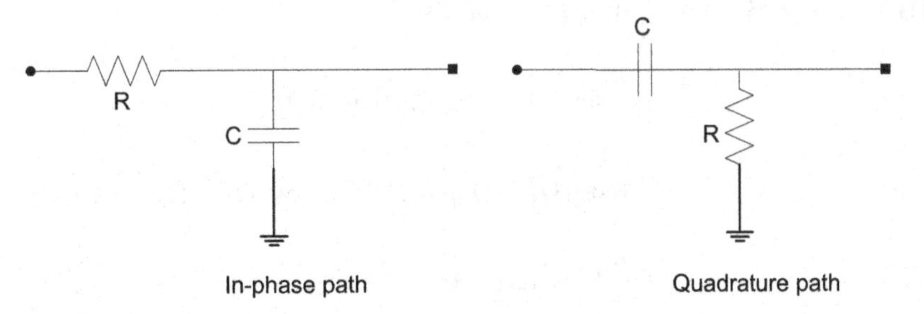

FIGURE 4.34

Basic *RC* and *CR* phase-shift circuits.

Finally, adding the result in Eqns (4.169)–(4.171) we obtain:

$$\begin{aligned}\widehat{I}_{\angle 90^\circ}(t) + \widehat{Q}(t) &= \frac{A}{2}\cos[(\Omega_{RF} - \Omega_{LO})t] - \frac{B}{2}\cos\left[\left(\Omega_{LO} - \Omega_{image}\right)t\right] \\ &\quad + \frac{A}{2}\cos[(\Omega_{RF} - \Omega_{LO})t] + \frac{B}{2}\cos\left[\left(\Omega_{image} - \Omega_{LO}\right)t\right] \\ &= A\cos[(\Omega_{RF} - \Omega_{LO})t] \end{aligned} \tag{4.172}$$

It is obvious that Eqn (4.172) is free from the image component of the signal. However, this is only true if the in-phase and quadrature paths are *perfectly* balanced. Any imbalance will create partial cancellation in the image. The imperfections are mainly due to the mixing process, inaccuracy in the resistor and capacitor parameters, as well as temperature variations.

In order to quantify the amount of image rejection, the image-reject ratio (IRR) is used as a metric:

$$IRR = \frac{\text{Desired signal level}}{\text{Image signal level}} \tag{4.173}$$

A derivation of the image reject ratio is found in Appendix B. Having said that, the amount of image rejection supplied by the mixer and coupled with front-end filtering, and provided we use high *IF*, could be sufficient depending on the image rejection dictated by the required performance. Furthermore, proper nonlinearity analysis of the mixer, which involves a weak received desired signal and a strong *LO* signal, cannot be accurately performed using the techniques presented above. For this reason, engineers resort to more sophisticated techniques such as harmonic balance, presented in Appendix C, to provide more accurate analysis of the circuit.

In order to mitigate the problems associated with the phase shift, we turn our attention to an alternate architecture, namely the Weaver architecture. In the Weaver architecture, a second quadrature mixing stage replaces the 90° phase shift. After the first down conversion by Ω_1 and lowpass filtering, the in-phase component is down-converted a second time by Ω_2 such that $\Omega_2 \ll \Omega_1$ to obtain:

$$\begin{aligned}\tilde{I}(t) = \widehat{I}(t)\sin(\Omega_2 t) &= \frac{A}{2}\sin[(\Omega_1 - \Omega_{RF})t]\sin(\Omega_2 t) \\ &\quad + \frac{B}{2}\sin\left[\left(\Omega_1 - \Omega_{image}\right)t\right]\sin(\Omega_2 t) \\ &= \frac{A}{4}\{\cos[(\Omega_1 - \Omega_{RF} - \Omega_2)t] - \cos[(\Omega_1 - \Omega_{RF} + \Omega_2)t]\} \\ &\quad + \frac{B}{4}\{\cos\left[\left(\Omega_1 - \Omega_{image} - \Omega_2\right)t\right] \\ &\quad - \cos\left[\left(\Omega_1 - \Omega_{image} - \Omega_2\right)t\right]\} \end{aligned} \tag{4.174}$$

Next, consider the quadrature component and examine the signal after it has undergone the first down-conversion, lowpass filtering, and then the second down-conversion:

$$\begin{aligned}\tilde{Q}(t) &= \widehat{Q}(t)\cos(\Omega_2 t) = \frac{A}{2}\cos[(\Omega_{RF}-\Omega_1)t]\cos(\Omega_2 t)\\&+\frac{B}{2}\cos\left[(\Omega_{image}-\Omega_1)t\right]\cos(\Omega_2 t)\\&=\frac{A}{4}\{\cos[(\Omega_{RF}-\Omega_1+\Omega_2)t]+\cos[(\Omega_{RF}-\Omega_1-\Omega_2)t]\}\\&+\frac{B}{4}\left\{\cos\left[(\Omega_{image}-\Omega_1+\Omega_2)t\right]+\cos\left[(\Omega_{image}-\Omega_1-\Omega_2)t\right]\right\}\end{aligned} \tag{4.175}$$

Subtracting the relation in Eqn (4.174) from Eqn (4.175), we obtain:

$$\tilde{Q}(t)-\tilde{I}(t) = \frac{A}{2}\cos[(\Omega_{RF}-\Omega_1-\Omega_2)t]+\frac{B}{2}\cos\left[(\Omega_{image}-\Omega_1-\Omega_2)t\right] \tag{4.176}$$

Define the *IF* as:

$$\Omega_{IF} = \Omega_{RF}-\Omega_1-\Omega_2 = \Omega_1+\Omega_2-\Omega_{image} \tag{4.177}$$

Substituting Eqn (4.177) into Eqn (4.176), we obtain:

$$\tilde{Q}(t)-\tilde{I}(t) = \frac{A}{2}\cos(\Omega_{IF}t)+\frac{B}{2}\cos[(\Omega_1+\Omega_2-\Omega_{IF})t] \tag{4.178}$$

The choice of Ω_1 and Ω_2 is not unique, however, it does present a secondary image problem as evident in the right hand of Eqn (4.178). To eliminate this problem, bandpass filters may be used in lieu of lowpass filters. However, this would imply filtering off-chip thus leading to more complexity in the design. Like the Hartley architecture, the weaver architecture suffers from any imbalance present between the in-phase and quadrature paths. Any gain and phase mismatch implies less than perfect image rejection. Both architectures depend on good layout techniques in order to ensure optimum matching. A further complexity of the Weaver architecture is the need for a second *LO* frequency.

4.2.6 Able-Baker spurs and intermodulation products

In order to understand how Able-Baker spurs are generated, let us consider the input signal in Eqn (4.164). Substitute Eqn (4.164) into the nonlinear memoryless polynomial provided in Eqn (4.12), we obtain:

$$\begin{aligned}y(t) &= \sum_{n=0}^{\infty}\beta_n x^n(t) = \beta_0+\beta_1 x(t)+\beta_2 x^2(t)+\beta_3 x^3(t)+\cdots\Big|_{x(t)=A\cos(\Omega_1 t)+B\cos(\Omega_2 t)}\\&=\beta_0+\beta_1\{A\cos(\Omega_1 t)+B\cos(\Omega_2 t)\}+\beta_2\{A\cos(\Omega_1 t)+B\cos(\Omega_2 t)\}^2\\&+\beta_3\{A\cos(\Omega_1 t)+B\cos(\Omega_2 t)\}^3+\cdots\end{aligned} \tag{4.179}$$

Table 4.4 Polynomial Expansion Using Eqn (4.180)

$n_1 \geq 0$	$n_2 \geq 0$	$n_1+n_2=n \triangleq 2$	Expression
0	0	Not valid	N/A
0	1		
1	0		
0	2	0+2	$\frac{2!}{0!2!}1 \times B^2\cos^2(\Omega_2 t) = B^2\cos^2(\Omega_2 t)$
1	1	1+1	$\frac{2!}{1!1!}A\cos(\Omega_1 t) \times B\cos(\Omega_2 t)$ $= 2A\cos(\Omega_1 t) \times B\cos(\Omega_2 t)$
2	0	2+1	$\frac{2!}{2!0!}A^2\cos^2(\Omega_1 t) \times 1 = A^2\cos^2(\Omega_1 t)$

The frequencies of the two-tone input signal $x(t)$ are generic expressed as Ω_1 and Ω_2. The two-tone input signal can be used to model an *RF* input signal mixed with an *LO* signal. If it is desired to predict the *IM* products due to two or more *RF* input signals mixed with an *LO* signal, a multitone signal must be used. If we assume the mixer, or for that matter any nonlinear memoryless device, to possess all the nonlinear components in Eqn (4.179), then a generalized formula is needed to predict the spur locations due to intermodulation or harmonics and *roughly* their respective levels. To do so, let us examine the relation $\{A\cos(\Omega_1 t)+B\cos(\Omega_2 t)\}^n$, which according to Eqn (4.241) can be expressed as:

$$\{A\cos(\Omega_1 t)+B\cos(\Omega_2 t)\}^n = \sum_{\substack{n_1,n_2\geq 0\\ n_1+n_2=n}} \frac{n!}{n_1!n_2!}A^{n_1}B^{n_2}\cos^{n_1}(\Omega_1 t)\cos^{n_2}(\Omega_2 t) \tag{4.180}$$

By way of a very simple example, let us compute $\{A\cos(\Omega_1 t)+B\cos(\Omega_2 t)\}^2$ as shown in Table 4.4 using the relation in Eqn (4.180) to further illustrate the notation.

For n_1 and n_2 even, the relationship in Eqn (4.180) can be further expressed with the aid of Eqn (4.236) as

$$\begin{aligned}\{A\cos(\Omega_1 t)+B\cos(\Omega_2 t)\}^n &= \sum_{\substack{n_1,n_2\geq 0\\ n_1+n_2=n}} \frac{n!}{n_1!n_2!}A^{n_1}B^{n_2}\cos^{n_1}(\Omega_1 t)\cos^{n_2}(\Omega_2 t)\\ &= \left(\frac{1}{2}\right)^{n-1}\sum_{\substack{n_1,n_2\geq 0\\ n_1+n_2=n}} \frac{n!}{n_1!n_2!}A^{n_1}B^{n_2}\\ &\quad\times\left\{\sum_{l=0}^{\frac{n_1-1}{2}}\binom{n_1}{l}\cos[((n_1-2l)\Omega_1 \pm n_2\Omega_2)t]+\cdots\right\}\end{aligned} \tag{4.181}$$

According to Eqn (4.181), it can be shown that in general a spur has the form:

$$A^{n_1} B^{n_2} C_{n_1,n_2 k_1,k_2} \cos[(k_1\Omega_1 \pm k_2\Omega_2)t]\Big|_{\substack{k_1 = 0\cdots n_1 \\ k_2 = 0\cdots n_2}} \tag{4.182}$$

where C_{n_1,n_2k_1,k_2} is a coefficient that can be computed via Eqn (4.181). The relationships in Eqns (4.180) and (4.181) show that higher order *IM* products tend to be more suppressed than lower order ones. These relationships, however, may be used to predict the power suppression trends of certain *IM* products rather than exact powers. For that specific purpose, intermodulation tables could be used to predict the theoretical *IM* levels [9,10]. In reality, however, either more accurate design models based on empirical data or measurements in the laboratory are utilized to predict the *IM* products accurately. The problem becomes more complicated when a multitone input signal is used. Although predicting the *IM* levels is not a trivial task, the prediction of the *IM* frequency remains a simpler procedure. Mixing a multitone input signal with the *LO* frequency results in *IM* products centered at:

$$\Omega_{IM} = \pm k_1\Omega_1 \pm k_2\Omega_2 \pm \cdots \pm k_N\Omega_N \pm k_{LO}\Omega_{LO} \tag{4.183}$$

Predicting the location of the *IM* products that occur in band may be done using any number of commercially available software such as SpurFinder™.[9] A spur or *IM* chart using this software is depicted in Figure 4.35. For two-tone analysis, the spurs and *IM* frequencies are given according to the general relation [11]:

$$F_{IF} = mF_{RF} + nF_{LO}, m, n \text{ are integers} \tag{4.184}$$

That is, the *IF* is produced by a certain combination of the input *RF* frequency and the *LO*.

Normalizing Eqn (4.184) by the *IF* leads to the relation:

$$m\frac{F_{RF}}{F_{IF}} + n\frac{F_{LO}}{F_{IF}} = 1, m, n \text{ are integers} \tag{4.185}$$

The formula in Eqn (4.185) can be used to construct a spur chart such as the one described in Figure 4.35. Based on Eqn (4.185), the desired *IF* occurs when $|m| = |n| = 1$, and hence at the output of the mixer, the desired response, based on first harmonics mixing, is either $\cos(\Omega_{RF} + \Omega_{LO})$ or $\cos(\Omega_{RF} - \Omega_{LO})$. All other *IM* products are not desirable and must be minimized via filtering or by design to acceptable levels. If an undesired *IM* product is in band and cannot be filtered out, then its level must be sufficiently low in order to meet the system's specifications. If that is not the case, then the designer is obliged to choose a different frequency plan. As a guiding principle, one must avoid in-band low-order *IM* products due to their large signal strength levels as projected by Eqn (4.180).

[9]SpurFinder™ is a graphical-based frequency planning tool that plots spurious frequencies.

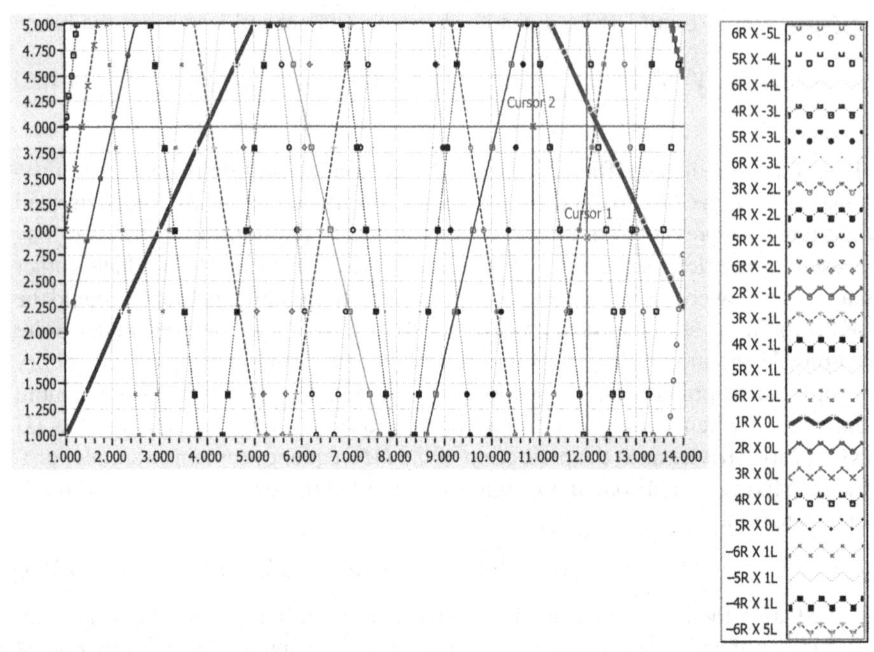

FIGURE 4.35

Spur chart obtained using the SpurFinder Software™: *RF* versus *LO* plot. *RF*, radio frequency. (For color version of this figure, the reader is referred to the online version of this book.)

EXAMPLE 4.4 CONSTRUCTION OF SPUR CHARTS

This is a rhetorical and very simplified example concerning the use of spur charts. A spur chart is a tool that a radio designer can use to develop a suitable frequency plan for his or her transceiver. The goal is to understand the utility and underlying principles of spur charts. In the following, all degradations other than *IM*s and spurs will be ignored.

Given an *RF* tone at 400 MHz that needs to be demodulated to an *IF* of 100 MHz using a 300 MHz *LO*. Using Eqn (4.185), draw the spurs (1×-1), (2×-1), and (3×-1). In a like manner, draw (1×1), (2×1), and (3×1). What is the impact of the $(4, -5)$ spur?

Next, consider a 20 MHz signal at 600 MHz *RF*. The signal is down-converted to 100 MHz *IF*. Assume that the *IF* filter is a 100-MHz-wide bandpass filter. What is the impact of having such a wideband filter? What is the impact of replacing the 100-MHz filter with a 50-MHz bandpass filter centered at *IF*?

Using the relationship in Eqn (4.185), it is obvious that

$$\frac{F_{RF}}{F_{IF}} - \frac{F_{LO}}{F_{IF}} = \frac{400}{100} - \frac{300}{100} = 1 \tag{4.186}$$

That is, Eqn (4.185) is true for (1×-1). The spurs for $(1 \times \pm 1)$, $(1 \times \pm 2)$, $(1 \times \pm 3)$, and (4×-5) are depicted in Figure 4.36. According to Figure 4.36, spur (4×-5) intersects

EXAMPLE 4.4 CONSTRUCTION OF SPUR CHARTS—cont'd

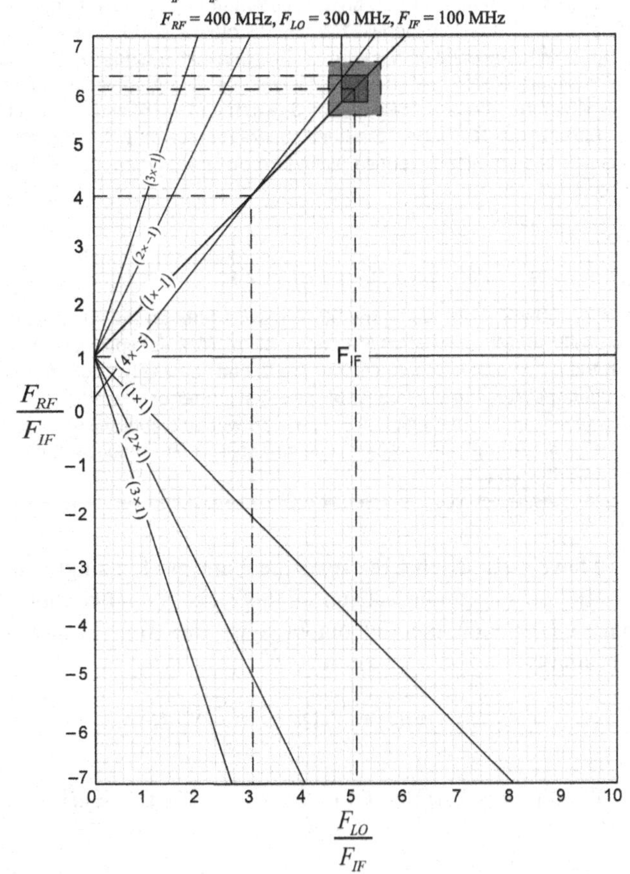

FIGURE 4.36

Spur chart for (1, ±1), (1, ±2), (1, ±3), and (4, −5) spurs. *IF*, intermediate frequency; *RF*, radio frequency. (For color version of this figure, the reader is referred to the online version of this book.)

(1 × −1). That is, spur (4,−5) appears at the *IF* and will degrade the quality of the desired signal. This can be verified simply using Eqn (4.185), that is

$$\frac{F_{RF}}{F_{IF}} - \frac{F_{LO}}{F_{IF}} = \frac{4 \times 400}{100} - \frac{5 \times 300}{100} = 1 \tag{4.187}$$

Continued

EXAMPLE 4.4 CONSTRUCTION OF SPUR CHARTS—cont'd

Note that, although we have used Eqn (4.185) to construct Figure 4.36, the coordinate axis used F_{RF}/F_{IF} and F_{LO}/F_{IF} is not unique. This chart depicts the impact of changing the *RF* and the *LO* while keeping the *IF* fixed.

Next, down-converting the 20 MHz signal from 600 MHz *RF* to 100 MHz *IF* implies that the *LO* frequency for this example is 500 MHz as shown in Figure 4.36. In order to assess the impact of having a 100-MHz-bandpass filter, let us draw a 100-MHz square centered at the locus point, that is the point on the (1 × −1) line that corresponds to 600 MHz *RF* and 500 MHz *IF*. The square is represented with a dashed line. It is graphically apparent that choosing a 100 MHz filter allows for degradation due to the (4 × −5) spur. This is true since the spur line intersects the 100-MHz dash-lined square representing the 100-MHz-bandpass filter. This can be numerically verified using Eqn (4.184) and an interferer say at 625 MHz as:

$$\left.\begin{aligned} F_{(4\times-5)} &= 4F_{RF} - 5F_{LO} \\ &= 4 \times 625 - 5 \times 475 = 125\text{ MHz} \end{aligned}\right\} m = 4, n = -5 \tag{4.188}$$

that is well within 100±50 MHz. Using the 50-MHz-wide filter on the other hand, depicted graphically as a solid square, shows that the impact of the (4 × −5) spur becomes peripheral. Therefore, with the aid of a well-populated spur chart, one can easily visualize the spurs and envision the frequency plan and the impact of the chosen filter bandwidths.

Other instantiations of the spur graph in Figure 4.36 will be to plot the spurs given the axis F_{RF}/F_{LO} and F_{IF}/F_{LO}, for example. In this case, one may observe the impact of the spurs as the *IF* and filter bandwidths vary.

Thus far, we have assumed that the spurs are comprised of single tones. The *RF* input signals to the *LO* may be broadband. In this case, given two input signals to a nonlinear device $x_1(t) + x_2(t)$, then the output signal due to first order nonlinearity, for example, is simply:

$$\begin{aligned} y(t) &= C_{1,1}x_1(t)x_2(t) \Leftrightarrow Y(\Omega) = C_{1,1}X_1(\Omega) * X_2(\Omega) \\ C_{1,1}X_1(\Omega) * X_2(\Omega) &\Leftrightarrow C_{1,1}\int_{-\infty}^{\infty} X_1(\xi) * X_2(\Omega - \xi)d\xi \end{aligned} \tag{4.189}$$

where $C_{1,1}$ is a constant due to first order nonlinearity. Higher order nonlinearities can also be expressed as:

$$\begin{aligned} y(t) &= C_{m,n}x_1^m(t)x_2^n(t) \\ Y(\Omega) &= C_{m,n}\underbrace{X_1(\Omega) * X_1(\Omega)\cdots * X_1(\Omega)}_{m-\text{times}} * \underbrace{X_2(\Omega) * X_2(\Omega)\cdots * X_2(\Omega)}_{n-\text{times}} \end{aligned} \tag{4.190}$$

where $C_{m,n}$ is a constant. Both Eqns (4.189) (4.190) show that a certain bandwidth expansion takes place when the input signals to a nonlinear system exhibit a certain bandwidth larger than 1 Hz. Therefore, in the case of Eqn (4.189), for example, if

$x_1(t)$ has a bandwidth B_1, for example, and $x_2(t)$ has a bandwidth B_2, then the resulting *IM* signal has a bandwidth that is the sum of $B_1 + B_2$.

4.2.7 Frequency multipliers and dividers

For certain frequencies of operations, for example, millimeter waves, it becomes difficult to build fundamental frequency oscillators that exhibit acceptable performance in terms of stability and noise characteristics,[10] Instead, a common approach would be to generate with the aid of a frequency multiplier a harmonic of the frequency oscillator. Frequency multipliers are also used to generate harmonically related signals used as building blocks in synthesizers.

A frequency multiplier is a nonlinear two-port device that takes in a single-tone input, for example, $x(t) = A\cos[\Omega t + \phi(t)]$ and produces an output tone of the form:

$$y(t) = B\cos\left(M[\Omega t + \phi(t)] + \theta\right) \tag{4.191}$$

where M is an integer. For $M = 2$, the device is called a frequency doubler, and for $M = 3$ it is called a frequency tripler. The phase θ is a variable independent of time.

There are two parameters of interest in Eqn (4.191), namely the phase $\theta(t)$ and the ratio between the output and input signal amplitudes. From our study on phase noise, if the phase $\phi(t)$ represents the phase noise of the input signal, then the output signal exhibits the same phase noise spectrum multiplied by 20 $\log_{10}(M)$. Thus, a frequency doubler will increase the phase noise by a factor of roughly 6 dB whereas a frequency tripler will increase the phase noise by a factor of 9.54 dB. Obviously, in a divider, the theoretical effect is exactly the opposite. The ratio between the output and input amplitude is referred to as the conversion gain or loss depending on whether one employs an active device using MESFET or HBT, for example, as opposed to using a passive device such as a diode. A single-diode frequency multiplier is shown in Figure 4.37 whereas a conceptual frequency multiplier based on an active device is depicted in Figure 4.38.

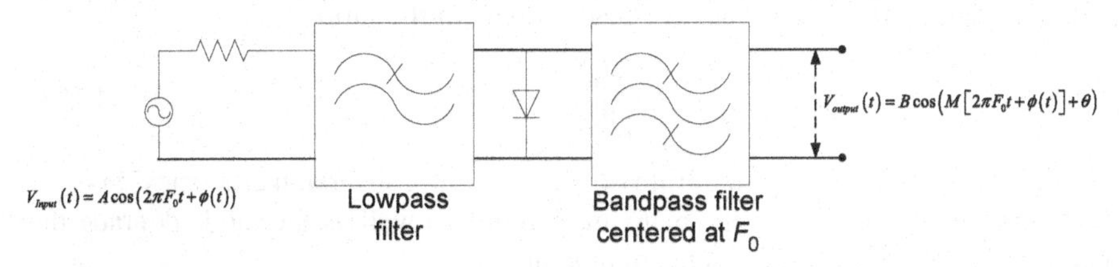

FIGURE 4.37

Passive frequency multiplier using single diode.

[10]The performance of solid state oscillators declines as a function of increasing the fundamental frequency of the oscillator. Therefore, designers rely on high-performance low-frequency oscillators followed by frequency multipliers to achieve the desired frequency.

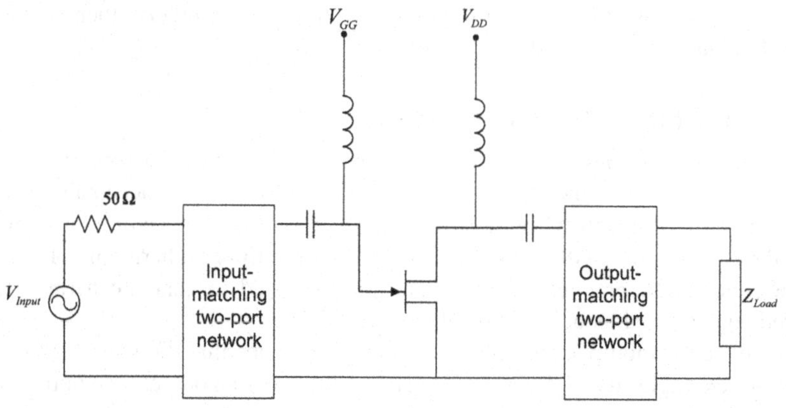

FIGURE 4.38

Active frequency multiplier with T-biased network.

Passive multipliers are simple and reliable. A passive multiplier is classified as resistive or reactive. The nonlinear element from which the various frequency harmonics are produced is typically a diode. The conversion loss can be compensated for using an amplifier in cascade with the multiplier circuit. Resistive multipliers have in theory infinite bandwidth [12]. This is true since resistive nonlinearities tend to be frequency independent. In this case, the bandwidth limitation is mainly due to the matching circuits [13]. Based on the Manley–Rowe relation [14,15], when a single-source frequency is used, a reactive multiplier satisfies the relation:

$$\sum_{m=1}^{\infty} P_m = 0, \text{ or } \sum_{m=2}^{\infty} P_m = P_1 \tag{4.192}$$

where P_m represents the average power of the mth harmonic. The average power at DC or P_0 is zero whereas the power of the first harmonic $P_1 > 0$ since it represents the average power delivered by the source. Terminating all the harmonics except the mth harmonic with lossless reactive loads results in the ratio

$$\left|\frac{P_m}{P_1}\right| = 1 \tag{4.193}$$

The implication, *in theory*, of Eqn (4.193) is 100% conversion efficiency. In practice, however, losses due to the matching circuits as well as the diode degrade the conversion efficiency by a significant amount.

The Manley–Rowe relations are not applicable for resistive multipliers using Schottky-barrier detector diode. This is true since a resistive multiplier is not lossless. Accordingly, it can be shown that

$$\sum_{m=0}^{\infty} m^2 P_m \geq 0 \tag{4.194}$$

Terminating all the harmonics except the *m*th harmonic implies that according to Eqn (4.194)

$$P_1 + M^2 P_m \geq 0 \tag{4.195}$$

or alternatively

$$\left|\frac{P_m}{P_1}\right| \leq \frac{1}{M^2} \tag{4.196}$$

Hence, the relationship in Eqn (4.196) demonstrates that a frequency multiplier's efficiency diminishes as a function of the square of the multiplication value M. Resistive multipliers become unpopular for high values of M as compared to reactive multipliers. Having said that, resistive multipliers tend to be more stable and operate at higher bandwidth than their reactive counterparts. Diode frequency multipliers can be improved upon by employing two diodes in balanced configuration, thus increasing the output power and improving the input impedance characteristics of the multiplier. The particular configuration of the diodes could lead to the rejection of all even or all odd harmonics [16–18]. A balanced frequency doubler configuration is shown in Figure 4.39. The purpose of the 180° coupler is to drive the two doublers with out-of-phase input signals. The doublers' outputs are then combined in phase, resulting in the fundamental and all odd harmonics to cancel out. The second harmonic and consequently all even order harmonics combine in phase. The input and output are thus isolated with the use of filters. The output power is obviously 3 dB higher due to coherent combining.

Next, we briefly turn our attention to frequency dividers. Frequency divider circuits are numerous and varied in design [11]. In this section, we will concentrate on regenerative dividers. The other type of divider relies on a phase-locked loop approach and is known as an oscillating divider. Regenerative dividers are based on passive circuits whereby the input signal at the fundamental frequency is divided in frequency to produce a signal at the desired fractional frequency. An example of a

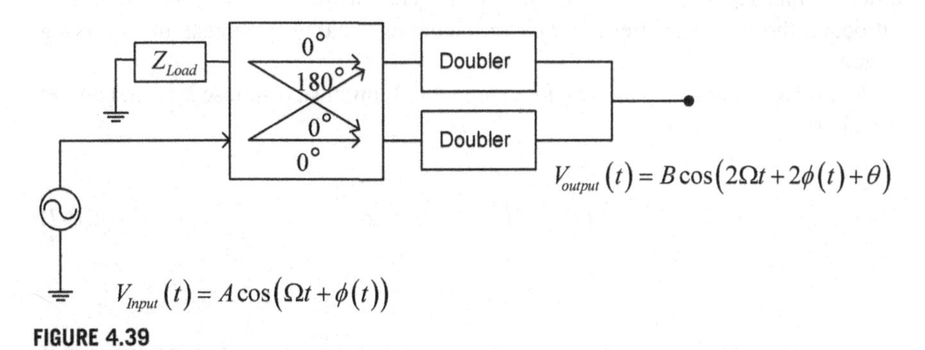

FIGURE 4.39

Balanced frequency doubler configuration.

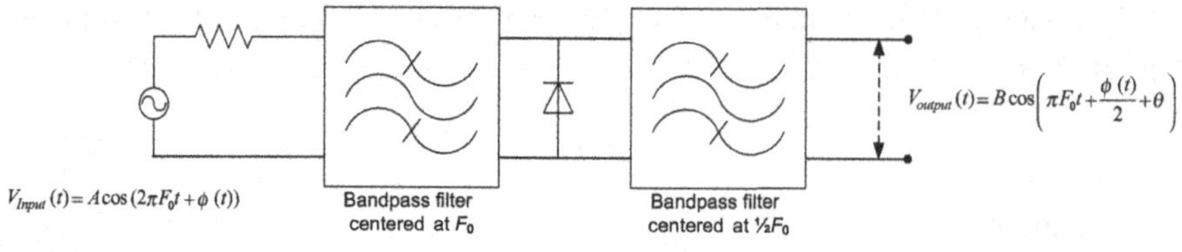

FIGURE 4.40

Passive frequency divide-by-2 using single diode.

regenerative divide-by-2 frequency divider is depicted in Figure 4.40. The reverse-biased diode acting as the nonlinear element is responsible for generating the desired subharmonic frequency. In [19], a balanced frequency divider configuration is presented using a balun.

Frequency multipliers and dividers usually involve very strong input signals and hence, in most cases, simple nonlinear techniques won't suffice to perform the analysis. In such scenarios, engineers resort to more advanced techniques such as harmonic balance, presented in Appendix C, to analyze the circuit.

4.3 Nonlinearity systems with memory: the Volterra series

In our analysis thus far, we have relied on the Taylor series expansion to analyze *weakly* nonlinear systems. However, a major drawback of Taylor series–based analysis techniques is their inability to model memory effects in the circuit. That is, the Taylor series can be used to analyze weakly nonlinear systems whose output strictly depends on the input signal at a particular time. In order to analyze the output of a nonlinear system at all other times, a new analysis tool is needed. Norbert Weiner was the first person to apply the Volterra[11] series to nonlinear circuit analysis. Weiner used the series to analyze the spectrum of FM signals in the presence of Gaussian noise. Volterra series can be employed to study the intermodulation properties of time-varying circuits such as mixers. The intent of this section is to briefly introduce the use of Volterra series in analyzing weakly nonlinear time-varying systems.

Consider a causal system with input signal $x(t)$, impulse response $h(t)$, and output signal $y(t)$ given as

$$y(t) = \int_{-\infty}^{\infty} h(\sigma)x(t-\sigma)d\sigma \tag{4.197}$$

[11]Introduced by Vito Volterra in 1887 in his *Theory of Functionals*.

Recall that a weakly memoryless nonlinear system can be represented by the Taylor series expansion:

$$y(t) = \sum_{n=1}^{\infty} \beta_n x^n(t) \tag{4.198}$$

where $\{\beta_n : n = 1, 2, \cdots\}$ are the Taylor series coefficients. A Volterra series combines Eqns (4.197) and (4.198) in order to incorporate memory to the model:

$$\begin{aligned} y(t) &= \sum_{n=1}^{\infty} \frac{1}{n!} \int_{-\infty}^{\infty} \cdots \int_{-\infty}^{\infty} k_n(u_1, u_2, \cdots u_n) \prod_{m=1}^{n} x(t - u_m) du_1 \cdots du_n \\ &= \int_{-\infty}^{\infty} k_1(u_1) x(t - u_1) du_1 + \frac{1}{2!} \int_{-\infty}^{\infty} \int_{-\infty}^{\infty} k_2(u_1, u_2) x(t - u_1) x(t - u_2) du_1 du_2 \\ &\quad + \frac{1}{3!} \int_{-\infty}^{\infty} \int_{-\infty}^{\infty} \int_{-\infty}^{\infty} k_3(u_1, u_2, u_3) x(t - u_1) x(t - u_2) x(t - u_3) du_1 du_2 du_3 + \cdots \end{aligned} \tag{4.199}$$

where $k_n(u_1, u_2, \cdots u_n)$ are called the Volterra kernels of the system and sometimes referred to as kernels for short. The variable $\{u_i : i = 1, \cdots, n\}$ is also a time variable to be distinguished from t. In the event where the nonlinearity is memoryless, the Volterra kernel reduces to a constant. Note the similarity between the first integral term in Eqn (4.199), namely $\int_{-\infty}^{\infty} k_1(u_1) x(t - u_1) du_1$ and the convolution integral presented in Eqn (4.197). Therefore, one could think of $k_2(u_1, u_2)$, $k_3(u_1, u_2, u_3)$, $\cdots$ as higher order impulse responses that are used to describe different orders of nonlinearities in the system [20]. The integral terms in Eqn (4.199) are comprised of multifold convolution integrals. Hence $y(t)$ consists of an infinite sum of n-fold convolution integrals.

The kernels in Eqn (4.199) can also be expressed in the frequency domain via the Fourier transform simply as:

$$K_n(F_1, \cdots, F_n) = \int_{-\infty}^{\infty} \cdots \int_{-\infty}^{\infty} k_n(u_1, \cdots, u_n) e^{-j2\pi F_1 u_1} \cdots e^{-j2\pi F_n u_n} du_1 \cdots du_n \tag{4.200}$$

In a similar manner, the kernels can also be expressed in the Laplace domain as:

$$K_n(F_1, \cdots, F_n) = \left. \int_{-\infty}^{\infty} \cdots \int_{-\infty}^{\infty} k_n(u_1, \cdots, u_n) e^{-js_1 u_1} \cdots e^{-js_n u_n} du_1 \cdots du_n \right|_{s_i = j2\pi F_i} \tag{4.201}$$

Using Eqn (4.200), the relationship in Eqn (4.199) can be expressed as:

$$Y(F) = K_1(F)X(F) + \frac{1}{2!}\int_{-\infty}^{\infty} K_2(F_1, F-F_1)X(F_1)X(F-F_1)\mathrm{d}F_1$$
$$+\frac{1}{3!}\int_{-\infty}^{\infty}\int_{-\infty}^{\infty} K_3(F_1, F_2, F-F_1-F_2)X(F_1)X(F_2)X(F-F_1-F_2)\mathrm{d}F_1\mathrm{d}F_2$$
$$+\cdots \tag{4.202}$$

The question now becomes: For a given nonlinearity, how do we determine the kernels? There are several techniques that can be used to explicitly determine the Volterra kernels, namely: the harmonic input method, the direct expansion method, and the power of transfer functions. In this chapter, we will elaborate on the harmonic input method.

4.3.1 The harmonic input method

The harmonic input method determines the Volterra series kernels in the frequency domain. Simply put, when the input signal $x(t)$ is the sum

$$x(t) = \sum_{p=1}^{n} e^{j\Omega_p t} \tag{4.203}$$

then the Kernel $K_n(F_1, \cdots, F_n)$ is none other than the coefficient of $e^{j(\Omega_1+\Omega_2+\cdots\Omega_n)t}$ as presented in Eqn (4.199) and the frequencies $\Omega_i = 2\pi F_i{:}i = 1, \cdots n$ are incommensurable. Hence we can claim

$$K_n(F_1, \cdots F_n) = \text{coefficient of } e^{j(\Omega_1+\Omega_2+\cdots\Omega_n)t} \tag{4.204}$$

as given in Eqn (4.199).

The proof is as follows. Substituting Eqn (4.203) into Eqn (4.199), we obtain

$$y(t) = \sum_{n=1}^{\infty}\frac{1}{n!}\int_{-\infty}^{\infty}\cdots\int_{-\infty}^{\infty} k_n(u_1, u_2, \cdots u_n)\prod_{m=1}^{n}\left[\sum_{p=1}^{n} e^{j\Omega_p(t-u_m)}\right]\mathrm{d}u_1\cdots\mathrm{d}u_n \tag{4.205}$$

According to Eqn (4.205), the nth term can be found after expanding the product term within Eqn (4.205), that is:

$$y_n(t) = \frac{1}{n!}e^{j\sum_{p=1}^{n}\Omega_p t}\int_{-\infty}^{\infty}\cdots\int_{-\infty}^{\infty} k_n(u_1, u_2, \cdots u_n)\sum_{n!} e^{-j\sum_{p=1}^{n}\Omega_p t u_{(p)}}\mathrm{d}u_1\cdots\mathrm{d}u_n$$
$$= \frac{1}{n!}e^{j\sum_{p=1}^{n}\Omega_p t}\sum_{n!}\int_{-\infty}^{\infty}\cdots\int_{-\infty}^{\infty} k_n(u_1, u_2, \cdots u_n)e^{-j\sum_{p=1}^{n}\Omega_p t u_{(p)}}\mathrm{d}u_1\cdots\mathrm{d}u_n \tag{4.206}$$

The relationship in Eqn (4.206) is none other than the n-fold Fourier transform and $\sum_{n!}(.)$ and $u_{(p)}$ denote the summation over $n!$ permutations of the subscripts of u_m [20]. In the frequency domain, the integral in Eqn (4.206) can be expressed according to Eqn (4.200) as:

$$y_n(t) = \frac{1}{n!}e^{j\sum_{p=1}^{n}\Omega_p t}\sum_{n!}K_n\left(F_{(1)},\cdots,F_{(n)}\right) \tag{4.207}$$

where again $F_{(i)}$ implies permutations over all $n!$ possibilities. For symmetric kernels, however, it turns out that (see [20])

$$\sum_{n!}K_n\left(F_{(1)},\cdots,F_{(n)}\right) = n!K_n(F_1,\cdots,F_n) \tag{4.208}$$

The Volterra kernels are said to be symmetric if and only if $k_n(u_1,\cdots u_n)$ has the same value regardless of the permutations of $u_1,\cdots u_n$. Substituting the result in Eqn (4.208) into Eqn (4.207), we obtain

$$y_n(t) = \frac{1}{n!}e^{j\sum_{p=1}^{n}\Omega_p t}K_n\left(F_{(1)},\cdots,F_{(n)}\right) \tag{4.209}$$

The method above is best illustrated via an example.

EXAMPLE 4.5 ANALYSIS OF FILTERED FM USING HARMONIC BALANCE

According to Bedrousian and Rice [20], certain forms of quasistatic approximations to filtered FM can be modeled according to the input–output relationship:

$$y(t) = x(t) + \mu\left[\frac{d}{dt}x(t)\right]^2\frac{d^2}{dt^2}x(t) \tag{4.210}$$

Compute the Volterra kernels of the system.

Following the analysis presented above, let us excite the system in Eqn (4.210) with the function $x(t) = e^{j\Omega_1 t}$, then the output signal according to Eqn (4.210) becomes

$$\begin{aligned} y(t) &= e^{j\Omega_1 t} + \mu\left[\frac{d}{dt}e^{j2\Omega_1 t}\right]^2\frac{d^2}{dt^2}e^{j2\Omega_1 t} \\ &= e^{j\Omega_1 t} + \mu\left[-\Omega_1^2 e^{j2\Omega_1 t}\right]\left[-\Omega_1^2 e^{j2\Omega_1 t}\right] \\ &= e^{j\Omega_1 t} + \mu\Omega_1^4 e^{j3\Omega_1 t} \end{aligned} \tag{4.211}$$

According to Eqns (4.199), (4.204), and (4.211), the kernel $K_1(F_1)$ is simply

$$K_1(F_1) = \text{coefficient of } e^{j\Omega_1 t} = 1 \tag{4.212}$$

Next consider the excitation input comprised of two tones, that is

$$x(t) = e^{j\Omega_1 t} + e^{j\Omega_2 t} \tag{4.213}$$

Continued

EXAMPLE 4.5 ANALYSIS OF FILTERED FM USING HARMONIC BALANCE—cont'd

Then the output $y(t)$ according to Eqn (4.210) becomes

$$\begin{aligned}y(t) &= e^{j\Omega_1 t} + e^{j\Omega_2 t} + \mu\left[j\Omega_1 e^{j\Omega_1 t} + j\Omega_2 e^{j\Omega_2 t}\right]^2\left[-\Omega_1^2 e^{j\Omega_1 t} - \Omega_2^2 e^{j\Omega_2 t}\right]\\ &= e^{j\Omega_1 t} + e^{j\Omega_2 t} + \mu\left[\Omega_1^2 e^{j2\Omega_1 t} + \Omega_2^2 e^{j2\Omega_2 t} + 2\Omega_1\Omega_2 e^{j(\Omega_1+\Omega_2)t}\right]\\ &\times\left[\Omega_1^2 e^{j\Omega_1 t} + \Omega_2^2 e^{j\Omega_2 t}\right]\\ &= e^{j\Omega_1 t} + e^{j\Omega_2 t} + \mu\left[\Omega_1^4 e^{j3\Omega_1 t} + \Omega_1^2\Omega_2^2 e^{j(\Omega_1+2\Omega_2)t} + 2\Omega_1^3\Omega_2 e^{j(2\Omega_1+\Omega_2)t}\right.\\ &\left.+\Omega_1^2\Omega_2^2 e^{j(\Omega_1+2\Omega_2)t} + \Omega_2^4 e^{j3\Omega_2 t} + 2\Omega_1\Omega_2^3 e^{j(\Omega_1+2\Omega_2)t}\right]\end{aligned} \tag{4.214}$$

In the expansion presented in Eqn (4.214), there are no $e^{j(\Omega_1+\Omega_2)t}$ terms, and hence we can declare that the kernel

$$K_2(F_1, F_2) = \text{coefficient of } e^{j(\Omega_1+\Omega_2)t} = 0 \tag{4.215}$$

And last, let the excitation input signal be comprised of three incommensurable tones, that is

$$x(t) = e^{j\Omega_1 t} + e^{j\Omega_2 t} + +e^{j\Omega_3 t} \tag{4.216}$$

Then the output signal $y(t)$ is given according to Eqn (4.210) as

$$\begin{aligned}y(t) &= e^{j\Omega_1 t} + e^{j\Omega_2 t} + e^{j\Omega_3 t}\\ &+\mu\left[j\Omega_1 e^{j\Omega_1 t} + j\Omega_2 e^{j\Omega_2 t} + j\Omega_3 e^{j\Omega_2 t}\right]^2\left[-\Omega_1^2 e^{j\Omega_1 t} - \Omega_2^2 e^{j\Omega_2 t} - \Omega_3^2 e^{j\Omega_3 t}\right]\end{aligned} \tag{4.217}$$

Expanding the terms in Eqn (4.217) shows that there will be three exponential terms like Eqn (4.216), and hence it can be deduced that the kernel is

$$\begin{aligned}K_3(F_1, F_2, F_3) &= \text{coefficient of } e^{j(\Omega_1+\Omega_2+\Omega_3)t}\\ &= 2\mu\Omega_1\Omega_2\Omega_3^2 + 2\mu\Omega_1\Omega_2^2\Omega_3 + 2\mu\Omega_1^2\Omega_2\Omega_3\\ &= 2\mu\Omega_1\Omega_2\Omega_3(\Omega_1 + \Omega_2 + \Omega_3)\end{aligned} \tag{4.218}$$

Furthermore, it can also be shown that the kernels for $n > 3$ will turn out to be zero.

It is clear from this example that the complexity of the harmonic input method increases quickly as n increases. Therefore, it becomes impractical to apply this method without the aid of a symbolic algebra program in order to assist in generating the kernels.

The harmonic input method can be easily applied to circuits where a nonlinear device is connected to a linear admittance with transfer function $H(F)$ as shown in Figure 4.41. The Volterra kernels as a function of $H(F)$ can be computed as shown by Bedrousian and Rice as a recurrence relation [21]. The first two terms are simply

$$\begin{aligned}K_1(F_1) &= \frac{H(F_1)}{a_1 + H(F_1)}\\ K_2(F_1, F_2) &= -\frac{2a_2 K_1(F_1)K_1(F_2)}{a_1 + H(F_1 + F_2)}\end{aligned} \tag{4.219}$$

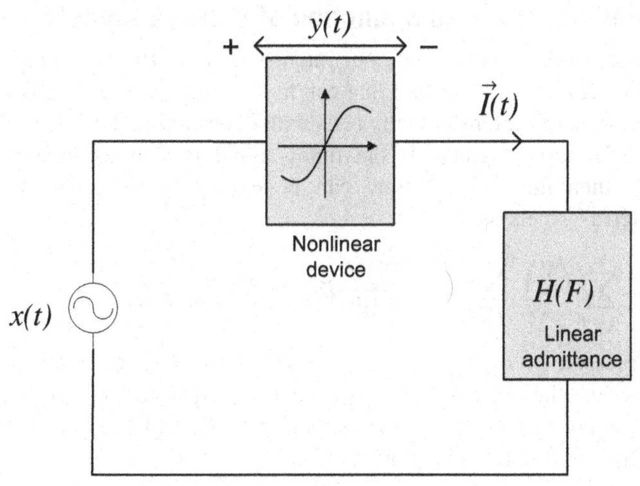

FIGURE 4.41

Circuit representing a nonlinear device in series with a linear admittance. (For color version of this figure, the reader is referred to the online version of this book.)

For example, let the linear admittance be represented by an inductance L. Furthermore, let the nonlinear device be represented by a square law detector where the current $\vec{I}(t)$ can be expressed in terms of the voltage $y(t)$ as:

$$\vec{I}(t) = a_2[y(t)]^2 \tag{4.220}$$

The differential equation representing the circuit becomes

$$La_2\frac{\mathrm{d}}{\mathrm{d}t}y^2(t) + y(t) = x(t) \tag{4.221}$$

The admittance function is $H(F) = 1/j\Omega L$ and the coefficient $a_2 \neq 0$. Then according to the relationship in Eqn (4.219), the Volterra kernels can be obtained as (see Ref. [21] Eqns (43) and (45)):

$$\begin{aligned}
K_1(F_1) &= \left.\frac{H(F_1)}{a_1 + H(F_1)}\right|_{a_1=0} = 1 \\
K_2(F_1, F_2) &= -\left.\frac{2a_2K_1(F_1)K_1(F_2)}{a_1 + H(F_1 + F_2)}\right|_{a_1=0} = -4a_2Lj\pi(F_1 + F_2) \\
K_3(F_1, F_2, F_3) &= 8\pi^2(-2ja_2L)^2(F_1 + F_2 + F_3)^2 \\
&\vdots
\end{aligned} \tag{4.222}$$

It is apparent from the previous analysis that the harmonic input method increased in complexity for kernel orders higher than 3. In this case, symbolic algebra programs can be used to aid in obtaining the higher order terms.

4.3.2 Output spectrum as a function of Volterra kernels

As far as this chapter is concerned, our interest in Volterra series remains in their ability to provide an analytical tool that can model and analyze a weakly nonlinear system. Therefore, our interest chiefly rests with determining the Volterra kernels by using sinusoidal input signals. If the input signal is a sinusoid namely $x(t) = V\cos(2\pi F_c t)$, then the output signal can be expressed in terms of frequency-domain Volterra kernels as

$$y(t) = \sum_{n=1}^{\infty}\left(\frac{V}{2}\right)^n \sum_{l=0}^{n} \frac{e^{j2\pi(2l-n)F_c t}}{l!(n-l)!} K_{l,n-l}(F_c)\Big|_{K_{l,n-l}(F_c)\,\cong\, K_n(F_1\cdots F_n)} \tag{4.223}$$

whereas expressed in Eqn (4.223) $K_{l,n-l}(F_c)$ is a simplified representation of $K_n(F_1 \cdots F_n)$ where the first l-frequency elements are equal to $+F_c$ and the remaining n-l elements equal to $-F_c$. To further elaborate on Eqn (4.223), let us expand the summation for the first few elements:

$$\begin{aligned}
y(t) = {} & \left\{\frac{V^2}{2}K_2(F_c, -F_c) + \cdots\right\} \\
& + e^{j2\pi F_c t}\left\{\frac{V}{2}K_1(F_c) + \frac{V^3}{16}K_3(F_c, F_c, -F_c) + \cdots\right\} \\
& + e^{-j2\pi F_c t}\left\{\frac{V}{2}K_1(-F_c) + \frac{V^3}{16}K_3(-F_c, -F_c, +F_c) + \cdots\right\} \\
& + e^{j4\pi F_c t}\left\{\frac{V^2}{8}K_2(F_c, F_c) + \cdots\right\} + e^{-j4\pi F_c t}\left\{\frac{V^2}{8}K_2(-F_c, -F_c) + \cdots\right\} \\
& + e^{j6\pi F_c t}\left\{\frac{V^3}{48}K_3(F_c, F_c + F_c) + \cdots\right\} \\
& + \cdots e^{-j6\pi F_c t}\left\{\frac{V^3}{48}K_3(-F_c, -F_c, -F_c) + \cdots\right\} + \cdots + \cdots
\end{aligned} \tag{4.224}$$

A careful examination of Eqn (4.224) reveals that the contributors to the fundamental frequency term $e^{j2\pi F_c t}$ as well as the odd harmonics are odd-numbered kernels, whereas the main contributors to the even harmonics are even-ordered kernels. In a weakly nonlinear system, given a small signal input power, the dominant term in Eqn (4.224) will be the first kernel $\frac{V}{2}K_1(F_c)$, which simply corresponds to the linear gain of the system or device under test. Increasing the input signal power results in an increase in the third Volterra kernel gain $\frac{V^3}{16}K_3(F_c, F_c, -F_c)$, thus signifying a certain compression in the system. The 1-dB compression point can be computed in a manner similar to Eqn (4.23). Likewise, the third order *IIP3* can be computed following the relationship in Eqn (4.77) by equating the terms $\frac{V}{2}K_1(F_c)$ and $\frac{V^3}{16}K_3(F_c, F_c, -F_c)$.

A similar expression could be obtained for a two-tone input say $x(t) = V_1 \cos(2\pi F_{c1}t) + V_2 \cos 2\pi F_{c2}t)$; however, as one might expect the complexity of the terms increases. For example, for $N,\ M \geq 0$, the term due to the frequency $NF_{c1} + MF_{c2}$ is [21]:

$$e^{j2\pi(NF_{c1}+MF_{c2})t}\sum_{l=0}^{\infty}\sum_{m=0}^{\infty}\frac{(V_1/2)^{2l+N}(V_1/2)^{2m+N}}{(N+l)!(M+m)!l!m!}K_{N+l,l;M+m.m}(F_{c1},F_{c2}) \quad (4.225)$$

where the kernel function in Eqn (4.225) is simply

$$K_{N+l,l;M+m.m}(F_{c1},F_{c2})\big|_{N+M+2(l+m)=n} = K_n(F_1,\cdots F_n)$$

and the first $N+l$ of the frequency terms are equal to $+F_{c1}$
The next l of the frequency terms are equal to $-F_{c1}$
The next $M+m$ of the frequency terms are equal to $+F_{c2}$ (4.226)
The next m of the frequency terms are equal to $-F_{c2}$
For $N < 0$ the signs of F_{c1} are reversed
And likewise for $M < 0$ the signs of F_{c2} are reversed

Obviously, albeit more complex, similar relationships can be found for a three-tone or higher input signals.

4.4 Appendix A: mathematical identities

The following trigonometric identities are found in more detail in [22]. The addition and subtraction identities are given via the relations

$$\begin{aligned}\cos(a+b) &= \cos a \cos b - \sin a \sin b\\ \cos(a-b) &= \cos a \cos b + \sin a \sin b\end{aligned} \quad (4.227)$$

and

$$\begin{aligned}\sin(a+b) &= \sin a \cos b + \sin b \cos a\\ \sin(a-b) &= \sin a \cos b - \sin b \cos a\end{aligned} \quad (4.228)$$

The sine relationship, for example, can be further manipulated as

$$\sin(a+b) = \sin a \cos b + \sin b \cos a = \{\sin a + \tan b \cos a\}\cos b \quad (4.229)$$

In most cases that we run into in this chapter, we encounter the products of two sines or cosines and are asked to relate them to the addition and subtraction identities of Eqns (4.227) and (4.228), that is

$$\begin{aligned}\cos a \cos b &= \frac{1}{2}(\cos(a+b) + \cos(a-b))\\ \sin a \sin b &= \frac{1}{2}(\cos(a-b) - \cos(a+b))\\ \sin a \cos b &= \frac{1}{2}(\sin(a-b) - \sin(a+b))\end{aligned} \quad (4.230)$$

Another identity that is encountered frequently are the function of double angles or sin $2a$ and cos $2a$. These functions can be expanded in terms of sines and cosines as

$$\begin{aligned} \cos 2a &= \cos^2 a - \sin^2 a = 2\cos^2 a - 1 = 1 - 2\sin^2 a \\ \sin 2a &= 2 \sin a \cos a \end{aligned} \tag{4.231}$$

Inversely, the squares of the sine and cosine can be related to cos $2a$ and sin $2a$ as

$$\begin{aligned} \cos^2 a &= \frac{\cos 2a + 1}{2} \\ \sin^2 a &= \frac{1 - \cos 2a}{2} \end{aligned} \tag{4.232}$$

In a similar manner, the sine and cosine functions of multiple angles can be expressed as

$$\begin{aligned} \cos 3a &= 4\cos^3 a - 3\cos a \Rightarrow \cos^3 a = \frac{\cos 3a + 3\cos a}{4} \\ \sin 3a &= 3\sin a - 4\sin^3 a \Rightarrow \sin^3 a = \frac{3\sin a - \sin 3a}{4} \end{aligned} \tag{4.233}$$

In general, a sinusoid to the nth power can be expressed as:

$$\begin{aligned} \cos^n(x) &= \left(\frac{1}{2}\right)^{n-1} \left\{ \cos(nx) + \binom{n}{1}\cos((n-2)x) + \binom{n}{2}\cos((n-4)x) \right. \\ &\quad \left. + \cdots + \binom{n}{\frac{n-1}{2}}\cos(x) \right\} \text{for } n \text{ odd} \\ \cos^n(x) &= \left(\frac{1}{2}\right)^{n-1} \left\{ \cos(nx) + \binom{n}{1}\cos((n-2)x) + \binom{n}{2}\cos((n-4)x) \right. \\ &\quad \left. + \cdots + \binom{n}{\frac{n-2}{2}}\cos(x) \right\} \text{for } n \text{ even} \end{aligned} \tag{4.234}$$

Assume that m and n are odd, then

$$\begin{aligned}
\cos^{m}(x)\cos^{n}(y) &= \left(\frac{1}{2}\right)^{m+n-2}\left\{\cos(mx) + \binom{m}{1}\cos((m-2)x)\right. \\
&\quad \left. + \binom{m}{2}\cos((m-4)x) + \cdots + \binom{m}{\frac{m-1}{2}}\cos(x)\right\} \\
&\quad \times \left\{\cos(ny) + \binom{n}{1}\cos((n-2)y) + \binom{n}{2}\cos((n-4)y)\right. \\
&\quad \left. + \cdots + \binom{n}{\frac{n-1}{2}}\cos(y)\right\} \\
&= \left(\frac{1}{2}\right)^{m+n-2}\left\{\cos(mx)\cos(ny) + \binom{m}{1}\cos((m-2)x)\cos(ny)\right. \\
&\quad + \binom{m}{2}\cos((m-4)x)\cos(ny) \\
&\quad \left. + \cdots + \binom{m}{\frac{m-1}{2}}\cos(x)\cos(ny) + \cdots\right\}
\end{aligned} \tag{4.235}$$

Further manipulations of Eqn (4.235) lead to:

$$\begin{aligned}
\cos^{m}(x)\cos^{n}(y) &= \left(\frac{1}{2}\right)^{m+n-1}\left\{\cos(mx \pm ny) + \binom{m}{1}\cos((m-2)x \pm ny)\right. \\
&\quad + \binom{m}{2}\cos((m-4)x \pm ny) \\
&\quad \left. + \cdots + \binom{m}{\frac{m-1}{2}}\cos(x \pm ny) + \cdots\right\} \\
&= \left(\frac{1}{2}\right)^{m+n-1}\sum_{l=0}^{\frac{m-1}{2}}\binom{m}{l}\cos((m-2l)x \pm ny) + \cdots
\end{aligned} \tag{4.236}$$

Similar universal expressions to Eqn (4.236) can be found for other combinations of m and n.

In general, we can express $\cos^n(x)$ as

$$\cos^n(x) = 2^{-n} \sum_{k=0}^{n} \binom{n}{k} \cos[(n-k)x] \tag{4.237}$$

Where $n \in \mathbb{N}^+$

$$\binom{n}{k} = \frac{n!}{k!(n-k)!} \tag{4.238}$$

The relationship in Eqn (4.238) represents the kth coefficient of the Newton binomial of degree n.

A further property of the sine and cosine functions are

$$\begin{aligned} \sin(-\alpha) &= -\sin(\alpha) \\ \cos(-\alpha) &= \cos(\alpha) \end{aligned} \tag{4.239}$$

A shift in $\pi/2$ in the argument of the sine function results in a cosine function, that is

$$\begin{aligned} \sin\left(\alpha + \frac{\pi}{2}\right) &= \sin(\alpha)\cos\left(\frac{\pi}{2}\right) + \sin\left(\frac{\pi}{2}\right)\cos(\alpha) = \cos(\alpha) \\ \sin\left(\alpha - \frac{\pi}{2}\right) &= \sin(\alpha)\cos\left(\frac{\pi}{2}\right) - \sin\left(\frac{\pi}{2}\right)\cos(\alpha) = -\cos(\alpha) \end{aligned} \tag{4.240}$$

Finally, a polynomial expansion can be expressed in terms of a multinomial series as

$$(x_1 + x_2 + \cdots + x_k)^n = \sum_{\substack{n_1, n_2, \cdots, n_k \geq 0 \\ n_1 + n_2 + \cdots + n_k = n}} \frac{n!}{n_1! n_2! \cdots n_k!} x_1^{n_1} x_2^{n_2} \cdots x_k^{n_k} \tag{4.241}$$

By way of an example, consider the polynomial

$$\begin{aligned} (x_1 + x_2 + x_3)^3 = {} & x_1^3 + 3x_1^2 x_2 + 3x_1 x_2^2 + x_2^3 + 3x_1^2 x_3 + 6x_1 x_2 x_3 + 3x_2^2 x_3 + 3x_1 x_3^2 \\ & + 3x_2 x_3^2 + x_3^3 \end{aligned} \tag{4.242}$$

4.5 Appendix B: effect of in-phase and quadrature imbalance on image rejection

The extent of image rejection in an image-reject mixer largely depends on the imbalance, or rather the lack of, in the quadrature demodulation. To illustrate this phenomenon, consider the *LO* signals demodulating an incoming desired signal along with its image as defined in Eqn (4.164) as

$$x_{LO,I} = \Gamma \sin(\Omega_{LO}t) \text{ and } x_{LO,Q} = (\Gamma + \delta)\cos(\Omega_{LO}t + \theta) \tag{4.243}$$

where $x_{LO,I}$ and $x_{LO,Q}$ are the *LO* signals demodulating the in-phase and quadrature components, Γ is the amplitude of the *LO* signal, and δ and θ are the amplitude and phase imbalance present in the *LO* signal. Consider by way of an example the Hartley image reject mixer depicted in Figure 4.32. The output of the phase shifter can be derived in a manner similar to Eqn (4.169) as:

$$\widehat{I}_{\angle 90^\circ}(t) = \frac{A\Gamma}{2}\cos[(\Omega_{RF} - \Omega_{LO})t] - \frac{B\Gamma}{2}\cos\left[\left(\Omega_{LO} - \Omega_{image}\right)t\right] \tag{4.244}$$

On the other hand, the quadrature component can be derived in a manner similar to Eqn (4.171) as:

$$\tilde{Q}(t) = \frac{A(\Gamma + \delta)}{2}\cos[(\Omega_{RF} - \Omega_{LO})t - \theta] + \frac{B(\Gamma + \delta)}{2}\cos\left[\left(\Omega_{image} - \Omega_{LO}\right)t - \theta\right] \tag{4.245}$$

At the output of the mixer, the *IF* signal is given as

$$\begin{aligned}
\widehat{I}_{\angle 90^\circ}(t) + \tilde{Q}(t) &= \underbrace{\frac{A\Gamma}{2}\cos[(\Omega_{RF} - \Omega_{LO})t]}_{\text{Signal component}} \underbrace{- \frac{B\Gamma}{2}\cos\left[\left(\Omega_{LO} - \Omega_{image}\right)t\right]}_{\text{Image component}} \\
&\quad + \underbrace{\frac{A(\Gamma + \delta)}{2}\cos[(\Omega_{RF} - \Omega_{LO})t - \theta]}_{\text{Signal component}} \\
&\quad + \underbrace{\frac{B(\Gamma + \delta)}{2}\cos\left[\left(\Omega_{\text{image}} - \Omega_{LO}\right)t - \theta\right]}_{\text{Image component}}
\end{aligned} \tag{4.246}$$

After simple rearrangements of the terms in Eqn (4.246), we obtain

$$\begin{aligned}
&\widehat{I}_{\angle 90^\circ}(t) + \tilde{Q}(t) \\
&= \underbrace{\frac{A\Gamma}{2}\cos[(\Omega_{RF} - \Omega_{LO})t] + \frac{A(\Gamma + \delta)}{2}\cos[(\Omega_{RF} - \Omega_{LO})t - \theta]}_{\text{Signal component}} \\
&\times \underbrace{- \frac{B\Gamma}{2}\cos\left[\left(\Omega_{LO} - \Omega_{image}\right)t\right] + \frac{B(\Gamma + \delta)}{2}\cos\left[\left(\Omega_{image} - \Omega_{LO}\right)t - \theta\right]}_{\text{Image component}}
\end{aligned} \tag{4.247}$$

Define the image-to-signal power ratio at the input of the mixer as

$$\Psi = \frac{B^2}{A^2} \tag{4.248}$$

Then the image-to-signal power ratio at the output of the mixer according to Eqns (4.247) and (4.248) is

$$\begin{aligned}\Upsilon &= \Psi\frac{\Gamma^2 + (\Gamma+\delta)^2 - 2\Gamma(\Gamma+\delta)\cos\theta}{\Gamma^2 + (\Gamma+\delta)^2 + 2\Gamma(\Gamma+\delta)\cos\theta} \\ &= \Psi\frac{1 + \left(1+\frac{\delta}{\Gamma}\right)^2 - 2\left(1+\frac{\delta}{\Gamma}\right)\cos\theta}{1 + \left(1+\frac{\delta}{\Gamma}\right)^2 + 2\left(1+\frac{\delta}{\Gamma}\right)\cos\theta}\end{aligned} \tag{4.249}$$

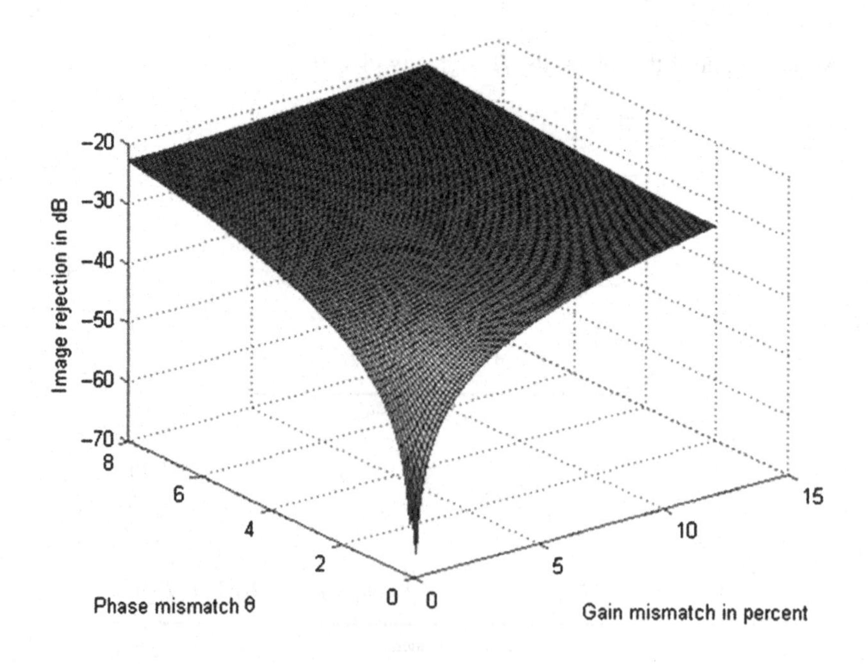

FIGURE 4.42

Image rejection as a function of phase mismatch θ and percentage of gain mismatch δ/Γ in percent. (For color version of this figure, the reader is referred to the online version of this book.)

The image rejection ratio is then defined as the ratio of Eqn (4.249) to Eqn (4.248) or in dB

$$IRR = 10\log_{10}\left(\frac{\Gamma^2 + (\Gamma+\delta)^2 - 2\Gamma(\Gamma+\delta)\cos\theta}{\Gamma^2 + (\Gamma+\delta)^2 + 2\Gamma(\Gamma+\delta)\cos\theta}\right) \tag{4.250}$$

The image rejection ratio expressed in Eqn (4.250) is a measure of the amount of rejection the image reject mixer provided to the system as shown in Figure 4.42. The mismatch presented herein is only due to the mixer. Possible mismatches could also be due to the filters as well as the 90° phase shift.

4.6 Appendix C: description of the harmonic balance method

Unlike the Volterra series, for example, harmonic balance analysis is a suitable technique for analyzing strongly nonlinear circuits in the presence of a *large* signal source [23]. The intent of this appendix is to give the reader a very high level overview of the harmonic balance method technique. By way of an example, the method is suitable for analyzing the steady state response of power amplifiers, mixers, and multipliers using solid state devices [24]. The harmonic balance method can also be applied to analyze weakly nonlinear circuits that employ single or multitone excitation [1].

In broad terms, the harmonic balance technique is used to convert a set of differential equations into a nonlinear algebraic system of equations that can be solved using numerical algorithms. Nakhla and Vlash [24] proposed using a gradient technique, whereas Filicori et al. [25] used the Newton–Raphson numerical method to solve harmonic balance problems. One method, of particular interest, is the piecewise harmonic balance technique. In this technique, the nonlinear circuit is separated into two subcircuits, namely, a linear circuit and a nonlinear circuit as shown in the simple example given in [1] and depicted in Figure 4.43(a). Assume that the source voltage is exciting the diode circuit with a given tone, namely $V_S(t) = A\cos(\Omega_0 t)$. The diode response will consist of all the harmonics of $V_S(t)$. The next step is to provide an *educated guess* of the diode voltage response at the various harmonic frequencies. This simplifies the diode circuit depicted in Figure 4.43(a) into the one depicted in Figure 4.43(b). Note that the impedance $Z(\Omega)$ is a function of frequency as well. Using Kirchhoff's voltage law, and knowing the source and diode voltage response at all the harmonics,[12] we can predict the value of the linear current $\underline{I}_{\text{linear}}(k\Omega_0)$.

Next, we compute the value of the nonlinear current $\vec{I}_{\text{nonlinear}}(t)$ using our best guess of the voltage across the diode. The voltage across the diode is obtained by

[12] Obviously, the voltage source has theoretically only one nonzero harmonic, namely at Ω_0.

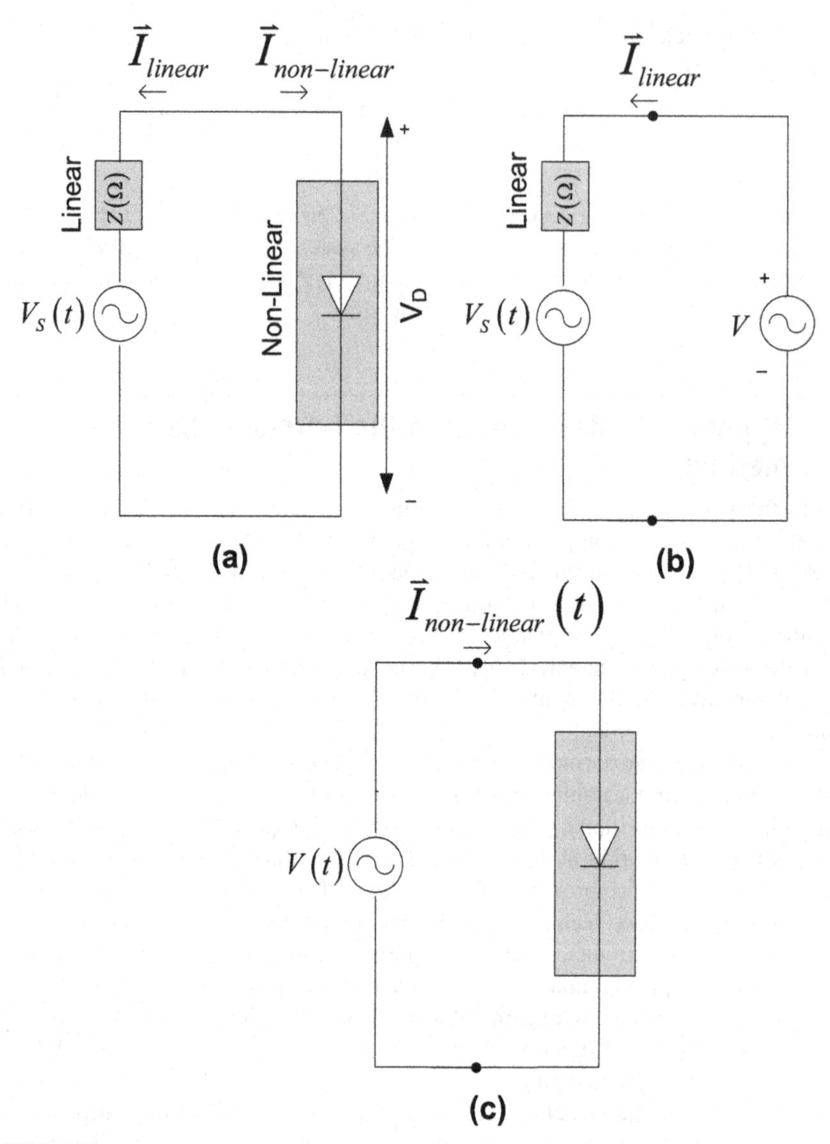

FIGURE 4.43

A simple nonlinear diode circuit: (a) linear and nonlinear instantiation, (b) linear instantiation, and (c) nonlinear instantiation. (For color version of this figure, the reader is referred to the online version of this book.)

taking the Fourier transform of the *assumed* voltage response at the various harmonics $V(t)$ using the diode junction equation

$$\vec{I}_{\text{nonlinear}}(t) = I_{\text{saturation}}\left(e^{qV(t)/\eta KT} - 1\right) \tag{4.251}$$

where q is the electron charge, η is a factor that accounts for the imperfections in the junction, T is the temperature in kelvin, and $K = 1.37 \times 10^{-23}$ J/K is Boltzmann's constant. Next, we test whether our guess of the harmonic voltages across the diode form an appropriate solution. This is done by applying Kirchhoff's current law to Figure 4.43(c), namely that the sum of all currents entering a node is zero or

$$\vec{I}_{\text{nonlinear}}(k\Omega_0) + \vec{I}_{\text{linear}}(k\Omega_0) = 0 \tag{4.252}$$

The harmonic currents $\vec{I}_{\text{nonlinear}}(k\Omega_0)$ are obtained from the Fourier transform of $\vec{I}_{\text{nonlinear}}(t)$. If the relationship in Eqn (4.252) is not satisfied to within a certain acceptable error, a new set of diode voltage values is proposed, and the process is repeated until an acceptable solution is found. The same basic technique is applied to more complex circuits whereby the circuit itself is subdivided into linear and nonlinear circuits as shown in Figure 4.44.

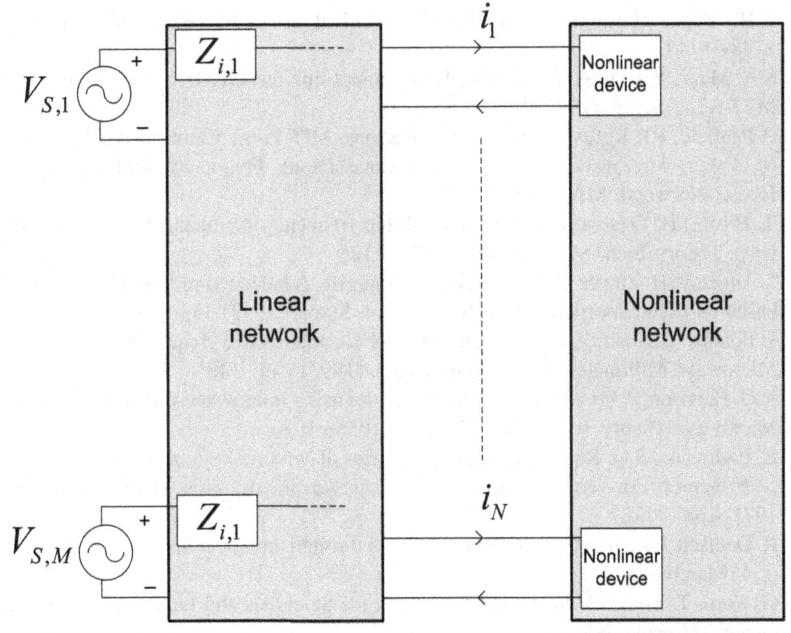

FIGURE 4.44

Partitioning of circuit into linear and nonlinear components. (For color version of this figure, the reader is referred to the online version of this book.)

References

[1] S. Maas, Nonlinear Microwave and RF Circuits, Artech House, Boston, MA, 2003.

[2] R. Watson, Receiver Dynamic Range: Part 1, Watkins-Johnson Technical Note, WJ Communications, San Jose, CA.

[3] R. Watson, Receiver Dynamic Range: Part 2, Watkins-Johnson Technical Note, WJ Communications, San Jose, CA.

[4] B. Razavi, Design considerations for direct-conversion receivers, IEEE Trans. Circuit Syst. II 44 (6) (June 1997).

[5] V. Aparin, E. Zeisel, P. Gazerro, Highly linear BiCMOS LNA and mixer for cellular CDMA/AMPS applications, IEEE Radio Freq. Integr. Circuits Symp. (2002) 129–132.

[6] T. Rouphael, RF and Digital Signal Processing for Software Defined Radio, Elsevier, Boston, MA, 2009.

[7] A.A.M. Saleh, Frequency-independent and frequency dependent non-linear models of TWT amplifiers, IEEE Trans. Commun. 29 (November 1981) 1715–1720.

[8] B. Razavi, RF Microelectronics, Prentice Hall, Saddle River, NJ, 1998.

[9] B.C. Henderson, Mixers part 1: characteristics and performance, Watkins-Johnson Tech Note 8 (2) (April 1981).

[10] B.C. Henderson, Predicting intermodulation suppression in double-balanced mixers, Watkins-Johnson Tech. Note 10 (4) (August 1983).

[11] V. Manassewitsch, Frequency Synthesizers Theory and Design, third ed., Wiley and Sons, New York, NY, 1987.

[12] C.H. Page, Harmonic generation with ideal rectifiers, Proc. IRE 46 (1958) 1738–1740.

[13] S.A. Maas, Y. Ryu, A broadband planar, monolithic resistive frequency doubler, IEEE MTT-S Int. Symp. (1994) 175–178.

[14] P. Penfield, P.P. Rafuse, Varactor Applications, MIT Press, Cambridge, MA, 1962.

[15] M. Faber, Microwave and Millimeter-wave Diode Frequency Multipliers, Artech House, Norwood, MA, 1995.

[16] T. Hirota, H. Ogawa, Uniplanar monolithic frequency doublers, IEEE Trans. Microwave Theory Tech. MTT-37 (1989) 1249–1254.

[17] I. Angelov, H. Zirath, N. Rorsman, H. Gronqvist, A balanced millimeter-wave doubler based on pseudomorphic HEMTS, IEEE Int. Symp. (1992) 353–356.

[18] J. Filkart, Y. Xuan, A new circuit structure for microwave frequency doublers, IEEE Microwave Millimeter Wave Circuit Symp. (1993) 145–148.

[19] R.G. Harrison, A broadband frequency divider using microwave varactors, IEEE Trans. Microwave Theory Tech. MTT-25 (1977) 1055–1059.

[20] E. Bedrosian, S.O. Rice, The output properties of Volterra systems (non-linear systems with memory) driven by harmonic and Gaussian inputs, Proc. IEEE 59 (December 1971) 1688–1707.

[21] R. Deutsch, On a method of Wiener for noise through nonlinear devices, IRE Conv. Rec. pt 4 (March 1955) 186–192.

[22] G. Korn, T. Korn, Mathematical Handbook for Scientists and Engineers, Dover Publications, Mineola, NY, 1968.

[23] S. El-Rabaie, V. Fusco, C. Stewart, Harmonic balance evaluation of nonlinear microwave circuits – a tutorial approach, IEEE Trans. Edu. 31 (3) (August 1988) 181–192.

[24] M.S. Nakhla, J. Vlach, A piecewise harmonic balance technique for determination of periodic response of nonlinear systems, IEEE Trans. Circuit Syst. CAS-23 (1976).

[25] F. Filicori, V.A. Monaco, C. Naldi, Simulation and design of microwave class-C amplifier through harmonic analysis, IEEE Trans. Microwave Theory Tech. MTT-27 (December 1979).

CHAPTER

Signal Sampling and Distortion

5

CHAPTER OUTLINE

In receivers, data conversion is the process of converting a signal from the analog domain to the digital domain. At the input of the analog-to-digital converter (ADC), the analog signal is sampled at uniform time intervals and the output is quantized into discrete amplitude levels. The sampled signal suffers from various distortions, namely quantization noise, jitter, the impact of phase noise on the sampling clock, and overloading and clipping.

This chapter is divided into four sections. Section 5.1 describes the sampling and reconstruction processes of baseband and bandpass signals. Section 5.2 delves into the various degradations caused by signal conversion and sampling imperfections. Topics such as quantization noise, sampling clock jitter, impact of phase noise on

Wireless Receiver Architectures and Design. http://dx.doi.org/10.1016/B978-0-12-378640-1.00005-6

the sampling clock, signal overloading, and clipping are discussed. The antialiasing filtering requirements and their impact on signal quality are discussed in some detail in Section 5.3. Finally, an exact formulation of the quantization noise and a derivation of the signal-to-quantization-noise ratio (*SQNR*) that takes into account the statistics of the input signal are presented in Section 5.4.

5.1 Analog and digital signal representation

This section serves as a theoretical prelude to practical analog-to-digital conversion and the various Nyquist and oversampling data converters. Its aim is to present the necessary mathematical background behind the sampling theorem. Topics such as signal representation, baseband and bandpass sampling, signal reconstruction, and out-of-band energy are discussed.

5.1.1 Sampling and reconstruction of lowpass signals

In this section, we discuss the sampling of lowpass signals as well as the required antialiasing filters needed to obtain a certain desired signal-to-noise ratio (*SNR*). The reconstruction of the sampled lowpass signal into an analog signal is then presented.

5.1.1.1 Lowpass sampling and filtering requirements

In order to *faithfully* reconstruct an analog signal from its digital counterpart, the sampling theorem dictates that the lowpass signal must be sampled at *least* at twice the highest frequency component of the analog bandlimited signal. As will be seen later, this condition simply implies that the spectral replicas that occur due to sampling do not overlap and hence cause no distortion to the reconstructed analog signal. This in turn implies that all the information in the original analog signal is preserved.

To proceed, consider the lowpass analog signal $x_a(t)$ that for the purposes of this discussion is *strictly* bandlimited with a frequency upper bound of $B/2$. Strictly bandlimited in this discussion implies that the highest frequency component of $x_a(t)$ is strictly less than $B/2$ or $X_a(F) \neq 0$ for $-B/2 < F < B/2$ where B is called the Nyquist rate.

Consider the Fourier transform of the analog signal $x_a(t)$

$$X_a(F) = \int_{-\infty}^{\infty} x_a(t)e^{-j2\pi Ft}dt \tag{5.1}$$

The frequency domain signal can also be used to recover the time-domain analog signal via the inverse Fourier transform:

$$x_a(t) = \int_{-\infty}^{\infty} X_a(F)e^{j2\pi Ft}dF \tag{5.2}$$

The analog time-domain signal can then be sampled at the sampling rate of T_s samples per seconds resulting in the *discrete-time* signal

$$x(n) = x_a(nT_s) \cong x_a(t)\big|_{t=nT_s} \tag{5.3}$$

In a similar manner to Eqn (5.2), the spectrum of $x(n)$ can be obtained using the Fourier transform of discrete aperiodic signals as [1,2]

$$\begin{aligned} X(f) &= \sum_{n=-\infty}^{\infty} x(nT_s)e^{-j2\pi fnT_s} \\ &= \sum_{n=-\infty}^{\infty} x(n)e^{-j2\pi fn} \end{aligned} \tag{5.4}$$

In a similar manner to the analog case, the discrete signal presented in Eqn (5.4) can be obtained by applying the inverse Fourier transform:

$$x(n) = \int_{-1/2}^{1/2} X(f)e^{j2\pi fn}df \tag{5.5}$$

According to the relationships presented in Eqns (5.2) and (5.5), we can relate the analog signal spectrum to the discrete signal spectrum as

$$x(n) = x_a(nT_s) = \int_{-\infty}^{\infty} X_a(F)e^{j2\pi\frac{F}{F_s}n}dF \tag{5.6}$$

From Eqn (5.6), we can imply that due to periodic sampling the analog and discrete frequencies are related according to the relationship

$$f = \frac{F}{F_s} \tag{5.7}$$

Thus if we compare Eqns (5.5) and (5.6), we obtain

$$\int_{-1/2}^{1/2} X(f)e^{j2\pi fn}df \Bigg|_{\substack{f=F/F_s \\ df=dF/F_s}} = \frac{1}{F_s}\int_{-F_s/2}^{F_s/2} X\left(\frac{F}{F_s}\right)e^{j2\pi\frac{F}{F_s}n}dF = \int_{-\infty}^{\infty} X_a(F)e^{j2\pi\frac{F}{F_s}n}dF \tag{5.8}$$

In order to further our understanding of the relationship presented in Eqn (5.8), the integral on the right-hand side can be expanded as an infinite sum of integrals

$$\int_{-\infty}^{\infty} X_a(F)e^{j2\pi\frac{F}{F_s}n}dF = \sum_{l=-\infty}^{\infty}\int_{(l-1/2)F_s}^{(l+1/2)F_s} X_a(F)e^{j2\pi\frac{F}{F_s}n}dF \tag{5.9}$$

Note that in Eqn (5.9), the signal $X_a(F)$ over the interval $[(l-1/2)F_s, (l+1/2)F_s]$ is equivalent to $X_a(F+lF_s)$ in the integration interval $[-F_s/2, F_s/2_s]$. The summation term in Eqn (5.9) can be further expanded as

$$\sum_{l=-\infty}^{\infty} \int_{(l-1/2)F_s}^{(l+1/2)F_s} X_a(F) e^{j2\pi \frac{F}{F_s} n} dF = \sum_{l=-\infty}^{\infty} \int_{-F_s/2}^{F_s/2} X_a(F+lF_s) e^{j2\pi \frac{F+lF_s}{F_s} n} dF \quad (5.10)$$

The term on the right-hand side of Eqn (5.10) can be further manipulated by swapping the integral and summation signs as

$$\sum_{l=-\infty}^{\infty} \int_{-F_s/2}^{F_s/2} X_a(F+lF_s) e^{j2\pi \frac{F+lF_s}{F_s} n} dF = \sum_{l=-\infty}^{\infty} \int_{-F_s/2}^{F_s/2} X_a(F+lF_s) e^{j2\pi \frac{F}{F_s} n} dF \quad (5.11)$$

where the exponential in Eqn (5.11) has been reduced since we can satisfy $e^{j2\pi \frac{F+lF_s}{F_s} n} = e^{j2\pi \frac{lF_s}{F_s} n} e^{j2\pi \frac{F}{F_s} n} = e^{j2\pi \frac{F}{F_s} n}$.

A comparison between Eqns (5.8) and (5.11) reveals that

$$\frac{1}{F_s} \int_{-F_s/2}^{F_s/2} X\left(\frac{F}{F_s}\right) e^{j2\pi \frac{F}{F_s} n} dF = \int_{-F_s/2}^{F_s/2} \sum_{l=-\infty}^{\infty} X_a(F+lF_s) e^{j2\pi \frac{F_s}{F_s} n} dF \quad (5.12)$$

A comparison between the right- and left-hand sides of Eqn (5.12) shows that

$$\frac{1}{F_s} X\left(\frac{F}{F_s}\right) = \sum_{l=-\infty}^{\infty} X_a(F+lF_s) \quad (5.13)$$

Or in terms of normalized frequency, the relationship in Eqn (5.13) can be written as

$$X(f) = F_s \sum_{l=-\infty}^{\infty} X_a[(f+l)F_s] \quad (5.14)$$

A close examination of the discrete spectrum $X(f)$ in Eqn (5.14) reveals that it is made up of replicas of the analog spectrum $X_a(F)$ scaled by the sampling frequency F_s and periodically shifted in frequency. The sampling process is illustrated in Figure 5.1.

Thus far, we have shown according to Eqn (5.14) that the discrete spectrum is made up of a series of periodic replicas of the analog spectrum with spacing dependent on the sampling frequency F_s. If the sampling frequency is chosen such that it is greater than the bandwidth of the signal, that is, $F_s > B$, where B is the intermediate

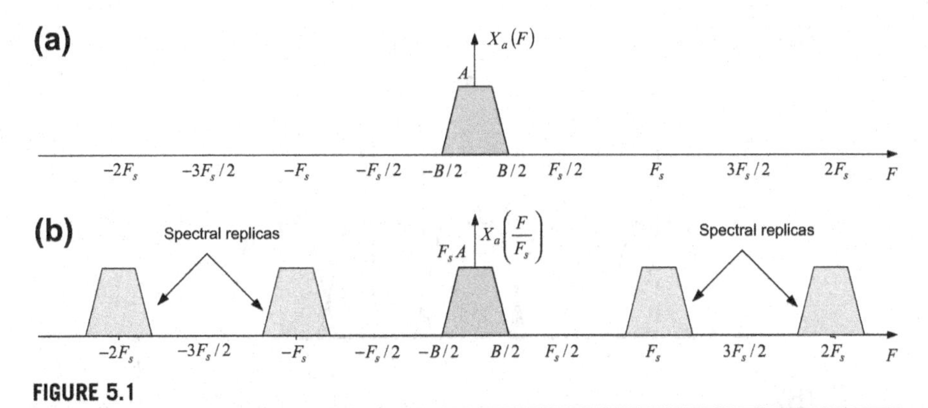

FIGURE 5.1

The sampling process of analog signals: (a) the analog signal bounded by the frequency $|B/2|$, and (b) the spectrum of discrete-time-sampled analog signal scaled by F_s and replicated at the sampling frequency. (For color version of this figure, the reader is referred to the online version of this book.)

frequency (IF) signal bandwidth, then the analog signal can in theory be *accurately* reconstructed without loss of information from the discrete-time samples.[1] As mentioned earlier, the particular frequency for which $F_{\text{Nyquist}} = B$ is known as the Nyquist rate. If, on the other hand, the sampling frequency is chosen such that $F_s < B$, then the aliased replicas of the spectrum overlap the desired signal spectrum as shown in Figure 5.2(a) and degradation ensues. In other words, the discrete spectrum is made up of overlapped replicas of the original spectrum scaled by the sampling frequency. In this case, the analog signal reconstructed from the aliased signal spectrum is not an exact replica of the original signal spectrum as shown in Figure 5.2(b). In this case, the sampled signal does not faithfully represent the information present in the analog signal.

In practice, in addition to the proper choice of sampling frequency, an important aspect of managing the amount of degradation due to aliasing is limiting the amount of out-of-band energy that is allowed to fold over onto the desired signal's band. Although the desired signal impinging at the antenna is bandlimited, it is typically accompanied by out-of-band signal components from within the desired signaling band as well as from outside of it. These signals are either blockers or interferers. The band definition filter that precedes the low noise amplifier (LNA) usually attenuates the signals and interferers that exist out of the signaling band. The received signals and interferers are further manipulated and conditioned throughout the receive chain by a variety of analog linear and nonlinear blocks. All throughout, noise and new signals due to nonlinearities may arise in or around the desired signal's band. The last line of defense on the analog side is the antialiasing filter.

[1]When converting a discrete-time signal to analog signal, proper analog filtering must be applied to prevent spectral images from corrupting the desired signal spectrum.

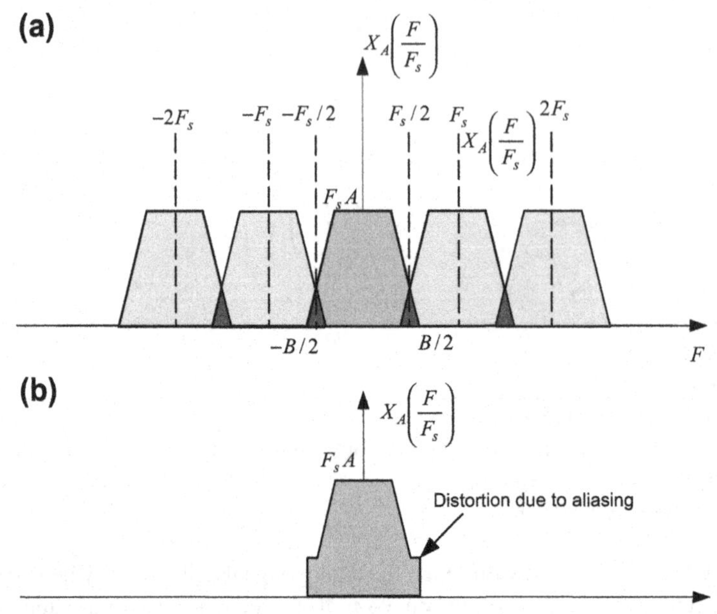

FIGURE 5.2

Aliased spectrum and its analog reconstructed counterpart: (a) aliased spectrum of discrete-time-sampled spectrum, and (b) spectrum of analog signal reconstructed from aliased discrete-time signal. (For color version of this figure, the reader is referred to the online version of this book.)

The effectiveness of the antialiasing filter is measured by its attenuation outside the desired signal band. Any blockers or interferers that are more than ½ least significant bit (LSB) in relative strength can fold over onto the desired signal's band and cause distortion to the desired signal's *SNR*. Whether the aliased energy itself falls within the signal's bandwidth or not depends on its frequency location and on the sampling rate of the ADC [3].

Depending on the number of bits in the ADC, it may not be reasonable to expect the antialiasing filter to attenuate all nondesired out-of-band signals to less than ½ LSB. This stringent filtering requirement may be alleviated by oversampling the ADC and further spreading apart the spectral replicas. Digital filtering can then be used to further attenuate some of the nondesired signal blockers and interferers. This in turn serves to reduce the complexity of the antialiasing filter at the cost of running the ADC and digital circuits at a higher clock rate and adding more complexity to the digital signal processing. All in all, an architectural trade-off between the analog and digital signal processing must be performed to properly manage the out-of-band degradation to provide an acceptable desired signal *SNR*.

5.1.1.2 Reconstruction of lowpass signals from discrete samples

Next, we examine the reconstruction of the analog lowpass signals from digital samples. Consider the Fourier transform of the analog signal $x_a(t)$

$$x_a(t) = \int_{-\infty}^{\infty} X_a(F)e^{j2\pi Ft}\mathrm{d}F \tag{5.15}$$

and the discrete-time signal sampled at the Nyquist rate $F_s = B$

$$X_a(F) = \begin{cases} \frac{1}{F_s}X\left(\frac{F}{F_s}\right) & -\frac{F_s}{2} < F < \frac{F_s}{2} \\ 0 & \text{otherwise} \end{cases} \tag{5.16}$$

The Fourier transform of $X\left(\frac{F}{F_s}\right)$ is given as

$$X\left(\frac{F}{F_s}\right) = \sum_{n=-\infty}^{\infty} x(nT_s)e^{-j2\pi\frac{F}{F_s}n} \tag{5.17}$$

Then the reconstructed analog signal $\hat{x}_a(t)$ can be obtained by substituting Eqn (5.17) into Eqn (5.16) and then substituting the result into Eqn (5.15), that is

$$\begin{aligned} \hat{x}_a(t) &= \frac{1}{F_s}\int_{-F_s/2}^{F_s/2} X\left(\frac{F}{F_s}\right)e^{j2\pi Ft}\mathrm{d}F \\ &= \frac{1}{F_s}\int_{-F_s/2}^{F_s/2}\left[\sum_{n=-\infty}^{\infty} x(nT_s)e^{-j2\pi\frac{F}{F_s}n}\right]e^{j2\pi Ft}\mathrm{d}F \end{aligned} \tag{5.18}$$

The order of the integral and the summation given in Eqn (5.18) can be further rearranged to obtain

$$\begin{aligned} \hat{x}_a(t) &= \sum_{n=-\infty}^{\infty} x(nT_s)\left[\frac{1}{F_s}\int_{-F_s/2}^{F_s/2} e^{j2\pi F\left(t-\frac{n}{F_s}\right)}\mathrm{d}f\right] \\ &= \sum_{n=-\infty}^{\infty} x(nT_s)\frac{\sin\left(\frac{\pi}{T_s}(t-nT_s)\right)}{\frac{\pi}{T_s}(t-nT_s)} \end{aligned} \tag{5.19}$$

The sine function divided by its argument is the sin *c* function shifted in time by the sampling period. Define the sin *c* function in Eqn (5.19) as $p(t)$ and its frequency domain equivalent as $P(F)$, or

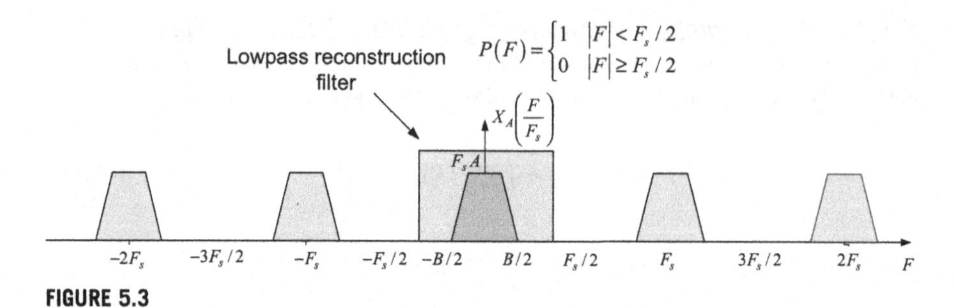

FIGURE 5.3

Recovered original signal spectrum from discrete-time periodic spectrum. (For color version of this figure, the reader is referred to the online version of this book.)

$$
\begin{aligned}
p(t) &= \frac{\sin\left(\frac{\pi}{T_s}(t-nT_s)\right)}{\frac{\pi}{T_s}(t-nT_s)} \quad \text{time domain} \\
P(F) &= \begin{cases} 1 & |F| < F_s/2 \\ 0 & |F| \geq F_s/2 \end{cases} \quad \text{frequency domain}
\end{aligned}
\tag{5.20}
$$

From Eqn (5.20), it is obvious that the sin *c* function plays the role of an ideal interpolation filter that when applied to the signal spectrum of a nonaliased discrete-time signal recovers the original analog signal *without degradation* or

$$
\widehat{X}\left(\frac{F}{F_s}\right) = X\left(\frac{F}{F_s}\right)P(F) \tag{5.21}
$$

This process is illustrated in Figure 5.3 and shows the ideal filter's role in recovering the spectrum of the original analog signal.

5.1.2 Sampling and reconstruction of bandpass signals

Bandpass sampling is prevalent in certain radio architectures that digitize the signal at IF.[2] There are certain advantages and disadvantages to sampling the signal at IF—for one, IF sampling does not suffer from DC offsets due to LO leakage and second order effects. Furthermore, IF sampling decreases the vulnerability of narrowband signals to $1/f$ noise. In this section, we will discuss bandpass sampling and signal reconstruction. Another potential advantage of IF sampling is the lack of imbalance between in-phase and quadrature components since the architecture may not require quadrature mixing in the receive path. However, this is

[2]Although in theory one could also sample the received signal at RF, there are numerous practical reasons, given today's technology, why an RF-sampling architecture is still not realizable for most frequency bands.

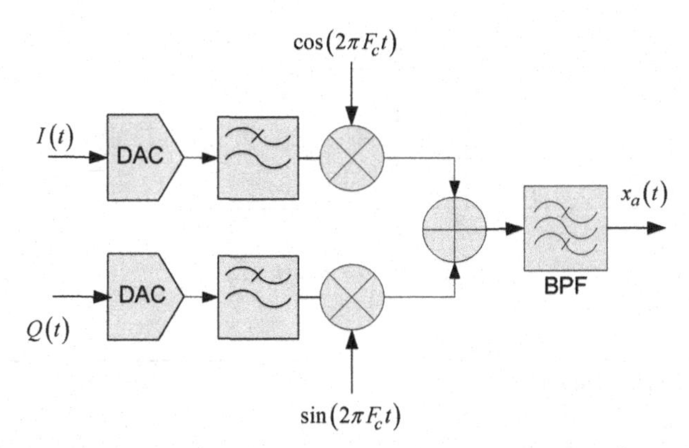

FIGURE 5.4 Quadrature up-conversion from baseband to intermediate frequency or radio frequency. BPF, bandpass filter; DAC, digital to analog converter. (For color version of this figure, the reader is referred to the online version of this book.)

accomplished at the expense of adding cost and increasing power consumption and complexity of the receiver.

5.1.2.1 Signal representation and bandpass sampling: integer positioning

The analog IF or radio frequency (RF) signal can, in its purest form, be expressed in terms of the in-phase $I(t)$ and quadrature $Q(t)$ signals. For example, given a direct conversion transmitter depicted in Figure 5.4, the in-phase and quadrature components are first converted from discrete-time signals to analog signals filtered and up-converted to RF. In this case, the in-phase and quadrature signals are analog, the *analog* signal $x_a(t)$ can be expressed as

$$\begin{aligned} x_a(t) &= I(t)\cos(2\pi F_c t) - Q(t)\sin(2\pi F_c t) \\ &= \mathrm{Re}\left\{(I(t) + jQ(t))e^{j2\pi F_c t}\right\} \\ &= \mathrm{Re}\left\{s_a(t)e^{j2\pi F_c t}\right\} \end{aligned} \tag{5.22}$$

An equivalent representation of Eqn (5.22) is in polar form where the signal takes the form $|r(t)|e^{j\phi(t)}$ or

$$x_a(t) = \sqrt{I^2(t) + Q^2(t)}\cos\left(2\pi F_c t + \tan^{-1}\left(\frac{-Q(t)}{I(t)}\right)\right) \tag{5.23}$$

Either Eqn (5.22) or Eqn (5.23) can be used when analyzing bandpass IF or RF signals.

When discussing bandpass sampling, a parameter of interest is the fractional bandwidth defined as the number of signal bandwidths separating the origin to the lower edge of the passband as defined in [2]. A special case occurs when the IF frequency (or RF frequency) F_c is related to the *noise equivalent* bandwidth B before sampling as

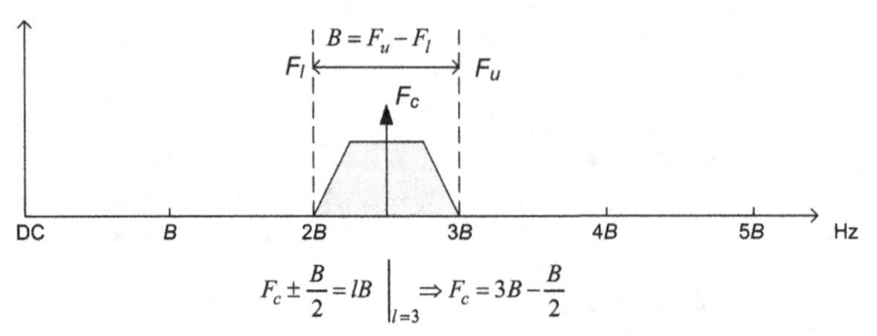

FIGURE 5.5

Integer band positioning for $l = 3$.

$$F_c \pm \frac{B}{2} = lB \quad l \text{ is a positive integer} \tag{5.24}$$

The special relationship in Eqn (5.24) is known as integer band positioning and is illustrated in Figure 5.5 for $l = 3$. The signal is bounded in the frequency domain *strictly* by an upper bound frequency F_u and a lower bound frequency F_l.

In order to observe the implication of integer band positioning on the sampling theorem, assume that the IF frequency is chosen such that $F_c = lB - B/2$, $l > 0$, then let the sampling rate be given as $T_s = 1/2B$, then the relationship in Eqn (5.22) becomes

$$\tilde{x}_a(t) = I(t)\cos(2\pi F_c t) - Q(t)\sin(2\pi F_c t)\Big|_{\substack{t=nT_s \\ F_c=(2l-1)\frac{B}{2}}} \Rightarrow$$

$$\tilde{x}_a(t)|_{t=nT_s} \triangleq \tilde{x}_a(nT_s) = I(nT_s)\cos\left(2\pi(2l-1)\frac{B}{2}nT_s\right) - Q(nT_s)\sin\left(2\pi(2l-1)\frac{B}{2}nT_s\right)$$

$$\tilde{x}_a(t)|_{t=nT_s} \triangleq \tilde{x}_a(nT_s) = I(nT_s)\cos\left(\frac{\pi}{2}(2l-1)n\right) - Q(nT_s)\sin\left(\frac{\pi}{2}(2l-1)n\right) \tag{5.25}$$

The notation ~ is used to designate the sampled version of the analog signal $x_a(t)$. Let us examine two scenarios in Eqn (5.25). For n even, say $n = 2m,\ m \in \mathbb{N}$, then the relationship in Eqn (5.25) becomes:

$$\begin{aligned} \tilde{x}_a(2mT_s) &= I(2mT_s)\cos((2l-1)m\pi) - Q(nT_s)\sin((2l-1)m\pi) \\ \tilde{x}_a(2mT_s) &= (-1)^m I(2mT_s),\ n \text{ is even} \end{aligned} \tag{5.26}$$

or simply

$$\tilde{x}_a(nT_s) = (-1)^{n/2} I(nT_s) \quad \text{for } n \text{ even} \tag{5.27}$$

The relationship in Eqn (5.27) implies that for n even, the quadrature samples disappear and the signal is made up of in-phase samples only.

For n odd, that is for $n = 2m - 1$, $m \in \mathbb{N}$, the relationship in Eqn (5.25) becomes

$$\begin{aligned}\tilde{x}_a((2m-1)T_s) &= I((2m-1)T_s)\cos\left(2\pi(2l-1)\frac{B}{2}(2m-1)T_s\right)\\&\quad - Q((2m-1)T_s)\sin\left(2\pi(2l-1)\frac{B}{2}(2m-1)T_s\right)\\ \tilde{x}_a((2m-1)T_s) &= I((2m-1)T_s)\cos\left(\frac{\pi}{2}(4lm-2l-2m+1)\right)\\&\quad - Q((2m-1)T_s)\sin\left(\frac{\pi}{2}(4lm-2l-2m+1)\right)\\ \tilde{x}_a((2m-1)T_s) &= (-1)^{l+m+1}Q((2m-1)T_s) \quad \text{for } n \text{ odd}\end{aligned} \tag{5.28}$$

In other words, the sampled analog signal in Eqn (5.28) is

$$\tilde{x}_a(nT_s) = (-1)^{l+\frac{n+1}{2}+1}Q(nT_s) \quad \text{for } n \text{ odd} \tag{5.29}$$

In a similar manner to the case for even values of n, for odd values of n the in-phase signal samples disappear and the sampled signal is made up of quadrature signal samples. In order to further elaborate on this special case, consider $T_s' = 2T_s = 1/B$, then substituting into Eqns (5.27) and (5.29), we obtain the relationship

$$\begin{aligned}\tilde{x}_a(mT_s') &= (-1)^m I(mT_s') && \text{for } n \text{ even}, \ n = 2m\\ \tilde{x}_a\left(mT_s' - \frac{T_s'}{2}\right) &= (-1)^{l+m+1}Q\left(mT_s' - \frac{T_s'}{2}\right) && \text{for } n \text{ odd}, \ n = 2m-1\end{aligned} \tag{5.30}$$

An interesting consequence for choosing the sampling rate such that the bandwidth of the signal is integer positioned in relationship to the IF is the simplified quadrature conversion (digital mixing) to baseband. In this case, the sampled in-phase component is obtained by multiplying the output of the ADC by $(-1)^{n/2}$ for n even and by zero for n odd. In a similar manner, the quadrature component is obtained by multiplying the output of the ADC by $(-1)^{l+\frac{n+1}{2}+1}$ for n odd and by zero for n even.

5.1.2.2 Reconstruction of bandpass signals based on integer positioning

Let us reconsider the discrete-time samples obtained in Eqn (5.30), that is $I(mT_s')$ and $Q\left(mT_s' - \frac{T_s'}{2}\right)$. These samples can be used to reconstruct the equivalent analog signal according to the interpolation relationship presented in Eqn (5.19), that is for the quadrature component, we have

$$I(t) = \sum_{m=-\infty}^{\infty} I(2mT_s) \frac{\sin\left(\frac{\pi}{2T_s}(t - 2mT_s)\right)}{\frac{\pi}{2T_s}(t - 2mT_s)} \tag{5.31}$$

And similarly, for the quadrature component we obtain a similar expression

$$Q(t) = \sum_{m=-\infty}^{\infty} Q(2mT_s - T_s) \frac{\sin\left(\frac{\pi}{2T_s}(t - 2mT_s + T_s)\right)}{\frac{\pi}{2T_s}(t - 2mT_s + T_s)} \tag{5.32}$$

Substituting Eqns (5.31) and (5.32) into Eqn (5.22) we obtain

$$x_a(t) = I(t)\cos(2\pi F_c t) - Q(t)\sin(2\pi F_c t)$$

$$x_a(t) = \sum_{m=-\infty}^{\infty} \left\{ (-1)^m \tilde{x}_a(2mT_s) \frac{\sin\left(\frac{\pi}{2T_s}(t - 2mT_s)\right)}{\frac{\pi}{2T_s}(t - 2mT_s)} + (-1)^{l+m} \tilde{x}_a(2mT_s - T_s) \frac{\sin\left(\frac{\pi}{2T_s}(t - 2mT_s + T_s)\right)}{\frac{\pi}{2T_s}(t - 2mT_s + T_s)} \right\} \tag{5.33}$$

The relationship in Eqn (5.33) can be further expressed in a yet simpler form as

$$x_a(t) = \sum_{m=-\infty}^{\infty} \tilde{x}_a(mT_s) \frac{\sin\left(\frac{\pi}{2T_s}(t - mT_s)\right)}{\frac{\pi}{2T_s}(t - mT_s)} \cos(2\pi F_c(t - mT_s)) \tag{5.34}$$

Bandpass sampling based on integer positioning is desired since it greatly simplifies the digital mixing process. However, in many cases, it is not practical to relate the IF center frequency to the signaling bandwidth and in turn to the antialiasing filter's bandwidth. Filters such as surface acoustic wave filters (SAWs) are designed with a certain *suitable* IF center frequency. Furthermore, the choice of the sampling frequency, and hence the sampling clock, the IF center frequency, and the antialiasing filter involve a trade study that encompasses the whole receiver chain, and the problem is not a mere simplification of the signal processing.

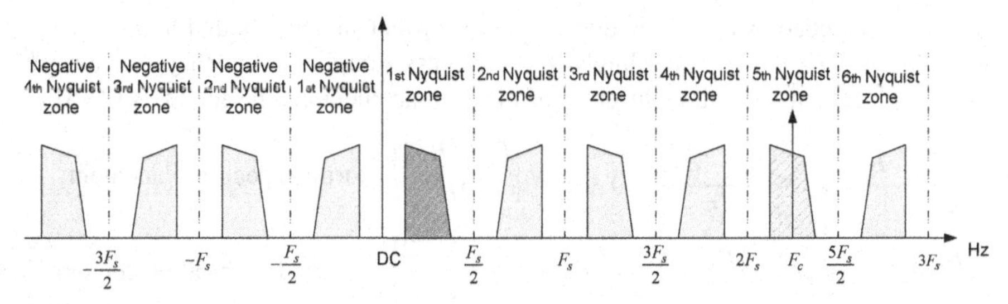

FIGURE 5.6

Images of the desired signal spectrum as they occur in various Nyquist zones. (For color version of this figure, the reader is referred to the online version of this book.)

5.1.2.3 General bandpass sampling and Nyquist zones

In general, when sampling a bandpass signal, it is imperative to know the region in the spectrum where the signal lies. That is, it is important to know in which Nyquist zone the signal lies. Nyquist zones divide the spectrum into uniformly spaced regions placed equidistantly at $F_s/2$ intervals. The Nyquist zones are divided into odd and even zones as shown in Figure 5.6. The odd Nyquist zones present exact replicas of the sampled signal, whereas the even Nyquist zones present exact *mirrored* replicas of the sampled signal.

The zone situated between DC and $F_s/2$ is known as the first Nyquist zone. On the other hand, the zone placed between $-F_s/2$ and DC is the first negative Nyquist zone. In general, if the signal is centered at F_C, where $F_C = 0$ as in the lowpass signal case, then exact spectral replicas appear at $F_C + kF_s$ for $k = 0, 1, 2, 3, \ldots$ It is obvious from Figure 5.6, that the first Nyquist zone corresponds to the case where $k = 0$. Consequently, first, third, and fifth Nyquist zones occur at $k = 1$, $k = 2$, and $k = 3$, respectively.

In a similar manner, the mirrored replicas of the signal's spectrum occur in even-numbered Nyquist zones centered at $kF_s - F_C$ for $k = 1, 2, 3, \ldots$ The second, fourth, and sixth Nyquist zones occur at $k = 1$, $k = 2$, and $k = 3$, respectively as presented in Figure 5.6. In general, the relationship that governs even and odd Nyquist zones and the signal's center frequency F_C is

$$F = \begin{cases} rem(F_C, F_s) & \left\lfloor \dfrac{F_C}{F_s/2} \right\rfloor \text{ is even} \Rightarrow \text{image is exact replica} \\ F_s - rem(F_C, F_s) & \left\lfloor \dfrac{F_C}{F_s/2} \right\rfloor \text{ is odd} \Rightarrow \text{image is mirrored replica} \end{cases} \tag{5.35}$$

where the function *rem* is the remainder after division. Furthermore, the notation $\lfloor . \rfloor$ represents the floor function, which rounds a number toward zero. F is the center frequency of the image in the first Nyquist zone.

Next, consider the center frequency of the signal to be F_C, then define in relation to the bandwidth B as $F_C = F_U - B/2 = F_L + B/2$, where the bandwidth of the signal or channel is bounded by a lower bound frequency F_L and an upper bound frequency F_U such that $B = F_U - F_L$. Accordingly, the bandpass sampling frequency for normal spectral placement as well as inverted or mirror placement can be found as [2–5]

$$\begin{aligned} &\frac{2F_c + B}{2n+1} \le F_s \le \frac{F_c - B/2}{n} \quad 0 \le n \le \left\lfloor \frac{F_c - B/2}{2B} \right\rfloor \quad \text{normal spectral placement} \\ &\frac{F_c + B/2}{n} \le F_s \le \frac{2F_c - B}{2n-1} \quad 1 \le n \le \left\lfloor \frac{F_c + B/2}{2B} \right\rfloor \quad \text{inverted spectral placement} \end{aligned} \tag{5.36}$$

The minimum sampling rates as given in Eqn (5.36) must not be taken at face value. Serious considerations must be given to imperfections in the receiver design and possible interference in the RF. For example, sampling clock jitter, carrier frequency drift, blockers, and other interferers in the vicinity of the desired signal, noise equivalent bandwidth of the antialiasing filter are just a few of the considerations the designer must take when deciding on the choice of the sampling frequency.

5.2 Signal distortion due to sampling and conversion imperfections

This section is concerned with signal degradations due to the sampling process. Topics ranging from quantization noise to jitter and clipping will be discussed. Performance measures such as signal-to-quantization ratio of oversampled signals will also be discussed. These parameters are specifically relevant to Nyquist converters. Similar parameters will be developed for oversampled converters in the next chapter.

5.2.1 Quantization noise

Given an analog signal $x_a(t)$ sampled at $t = nT_s$ results in the discrete-time signal $x_a(nT_s)$. The Nyquist converter has a finite resolution depending on the total number of bits that represent the analog signal. This in turn implies that the signal $x_a(nT_s)$ will be mapped to discrete signal $x_k(nT_s)$, which belongs to a set of finite values representing the range of the ADC, that is

$$\cdots x_{k-1}(nT_s) \le x_k(nT_s) \le x_{k+1}(nT_s) \cdots \text{ for } k = 1, \cdots, K \tag{5.37}$$

The quantization process as presented above implies that the process itself is both memoryless and nonlinear. Furthermore, the error due to quantization *seems* to be uniformly distributed. This is true given that the quantization step size is the same between any two consecutive levels.

Define the quantization step as

$$\delta = \cdots x_k(nT_s) - x_{k-}(nT_s) \triangleq x_{k+1}(nT_s) - x_k(nT_s) \cdots \text{ for all } k = 1, \cdots, K-1 \tag{5.38}$$

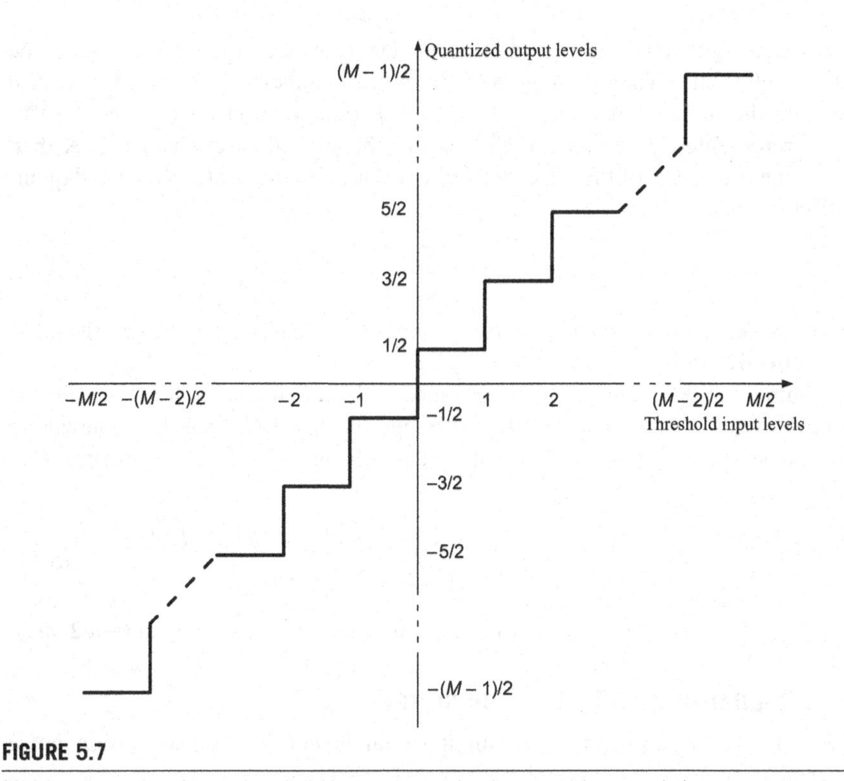

FIGURE 5.7

Midriser quantizer input threshold versus quantized output characteristics.

The digital mapping of the analog signal results in the *quantizer* being a midtread quantizer if the DC value is assigned a quantization level, otherwise the quantizer is known as a midriser, which obviously implies that DC is not represented by a quantization step. An example of a midriser quantizer is depicted in Figure 5.7. Given a b-bit quantizer, the number of threshold input levels versus the quantized output levels for a midriser quantizer is simply $M = 2^b$. On the other hand, a midtread quantizer has $M = 2^b - 1$ input threshold levels.

Next, we define the uniform quantization error as the difference between the discrete time-sampled analog signal and its equivalent digital representation, that is

$$-\frac{\delta}{2} < e_k(nT_s) = x_a(nT_s) - x_k(nT_s) < \frac{\delta}{2} \tag{5.39}$$

According to Eqn (5.39), the quantization error is bounded by half the step size provided that $x_a(nT_s)$ lies within two quantization levels. Define the peak-to-peak voltage or full-scale range of the ADC as V_{F_s}. The relationship in Eqn (5.39) is true only if the analog signal level is less than the peak voltage $V_{x,F_s}/2$, otherwise the quantization error could exceed half the quantization step.

In this case, signal clipping is said to occur. The step size of the ADC can also be used to define the dynamic range as well as the number of effective bits. Let b designate the number of effective bits, then K quantization levels can be sufficiently represented *if and only if* $2^b \geq K$. For the special case where $2^b = K$ then let δ be the resolution of the ADC defined in relation to the peak voltage and quantization levels as

$$\delta = \frac{V_{x,F_s}}{2^b} \tag{5.40}$$

For a fixed peak voltage, the resolution of the ADC increases by increasing the number of effective bits.

Assume that the input signal is zero-mean white Gaussian distributed and that the input signal is always within the full range of the ADC, then the quantization errors are statistically uniformly distributed and the error samples are uncorrelated, that is:

$$\begin{aligned} E\{e_k(nT_s)e_k(nT_s+T_s)\} &= E\{e_k(nT_s)\}E\{e_k(nT_s+T_s)\} \\ e_k(nT_s) &\in (-\delta/2, \delta/2) \end{aligned} \tag{5.41}$$

From Eqn (5.41), it is inferred that the error is confined to the interval $(-\delta/2, \delta/2)$.

5.2.2 Signal-to-quantization-noise ratio

Define the *SQNR* as the ratio of input signal power P_x to quantization noise power P_e

$$SQNR_{dB} = 10\log_{10}\left(\frac{P_x}{P_e}\right) \tag{5.42}$$

By way of an example, let the input to the ADC be an additive white Gaussian noise (AWGN) signal, then the distribution of the signal is the familiar bell-shaped curve depicted in Figure 5.8. At the output of the ADC, the quantized signal is somewhat a flawed reproduction of the original input signal. The difference between the original signal and the quantized signal is the quantization error. The quantization error is uniformly distributed as shown in Figure 5.9.

Given that the error signal is zero mean, that is $E\{e_k(t)\} = 0$, then the quantization noise power is the variance of the quantization error signal σ_e^2. The variance can be simply computed according to the relation

$$\begin{aligned} P_e &= E\{e_k^2(t)\} - \underbrace{E^2\{e_k(t)\}}_{=0} \\ &= E\{e_k^2(t)\} = \sigma_e^2 \end{aligned} \tag{5.43}$$

The theoretical probability distribution function (PDF) of the quantization noise is shown in Figure 5.10. Next, let us compute the error variance σ_e^2

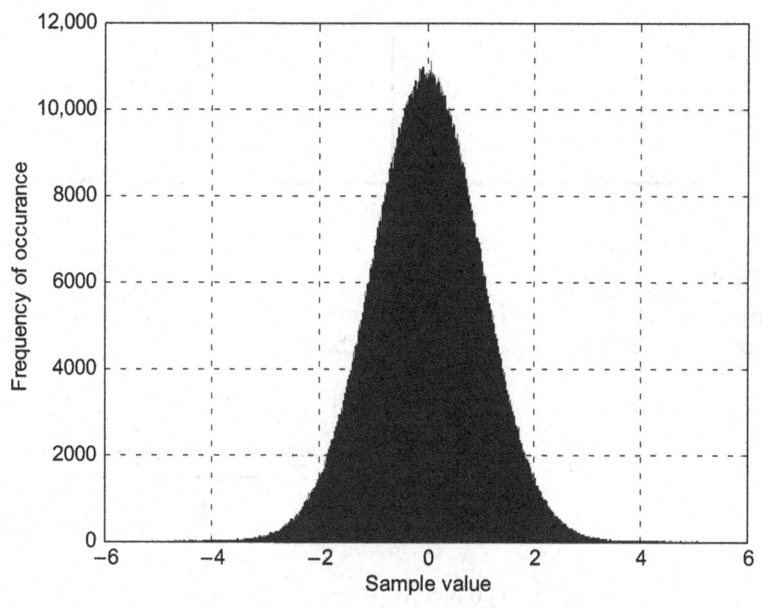

FIGURE 5.8

Histogram of additive white Gaussian noise signal used as input to an analog-to-digital converter device. (For color version of this figure, the reader is referred to the online version of this book.)

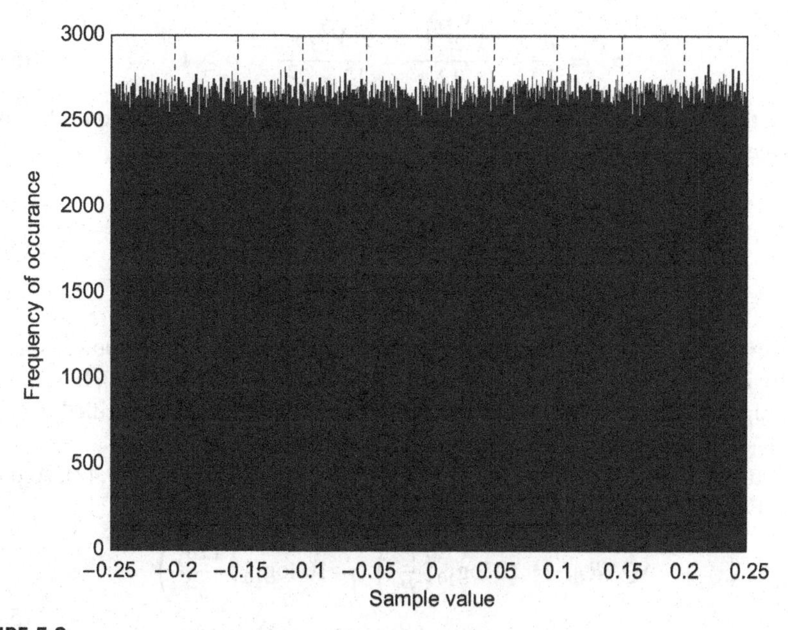

FIGURE 5.9

Histogram of quantization error signal showing a uniform distribution. (For color version of this figure, the reader is referred to the online version of this book.)

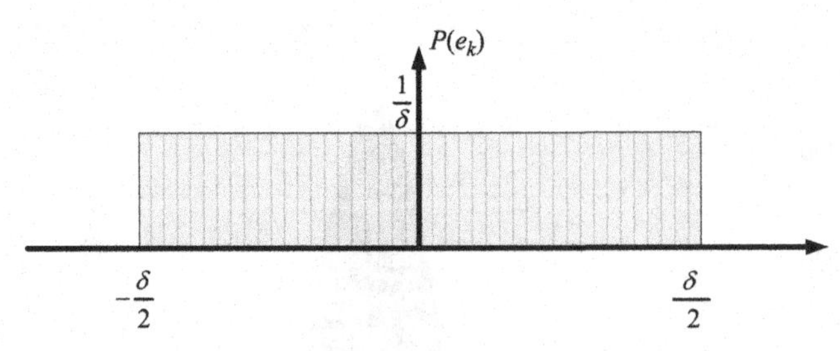

FIGURE 5.10

Probability distribution function of quantization error signal. (For color version of this figure, the reader is referred to the online version of this book.)

$$\begin{aligned} P_{e_k} &= \sigma_e^2 = E\{e_k^2(t)\} = \int_{-\delta/2}^{\delta/2} e_k^2(t)\underbrace{P(e_k(t))}_{=\frac{1}{\delta}}de_k(t) \\ &= \frac{1}{\delta}\int_{-\delta/2}^{\delta/2} e_k^2(t)de_k(t) = \frac{\delta^2}{12} \end{aligned} \tag{5.44}$$

The relationship in Eqn (5.44) can be further expressed in terms of the number of effective bits and full-scale voltage by substituting Eqn (5.40) for δ

$$\sigma_e^2 = \frac{\delta^2}{12} = \frac{\left(\frac{V_{x,Fs}}{2^b}\right)^2}{12} = \frac{V_{x,Fs}^2}{12 \times 2^{2b}}, \quad b \geq 1 \tag{5.45}$$

Define the power spectral density of the quantization noise as the ratio of the noise variance to the sampling frequency or

$$\begin{aligned} S_e(f) &= \frac{2P_e}{F_s}; \text{ sampling at baseband} \\ S_e(f) &= \frac{P_e}{F_s}; \text{ sampling at IF or bandpass sampling} \end{aligned} \tag{5.46}$$

In the event where the signal is sampled at baseband, traditionally the power spectral density is restricted to the frequency band $[0, F_s/2]$. If on the other hand, IF sampling or bandpass sampling is employed, the power spectral density is specified over the band $[-F_s/2, F_s/2]$.

Next, let us further examine the *SQNR* relationship given in Eqn (5.42). According to Eqns (5.44) and (5.45), we can express Eqn (5.42) as

$$\begin{aligned} SQNR_{\text{dB}} &= 10\log_{10}\left(\frac{P_x}{P_e}\right) = 10\log_{10}\left(\frac{12\sigma_x^2}{\delta^2}\right) \\ &= 10\log_{10}\left(\frac{12 \times 2^{2b} \times \sigma_x^2}{V_{x,Fs}^2}\right), \quad b \geq 1 \end{aligned} \tag{5.47}$$

Further expansion of Eqn (5.47) reveals the classic equation for $SQNR$ as

$$\begin{aligned} SQNR_{\text{dB}} &= 10\log_{10}(12) + b \times 10\log_{10}(4) + 10\log_{10}\left(\frac{\sigma_x^2}{V_{x,F_s}^2}\right) \\ &= 10.8 + 6.0206b + 10\log_{10}\left(\frac{\sigma_x^2}{V_{x,F_s}^2}\right), \quad b \geq 1 \end{aligned} \tag{5.48}$$

Define the peak power-to-average-power ratio ($PAPR$) as the ratio of the full-scale peak signal power $P_{\text{peak}} = (V_{x,F_s}/2)^2$ to the average signal power or σ_x^2

$$\begin{aligned} PAPR &= 10\log_{10}\left(\frac{P_{\text{peak}}}{P_s}\right) \\ &= 10\log_{10}\left(\frac{V_{x,F_s}^2}{4\sigma_x^2}\right) = 10\log_{10}\left(\frac{1}{\kappa^2}\right) \end{aligned} \tag{5.49}$$

The variable κ is known as the waveform loading factor. It is given as the ratio of the input signal root mean square (*rms*) voltage to the full-scale voltage of the ADC.

Next, we examine the impact of oversampling on quantization noise. The bandwidth of the uniformly distributed quantization noise ranges from $-F_s/2$ to $F_s/2$ as depicted in Figure 5.11. Therefore, increasing the sampling rate serves to spread the quantization noise across the band while keeping the quantization noise power the same. The quantization noise within the signal band, however, diminishes as a function of oversampling. Therefore, the noise power within the sampling band gets scaled by the oversampling ratio defined simply as

$$OSR = \frac{F_s}{B} \tag{5.50}$$

The bandwidth B refers to the IF bandwidth as shown in Figure 5.11. The spreading of the noise due to oversampling causes the noise power within the desired

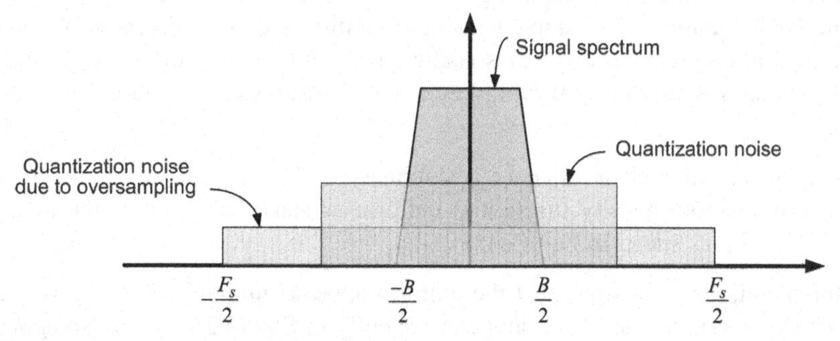

FIGURE 5.11

Power spectral density of quantization noise spread between $-F_s/2$ and $F_s/2$. (For color version of this figure, the reader is referred to the online version of this book.)

bandwidth to be scaled by the oversampling ratio σ_e^2/OSR and hence the $SQNR$ to be expressed as

$$\begin{aligned} SQNR &= 10\log_{10}\left(\frac{\sigma_x^2}{\sigma_e^2}OSR\right) \\ &= 10\log_{10}\left(\frac{12\times 2^{2b}\times\sigma_x^2}{V_{x,F_s}^2}\frac{F_s}{B}\right) = 10\log_{10}\left(\frac{3\kappa^2 2^{2b}F_s}{B}\right) \\ &= 4.7712 + 6.02b + 10\log_{10}\left(\frac{F_s}{B}\right) - PAPR \\ &= 4.7712 + 6.02b + 10\log_{10}(OSR) - PAPR \end{aligned} \tag{5.51}$$

The quantization noise power after oversampling can then be computed via the relation

$$\int_{-B/2}^{B/2} S_e(f)df = \frac{BP_e}{F_s} = \frac{1}{OSR}P_e \tag{5.52}$$

The improvement due to oversampling can be seen in the $SQNR$ only after digital filtering the signal to its occupied bandwidth. This gain in $SQNR$ is sometimes referred to as processing gain.

At this point, it is important to clarify that the *quantization noise* discussed thus far is not statistically independent in nature but rather depends completely on the input signal. Historically, quantization noise has been modeled as a uniform random variable—an approach due in large to the fact that the quantization error varies uniformly between $-\delta/2 < e_k(nT_s) < \delta/2$ as the input varies over the peak-to-peak values of the ADC [6,7]. This assumption is particularly true provided that the input signal assumes arbitrary values within the peak-to-peak range of the ADC, thus causing the output signal to crisscross the various quantization steps as the input changes from sample to sample. In this case, the uniform noise model described in Figure 5.12(a) is a reasonable quantization model to use. The validity of the model rests on the following key assumptions concerning the quantization error:

1. The quantization error sequence is stationary
2. The probability density function is uniform as stated earlier over the interval $(-\delta/2, \delta/2)$ as stated in Eqn (5.41)

At this point, we must stress that the uniform noise assumption used to derive the $SQNR$ expression in Eqn (5.48) and consequently in Eqn (5.51) is only an *approximation*. For example, we shall see in Section 5.2.5 that as the signal exceeds the peak of the ADC, clipping occurs. As the *rms* of the input signal increases, the mean square error due to clipping increases, eventually dominating the uniform

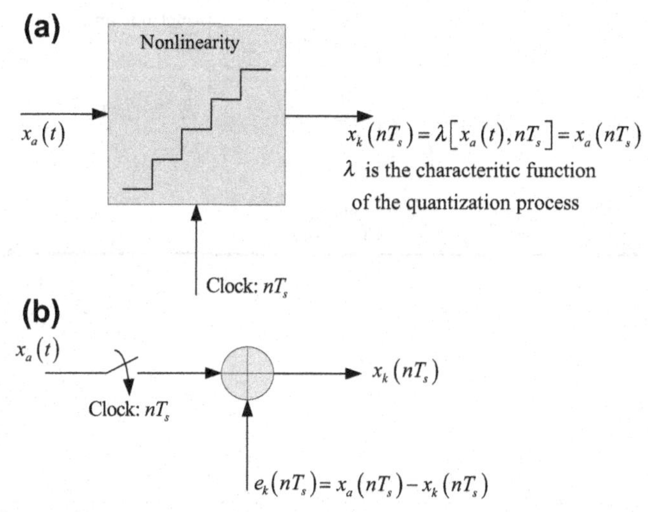

FIGURE 5.12

Quantization model based on: (a) nonlinear deterministic model, or (b) statistical additive model. (For color version of this figure, the reader is referred to the online version of this book.)

quantization noise. A spectral approach to quantization noise will be presented in the appendix.

5.2.3 Effect of clock jitter on sampling

Clock jitter is typically an ambiguous specification in data converters that describes the degradation due to timing errors in the *uniform*[3] sampling process. Clock jitter introduces a certain amount of uncertainty to the position of the samples taken at uniform intervals in time.

To further understand the impact of clock jitter, consider the tone $x_a(t) = A\sin(2\pi Ft)$ when sampled at F_s expressed as

$$\begin{aligned} x_a(t) &= A\sin(2\pi Ft) \xrightarrow{\text{Sample at } F_s} x(n) \cong x_a(t = nT_s)|_{T_s=1/F_s} \\ x(n) &= A\sin\left(2\pi\frac{F}{F_s}n\right) = A\sin(2\pi fn) \end{aligned} \tag{5.53}$$

[3]In this discussion, we are concerned with uniform sampling only versus, say, nonuniform sampling. Incidentally, the Nyquist sampling theorem assumes that the samples are obtained periodically and hence uniformly in time.

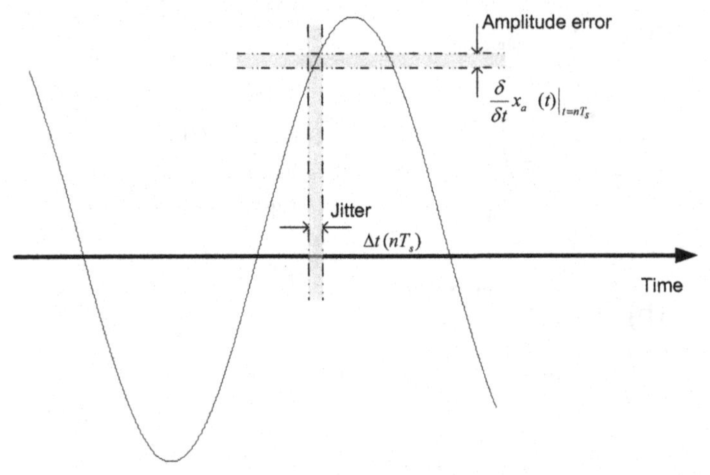

FIGURE 5.13

Effect of sampling jitter on a single tone. (For color version of this figure, the reader is referred to the online version of this book.)

where the normalized frequency is given as $f = F/F_s$. Let the jitter value $\Delta t(.)$ denote the timing jitter at the sampling instant nT_s, and then the sampling error is proportional to the derivative of the input signal $x_a(t)$ or [8]

$$e(t) = \Delta t(nT_s) \frac{\delta}{\delta t} x_a(t) \bigg|_{t=nT_s} \tag{5.54}$$

For a sinusoidal input, the sampled signal becomes

$$x(n) \approx x_a(nT_s) + \Delta t(nT_s) \frac{\delta}{\delta t} x_a(t)|_{t=nT_s} \tag{5.55}$$

The effect of jitter on a sinusoid is depicted in Figure 5.13. The maximum error occurs at the zero crossings, that is

$$\max\{e(t)\} = 2\pi F A \Delta t(nT_s) \tag{5.56}$$

where according to Eqn (5.55) the error function is given as $e(t) = \Delta t(nT_s) \frac{\delta}{\delta t} x_a(t)|_{t=nT_s}$.

EXAMPLE 5.1 JITTER OF AN *N*-BIT DATA CONVERTER

Let the amplitude range of a certain sinusoid vary between $\pm A$. What is the maximum jitter of an N-bit converter[4] such that the error is constrained to less than half LSB?

Let the ADC sampling clock be 20 MHz, then given an input tone at 10 MHz, find the maximum jitter value as you increase N from 7 to 12 bits. Determine the trend.

Next, consider fixing the number of bits to 10. What is the impact on maximum jitter as you increase the tone frequency from 10 to 60 MHz in 10 MHz steps and *changing* the sampling clock accordingly? Determine the trend.

According to the relationship presented in Eqn (5.56), the error magnitude is constrained to be less than ½ LSB:

$$e(t) < \frac{A}{2^N} \tag{5.57}$$

In other words, the jitter $\Delta t(.)$ is also constrained to

$$\Delta t(nT_s) < \frac{1}{2\pi F 2^N} \tag{5.58}$$

It is important to note that in Eqn (5.58), the frequency F is half the sampling clock frequency and represents an *extreme* scenario.

Given a 10-MHz tone and increasing the number of bits from 7 to 12 bits seems to decrease the maximum jitter as shown in Table 5.1. The trend depicted in Figure 5.14 shows that the jitter decays as N increases, varying from 124 to 3.89 ps.

Table 5.1 Maximum Jitter as a Function of Increasing Number of Bits

Parameter	Value					
Number of bits	7	8	9	10	11	12
Frequency (MHz)	10	10	10	10	10	10
Sampling clock (MHz)	20	20	20	20	20	20
Jitter (ps)	124.34	62.17	31.08	15.54	7.77	3.89

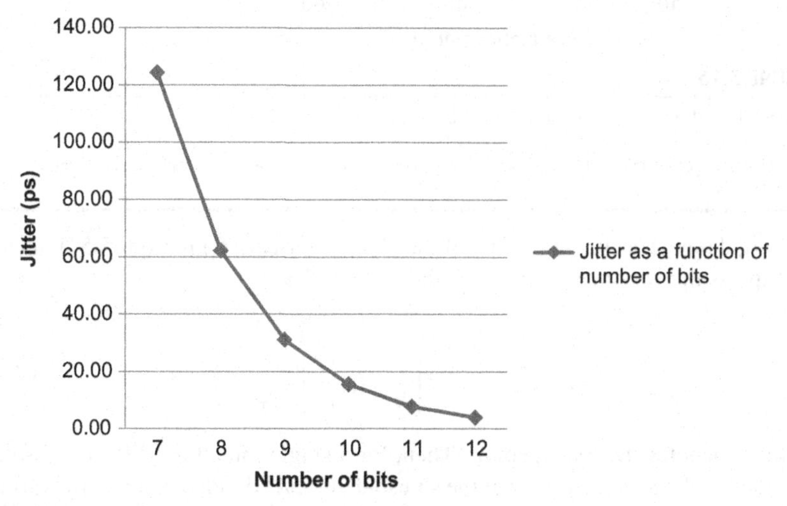

FIGURE 5.14

Maximum jitter as a function of increasing number of bits.

Continued

EXAMPLE 5.1 JITTER OF AN *N*-BIT DATA CONVERTER—cont'd

Next, given a 10-bit data converter, increasing the frequency of the tone, along with the sampling clock, seems to also decrease the maximum jitter as numerically shown in Table 5.2 and depicted in Figure 5.15.

Table 5.2 Maximum Jitter as a Function of Increasing Sinusoidal Frequency

Parameter	Value					
Number of bits	10	10	10	10	10	10
Frequency (MHz)	10	20	30	40	50	60
Sampling clock (MHz)	20	40	60	80	100	120
Jitter (ps)	15.54247491	7.77123746	5.18082497	3.88561873	3.108495	2.5904125

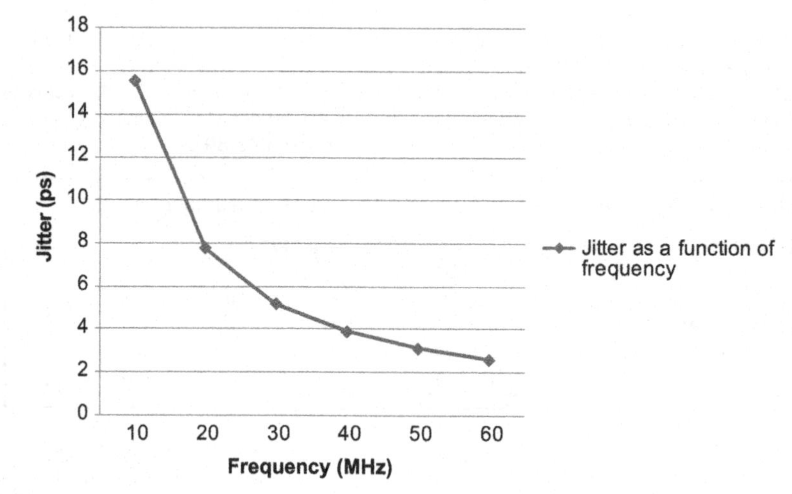

FIGURE 5.15

Maximum jitter a function of input frequency.

[4] *N* is sometimes used in lieu of *b* to denote the total number of bits as opposed to *b*, which designates the effective number of bits.

Next, define Υ as the *rms* value of the clock jitter $\Delta t(nT_s)$ in Eqn (5.54), then we can express the variance of the error in Eqn (5.54) as

$$\sigma_{rms}^2 = \Upsilon^2 \left[\widehat{\frac{\delta}{\delta t} x_a(t)} \right]^2 \Bigg|_{t=nT_s} \tag{5.59}$$

where $\frown$ denotes the *rms* operator. Then, for a single tone, the *SNR* can be simply expressed as the average power of the sinusoid or $0.5A^2$ divided by the error variance or in dB as

$$SNR_{jitter} = 10\log_{10}\left(\frac{\frac{1}{2}A^2}{\sigma_{rms}^2}\right) = 10\log_{10}\left(\frac{\frac{1}{2}A^2}{\Upsilon^2\left[\frac{\delta}{\delta t}x_a(t)\right]^2\Big|_{t=nT_s}}\right) \tag{5.60}$$

Recall, however, that according to Eqn (5.54), for a single tone we can express the standard deviation of the error in terms of the *rms* value of $\Delta t(nT_s)$, or Υ as

$$\sigma_{rms} = \frac{1}{\sqrt{2}}A2\pi F\Upsilon \tag{5.61}$$

Substituting Eqn (5.61) into the *SNR* expression for jitter in Eqn (5.60), we obtain:

$$SNR_{jitter} = 10\log_{10}\left(\frac{\frac{1}{2}A^2}{4\pi^2F^2\Upsilon^2\frac{1}{2}A^2}\right) = -20\log_{10}(2\pi F\Upsilon) \tag{5.62}$$

The *SNR* relationship in Eqn (5.62) shows that the degradation due to jitter at low frequency has minimal impact on the overall *SNR*, while the impact of jitter is noticeable at high frequency.

Despite the fact that Eqn (5.62) is representative of sinusoids sampled at *half the sampling clock only*, the expression is still used as a performance metric of clock jitter and its impact on the performance of digital converters. However, this metric is limited since it only describes *SNR* performance of a *full-scale tone* presented at the highest possible frequency, which is half the sampling clock. In most cases, however, the desired signal in question has a certain finite bandwidth that is the signal and is not a tone. This signal may also be oversampled implying that its bandwidth is less than the Nyquist frequency.[5]

[5] Recall that in signal processing, the Nyquist rate is the sampling rate that is equivalent to the IF signal bandwidth for an IF sampling receiver. In a similar vein, it is *conventionally twice the baseband bandwidth of the signal,* which is half the IF signal bandwidth [9].

EXAMPLE 5.2 IMPACT OF JITTER ON *SNR*

Consider a 12-bit converter sampled at 20 MHz. Assume that the total desired *SNR* at the output of the converter is 58 dB and that the input is a single tone given at a frequency of slightly less than 10 MHz. Compute the *rms* jitter assuming that the only other degradation in the converter is due to quantization. What is the impact on *rms* jitter if you vary the total *SNR* from 58 to 64 dB? What is the impact on jitter if you change the number of bits from 10 to 12 bits while varying the total *SNR* between 58 and 64 dB? Display the results graphically and draw conclusions.

Assuming that the noise is additive, the total *SNR* in dB at the output of the converter is simply given as

$$SNR_{Total} = -20\log_{10}\sqrt{10^{\frac{-SQNR}{10}} + 10^{\frac{-SNR_{jitter}}{10}}} \tag{5.63}$$

where SNR_{Total} is the total *SNR* of the converter, *SQNR* is the signal to quantization ratio, and SNR_{jitter} is the *SNR* due to jitter all expressed in decibels.

In linear scale, this relationship may be expressed as

$$\frac{1}{SNR_{Total,\ linear}} = \frac{1}{SQNR_{linear}} + \frac{1}{SNR_{jitter,\ linear}} \tag{5.64}$$

which simply implies that the dB-scaled *SNR* due to jitter can be expressed in terms of *SQNR* and total *SNR* as

$$SNR_{jitter} = 10\log_{10}\left(\frac{SNR_{Total,\ linear} \times SQNR_{linear}}{SQNR_{linear} - SNR_{Total,\ linear}}\right) \tag{5.65}$$

According to Eqn (5.65), in order to compute the jitter *SNR*, we must first compute the *SQNR*. The *SQNR* is given in Eqn (5.51) and can be computed for a tone sampled at ever so slightly less than the sampling clock rate as

$$\begin{aligned} SQNR &= 4.7712 + 6.02b + 10\log_{10}\left(\frac{F_s}{B}\right) - PAR \\ &= 4.7712 + 6.02 \times 12 + 10\log_{10}\left(\frac{20\text{ MHz}}{10\text{ MHz}}\right) - 3\text{ dB} = 77.012\text{ dB} \end{aligned} \tag{5.66}$$

The *PAPR* of the tone is 3 dB and can be obtained via the relationship

$$\begin{aligned} PAPR &= 10\log_{10}\left(\frac{P_{peak}}{P_{avg}}\right) = 10\log_{10}\left(\frac{\max\left\{|x(t)|^2\right\}}{F\int_0^{1/F} x^2(t)dt}\right) \\ &= 10\log_{10}\left(\frac{\max\left\{\left|A^2\cos^2(2\pi Ft)\right|\right\}}{F\int_0^{1/F} A^2\cos^2(2\pi F_c)dt}\right) = 10\log_{10}\left(\frac{A^2}{\frac{1}{2}A^2}\right) = 3\text{ dB} \end{aligned} \tag{5.67}$$

Next, given the total *SNR* of 58 dB and the *SQNR* of 77.012 dB, we can obtain the jitter *SNR* according to Eqn (5.65) as

EXAMPLE 5.2 IMPACT OF JITTER ON *SNR*—cont'd

$$SNR_{jitter} = 10\log_{10}\left(\frac{SNR_{Total,\ linear} \times SQNR_{linear}}{SQNR_{linear} - SNR_{Total,\ linear}}\right)$$
$$= 10\log_{10}\left(\frac{10^{58/10} \times 10^{77.02/10}}{10^{77.02/10} - 10^{58/10}}\right) = 58.0548\ \text{dB} \tag{5.68}$$

The *rms* jitter can then be found via the expression given in Eqn (5.62) as

$$SNR_{jitter} = -20\log_{10}(2\pi F\Upsilon) \Rightarrow \Upsilon = \frac{10^{-SNR_{jitter}/20}}{2\pi F}$$
$$\Upsilon = \frac{10^{-SNR_{jitter}/20}}{2\pi F} = \frac{10^{-58.054/20}}{2\times\pi\times 10\ \text{MHz}} = 19.91\ \text{ps} \tag{5.69}$$

Next, if we vary the total *SNR* from 58 to 64 dB while maintaining the number of effective bits in the converter at 12 bits, we observe that the jitter variance decreases as the total *SNR* increases. This is evident as can be seen in Table 5.3 where the *rms* jitter has decreased from 19.9 to 9.8 ps. Decreasing the number of bits, however, from 12 to 10, the *rms* jitter decreases from that of 12 bits for the same signal conditions. This effect is more pronounced at high *SNR* as can be clearly seen in Figure 5.16.

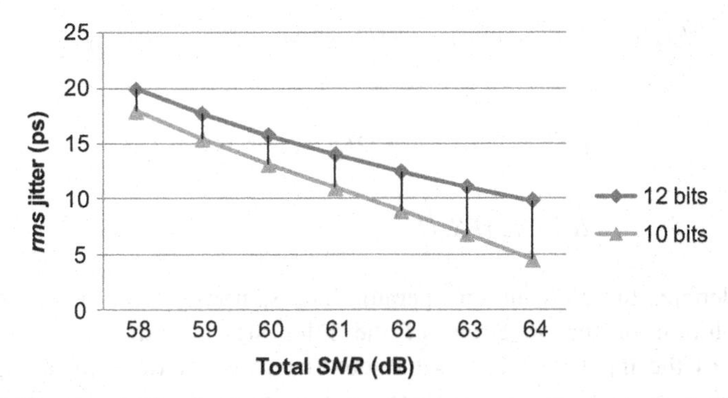

FIGURE 5.16

Impact of root mean square clock jitter on total analog-to-digital converter signal-to-noise ratio for 10- and 12-bit converters. (For color version of this figure, the reader is referred to the online version of this book.)

Table 5.3 Root Mean Square (*rms*) Jitter as a Function of *SQNR* and Total *SNR*

Parameter	Value						
Total *SNR* (dB)	58	59	60	61	62	63	64
PAPR (dB)	3	3	3	3	3	3	3
Number of bits	12	12	12	12	12	12	12
SQNR (dB)	77.02149996	77.0215	77.02149996	77.0215	77.0215	77.0215	77.0215
Jitter *SNR* (dB)	58.05474807	59.0690369	60.08709268	61.10993	62.13885	63.17554	64.22218
Frequency (MHz)	10	10	10	10	10	10	10
rms jitter (ps)	19.91052575	17.7161066	15.75670839	14.00631	12.44163	11.04188	9.788392

PAPR, *peak to average power ratio;* SQNR, *signal-to-quantization-noise ratio;* SNR, *signal-to-noise ratio.*

In the white noise model [10], assume that the analog signal $x_a(t)$ is sampled at the rate $F_s = 1/T_s$ and the sampled signal is given according to the relationship in Eqn (5.55). In the white noise model the *rms* value of $\Delta t(nT_s)$ or Υ can be found provided that the signal occupies the full scale of the ADC at its peak:

$$\Upsilon = \frac{1}{2^N \pi B}\sqrt{\frac{2}{3}OSR} \tag{5.70}$$

where again N is the number of bits, B is the IF bandwidth, and OSR is the oversampling ratio. It is important to note that in this case, as is in the case presented above concerning a sinusoidal input, the error signal due to sampling jitter is considered to be white. In the event where the desired signal is significantly oversampled such that it occupies a small portion of the converted bandwidth, then the power spectral density of the error signal becomes very relevant [10].

To obtain a better understanding of the spectral distribution of the error, consider the discrete-time Fourier transform (*DTFT*) of $e(t)$ as:

$$\begin{aligned} E(F) &= DTFT\{e(t)\} = DTFT\left\{\Delta t(nT_s)\frac{\delta}{\delta t}x_a(t)\big|_{t=nT_s}\right\} \\ &= DTFT\{\Delta t(nT_s)\} * DTFT\left\{\frac{\delta}{\delta t}x_a(t)\big|_{t=nT_s}\right\} \\ &= \Delta t(F) j2\pi F X_a(F) \end{aligned} \tag{5.71}$$

where * denotes the convolution operator. The sampling error spectrum is then the convolution of the spectrum of the jitter signal with that of the time-derivative of the input signal. It is obvious from the relationship in Eqn (5.71) that in order to compute the spectrum of the sampling error signal $e(t)$, both the input signal and the jitter must be known [11]. The sampling jitter can be computed or measured in the lab. The input signal, on the other hand, is a combination of a desired *known* input signal corrupted by noise and possibly interferers or blockers.

Allowing the input signal to be a single tone of magnitude A and centered at F_c, then the spectrum of the error signal is none other than the spectrum of the jitter signal scaled and centered at F_c or

$$E(F) = \sqrt{2}\pi A F_c \Delta t(F - F_c) \tag{5.72}$$

If the clock is generated via a phase-locked loop (PLL), then the output spectrum is similar to the one depicted in Figure 5.17. According to Eqns (5.71) and (5.72), the convolution operation of $\Delta t(F)$ with a sinusoidal input causes the jitter signal spectrum to be centered around F_c. This process is illustrated in Figure 5.18.

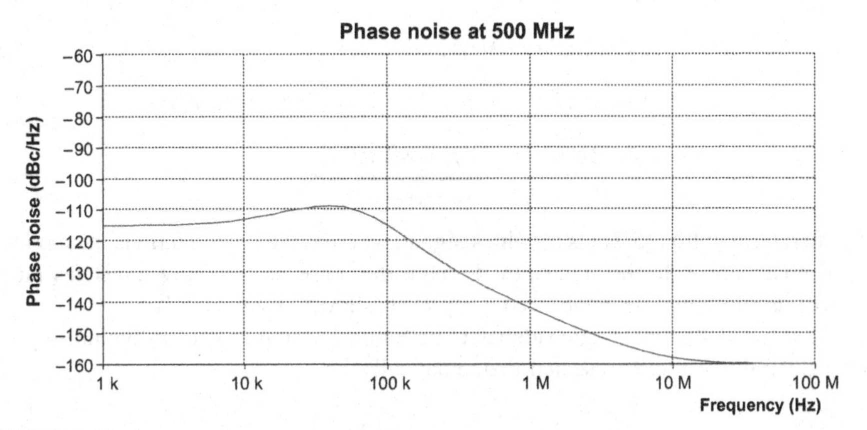

FIGURE 5.17

Typical clock jitter spectrum generated by a phase-locked loop. (For color version of this figure, the reader is referred to the online version of this book.)

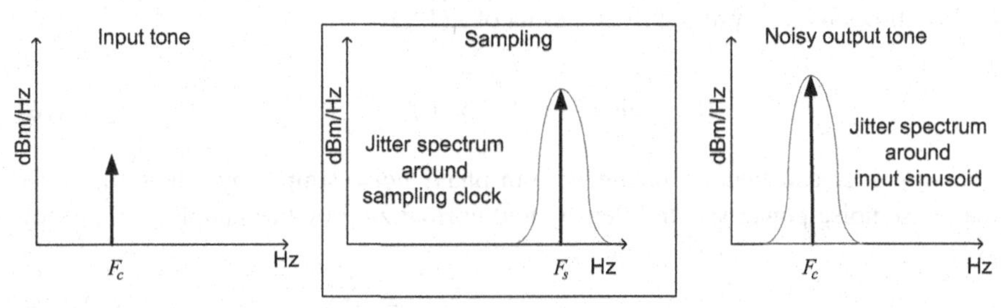

FIGURE 5.18

Impact of phase-locked loop induced jitter on sinusoidal input. (For color version of this figure, the reader is referred to the online version of this book.)

5.2.4 Impact of phase noise on clock jitter

Phase noise was discussed in some detail in Chapter 3. In this section, we will discuss the impact of phase noise on the jitter of the sampling clock. Consider the clock signal to be a sinusoid obtained at the sampling frequency F_s:

$$\begin{aligned} V_{clock}(t) &= A \sin\left(2\pi F_s t + \underbrace{2\pi F_s \Delta t(t)}_{\text{Phase noise}}\right) \\ &= A \sin\left(2\pi F_s t + \phi(t)\right) \end{aligned} \tag{5.73}$$

where $\phi(t) = 2\pi F_s \Delta t(t)$ is the phase noise. Assuming small signal perturbation, that is $\sin(\phi(t)) \approx \phi(t)$ and $\cos(\phi(t)) \approx 1$, then the relationship in Eqn (5.73) becomes

$$
\begin{aligned}
V_{clock}(t) &= A \sin(2\pi F_s t + \phi(t)) \\
&= A \sin(2\pi F_s t) \cos(\phi(t)) + A \cos(2\pi F_s t) \sin(\phi(t)) \\
&\approx A \sin(2\pi F_s t) + \underbrace{A\phi(t) \cos(2\pi F_s t)}_{\text{Modulated noise}}
\end{aligned}
\tag{5.74}
$$

The first term in Eqn (5.74) is the clock signal, whereas the second term is the modulated phase noise. In the frequency domain, the noise term simply translates into sidebands adjoining the center frequency F_s as shown in Figure 5.18.

Next, we define the single sideband phase noise spectrum $L(F)$ as the noise power spectral density, represented mathematically as

$$
L(F) = 10 \log_{10}\left(\frac{\Phi^2(F)}{2}\right) \text{ dBc/Hz} \tag{5.75}
$$

where $\Phi(F)$ is phase noise spectrum of $\phi(t)$. Conversely, the relationship in Eqn (5.75) allows us to express $\Phi(F)$ in terms of $L(F)$ as

$$
\Phi(F) = \sqrt{2 \times 10^{\frac{L(F)}{10}}} \tag{5.76}
$$

The *rms* jitter can then be obtained from phase noise simply by integrating over the phase noise power spectral density and normalizing by the sampling frequency [12]

$$
\begin{aligned}
\Upsilon &= \frac{1}{2\pi F_s}\sqrt{2\int_0^{\infty} 10^{\frac{L(F)}{10}} dF} \\
&= \frac{1}{2\pi F_s}\sqrt{\int_0^{\infty} \Phi^2(F) dF}
\end{aligned}
\tag{5.77}
$$

The *SNR* due to jitter can be obtained as the ratio of the tone-signal power to the error variance over the frequency band (F_{min}, F_{max}), or

$$
SNR_{jitter} = 10 \log_{10}\left(\frac{\frac{A^2}{2}}{\int_{F_{min}}^{F_{max}} E^2(F) dF}\right) \tag{5.78}
$$

The error spectrum can be further simplified by substituting the clock phase noise $\Phi(F) = 2\pi F_s \Delta t(F)$ into the error spectrum $E(F) = j2\pi F \Delta t(F) X_a(F)$. We can rewrite Eqn (5.78) as[6]

$$SNR_{jitter} = -20 \log_{10}\left(\frac{F_{input}}{F_s}\sqrt{\int_{F_{min}}^{F_{max}} 10^{\frac{L(F-F_{input})}{10}} dF}\right) \tag{5.79}$$

where F_{input} is the input frequency of the tone.

Next, we define period jitter or cycle-to-cycle jitter as the variation in the clock period and which can be expressed as [12]

$$\Delta t_{period}(kT) = \Delta t(kT) - \Delta t((k-1)T) \tag{5.80}$$

When normalized to the sampling frequency, in the discrete-time domain, the relationship in Eqn (5.80) can be expressed as

$$\Delta t_{period}(z)\big|_{f=F/F_s} = \Delta t(z)\left(1 - z^{-1}\right) \tag{5.81}$$

Substituting $z = e^{j2\pi F/F_s}$ in Eqn (5.81), we obtain

$$\begin{aligned} \Delta t_{period}(f)\big|_{f=F/F_s} &= \Delta t\left(\frac{F}{F_s}\right)\left(1 - e^{-j2\pi F/F_s}\right) \\ &= 2j\Delta t\left(\frac{F}{F_s}\right)e^{-j\pi F/F_s}\left(\frac{e^{j\pi F/F_s} - e^{-j2\pi F/F_s}}{2j}\right) \\ &= 2j\Delta t\left(\frac{F}{F_s}\right)e^{-j\pi F/F_s}\sin\left(\frac{\pi F}{F_s}\right) \end{aligned} \tag{5.82}$$

The period jitter can then be related to the phase noise

$$\Delta t_{period}(F) = \frac{1}{2\pi F_s}\sqrt{8\int_0^{\infty} 10^{\frac{L(F)}{10}} \sin^2\left(\frac{\pi F}{F_s}\right) dF} \tag{5.83}$$

Period jitter can be measured in the laboratory, thus verifying important timing information concerning circuit settling time within a clock period.

5.2.5 Overloading and clipping

Thus far, we have looked at quantization noise as a source of signal degradation in analog-to-digital conversion. Quantization noise is characterized as white and uniformly distributed. In our analysis, we have assumed that the desired signal is

[6]Recall that for an input sinusoid at a given frequency F_{input} the error spectrum is simply shifted in frequency, that is $E(F) = j\frac{A}{\sqrt{2}}2\pi F_{input}\Delta t(F)$.

received within the dynamic range of the converter. However, under certain circumstances, the received signal may exceed the full scale or range of the ADC thus causing the signal to *clip*. Clipping is a form of degradation that occurs when the received signal exceeds the dynamic range of the ADC [13,14].

To further quantify the clipping process, assume the input signal $x(t)$ to be white Gaussian distributed. The PDF of the signal can be expressed as

$$p(x(t)) = \frac{1}{\sigma_x\sqrt{2\pi}} e^{-\frac{x^2(t)}{2\sigma_x^2}} \tag{5.84}$$

where $x(t)$ is assumed to be zero mean, that is $E\{x(t)\} = 0$, and has a variance of σ_x^2. Furthermore, suppose that the signal clips at half the full-scale voltage $V_{x,F_s}/2$, then the total power of the clipped signal is derived in Refs. [13,14] to be

$$P_{clipping} = 2 \int_{V_{F_s}/2}^{\infty} \left(x(t) - \frac{V_{x,F_s}}{2} \right)^2 p(x(t)) \mathrm{d}x(t) \tag{5.85}$$

Define the ratio $\mu = V_{x,F_s}/(2\sigma_x)$, then the power of the clipped signal as given in Eqn (5.85) simply becomes

$$P_{clipping} = \sqrt{\frac{8}{\pi}} \frac{\sigma_x^2}{\mu^3} e^{-\mu^2/2} \tag{5.86}$$

If we assume the signal to be zero-mean Gaussian distributed, then the signal power is simply the variance σ_x^2. The signal-to-clipping-degradation ratio can then be defined as [13,14]:

$$SCR = \frac{1}{2} \sqrt{\frac{\pi}{2}} \mu^3 e^{\mu^2/2} \tag{5.87}$$

If we assume the noise due to quantization, clipping, and jitter to be independent, the total signal-to-noise ratio can be estimated:

$$\begin{aligned} \frac{1}{SNR_{Total}} &= \frac{1}{SNR_{jitter}} + \frac{1}{SQNR} + \frac{1}{SCR} \\ &= \frac{SQNR \times SCR + SNR_{jitter} \times SCR + SNR_{jitter} \times SQNR}{SNR_{jitter} \times SQNR \times SCR} \\ SNR_{Total} &= \frac{SNR_{jitter} \times SQNR \times SCR}{SQNR \times SCR + SNR_{jitter} \times SCR + SNR_{jitter} \times SQNR} \end{aligned} \tag{5.88}$$

Clipping is allowed for certain modulation schemes in order to lower the *PAPR* and allow the ADC to provide better statistical representation of the signal. This is especially true for OFDM where a certain amount of clipping is tolerated for lower overlaying constellations such as QPSK.

5.3 Antialiasing filtering requirements

The ADC is typically preceded by an analog filter known as the antialiasing filter. The purpose of this filter is really twofold: (1) to eliminate or attenuate any signal component higher than $F_s/2$, and (2) a last-resort filter to eliminate any unwanted blockers or interferers before digital conversion. The aim is to be able to remove any remaining unwanted signal components after digital conversion via digital filtering as shown in Figure 5.19. The level of desired filtering required to remove all unwanted signals, however, may not be totally practical under certain scenarios. For example, in order to maximize the dynamic range of the ADC, the distortion power due to aliasing cannot exceed that of ½ LSB. The power present in ½ LSB depends on the number of bits in a digital converter, or more specifically,

$$P_{1/2\ \mathrm{LSB}} = 10\log_{10}\left(\frac{1}{2^{2(b+1)}}\right)\mathrm{dBc} \tag{5.89}$$

where b is the effective number of bits defined as

$$b = \frac{SNR_{Total} - 1.76}{6.02} \tag{5.90}$$

Both relationships cited in Eqns (5.89) and (5.90) refer to the ADC's full-scale dynamic range. In this case, the designer may resort to oversampling and digital filtering to lighten the requirements on the analog filter. Therefore, a careful analysis is required in order to determine the optimum solution in terms of cost and performance that would present a balanced approach between filtering and sampling.

Another very important aspect that must be taken into account in determining the filtering and sampling strategy is the radio architecture. For example, in an IF-sampling architecture, the analog antialiasing filter is most likely external to the chip. IF sampling requires only one ADC that samples the received signal at IF. The choice of IF frequency used will also determine the filter technology. On the other extreme, consider a direct-conversion receiver where the signal is split into analog in-phase and quadrature components after mixing. In this case, there

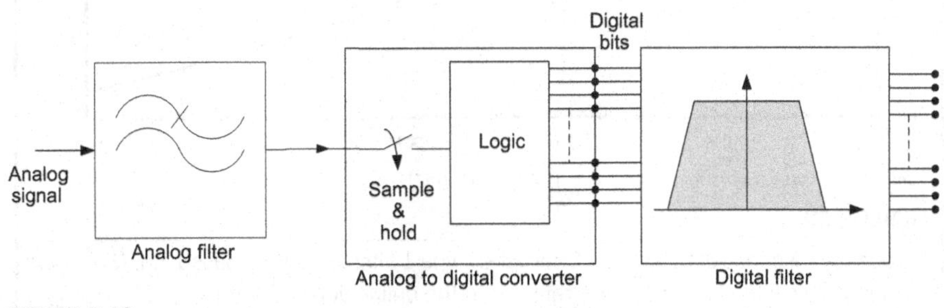

FIGURE 5.19

Analog antialiasing filter preceded by analog-to-digital converter and digital filter.

are two analog lowpass antialiasing filters that are integrated on the chip. The sampling rate and resolution of the in-phase and quadrature ADCs is typically less than that used for the IF-sampling ADC. Therefore, once the architecture is determined, the required instantaneous ADC dynamic range and blocking requirements will determine to a great extent the minimum sampling frequency of the ADC, its resolution, and the antialiasing filter requirements.

EXAMPLE 5.3 ANTIALIASING FILTER COMPLEXITY AND SAMPLING RATE TRADE-OFF

Consider a 5-MHz OFDM signal and a direct conversion receiver. Both data converters are 9-bit Nyquist ADCs. Furthermore, assume that a single narrowband blocker could exist at the following frequencies:

- Slightly higher than 10 MHz, or
- Slightly higher than 20 MHz, or
- Slightly higher than 40 MHz.

Given the continuous frequency responses of three possible antialiasing Chebyshev type 1 filters depicted in Figures 5.20–5.22 simultaneously. In each case, choose the minimum sampling frequency such that the degradation due to the blocker is less than ½ LSB. Assume that the blocker power does not exceed 0 dBc relative to the signal power.

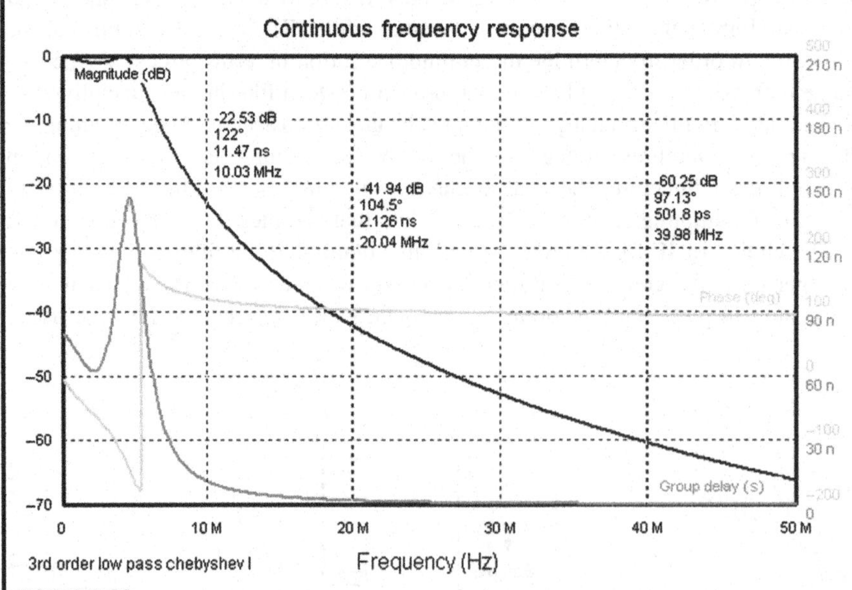

FIGURE 5.20

Frequency response of third order Chebyshev type 1 filter for 5 MHz passband. (For color version of this figure, the reader is referred to the online version of this book.)

EXAMPLE 5.3 ANTIALIASING FILTER COMPLEXITY AND SAMPLING RATE TRADE-OFF—cont'd

First, given a 9-bit Nyquist ADC, let us compute the necessary rejection due to the power present in ½ LSB of dynamic range

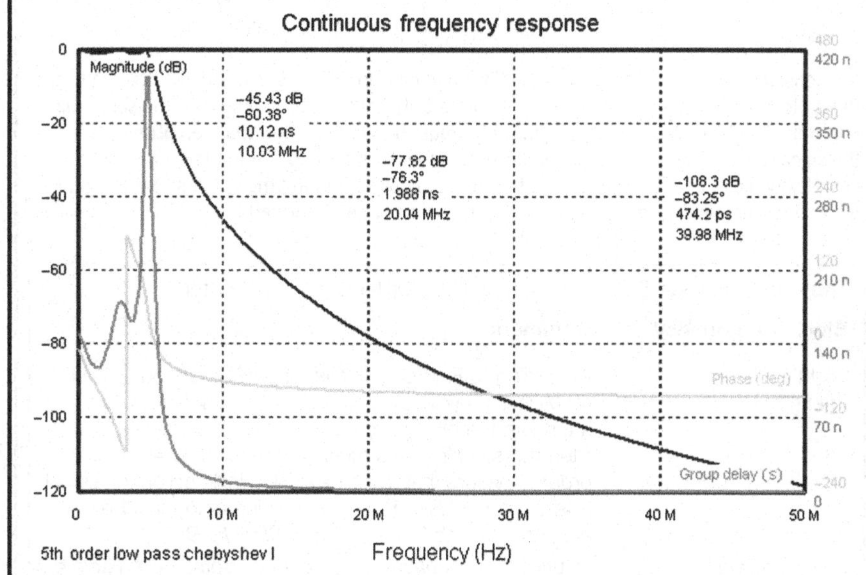

FIGURE 5.21

Frequency response of fifth order Chebyshev type 1 filter for 5 MHz passband. (For color version of this figure, the reader is referred to the online version of this book.)

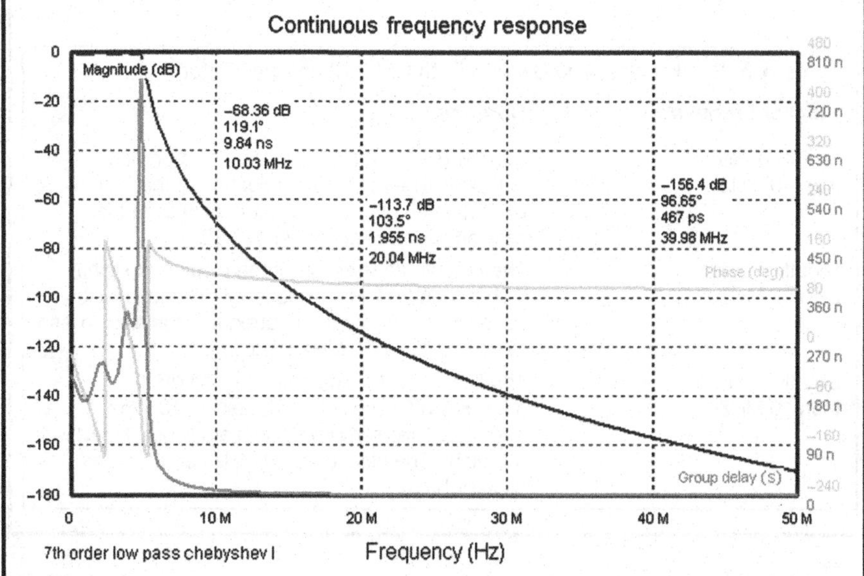

FIGURE 5.22

Frequency response of seventh order Chebyshev type 1 filter for 5 MHz passband. (For color version of this figure, the reader is referred to the online version of this book.)

EXAMPLE 5.3 ANTIALIASING FILTER COMPLEXITY AND SAMPLING RATE TRADE-OFF—cont'd

$$P_{1/2\,\mathrm{LSB}} = 10\log_{10}\left(\frac{1}{2^{2(b+1)}}\right) = 10\log_{10}\left(\frac{1}{2^{2(9+1)}}\right) = -60.2\ \mathrm{dBc} \quad (5.91)$$

Consider the third order Chebyshev filter depicted in Figure 5.20. The results due to the three blocker scenarios is summarized in Table 5.4. According to Table 5.4, the sampling rate must be chosen such that a digital filter must be implemented that can attenuate the signal post data conversion to an acceptable level. A trade-off of implementation complexity, cost, and power consumption to justify oversampling the clock and the digital filter versus employing higher order analog filters must be performed.

Table 5.4 Blocker Scenario Given a Third Order Chebyshev Filter

Blocker Frequency	Comment
Slightly greater than 40 MHz	According to Figure 5.20, the *magnitude* attenuation at 40 MHz is roughly 60 dBc, which is not sufficient to attenuate the blocker to below ½ LSB. Therefore, for this filter, the sampling rate must be in excess of 40 MHz in order to ensure that the degradation inflicted on the signal is less than ½ LSB. The oversampling ratio must be in accordance with Eqn (5.50) or $OSR = F_s/B$.
Slightly greater than 20 MHz	Insufficient attenuation. See note on oversampling above.
Slightly greater than 10 MHz	Insufficient attenuation. See note on oversampling above.

Table 5.5 Blocker Scenario Given a Fifth Order Chebyshev Filter

Blocker Frequency	Comment
Slightly greater than 40 MHz	Sampling at 10 MHz in this case is more than sufficient. The magnitude rejection of the filter exceeds 108 dB, which is far in excess of the needed power rejection for ½ LSB.[7]
Slightly greater than 20 MHz	The magnitude rejection of the blocker at slightly higher than 20 MHz exceeds 77 dB and is well beyond what is needed to attenuate the alias to less than ½ LSB for 10 MHz sampling.
Slightly greater than 10 MHz	The filter provides approximately 45 dB of attenuation for a blocker situated at 10 MHz. Therefore, oversampling signal at 15 or 20 MHz is sufficient. The blocker at 10 MHz can then be digitally filtered.

EXAMPLE 5.3 ANTIALIASING FILTER COMPLEXITY AND SAMPLING RATE TRADE-OFF—cont'd

Next consider the fifth order Chebyshev filter depicted in Figure 5.21. The results for the three blocker scenarios are summarized in Table 5.5. The filter is sufficient to attenuate the blockers at 20 and 40 MHz such that the degradation power in each case is below that of ½ LSB. However, that is not the case for the 10 MHz blocker, and oversampling and digital filtering are necessary to meet the desired performance.

Finally, given the seventh order Chebyshev filter depicted in Figure 5.22, it can be easily seen that the filter's magnitude response in all three cases provides sufficient rejection. After filtering and sampling, the degradation due to the blocker signal is certainly less than the relative power contained in ½ LSB.

[7] The filter resolution obtained via simulation represents only theoretical results. In practice, analog rejection of 108 dB at IF may not be easily achievable using conventional filter design! For this reason, proper analog filter simulation that takes into account all the relevant practical design and process parameters must be performed to obtain reasonable rejection numbers.

5.4 Quantization noise based on spectral analysis

In this section, an exact expression for quantization noise and *SQNR* is derived. The analysis herein does not make any assumptions about the signal nor its nature. Furthermore, there is no assumption made concerning the loading of the converter. Consider the sinusoidal input signal plus noise

$$\begin{aligned} x(t) &= s(t) + \eta(t) \\ &= A\cos(\Omega t + \phi) + \eta(t) \end{aligned} \tag{5.92}$$

where Ω and ϕ are the frequency and phase of the sinusoid, and $\eta(.)$ is white Gaussian with variance σ_η^2. According to Ref. [15], an arbitrary nonlinearity $f[.]$ acting on a sinusoidal input plus noise can be generally conveyed as[8]

$$f[x] \triangleq f[s+\eta] = \sum_{k=0}^{\infty}\sum_{m=0}^{\infty}\frac{1}{k!}\Psi_m h_{km} He_k\left(\frac{\eta}{\sigma}\right)\cos(m(\Omega t + \phi)) \tag{5.93}$$

where Ψ_m is a constant used to simplify the mathematics

$$\Psi_m = \begin{cases} 1 & m = 0 \\ 2 & \text{otherwise} \end{cases} \tag{5.94}$$

and $He_k\left(\frac{\eta}{\sigma}\right)$ are the Hermite polynomial defined as

$$He_k\left(\frac{\eta}{\sigma}\right) = (-\sigma)^k e^{\eta/2\sigma^2}\frac{d^k}{d\eta^k}\left[e^{-\eta/2\sigma^2}\right] \tag{5.95}$$

[8] The time variable has been dropped for notational convenience.

and finally the coefficients of the expansion h_{km} are given as

$$h_{km} = \int_{-A}^{A} \int_{-\infty}^{\infty} f[s+\eta] He_k\left(\frac{\eta}{\sigma}\right) \frac{e^{-\eta^2/2\sigma^2}}{\sqrt{2\pi\sigma^2}} \frac{T_m(s/A)}{\pi\sqrt{A^2-s^2}} d\eta ds \qquad (5.96)$$

The function $T_m(.)$ denotes the Chebyshev polynomial defined as

$$T_m(s/A) = \cos\left[m \cos^{-1}(s/A)\right] \qquad (5.97)$$

The Hermite polynomials $He_k(.)$ form an orthogonal set in the interval $(-\infty, \infty)$ with regard to the weighting function $e^{-x^2/2}$, that is

$$\int_{-\infty}^{\infty} e^{-x^2/2} He_k(x) He_m(x) dx = \begin{cases} k!\sqrt{2\pi} & k = m \\ 0 & \text{otherwise} \end{cases} \qquad (5.98)$$

The orthogonality property of Hermite polynomials makes them ideal for series expansions. Furthermore, the weighting function of the Hermite polynomials makes them appropriate for representing nonlinear functions acting on zero-mean Gaussian noise. This is particularly true since the cross-correlation of the various terms in a series expansion using Hermite polynomials of the function $f[\eta]$ is zero. The Hermite polynomial can be expressed in recursive form as [16]

$$\begin{aligned} He_{k+1}(x) &= xHe_k(x) - kHe_{k-1}(x) \\ He_0(x) &= 1 \\ He_1(x) &= x \end{aligned} \qquad (5.99)$$

The relationship in Eqn (5.96) can be further simplified as

$$h_{km} = j^{(k+m)} \sigma_\eta^k \int_{-\infty}^{\infty} F[\xi] \xi^k e^{-\frac{\sigma^2\xi^2}{2}} J_m(A\xi) d\xi \qquad (5.100)$$

where j denotes the complex number $j = -1$ and $J_m(.)$ denotes the mth order Bessel function and $F[\xi]$ is the Fourier transform of $f[x]$, then the Fourier transform pair of $f[x]$ and $F[\xi]$ is given as

$$F[\xi] = \int_{-\infty}^{\infty} f[x] e^{-j\xi x} dx \Leftrightarrow f[x] = \int_{-\infty}^{\infty} F[\xi] e^{j\xi x} d\xi \qquad (5.101)$$

Consider the ideal midriser quantization characteristics depicted in Figure 5.7, then $F[\xi]$ can be expressed as

$$F[\xi] = \frac{\delta}{2\pi j\xi} \sum_{i=1}^{2^b-1} \cos(\xi\tau_i) \qquad (5.102)$$

where δ is the quantization step, b is the number of effective bits, and $\{\tau_i\}_{i=1}^{2^b-1}$ is a set of threshold levels, depicted as the values representing the x axis in Figure 5.7 and ideally represented as

$$\tau_i = \begin{cases} -\infty & \text{for } i = 0 \\ \left(i - \frac{M}{2}\right)\delta & \text{for } i = 1, \cdots, M-1 \\ \infty & \text{for } i = M \end{cases} \tag{5.103}$$

where, as defined earlier, $M+1$ represents the number of thresholds in the quantizer.

Substituting the relationship in Eqn (5.102) into Eqn (5.100), we obtain a new expression for the Hermite coefficients given as

$$h_{km} = \frac{j^{(k+m-1)}\delta\sigma_\eta^k}{2\pi}\sum_{i=1}^{2^b-1}\int_{-\infty}^{\infty}\cos(\xi\tau_i)\xi^{k-1}e^{-\frac{\sigma^2\xi^2}{2}}J_m(A\xi)d\xi \tag{5.104}$$

The series expansion expressed in Eqn (5.93) implies that $f[s+\eta]$ is comprised of many terms. The majority of these terms are harmonics. Two terms, however, corresponding to the signal and noise, hold particular importance, namely

$$\begin{aligned} s_0(t) &= 2h_{01}\cos(\Omega t + \phi) \\ \eta_0(t) &= h_{10}He_1\left(\frac{\eta}{\sigma}\right) = h_{10}\frac{\eta(t)}{\sigma} \end{aligned} \tag{5.105}$$

According to Eqn (5.105), the signal power P_s and the noise power P_η can be computed, respectively, as

$$\begin{aligned} P_s &= 2h_{01}^2 \\ P_\eta &= h_{10}^2 \end{aligned} \tag{5.106}$$

Next, compute the total output power P_{Total} of the quantizer determined by the probability density function of the input $x(t) = s(t) + \eta(t)$

$$P_{Total} = \sum_{i=1}^{2^b-1}\delta_i^2 \Pr(\tau_i < x \le \tau_{i+1}) \tag{5.107}$$

The quantization error power can then be derived via the simple relationship

$$\begin{aligned} P_{Quantization} &= P_{Total} - P_s - P_\eta \\ &= \sum_{i=1}^{2^b-1}\delta_i^2 \Pr(\tau_i < x \le \tau_{i+1}) - 2h_{01}^2 - h_{10}^2 \end{aligned} \tag{5.108}$$

The exact *SQNR* can then be expressed as the ratio

$$\begin{aligned} SQNR_{Exact} &= \frac{P_s}{P_{Quantization}} = \frac{P_s}{P_{Total} - P_s - P_\eta} \\ &= \frac{P_s}{\sum_{i=1}^{2^b-1}\delta_i^2 \Pr(\tau_i < x \le \tau_{i+1}) - 2h_{01}^2 - h_{10}^2} \end{aligned} \tag{5.109}$$

The importance of this derivation of *SQNR* is the fact that one can obtain the exact power of the harmonics and intermodulation products present at the output of the quantizer. The quantization error itself is correlated with the signal mostly in the form of odd harmonics. Furthermore, the relationships in Eqns (5.108) and (5.109) indicate that the nature of the quantization noise and hence the *SQNR* depends on the statistical nature of noise present in the signal. More precisely, the quantization noise largely depends on the amplitude of the signal and noise present at the input of the quantizer.

References

[1] A. Oppenheim, R. Schafer, Discrete-Time Signal Processing, Prentice Hall, Englewood Cliffs, NJ, 1989.

[2] R. Vaughn, et al., The theory of bandpass sampling, IEEE Trans. Signal Process. 39 (9) (September 1991).

[3] T. Rouphael, RF and Digital Signal Processing for Software Defined Radio, Elsevier, Boston, MA, 2009.

[4] J. Liu, X. Zhou, Y. Peng, Spectral arrangements and other topics in first order bandpass sampling theory, IEEE Trans. Signal Process. 49 (6) (June 2001) 1260–1263.

[5] M. Choe, H. Kang, K. Kim, Tolerable range of uniform bandpass sampling for software defined radio, in: 5th International Symposium on Wireless Personal Multimedia Communications, vol. 2, October 27–30, 2002, pp. 840–842.

[6] W.R. Bennet, Spectral of quantization signals, Bell Syst. Tech. J. 27 (3) (June 1948) 446–472.

[7] A. Gersho, Principles of quantization, IEEE Trans. Circuits Syst. CAS-25 (7) (July 1978) 427–436.

[8] B. Brannon, A. Barlow, Aperture Uncertainty and ADC System Performance, Analog Devices Application Note AN-501 (2006).

[9] J. Proakis, D.G. Manolakis, Digital Signal Processing, Principles, Algorithms, second ed., Macmillan, New York, 1992.

[10] V. Arkesteijn, E. Klumperink, B. Nauta, Jitter requirements of the sampling clock in software radio receivers, IEEE Trans. Circuits Syst. 53 (2) (February 2006) 90–94.

[11] D.H. Shen, C.M. Hwang, B.B. Lusignan, B. Wooley, A 900 MHz RF front-end with integrated discrete-time filtering, IEEE J. Solid-State Circuits 31 (12) (December 1996) 1945–1954.

[12] C. Azeredo-Leme, Clock jitter effects on sampling: a tutorial, IEEE Circuits Syst. Mag. third quarter (2011) 26–37.

[13] D. Mestdagh, P. Spruyt, B. Brian, Effect of amplitude clipping in DMT-ADSL transceivers, IEEE Trans. Electron. Lett. 29 (15) (July 1993) 1354–1355.

[14] N. Al-Dahir, J. Cioffi, On the uniform ADC bit precision and clip level computation for a Gaussian signal, IEEE Trans. Signal Process. 44 (2) (February 1996) 434–438.

[15] N.M. Blachman, The effect of non-linearity upon signals in the presence of noise, IEEE Trans. Comm. COM-21 (2) (February 1973) 152–154.

[16] G. Korn, T. Korn, Mathematical Handbook for Scientists and Engineers, Dover Publications, Mineola, NY, 1968.

Further reading

[17] R. Van de Plassche, Integrated Analog-to-Digital-to-Analog Converters, Kluwer Academic Publishers, 1994.

[18] B. Song, High-speed pipelined ADC, in: Tutorial European IEEE Solid-State Circuits Conf., September 2002.

CHAPTER

Data Conversion

6

CHAPTER OUTLINE

Wireless Receiver Architectures and Design. http://dx.doi.org/10.1016/B978-0-12-378640-1.00006-8

Analog to digital conversion is the process of transforming the signal from the analog domain to the digital domain. This process could take place at baseband, as is the case of direct conversion receivers, or at intermediate frequency (IF) or low IF depending on the requirements and consequently on the receiver architecture pursued by the designers. In this chapter, we discuss the various hardware architectures in which an analog-to-digital converter (ADC) can be implemented. The chapter is divided into five sections. Section 6.1 discusses the main building blocks of ADCs, namely track-and-hold amplifiers and comparators. Topics such as aperture time accuracy, clock feedthrough, and charge injection and their impact on the signal-to-noise ratio (SNR) are discussed. In Section 6.2, we introduce the Nyquist converter. We delve into the architectural details of the FLASH, pipelined, and folding ADC architectures. In this section, we also discuss the impact of key performance parameters such as dynamic range, harmonic distortion, and thermal noise on the performance of the converter. Section 6.3 presents the concept of oversampled converters and $\Delta\Sigma$ modulation. The basic loop dynamics are derived. The architectures of continuous-time and discrete-time $\Delta\Sigma$ modulators are presented and the advantages and disadvantages of each design architecture are given in some detail. Next, the signal processing of the MASH architecture, which is comprised of a cascade of basic $\Delta\Sigma$ modulators, is studied. The pros and cons for using a MASH converter versus a single-loop converter are given. The chapter concludes with a small discussion on further nonidealities of $\Delta\Sigma$ modulators, which are largely outside the scope of this text.

6.1 Basic building blocks

In this section, the two main blocks of data converters, namely the sample and hold amplifier (SHA) or the track and hold amplifier (THA) and the comparator are discussed. The basic architecture of each block is presented. The performance and sources of degradations are discussed.

6.1.1 Track-and-hold and sample-and-hold amplifiers

In data conversion, an SHA or a THA is typically needed in order to capture the analog signal and hold it during the conversion process from the analog domain to the digital domain. A THA circuit tracks the analog signal during the *track* mode for roughly 50% of the time, and the analog value is captured as the circuit switches into the *hold* mode for the remaining 50% of the time. An SHA circuit, on the other hand, is comprised of two THA circuits placed back to back. The second THA circuit in this case is clocked on the opposite phase of the first THA circuit. With the exception of the short transition periods at the rising clock edges, an SHA produces a *held* signal during the full sampling clock period. In this case, in order to limit the transition between hold and track phases to a small proportion of the clock period, the bandwidth of the second THA must be sufficiently large.

6.1.1.1 Basic architecture

A basic open-loop functional THA is depicted in Figure 6.1. The performance of a typical ADC is gravely influenced by the performance of track-and-hold (T/H) circuit. Parameters such as spur-free dynamic range (SFDR), noise, dynamic range, and other types of distortion are largely influenced by the performance of the T/H circuit. For example, the choice of a single-ended THA or SHA, as opposed to differential THA or SHA, is limited by charge feedthrough that can significantly degrade the performance of the ADC. THA circuits have been implemented in gallium-arsenide (GaAs) technology for applications requiring very high speed, Silicon bipolar, complementary metal-oxide-semiconductor (CMOS) at lower frequency of operation, indium-phosphide (InP), and silicon-germanium (SiGe). All of these process technologies offer certain advantages and disadvantages. While CMOS for example is relatively cheap to fabricate at the expense of lower speed, InP and GaAs process technologies are more expensive but tend to provide much faster devices.

Apart from the design details or types of THA or SHA used, all circuits have four basic components common to all configurations. According to Figure 6.1, a THA consists of an input amplifier, a hold capacitor C_{Hold} serving as a storage device, a switching circuit, and an output buffer. Operationally, the input amplifier presents high impedance to the signal source, thus buffering the input signal, and provides sufficient current gain in order to charge the hold capacitor C_{Hold}. During track mode, the voltage across C_{Hold} follows the input signal with a certain delay. During the hold mode, the switch opens and the hold capacitor retains the same voltage that is theoretically present at the last instant of the track period. The output buffer, on the other hand, presents a high impedance to the hold capacitor

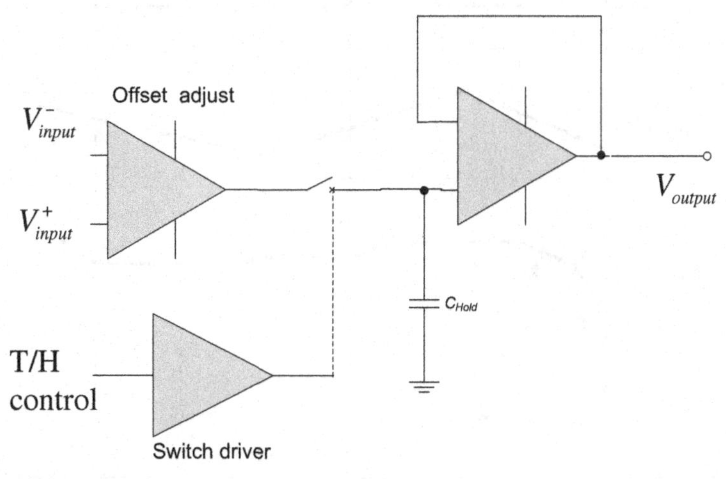

FIGURE 6.1

Functional block diagram of an open-loop THA circuit.

C_{Hold} in order to maintain the held voltage from rapid discharge. The switching circuit consists of a switch driver and the switch itself. The basic operation of a THA is illustrated in Figure 6.2. When the clock ϕ is high, then the switch is closed and the capacitor C_{Hold} charges its voltage to V_{input}. The capacitor used is assumed to be a low-leakage capacitor in order to avoid voltage droop. Voltage droop is

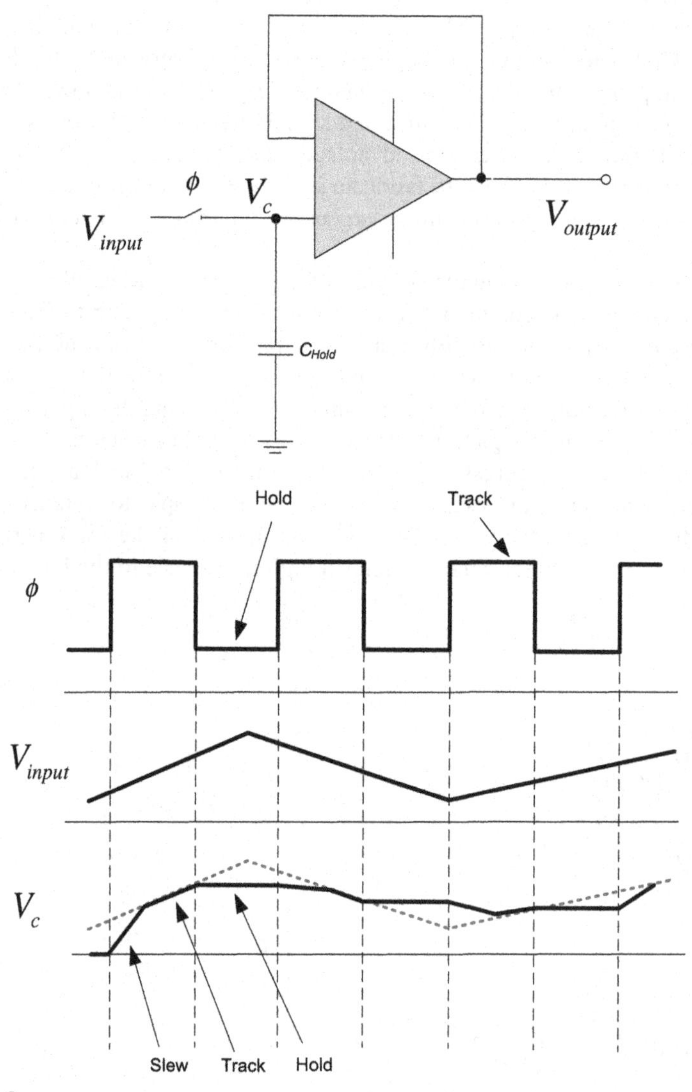

FIGURE 6.2

Basic track-and-hold circuit excluding the input amplifier. (For color version of this figure, the reader is referred to the online version of this book.)

defined in terms of the hold capacitor leakage current as rate of change in voltage across C_{Hold} as

$$\frac{\mathrm{d}}{\mathrm{d}t}V_c(t) = \frac{I_{leakage}}{C_{Hold}} \tag{6.1}$$

In data conversion, it is greatly desired that the THA output voltage does not droop by more than ±½ LSB during conversion time. In this case, the maximum allowable droop is given in terms of full-scale voltage V_{FS} as

$$\max\left\{\frac{\mathrm{d}}{\mathrm{d}t}V_c(t)\right\} = \frac{V_{FS}}{2^{b+1}T_{conversion}} \tag{6.2}$$

where T_c is the data conversion time.

When the clock ϕ is low, the switch is open and the voltage is held. Let I_{source} be the maximum current the supply can provide in order to charge the hold capacitor C_{Hold}, and let R_{switch} be the ON resistance of the switch, then the maximum rate of change during slew is given as the ratio

$$\frac{\mathrm{d}}{\mathrm{d}t}V_c(t) = \frac{I_{source}}{R_{switch}} \tag{6.3}$$

Slew rate is then defined as the maximum rate of change of the output voltage during track mode.

The 3-dB bandwidth, on the other hand, is a function of the switch resistance and the hold capacitor simply expressed as

$$BW_{3-dB} = \frac{1}{2\pi R_{\mathrm{switch}} C_{\mathrm{Hold}}} \tag{6.4}$$

The relationship in Eqn (6.4) is commonly referred to as the small signal bandwidth and is simply the frequency at which the voltage gain of the THA drops by 3 dBs relative to the gain at DC. Small signal refers to an input signal, which is much smaller than full scale by as much as 20 or even 40 dBs. The small signal bandwidth is much smaller than the full power bandwidth, which is defined as the frequency for which the THA drops by 3 dB relative to the gain at DC given a full-scale input signal.[a] Generally speaking, the small signal bandwidth tends to be larger than the full-scale bandwidth. This is true if the input driver cannot provide sufficient current or the ON resistance of the switch is dependent on the input signal voltage.

The switch resistance can be further manipulated and increased with the aid of the output impedance of the source driving the circuit. From basic circuit theory, we can also compute the acquisition time due to a step voltage as

$$V_{output}(t) = \left(1 - e^{-t/R_{switch}C_{Hold}}\right)V_{input}(t) \tag{6.5}$$

[a]Full power and small signal bandwidths are usually measured using an input tone.

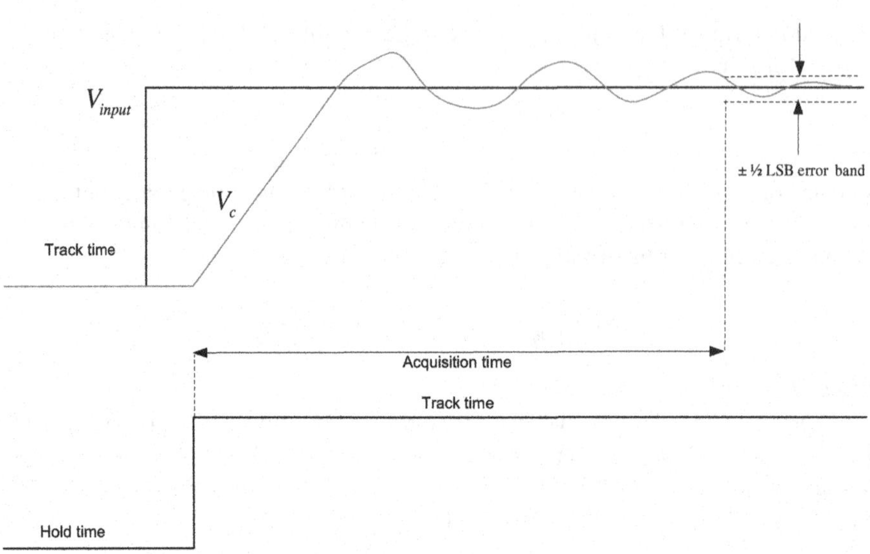

FIGURE 6.3

Acquisition time. (For color version of this figure, the reader is referred to the online version of this book.)

Acquisition time by definition is the maximum time needed to attain the new level of the input voltage $V_{input}(t)$ once the THA switches to track mode. A signal is said to be acquired once the THA output voltage $V_{output}(t)$ has settled within a certain error band of its final voltage as shown in Figure 6.3. This error is typically bounded by ±½ LSB. The acquisition time obviously depends on the hold capacitor C_{Hold} as expressed in Eqn (6.5). Therefore, maximum acquisition time occurs when C_{Hold} must fully charge to a full-scale voltage change. The acquisition time can be reduced by simply choosing a smaller capacitance for C_{Hold}. The implication of choosing a smaller capacitance, however, results in increase in the hold step as an increase in droop rate.

6.1.1.2 Aperture time accuracy

Another parameter of interest is aperture time accuracy. An ideal T/H circuit samples the incoming signal instantaneously. In practice, however, the T/H circuit requires a certain aperture turn-off time $\tau/2$, which is defined as the fall time of the sampling pulse [1]. In a similar vein, a practical sampling pulse exhibits a rising time $\tau/2$ resulting in aperture timing error τ. Figure 6.4 shows that the ideal sample should instantaneously occur at time t, but instead it occurs within the time interval $(t - \tau/2, t + \tau/2)$. Let the weighting function that defines the aperture window be $g(.)$, then the T/H circuit samples the input signal $V_{in}(t)$ resulting in the output signal $V_{output}(t)$ [2]:

$$V_{output}(t) = \int_{-\tau/2}^{\tau/2} g(\varepsilon) V_{input}(t+\varepsilon) d\varepsilon \tag{6.6}$$

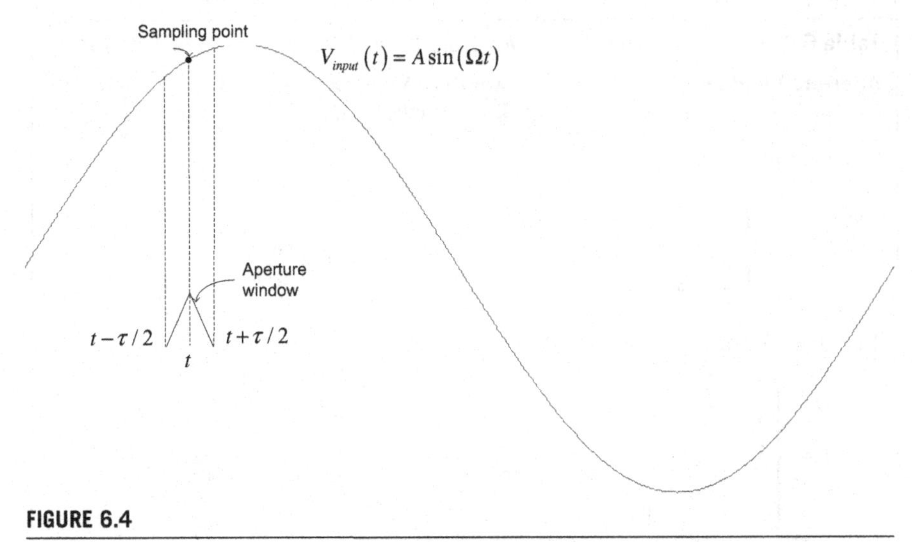

FIGURE 6.4

Output sample with finite aperture time τ. (For color version of this figure, the reader is referred to the online version of this book.)

For the sake of this analysis, consider the weighting function to be normalized and hence satisfy the relationship:

$$\int_{-\tau/2}^{\tau/2} g(\varepsilon)d\varepsilon = 1 \tag{6.7}$$

Next, we compute the error power due to aperture time accuracy as shown in [2]:

$$\begin{aligned} P_{error} &= E\left\{\left[V_{output}(t) - V_{input}(t)\right]^2\right\}\Big|_{t=nT_s} \\ &= \lim_{N\to\infty} \frac{1}{N}\sum_{n=0}^{N-1}\left[V_{output}(nT_s) - V_{input}(nT_s)\right]^2 \end{aligned} \tag{6.8}$$

For a sinusoidal input $V_{input}(t) = A\sin(2\pi F_c t)$, the resulting error power for a finite aperture time is given as

$$P_{error} = \frac{A^2}{2}|1 - G(j\Omega)|^2 \tag{6.9}$$

The authors in [2] show the error power Eqn (6.8) for three aperture windows as summarized in Table 6.1.

Table 6.1 Error Power Due to Finite Aperture Time for Various Aperture Windows

Aperture Window $g(\tau)$	Aperture Window $G(j\Omega)$ and Error Power P_{error}
Rectangular window: $g(\varepsilon) = \begin{cases} \frac{1}{\varepsilon} & -\frac{\tau}{2} \le \varepsilon \le \frac{\tau}{2} \\ 0 & \text{otherwise} \end{cases}$	$G(j\Omega) = \sin c\left(\frac{\Omega\tau}{2}\right)$ $P_{error} = \frac{A^2}{2}\left\vert 1 - \sin c\left(\frac{\Omega\tau}{2}\right)\right\vert^2$
Triangular window: $g(\varepsilon) = \begin{cases} \frac{2}{\tau} + \frac{4}{\tau^2}\varepsilon & -\frac{\tau}{2} \le \varepsilon \le 0 \\ \frac{2}{\tau} - \frac{4}{\tau^2}\varepsilon & 0 \le \varepsilon \le \frac{\tau}{2} \\ 0 & \text{otherwise} \end{cases}$	$G(j\Omega) = \frac{8}{\Omega^2\tau^2}\left(1 - \cos\left(\frac{\Omega\tau}{2}\right)\right)$ $P_{error} = \frac{A^2}{2}\left\vert 1 - \frac{8}{\Omega^2\tau^2}\left(1 - \cos\left(\frac{\Omega\tau}{2}\right)\right)\right\vert^2$
$g(\varepsilon) = \begin{cases} \frac{2}{\tau}\cos^2\left(\frac{\pi\varepsilon}{\tau}\right) & -\frac{\tau}{2} \le \varepsilon \le \frac{\tau}{2} \\ 0 & \text{otherwise} \end{cases}$	$G(j\Omega) = \left[1 - \left(1 - \frac{4\pi^2}{\Omega^2\tau^2}\right)^{-1}\right]\sin c\left(\frac{\Omega\tau}{2}\right)$ $P_{error} = \frac{A^2}{2}\left\vert 1 - \left[1 - \left(1 - \frac{4\pi^2}{\Omega^2\tau^2}\right)^{-1}\right]\sin c\left(\frac{\Omega\tau}{2}\right)\right\vert^2$

6.1.1.3 Charge injection and clock feedthrough

At this juncture, it is important to address the various architectures of T/H circuits. For low-speed low-power applications such as pipelined ADCs, a switched capacitor circuit, conceptually depicted in Figure 6.5, is typically used [3]. During hold, there is limited voltage droop due to limited leakage paths for the hold capacitor C_{Hold}. According to Eqn (6.3), the voltage change across C_{Hold} is determined by the current passing through the MOS device. In turn, the current capacity of the MOS device depends on the biasing characteristics of the transistor.

In practice, the performance of the T/H circuit depicted in Figure 6.5 suffers from two major types of degradations: charge injection and clock feedthrough. In order to understand the problem associated with charge injection, consider the metal-oxide-semiconductor (MOS) device depicted in Figure 6.5. When the transistor is on, the drain-to-source voltage V_{DS} is roughly zero and the transistor is said to operate in the triode region. When the MOS device is turned off, the mobile charges flow from

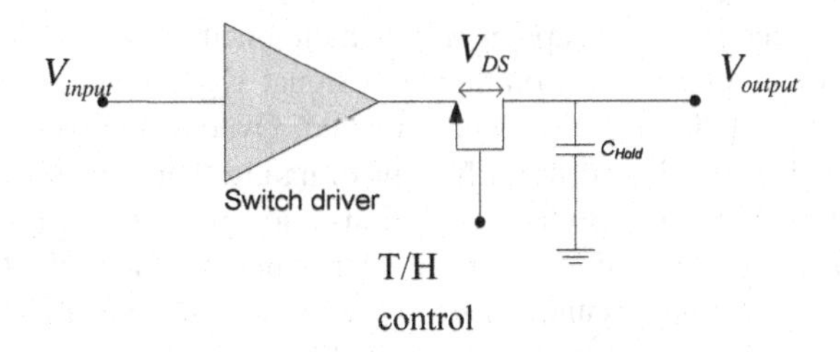

FIGURE 6.5

T/H switched capacitor circuit.

the channel region and into the drain and source junctions [4]. For an NMOS transistor, the amount of channel charge is directly proportional to the channel width W and the channel length L or:

$$Q_{channel} = -WLC_o(V_{dd} - V_{threshold} - V_{input}) \tag{6.10}$$

where $V_{threshold}$ is the threshold voltage of the NMOS device and C_o is the gate oxide capacitance. Once the NMOS switch is turned off, a certain amount of channel charge transfers into the input V_{input} while the rest transfers into the hold capacitor C_{Hold}

$$\Delta Q_{channel} = \kappa Q_{channel} = -\kappa WLC_o(V_{dd} - V_{threshold} - V_{input}) \tag{6.11}$$

where κ is the fraction of the channel charge that reverted to the hold capacitor. The change in the output voltage ΔV_{output} due to charge injection is simply the ratio

$$\begin{aligned}\Delta V_{output} &= \frac{\Delta Q_{channel}}{C_{Hold}} = \frac{-\kappa WLC_o(V_{dd} - V_{threshold} - V_{input})}{C_{Hold}} \\ &= \text{Constant} + \frac{\kappa WLC_o}{C_{Hold}} f(V_{input})\big|_{=V_{threshold}} + \frac{\kappa WLC_o}{C_{Hold}} V_{input}\end{aligned} \tag{6.12}$$

where the nonlinear function $f(V_{input})$ indicates that $V_{threshold}$ is nonlinearly related to the input voltage V_{input}. Hence, the relationship in Eqn (6.12) indicates that the output voltage change ΔV_{output} has a linear dependency on V_{input} as well as $V_{threshold}$, which in turn has a nonlinear dependency on the input voltage V_{input} [4,5]. In short, charge injection introduces a nonlinear dependency on the input signal to the output of the T/H circuit.

Another prominent issue in T/H circuit design is clock feedthrough. Clock feedthrough is attributed to the gate-to-source capacitance of the NMOS device. The resulting voltage change at the output of the T/H circuit is the ratio

$$\Delta V_{output} = -\frac{C_{parasitic}}{C_{parasitic} + C_{Hold}}(V_{DD} + V_{SS}) \tag{6.13}$$

where $C_{parasitic}$ is the parasitic capacitance as mentioned in [4]. The relationship in Eqn (6.13) shows no dependency on the input signal V_{input}, and hence the resulting voltage offset is predictable and can be removed. Overall, the degradation due to clock feedthrough is small and more benign in nature than the degradation introduced by charge injection. Both degradations, however, are byproducts of the intrinsic limitations of MOS devices. Both charge injection and clock feedthrough adversely affect the resolution and performance of the data converter [5,6]. In order to mitigate both problems, researchers in the field resorted to alternative T/H circuit designs.

Thus far, we have only shown the open-loop architecture depicted in Figure 6.1. Despite its speed, simplicity, stability, and high linearity, this particular architecture is prone to input-dependent degradation due to charge injection [1]. In order to better manage the errors due to charge injection, a closed-loop THA architecture, shown in Figure 6.6, may be employed. Comparing the open-loop THA of Figure 6.1 to the closed-loop THA of Figure 6.6, we note that in the latter, the output of the THA is fed back to the input or transconductance amplifier. The feedback configuration limits voltage swings across the switch as compared to the input and output of the THA.

During acquisition, the switch is closed and the output follows the input. The MOS switch is maintained at a virtual ground during the entire sampling phase, thus ensuring that the charge injection is independent of the input signal. Meanwhile, during this process, the hold capacitor C_{Hold} tracks the input signal. In this case, the error due to clock feedthrough from the switch is effectively removed [1,7]. When the switch is open, the voltage across the output is that of the hold capacitor, and hence the sampled or tracked voltage is retained during the hold period. The disadvantages of the closed-loop T/H circuit are lower speed when compared to the open-loop architecture, limited bandwidth, and increased design complexity. Other architectures such as the open-loop architecture with Miller capacitance, the multiplexed input architecture, the Schottky diode-bridge T/H

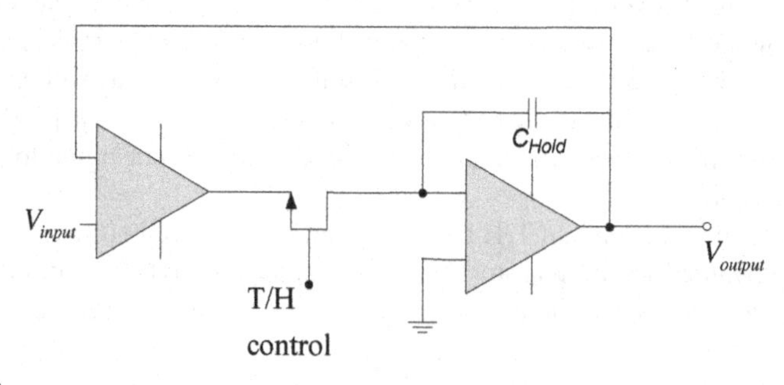

FIGURE 6.6

Closed-loop THA architecture.

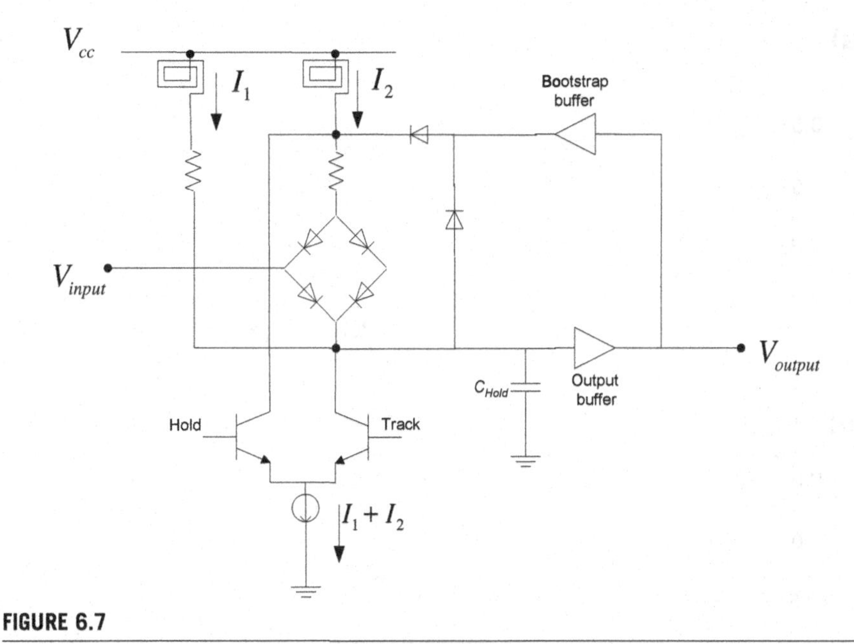

FIGURE 6.7

High-speed Schottky diode-bridge THA circuit.

depicted in Figure 6.7, and others that present certain performance advantages over the traditional open- and closed-loop architectures presented herein can be found in [1]. Details concerning certain implementation of the Schottky diode-bridge THA of Figure 6.7 can be found in [8–10].

6.1.1.4 Impact of voltage droop on performance

Before we move on to discuss the impact of the track-and-hold function on the signal itself, we will discuss one more type of degradation, namely droop. Given the basic architecture depicted in Figure 6.1, in THA, droop is simply defined as change in amplitude during the hold mode resulting from discharge in the hold capacitor. For a Nyquist converter, be it serial or parallel, let the number of bits be N, then there exists $2^N - 1$ total thresholds in the data converter. Furthermore, let n be the number of bits converted each cycle ($n = N$ for Flash converters), then the mean square level error resulting from droop is:

$$\Delta^2 = \left(\frac{2^n - 1}{2^N - 1}\right) d^2 \tag{6.14}$$

where d is the level error at the output of the comparator due to droop. According to Eqn (6.14), and as will become obvious later on in the chapter, the mean square

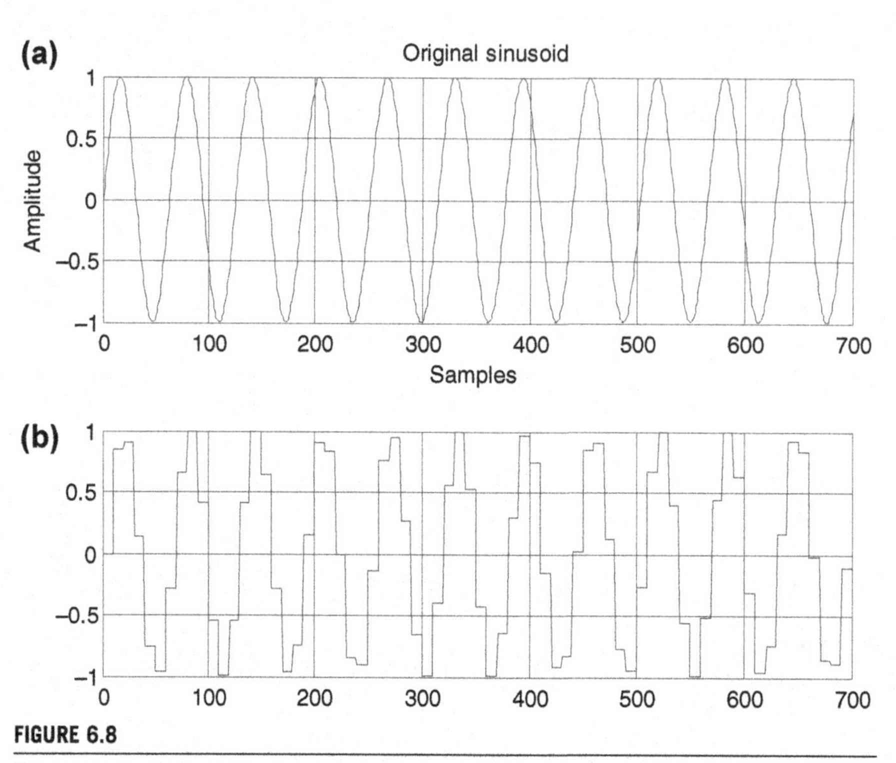

FIGURE 6.8

Time-domain effects of T/H circuit on input signal: (a) input tone, and (b) output of T/H circuit. (For color version of this figure, the reader is referred to the online version of this book.)

level error due to droop in the FLASH converter is d^2. In the same vein, a similar expression can be obtained for another popular Nyquist converter architecture, the successive approximation ADC, expressed as the sum of errors of L conversion cycles:

$$\Delta^2 = \frac{2^n - 1}{2^N - 1} d^2 \sum_{l=1}^{L-1} 2^{n(l-1)} \left(\frac{L-l}{L-1} \right)^2 \tag{6.15}$$

The relationship in Eqn (6.15) accounts for the total droop that takes place over the total number of conversion cycles. The successive approximation ADC architecture will be discussed later on in the chapter.

Finally, we examine the impact of droop on the SNR of the converter. The SNR due to droop is given as

$$SNR_{droop} = 10 \log_{10} \left(\frac{2^{2b} \kappa^2}{4\Delta^2} OSR \right) \tag{6.16}$$

again where OSR is the oversampling ratio and κ is the loading factor.

6.1.1.5 Impact of track and hold on signal quality

Finally, it is important to note the impact of the track-and-hold function on the input signal itself from a pure signal processing perspective. We will discuss the signal processing aspects of THA in more detail later on in the chapter, but at this point it is important to point out some of the time domain and frequency domain aspects of the T/H function.

Consider an analog tone input to the T/H function as depicted in Figure 6.1. Let the frequency of the input tone to the T/H be $F_c \approx 5$ MHz and the output of the T/H be as shown in Figure 6.8(a) and (b), respectively. Note that the output signal of the T/H is a degraded version of the input signal. In the frequency domain, the input tone to the T/H and its output are depicted in Figure 6.9(a) and (b), respectively. The degradation due to the hold process manifests itself as replicas of the input tone centered around multiples of the sampling frequency F_s or more precisely at $mF_s - F_c$ and $mF_s + F_c$, $m \in N$.

6.1.2 Comparators

A variety of high speed ADCs, such as the FLASH converter for example, use comparators in their circuits to compare an analog signal to a certain reference voltage as will be demonstrated in Section 6.2.1. A comparator must be able to amplify and compare the input signal at significantly higher speed than the sampling rate of the incoming signal. The basic function of a comparator is to compare a varying input signal to a fixed reference signal resulting in a logic 0 or logic 1 at the output, in effect acting as a single-bit ADC.

A traditional latched comparator circuit is shown in Figure 6.10. In a latched comparator, for example, the comparison is performed at time instants controlled by the latch signal. Once the track signal voltage goes high, the input signal voltage gets amplified. Once the latch signal goes high, the voltage differences at the output will force the positive feedback transistor pair Q_5 and Q_6 to latch, thus resulting in a digital output signal. At high speed, this basic design is prone to certain transient noise at the input during latch mode mainly due to Q_3 and Q_4 abruptly shutting off. This type of degradation is known as *kickback noise*. Kickback noise could take a significant amount of time before it decays to less than 1 LSB [1].

Another important conversion error is metastability. Metastability occurs when the difference between the input signal and the threshold is not adequate enough to drive the comparator into saturation within the allotted time [11], thus causing the output of the comparator to be undefined or completely erroneous. This phenomenon is illustrated in Figure 6.11. To further illustrate metastability, we can estimate the output of the comparator $V_{output}(t)$ in response to an input voltage $V_{input}(t)$ at the latching instance as

$$V_{output}(t) = GV_{input}(t)e^{t_d/\tau} \tag{6.17}$$

where G is the gain of the preamplifier. The time τ is the regeneration time constant of the latch, and the time t_d is the elapsed time after the comparator output has latched.

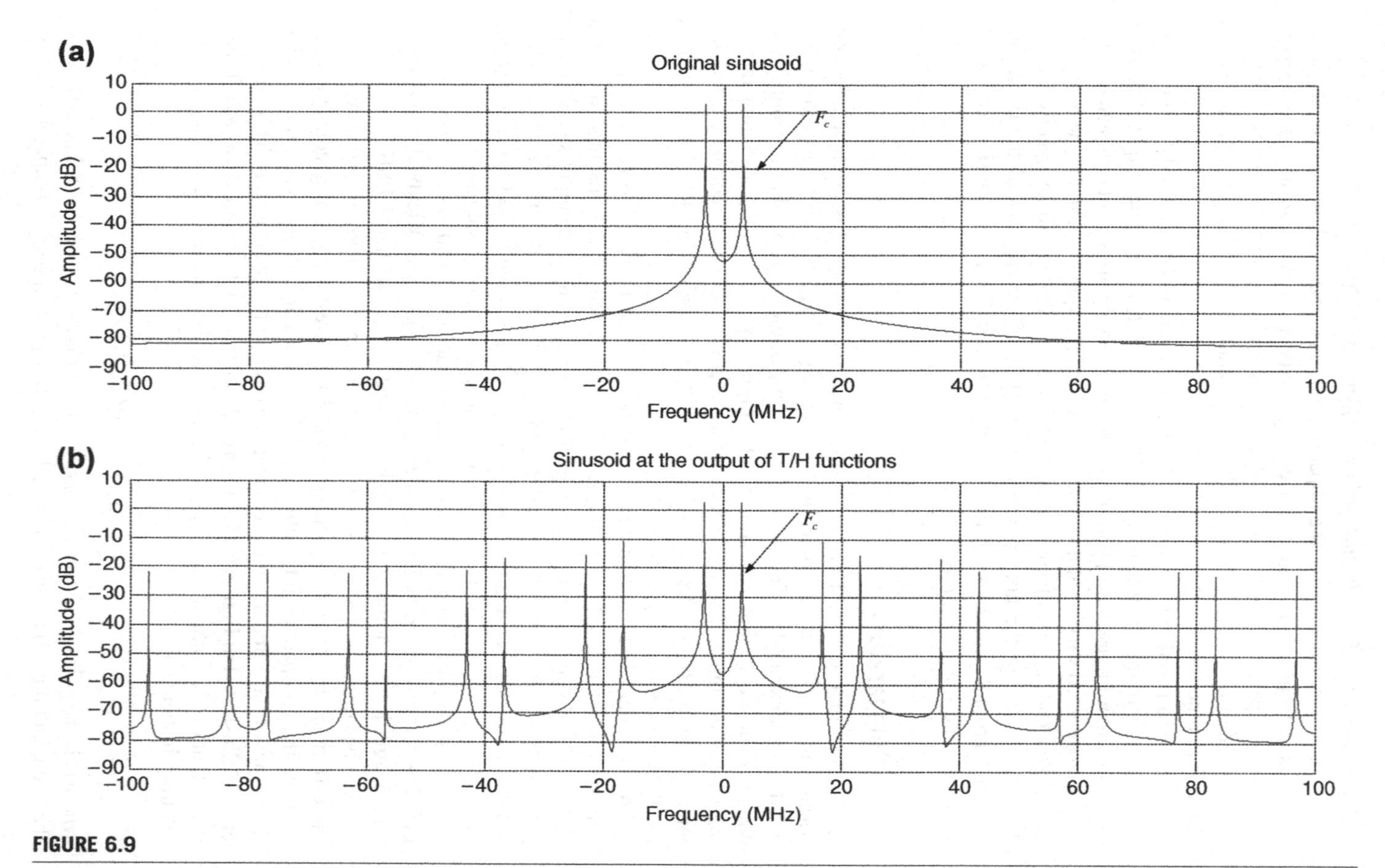

FIGURE 6.9

Spectrum of sinusoid with frequency F_c at the input and output of THA: (a) spectrum of input tone, and (b) spectrum of degraded tone at the output of THA. (For color version of this figure, the reader is referred to the online version of this book.)

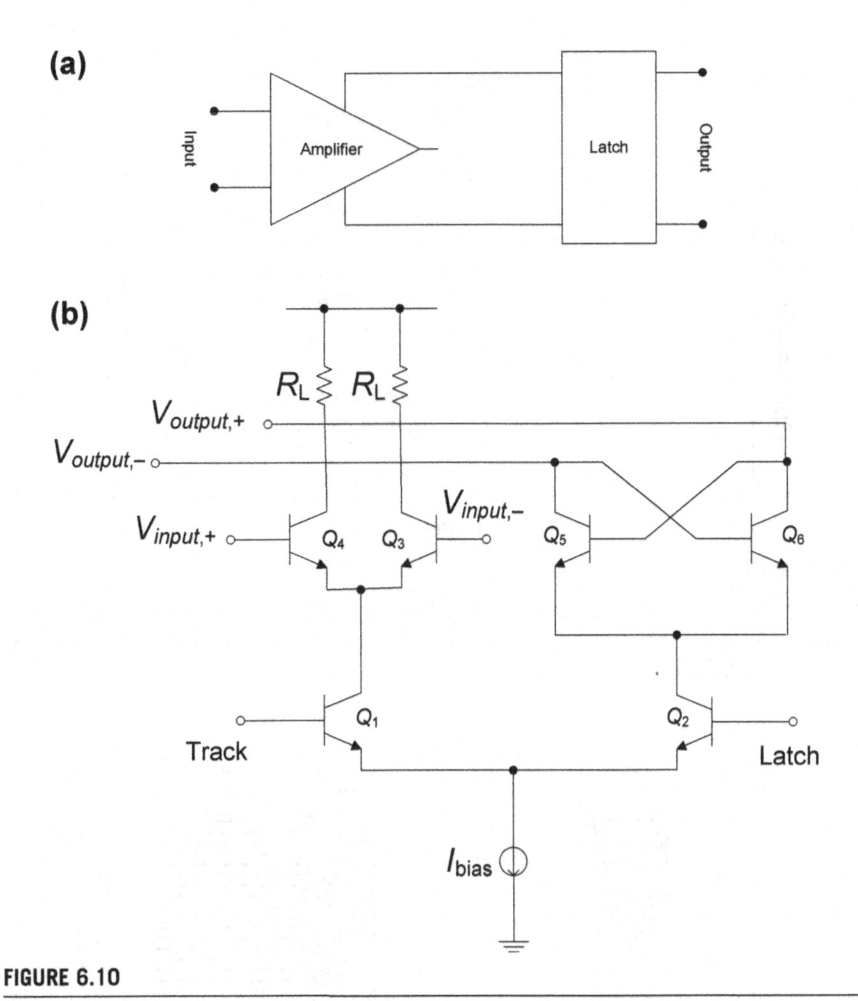

FIGURE 6.10

Basic comparator: (a) block diagram, and (b) classical track-latch comparator design.

Time regeneration is defined as the total time the latch needs in order to produce a legitimate digital signal. To gain a better understanding of metastability, consider the three scenarios depicted in Figure 6.11, namely scenarios A, B, and C. In scenario A, the comparator output reached a logic value 1 from a logic value 0 before the hold-mode signal is asserted. The output of the comparator is an unmistakable logic value 1 in response to a large differential input voltage. Next, consider scenario B where the differential input voltage is small. In this case, at the time of the assertion of the hold signal, the output of the comparator is in between logic level 0 and logic level 1 and consequently it has not reached logic level 1. In this case, the comparator has resulted in an erroneous output logic level. In scenario C, the differential input level to the comparator is very small, and hence the output of the comparator at the time that the hold signal is asserted is at the erroneous logic output level 0.

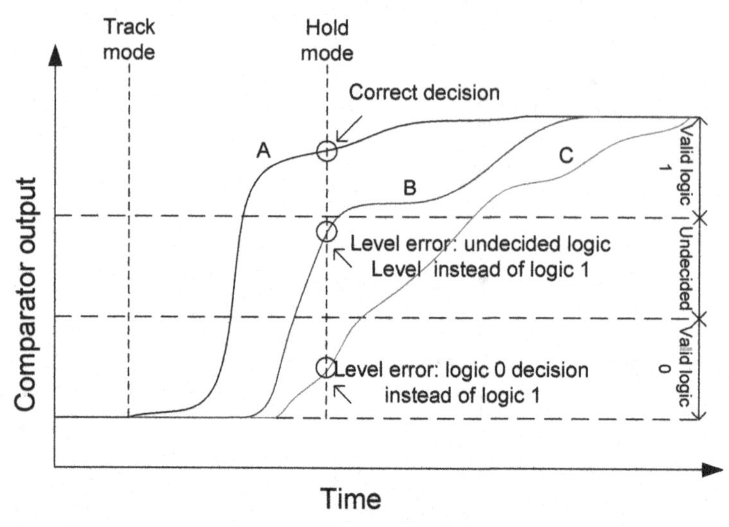

FIGURE 6.11

Metastability—an error that occurs when the latch does not produce the desired output level within a certain time [11]. (For color version of this figure, the reader is referred to the online version of this book.)

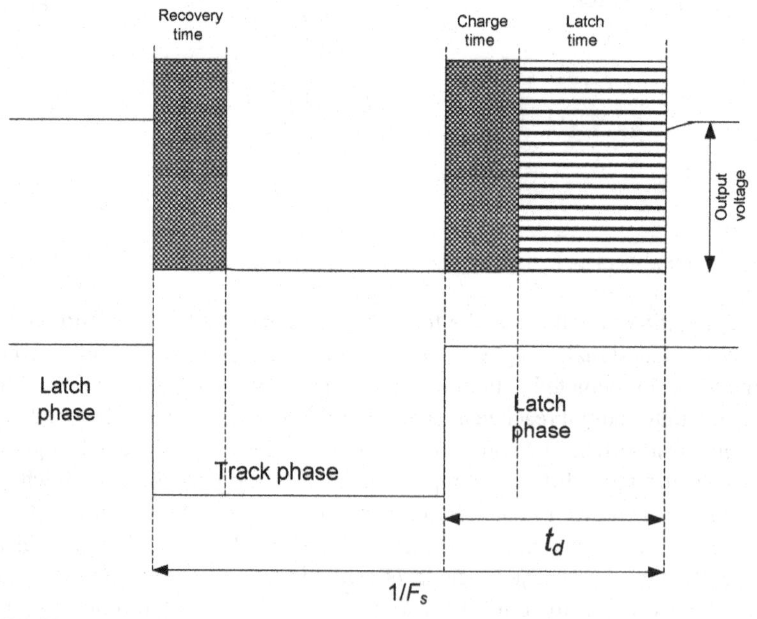

FIGURE 6.12

Comparator track, delay, and latch time during track and latch phase transitions.

The probability that a metastate has occurred after the decision time t_d has elapsed is [12,13]:

$$P_e(t > t_d) = e^{-\frac{G-1}{\tau}t_d} \quad (6.18)$$

Typically, for symmetrical clocking, the latch period t_d is $T_s/2 = 1/(2F_s)$, where F_s is the sampling speed of the comparator as illustrated in Figure 6.12. Realistically, if we account for a certain charge time, say t_{charge}, then the decision time becomes $t_d = 1/(2F_s) - t_{charge}$. Obviously, reducing charge time allows more time for the comparator to reach a digital decision and hence reduces the occurrence of metastable points. For a sampling rate F_s, and a moderate gain G, the probability of metastable states per second can be implied from Eqn (6.18) simply as

$$\Upsilon = F_s e^{-\frac{G-1}{\tau}t_d} = F_s e^{-\frac{G-1}{\tau}(T_s/2 - t_{charge})} \quad (6.19)$$

According to Eqn (6.18), the occurrence of metastability can be minimized by increasing the unity gain bandwidth G of the comparator [13].

EXAMPLE 6.1 METASTABILITY IN A LATCHING COMPARATOR

A simplified model of the comparator latching process is depicted in Figure 6.13. Let V_O be the voltage across the capacitor at the time of regeneration t_O, then compute the regeneration time constant $\tau = RC$.

Solving for the current $I(t)$ in Figure 6.13 results in the well-known equation

$$-V_{output}(t) + I(t)R + V(t) = 0 \quad (6.20)$$

Substituting $V_{output}(t) = GV(t)$ into Eqn (6.21) and solving for the current $I(t)$, we obtain

$$I(t) = \frac{G-1}{R}V(t) \quad (6.21)$$

Furthermore, the voltage across the capacitor C can be further expressed as

$$I(t) = C\frac{dV(t)}{dt} \quad (6.22)$$

Substitute the result in Eqn (6.20) into the current relationship in Eqn (6.22) and rearrange in terms of the voltage $V(t)$

$$\frac{dV(t)}{dt} = \frac{G-1}{RC}V(t) = \frac{G-1}{\tau}V(t) \quad (6.23)$$

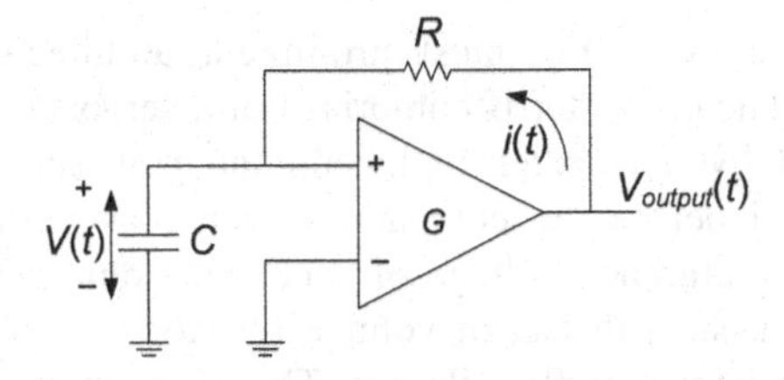

FIGURE 6.13

Basic model of comparator during the latching process.

Continued

EXAMPLE 6.1 METASTABILITY IN A LATCHING COMPARATOR—cont'd

Integrating both sides of Eqn (6.23) relates the voltage $V(t)$, the gain G, and time to the regeneration time constant

$$\ln(V(t)) + K = \frac{G-1}{\tau}t \tag{6.24}$$

Define the integration constant as $K = \ln(V_0)$, then Eqn (6.24) can be rewritten in the form

$$\ln\left(\frac{V(t)}{V_0}\right) = \frac{G-1}{\tau}t \Rightarrow V(t) = V_0 e^{\frac{G-1}{\tau}t} \tag{6.25}$$

Next, in order to determine the regeneration time constant, at an arbitrary instant time t_{init} measure the initial voltage V_{init}. After a certain time ΔT has elapsed such that $t_{sample} = t_{init} + \Delta T$, measure the voltage at the sampling instant V_{sample}. According to Eqn (6.25), the ratio of the sample voltage $V_{sample} \cong V(t_{sample}) = V_0 e^{\frac{G-1}{\tau}t_{sample}}$ to the initial voltage $V_{init} \cong V(t_{init}) = V_0 e^{\frac{G-1}{\tau}t_{init}}$ is

$$\frac{V_{sample}}{V_{init}} = \frac{V_0 e^{\frac{G-1}{\tau}t_{sample}}}{V_0 e^{\frac{G-1}{\tau}t_{init}}} = e^{\frac{G-1}{\tau}\Delta T} \tag{6.26}$$

Thus implying that the regeneration time constant is

$$\tau = \frac{G-1}{\ln\left(\frac{V_{sample}}{V_{init}}\right)}\Delta T \tag{6.27}$$

It is interesting to note that, according to Eqn (6.27), the regeneration time constant depends on the gain G as well as $\Delta T = t_{sample} - t_{init}$.

6.2 Nyquist converters

In this section, we present some popular Nyquist converter architectures, namely the FLASH converter, the pipelined and subranging converters, and finally the folding converter. Performance degradation relevant to these types of converters is discussed in some detail.

6.2.1 The FLASH architecture

The FLASH architecture is the most prominent architecture of high speed Nyquist converters. The converter is comprised of a series of comparators strung in parallel followed by a decoder logic and latch as shown in Figure 6.14. The resistors situated before the comparators serve as voltage dividers in an ascending order. The reference voltage of each individual comparator is tapped onto the resistor bank such that each voltage resistor is 1 LSB higher than the voltage across the resistor directly below it. Therefore, starting from the bottom comparator as comparator $m = 0$, if the analog input voltage of the signal is higher

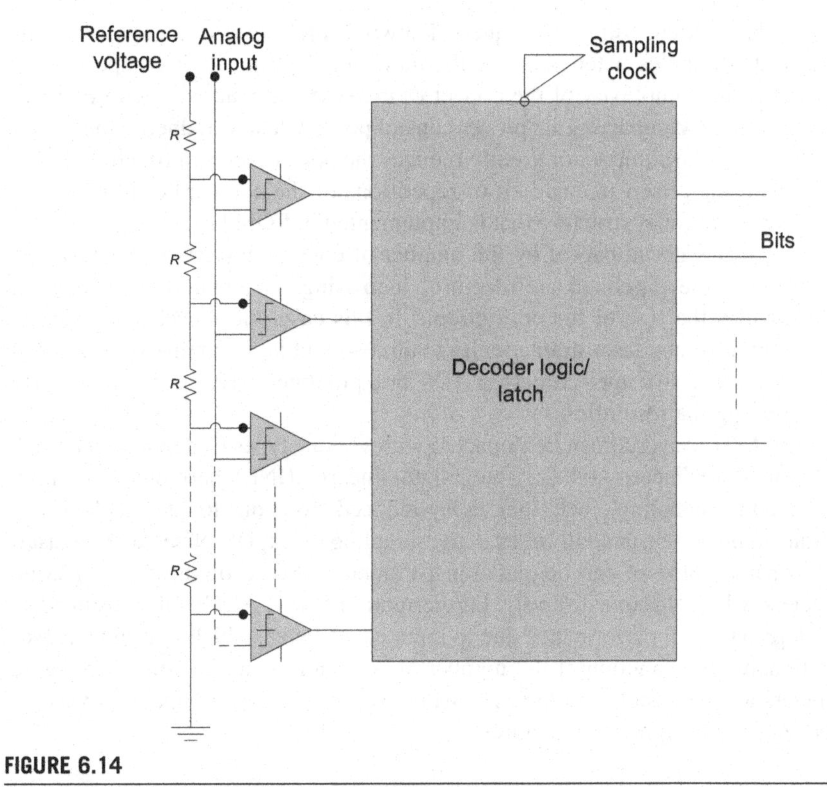

FIGURE 6.14

The FLASH or parallel ADC.

than the voltage at the mth comparator and lower than the voltage at the $(m+1)^{th}$ comparator, then the output corresponding to the mth comparator is logic "1", whereas the output corresponding to the $(m+1)^{th}$ comparator is logic "0". The structure of the comparator/resistor ladder resembles a thermometer logic that is the trademark of FLASH ADCs. The output of the thermometer logic is further decoded resulting in the proper binary output. This output could be represented in Gray code or binary weighted code if so desired. The output of the thermometer is postponed by one comparator delay after which the results are processed by the decoder logic and switch. The latter have an insignificant delay in contrast to the comparator's delay. Given a reasonably matched set of comparators, the comparator delay could act as an S/H circuit, thus enabling the designer to forgo the design of an SHA. Once the input voltage is applied to the resistor ladder, the comparators process the respective signal with certain inherent delay as mentioned earlier in the chapter.

Given an N-bit FLASH converter simply implies that the number of resistors present in the ladder structure is 2^N resistors whereas the number of comparators is $2^N - 1$. This bit-encoding technique is known as thermometer logic since in its operation it resembles a temperature thermometer. Despite its high speed, it is obvious that the complexity of the converter increases dramatically with its resolution thus directly impacting its power consumption. Clearly, reducing the power consumption of the comparator greatly reduces the power consumption of the converter. The design itself is made up of repetitions of the comparator block along with the single decoder structure that is implemented in ROM [13]. The conversion time, however, is not impacted by the number of comparators, as is the case, for example, with the pipelined architecture. Increasing the resolution by one bit almost doubles the size of the core circuit. In this case, the power consumption of the circuit also increases dramatically to almost double. In comparison, the resolution of a successive approximation ADC or a pipelined ADC increases linearly with increasing the resolution.

The FLASH converter can be impacted with various types of degradations such as differential nonlinearity (DNL), integral nonlinearity (INL), harmonic distortion, kickback noise, sampling clock jitter, delay-induced errors due to uneven clock distribution, nonideal rise and fall times in the sampling clock, DC offset, and metastability to name some of the degradation parameters. Some of these degradation parameters will be discussed shortly. Furthermore, FLASH ADCs suffer from additional degradation in performance due to input capacitance, which in turn increases proportionally as a function of the number of comparators in the converter. These parameters are not peculiar to the FLASH converter but can be applied in varying degrees to other Nyquist architectures.

6.2.1.1 Degradation due to differential and integral nonlinearities

DNL is a static parameter that denotes the maximum deviation of the converter step width and the ideal value of a single LSB. The ADC specification calls out for a DNL of strictly less than an LSB in order to avoid having any missing codes at the output of the converter. This, in essence, ensures that the input/output transfer function of the ADC is monotonic. In order to represent DNL mathematically, let the analog input to the converter be $x_a(t,i)$, which is obtained at multiples of the sampling period T_s be given as:

$$\begin{aligned} x_a(t,i)|_{t=nT_s} \cong x(nT_s) + e_q(nT_s) &= x_{ref} \sum_{m=0}^{M-1} b_m(nT_s)2^m \\ &\quad + e_q(nT_s), \quad \text{for } b_m(.) \in \{0,1\} \\ &= x_{ref}\left[b_0(nT_s) + 2b_1(nT_s) + \cdots \right. \\ &\quad \left. + 2^{M-1}b_{M-1}(nT_s)\right] \\ &\quad + e_q(nT_s), \quad \text{for } b_m(.) \in \{0,1\} \end{aligned} \tag{6.28}$$

In Eqn (6.28), $\{b(nT_s)\}_{m=0}^{M-1}$ is a binary sequence with b_0 being the LSB and b_{M-1} being the most significant bit or MSB. The input reference is represented by x_{ref}

whereas the quantization noise is associated with $e_q(.)$. The variable i associated with the analog input signal $x_a(t,i)$ signifies the quantization signal level at the output of the converter.

Next we define the DNL in terms of the difference between two analog inputs, resulting in two adjacent quantization values and the resolution of the ADC as:

$$DNL(i) = \frac{x_a(t,i) - x_a(t,i-1) - \delta}{\delta} = \frac{x_a(t,i) - x_a(t,i-1)}{\delta} - 1 \tag{6.29}$$

DNL is illustrated in Figure 6.15. In Eqn (6.29), we define the resolution of the converter or single LSB in terms of the full-scale voltage $V_{x,Fs}$ as the ratio (Figure 6.15)

$$\delta = V_{x,Fs}/2^b \tag{6.30}$$

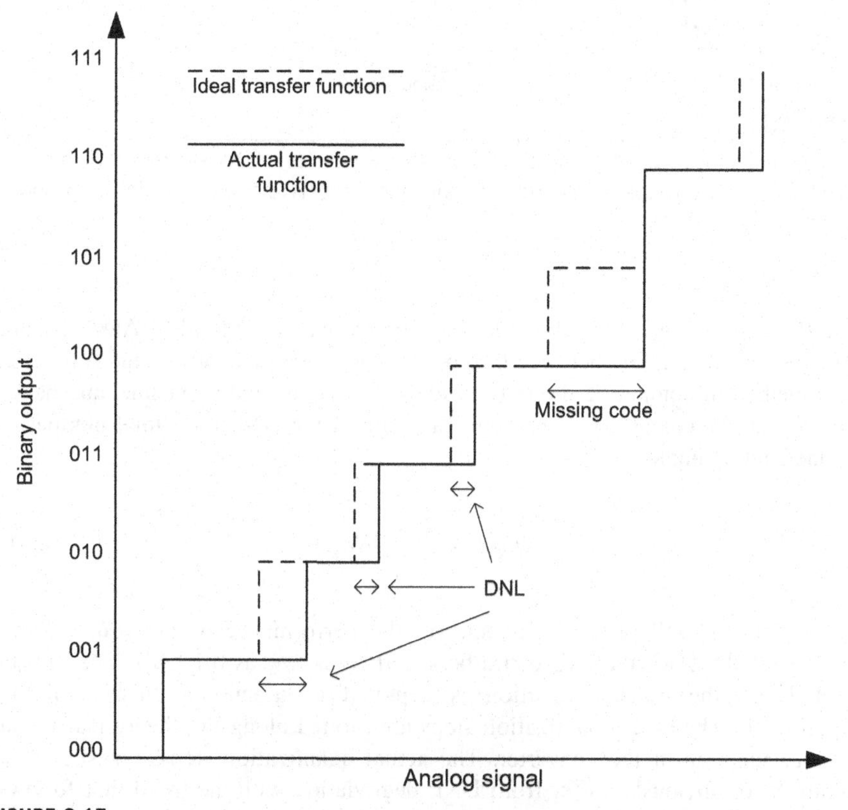

FIGURE 6.15

Input/output transfer function of 3-bit ADC showing both ideal and actual characteristics and missing codes, thus illustrating the effect of DNL.

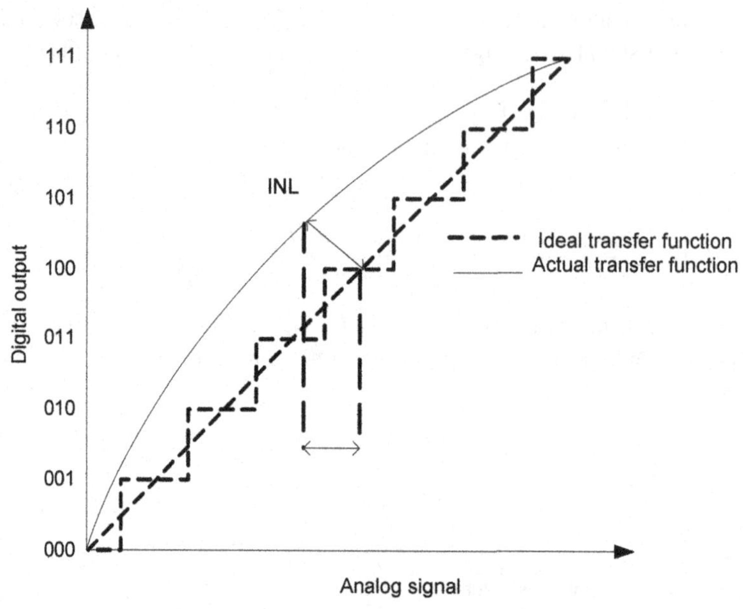

FIGURE 6.16

Input/output transfer function of 3-bit ADC displaying the effect of INL with respect to an ideal transfer function. (For color version of this figure, the reader is referred to the online version of this book.)

The INL can now be defined as the deviation curve from the ideal ADC transfer function, which is a line between 0, provided there is no DC offset, and full scale. This method of computing the INL is referred to as the end-point line, and hence the INL can be computed as the aggregate sum of the DNL for a total number of k codes, for example:

$$INL(k) = \sum_{i=0}^{k-1} DNL(i) \tag{6.31}$$

Both DNL and INL are used to specify the performance of a Nyquist ADC. For a FLASH ADC, the INL could be specified as low as 0.1 LSB. The impact of INL on the transfer function is depicted as a smooth curved line in Figure 6.16. The ideal quantization steps are plotted along the theoretical linear transfer function of the converter. The actual quantization steps, not shown in Figure 6.16, obviously suffer from DNL degradation with the trend that follows the INL curve.

EXAMPLE 6.2 IMPACT OF INL ON SIGNAL QUALITY

The purpose of this example is to simply illustrate the impact of INL on signal quality. Assume that the nonlinear transfer function of an ADC due to INL can be modeled as a third order polynomial of the form

$$y(t) = \beta_1 x(t) + \beta_2 x^2(t) + \beta_3 x^3(t)\Big|_{t=nT_s} \tag{6.32}$$

where x(t) is the input signal to the ADC and y(t) is the equivalent output signal, and T_s is the sampling period. The INL transfer function depicted in Figure 6.17 is generated using the coefficients $\beta_1 = 1.4$, $\beta_2 = -0.1$, and $\beta_3 = -0.3$. Determine the IIP2 and IIP3 points.

Assume the input signal after the SHA to be $x(nT_s) = \alpha\cos(\omega nT_s)$, where ω is the normalized frequency. Determine the output signal $y(nT_s)$, and plot its respective normalized spectrum.

In order to determine the output signal $y(nT_s)$, let us use the analysis developed in Chapter 4 to determine the various nonlinear signal components. That is, let

$$y(nT_s) = \underbrace{\beta_1 \alpha \cos(\omega nT_s)}_{\text{Linear term}} + \underbrace{\beta_2 \alpha^2 \cos^2(\omega nT_s)}_{\text{Second order term}} + \underbrace{\beta_3 \alpha^3 \cos^3(\omega nT_s)}_{\text{Third order term}} \tag{6.33}$$

The second order term in Eqn (6.33) can simply be expressed as:

$$\beta_2 \alpha^2 \cos^2(\omega nT_s) = \frac{\beta_2 \alpha^2}{2}\left[\underbrace{\cos(2\omega nT_s)}_{\text{Second harmonic}} + \underbrace{1}_{\text{DC term}}\right] \tag{6.34}$$

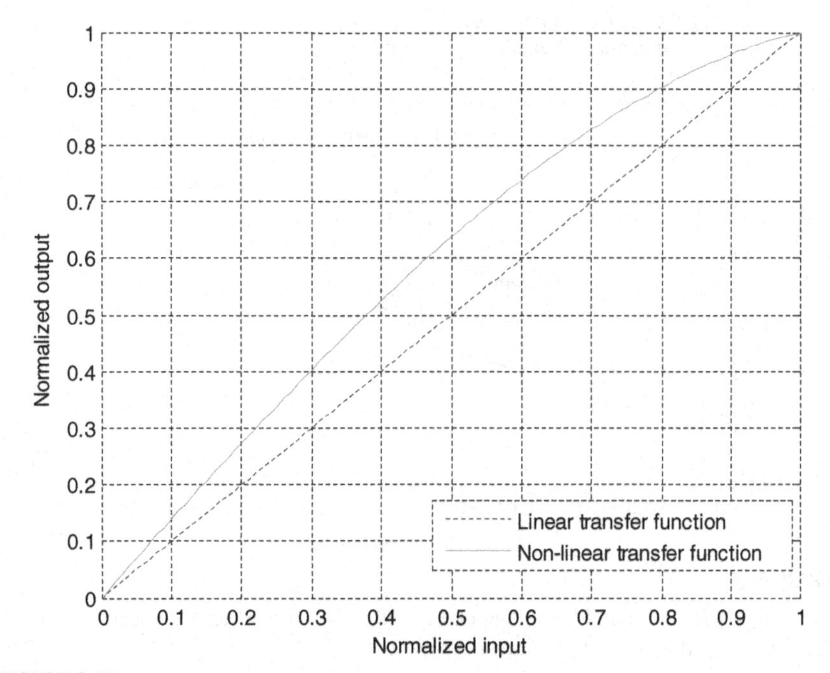

FIGURE 6.17

Linear and nonlinear transfer function of an ADC. (For color version of this figure, the reader is referred to the online version of this book.)

Continued

EXAMPLE 6.2 IMPACT OF INL ON SIGNAL QUALITY—cont'd

Similarly, the third order term in Eqn (6.33) can be written as:

$$\beta_3\alpha^3\cos^3(\omega nT_s) = \underbrace{\frac{3}{4}\beta_3\alpha^3\cos(\omega nT_s)}_{\text{Distortion to linear term}} + \underbrace{\frac{1}{4}\beta_3\alpha^3\cos(3\omega nT_s)}_{\text{Third harmonic}} \tag{6.35}$$

Finally, the output signal is given as the sum of the linear term plus the results expressed in Eqns (6.34) and (6.35), or as

$$y(\omega nT_s) = \frac{\beta_2\alpha^2}{2} + \left(\beta_1\alpha + \frac{3}{4}\beta_3\alpha^3\right)\cos(\omega nT_s) + \frac{\beta_2\alpha^2}{2}\cos(2\omega nT_s) + \frac{1}{4}\beta_3\alpha^3\cos(3\omega nT_s) \tag{6.36}$$

The spectrum of the ideal tone along with that of the degraded output signal is depicted in Figure 6.18.

The *IIP2* due to INL can be computed for $\beta_1 = 1.4$, and $\beta_2 = -0.1$ according to Chapter 4 as:

$$IIP2_{dBm} = 10\log_{10}\left(\frac{|\beta_1|}{|\beta_2|}\right) + 30 = 10\log_{10}\left(\frac{|1.4|}{|-0.1|}\right) + 30 = 41.5dBm \tag{6.37}$$

In a similar manner, the *IIP3* due to INL can be computed for $\beta_1 = 1.4$ and $\beta_3 = -0.3$

$$IIP3_{dBm} = 10\log_{10}\left(\sqrt{\frac{4}{3}\frac{|\beta_1|}{|\beta_3|}}\right) + 30 = 10\log_{10}\left(\sqrt{\frac{4}{3}\frac{|1.4|}{|-0.3|}}\right) + 30 = 34dBm \tag{6.38}$$

By employing the same techniques developed in Chapter 4, the harmonic distortion due to the second and third harmonic can also be computed.

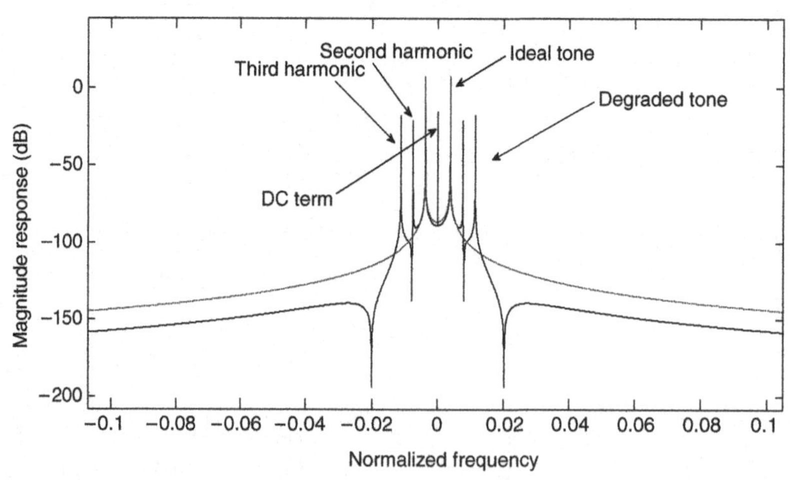

FIGURE 6.18

Spectrum of ideal tone and tone degraded by INL. (For color version of this figure, the reader is referred to the online version of this book.)

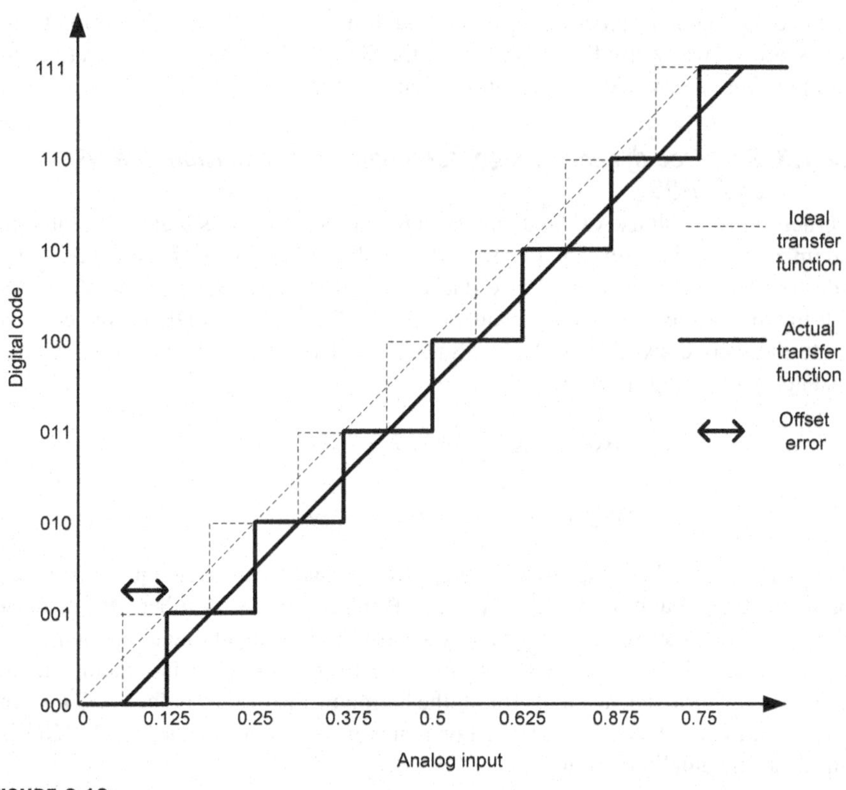

FIGURE 6.19

Impact of voltage offset error on ADC transfer function.

6.2.1.2 Offset errors

An offset error is a common error that impacts all the output codes of a Nyquist ADC as shown, by way of an example, in Figure 6.19. The voltage or current offset referred to here is not due to imperfections in the analog receiver but rather due to the ADC circuitry. In this case, an offset voltage, offset current, or in certain cases a digital code is present at the output of the ADC despite the fact that there is no signal applied to the input. Offset errors can also manifest themselves as gain errors. A gain error is a change or deviation in slope from the ADC's ideal transfer function. The offset may be present in the input or output amplifiers as well as comparators. In order to ensure that there are no missing codes, it is imperative that the maximum value of the offset does not exceed ½ LSB. In comparators, offsets are mainly due to the preamplifier, and hence extra care is necessary in the circuit design in order to minimize their impact. This implies, for example, optimizing the layout in order to mitigate the random offset caused by the transistor fabrication process. The offset is also dependent on the process itself. For example, offsets present in CMOS comparators are higher than offsets present in their bipolar junction transistor

(BJT) counterparts. In order to ensure a good design yield, offset cancellation techniques are typically used, thus reducing the offset of a CMOS comparator, for example, from a few mVs to a small portion of an mV.

6.2.1.3 Spur-free dynamic range, total harmonic distortion, SINAD, and ENOB

In order to make full use of the dynamic range of the ADC, it is highly desirable to maximize the ratio between the desired signal, say an input sinusoid, and any undesired harmonics that may arise due to circuit nonlinearity. In general, ADCs with low distortion due to INL result in a large SFDR [13]. SFDR in this context due to the second and third order nonlinearity is defined exactly as it was defined in Chapter 4 and repeated here as

$$\begin{aligned} \Delta_{SFDR-IIP2} &= \frac{1}{2}(IIP2 - P_{ADC\ noise}) \\ \Delta_{SFDR-IIP3} &= \frac{2}{3}(IIP3 - P_{ADC\ noise}) \end{aligned} \tag{6.39}$$

where $P_{ADC\ noise}$ is the noise power of the converter and not restricted just to thermal noise. The typical unit used to describe SFDR in this context is either dB full scale (dBFS) or dBc. Harmonics are typically measured with input signal applied at or near full scale of the ADC. Harmonics can be further specified in general via an encompassing term that accounts for all the harmonic distortions in the ADC known as total harmonic distortion (*THD*). For a single tone input to the ADC, *THD* is defined as the amplitude ratio:

$$THD = \frac{\sqrt{\sum_{n=2}^{N} \widehat{v_n^2}}}{\widehat{v_1^2}} = \frac{D}{S} \tag{6.40}$$

where $\widehat{v_1^2}$ and $\{\widehat{v_n^2}\}_{n=2}^{N}$ are the RMS voltage of the fundamental and second to *Nth* harmonic, respectively. The variable S and D signify the signal and distortion, respectively. Another figure of merit is *THD* plus noise or commonly referred to as *THD* + N. *THD* + N is defined as the sum of harmonic powers plus noise power divided by the fundamental power.

A further specification of ADCs in general and FLASH converters in particular is the signal-to-noise-and-distortion ratio (SINAD). SINAD is another encompassing specification that describes the general performance of a converter. For a single-tone input, SINAD is defined as the ratio of the RMS desired signal amplitude to the mean of the root-sum-square of all nonfundamental spectral components except DC. SINAD provides a good indication of the performance of an ADC converter since it includes the effects of almost all degradations present at the output of the converter. SINAD can be mathematically expressed as

$$SINAD = 20 \log_{10}\left(\frac{S}{N + D}\right) \tag{6.41}$$

It is interesting to further dissect the ratio $S/(N + D)$ by looking at its inverse, or

$$\frac{N+D}{S} = \frac{N}{S} + \frac{D}{S} = 10^{-\frac{SNR}{20}} + 10^{\frac{THD}{20}} \tag{6.42}$$

Hence, SINAD can be expressed in terms of *SNR* and *THD* simply as

$$SINAD = 20\log_{10}\left(\frac{1}{10^{-\frac{SNR}{20}} + 10^{\frac{THD}{20}}}\right) \tag{6.43}$$

Note that in this case, SNR was used and not SQNR in order to account for input-referred noise if the latter becomes significant.

Another measure of dynamic performance of the FLASH ADCs, as well as other ADCs, is the effective number of bits (ENOB). For an ideal ADC, ENOB is related to SINAD via the relationship as

$$ENOB = \frac{SINAD - 1.76 + 10\log_{10}\left(\frac{V_{FS}^2}{V_{input}^2}\right)}{6.02} \tag{6.44}$$

where V_{FS} is the full-scale voltage of the converter and V_{input} is the input signal voltage amplitude. When the input signal is applied at full scale, the result in Eqn (6.44) becomes the familiar number of bits derived from the theoretical expression $ENOB \cong N = (SNR - 1.76)/6.02$. Note that both ENOB and SINAD are figures of merit that measure the dynamic performance of the ADC.

6.2.1.4 Impact of thermal noise on FLASH ADCs

The FLASH ADC architecture that is depicted in Figure 6.14 relies on a series of amplifiers and resistors in order to create the thermometer circuit discussed earlier. Thermal noise, as discussed in Chapter 3, is white in nature and manifests itself around the nominal code transitions as shown in Figure 6.20. Code transition noise, also known as input-referred noise, causes dithering around the code transition levels. Thermal noise is additive to quantization noise. Assuming the total noise power due to quantization and thermal noise to be $N_{\text{total}}^2(.)$, then the total SNR can be expressed as the ratio of the signal power to noise power or

$$SNR_{total} = \frac{P_{signal}}{E\{N_{total}^2(t)\}} \tag{6.45}$$

where P_{signal} is the signal power.

The total noise power can be further expressed in terms of thermal and quantization noise simply as

$$\begin{aligned} E\{N_{total}^2(t)\} &= E\left\{\left[N_{thermal} + N_{quantization}\right]^2\right\} \\ &= E\left\{N_{thermal}^2(t)\right\} + E\left\{N_{quantization}^2(t)\right\} + 2E\left\{N_{thermal}N_{quantization}\right\} \end{aligned} \tag{6.46}$$

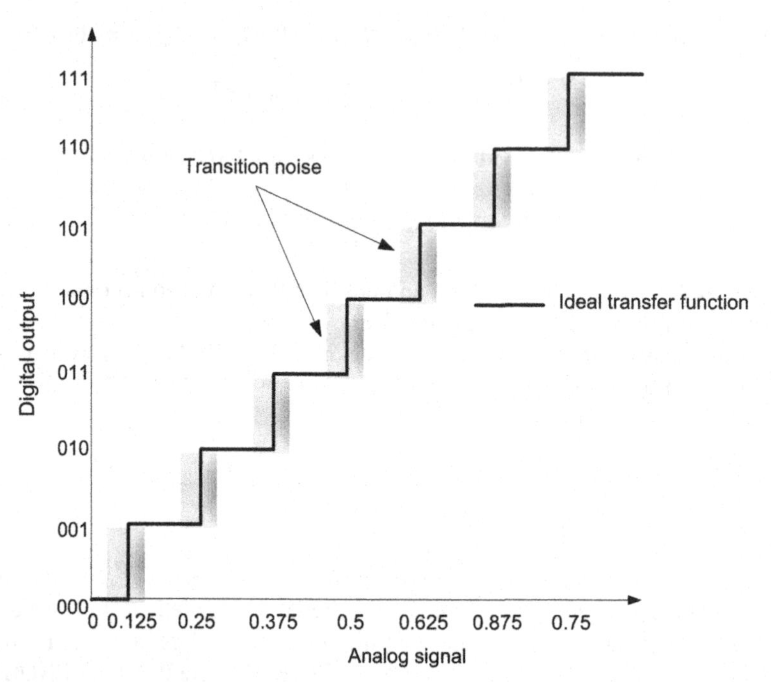

FIGURE 6.20

Degradation caused by transition noise in FLASH ADC.

where $N_{thermal}(.)$ and $N_{quantization}(.)$ are the thermal and quantization noise, respectively. Assume both $N_{thermal}(.)$ and $N_{quantization}(.)$ to be zero-mean white and uncorrelated [13], that is

$$E\{N_{\text{thermal}}N_{\text{quantization}}\} = 0 \tag{6.47}$$

then the expression in Eqn (6.46) can be reduced to the sum

$$\begin{aligned} E\{N_{total}^2(t)\} &= E\left\{\left[N_{thermal} + N_{quantization}\right]^2\right\} \\ &= E\left\{N_{thermal}^2(t)\right\} + E\left\{N_{quantization}^2(t)\right\} = \sigma_{thermal}^2 + \sigma_{quantization}^2 \end{aligned} \tag{6.48}$$

In Eqn (6.48), $\sigma_{thermal}^2$ and $\sigma_{quantization}^2$ are the thermal and quantization noise variances. Substitute the results obtained in Eqn (6.48) into the total signal-to-noise-ratio expression obtained in Eqn (6.45), then we obtain

$$SNR_{total} = \frac{P_s}{\sigma_{thermal}^2 + \sigma_{quantization}^2} \tag{6.49}$$

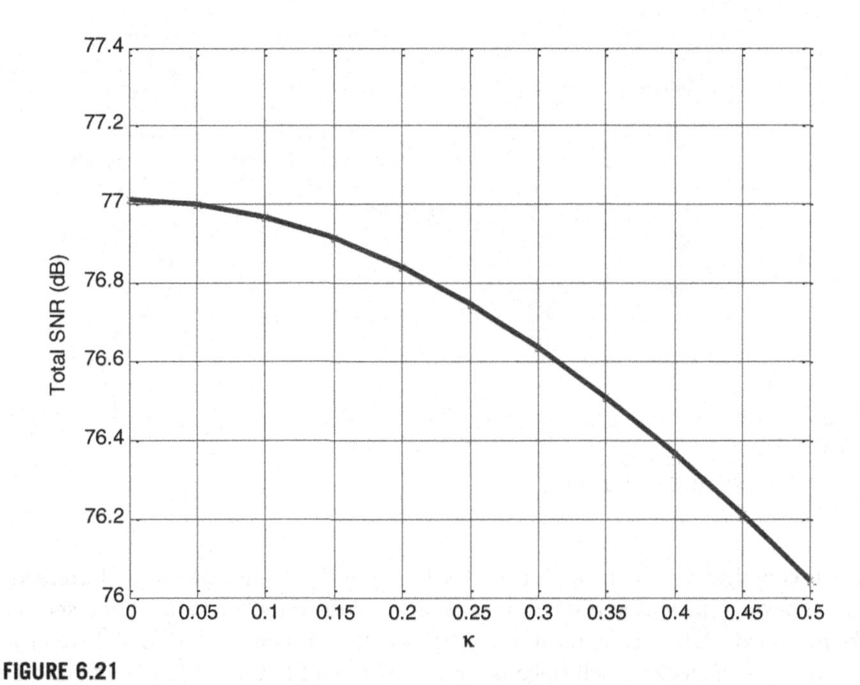

FIGURE 6.21

Impact of additive thermal noise on total SNR for a 12-bit ADC. (For color version of this figure, the reader is referred to the online version of this book.)

The relationship in Eqn (6.49) can be further expressed in terms of the signal-to-quantization ratio $SQNR = P_s/\sigma^2_{quantization}$ via simple algebraic manipulation, resulting in the total linear SNR:

$$SNR_{total} = \frac{P_s}{\sigma^2_{quantization}\left(\frac{\sigma^2_{thermal}}{\sigma^2_{quantization}} + 1\right)} = \frac{SQNR}{\left(\frac{\sigma^2_{thermal}}{\sigma^2_{quantization}} + 1\right)} = \frac{SQNR}{\kappa^2 + 1}, \quad \kappa > 0 \text{ and } \kappa^2 \ll 1 \tag{6.50}$$

The impact of additive thermal noise on the total SNR degrades the SQNR as depicted in Figure 6.21. The degradation is inversely proportional to the denominator of Eqn (6.50) where $\kappa^2 \ll 1$ and $\kappa > 0$.

6.2.2 Pipelined and subranging ADC architectures

The block diagram of a typical pipelined ADC architecture is depicted in Figure 6.22. The main advantage of this architecture is high resolution with reduced

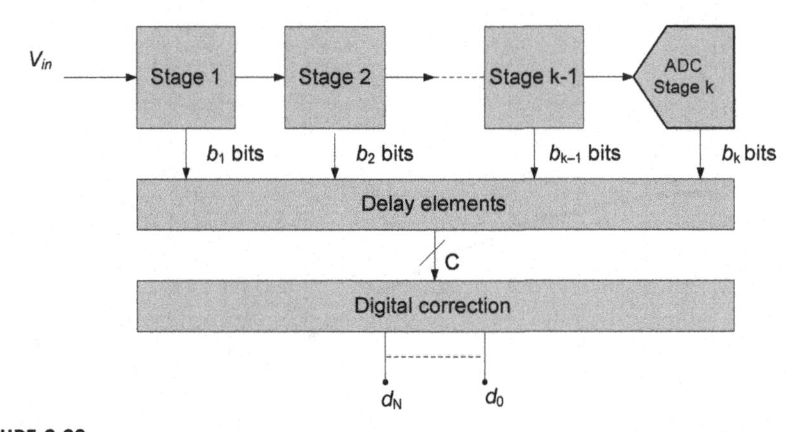

FIGURE 6.22

Typical pipelined ADC architecture.

design complexity compared to a FLASH ADC. These advantages of decreased complexity and increased resolution come at the cost of reduced conversion speed. The pipelined ADC is comprised of multiple stages set up in cascade and driven by nonoverlapping clocks. Each stage is comprised of an SHA, a FLASH ADC, and a corresponding DAC of the same resolution, a subtractor circuit, followed by an amplifier as shown in Figure 6.23. In the kth stage, the output of the SHA is digitized by the FLASH ADC, which produces b_k bits. These very same bits are then converted to an analog signal via a DAC of the same resolution. The output of the DAC is then subtracted from the output of the SHA and after amplification serves as input to the following stage. The amplifier circuit must settle in less than ½ clock

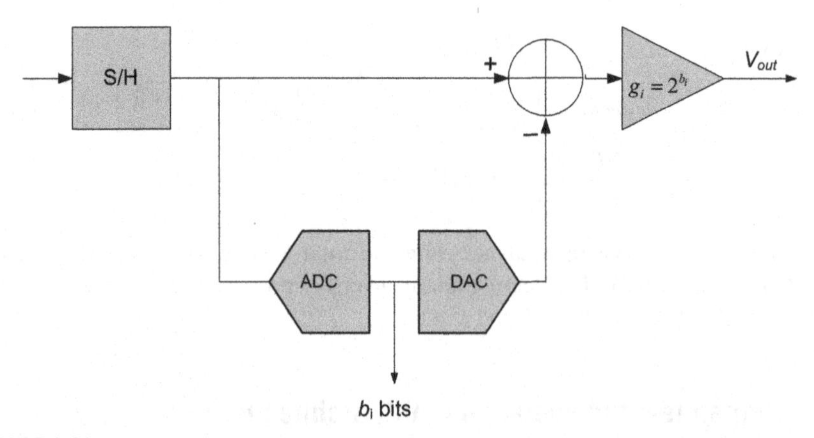

FIGURE 6.23

Single stage of a pipelined ADC.

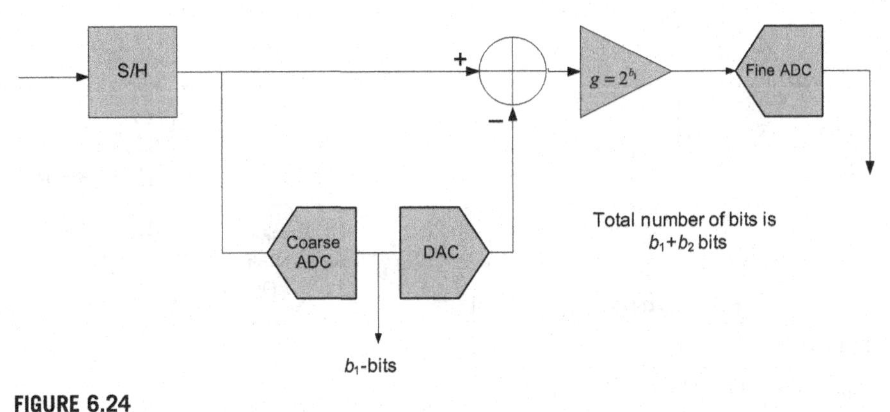

FIGURE 6.24

Subranging two-step ADC architecture.

cycle to the desired gain. The gain itself must be accurate especially in the early stages in order to achieve the desired resolution. To ensure the accuracy of the linear processing of the analog signal, pipelined architectures often rely on calibration techniques to compensate for capacitor mismatches and finite operational amplifier gains [14].

A subranging ADC is comprised of a single-stage circuit containing a *coarse* ADC followed by a *fine* ADC as shown in Figure 6.24. The complexity of a subranging architecture is inferior to that of a FLASH converter. For example, for a 10-bit subranging ADC, assume that the fine ADC is comprised of 6 bits requiring $2^6 - 1 = 63$ comparators and a coarse 4-bit ADC requiring $2^4 - 1 = 15$ comparators resulting in a total of 78 comparators, which is still far less than the 1023 comparators needed for a FLASH converter. In the event where the number of bits is split evenly between the coarse and the fine ADC, the number of comparators is proportional to $2^{b/2}$ where b is the resolution [15]. This results in a significant reduction in die area and power consumption compared to a FLASH converter. The subranging ADC, however, requires a minimum of two clock periods to produce its digital output as compared to the single clock period required by the FLASH converter. The accuracy of this converter depends amongst other things on the accuracy of the gain errors, offset errors, as well as the input-referred noise of the SHA.

6.2.3 Folding ADC architecture

Similar to a subranging ADC, a folding ADC, depicted in Figure 6.25, is comprised of a coarse ADC used to generate the MSBs, a fine ADC used to generate the LSBs, and a folding circuit. The two distinct differences between the subranging ADC and the folding ADC is the absence of a THA circuit as well as the DAC [16]. Both outputs of the coarse and fine ADCs are fed to a bit encoder circuit in order to generate the final digital output. The analog input, however, is sent simultaneously to the

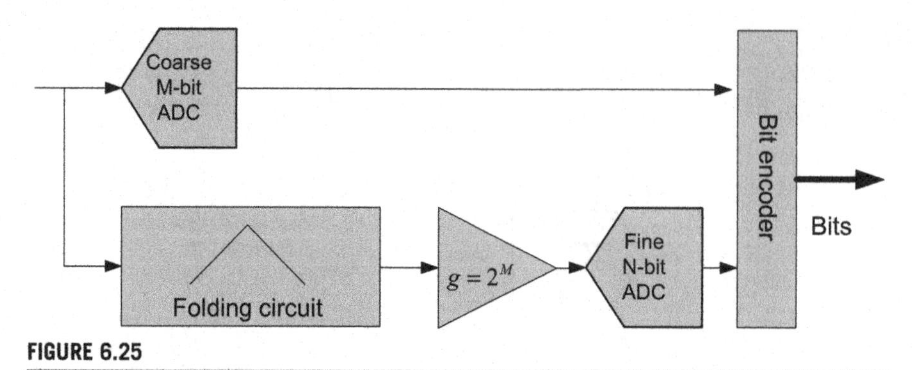

FIGURE 6.25

Folding ADC architecture.

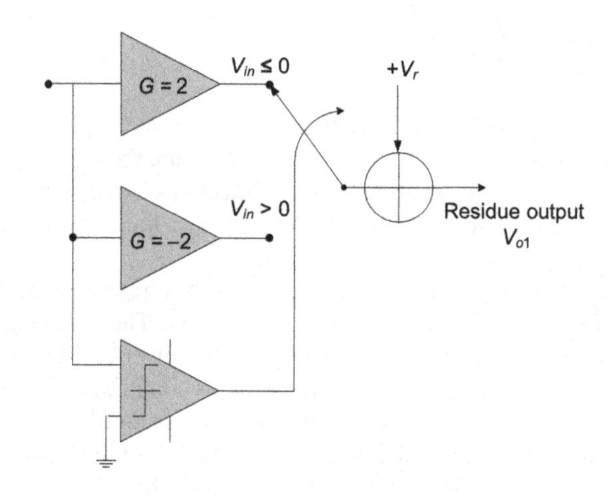

FIGURE 6.26

Basic Gray code folding circuit as presented in [17].

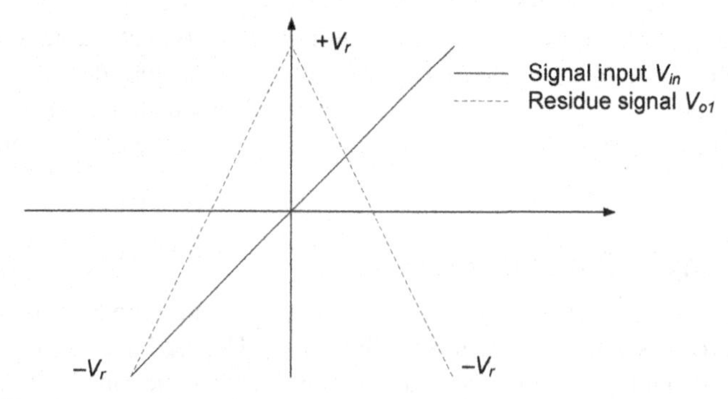

FIGURE 6.27

Output of folding circuit depicted in Figure 6.26. (For color version of this figure, the reader is referred to the online version of this book.)

coarse ADC that generates the MSBs and to the folding circuit. The number of comparators is similar to the number of comparators needed in the subranging ADC, thus resulting in a lower component count than the FLASH converter and smaller die area.

In order to understand the folding operation, consider the folding circuit presented in [17] and depicted in Figure 6.26. Let the input signal V_{in} vary within

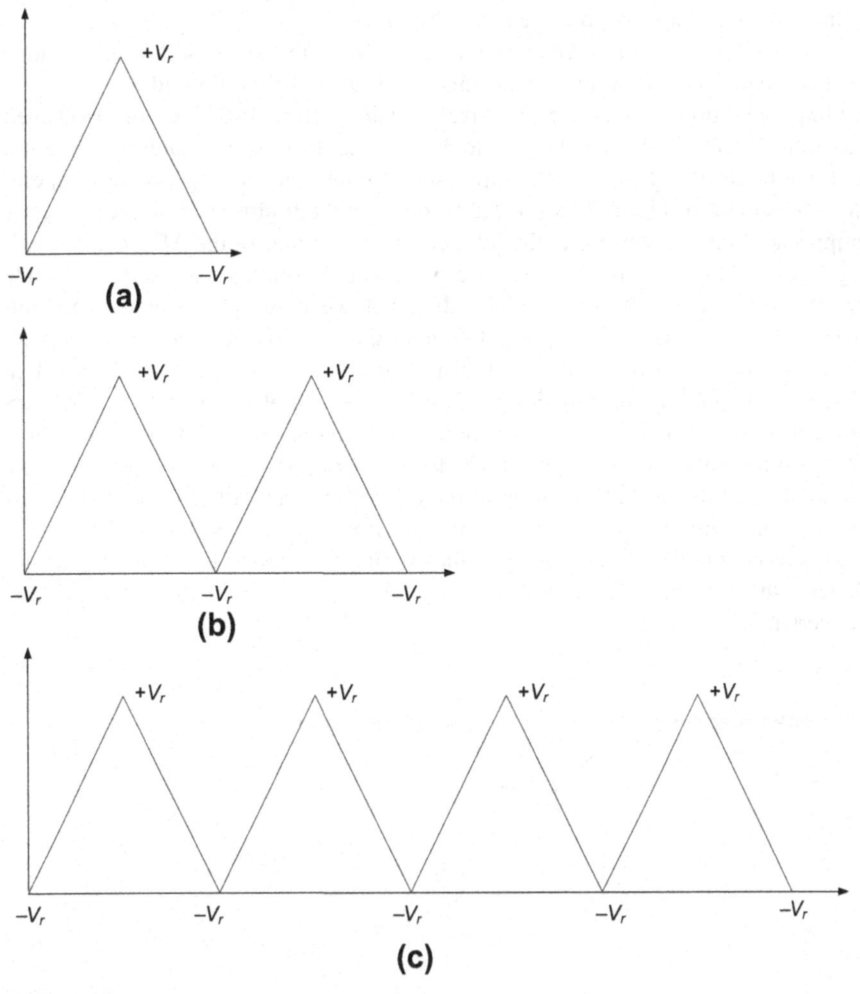

FIGURE 6.28

Folding circuit response with single- and multiple-folding operations: (a) single-folding, (b) double-folding, and (c) quadruple-folding. (For color version of this figure, the reader is referred to the online version of this book.)

the dynamic range $[-V_r, +V_r]$, then in the extreme case where the input signal is $V_{in} = -V_r$, the comparator enables the output of the $G = 2$ amplifier and the residue output signal is simply $V_{o1} = 2V_{in} + V_r = -V_r$. Next consider the other extreme of the dynamic range where the input signal is $V_{o1} = -2V_{in} + V_r = -2V_r + V_r = -V_r + V_r$, the comparator enables the output of the $G = -2$ amplifier and the residue output signal is again $V_{o1} = -2V_{in} + V_r = -2V_r + V_r = -V_r$. Next, assume that the input signal is in the middle of the dynamic range, that is $V_{in} = 0$, then the output of the folding circuit is at its maximum or $V_{o1} = -2V_{in} + V_r = -2 \times 0 + V_r = V_r$. The resulting transfer function is triangular in shape as portrayed in Figure 6.27.

The folding operation takes place at one-half the full-scale voltage range of the ADC. The folding circuit discussed thus far is limited to a single-folding operation and hence produces a single-bit (MSB) output. However, the folding circuit does not have to be limited to a single-folding operation and can be designed to split the input signal range into multiple-folding operations as shown in Figure 6.28. In the case where the folding is doubled, the peak amplitude occurs at one-fourth the full-scale voltage range of the ADC corresponding to two MSBs, whereas in the case of quadruple folding, the peak amplitude occurs at one-eighth the full-scale-folding voltage range corresponding to four MSBs. At this point, it is important to note that the folding operation does not occur without a certain amount of distortion around the edges as depicted in Figure 6.29. Folding circuits designed in MOS or bipolar will exhibit a certain amount of distortion in the form of rounded corners as opposed to the theoretical sharp corners depicted in Figure 6.28, for example. This nonideal response can cause degradation in the resolution of the converter especially if the folding operation occurs multiple times over the entire dynamic range. This type of distortion is more severe at higher frequencies and thus limits the speed of the converter. Nonetheless, the resulting die area is still significantly less than that of a FLASH converter.

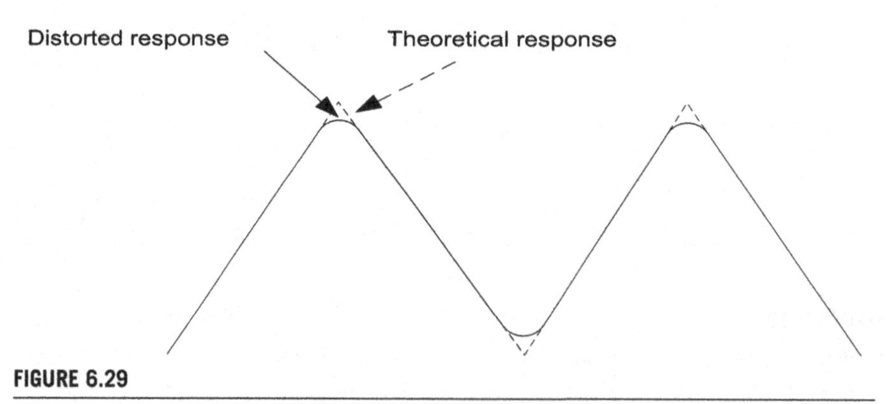

FIGURE 6.29

Theoretical versus actual distorted response of folding circuit.

6.3 ΔΣ converters

Oversampled converters in this chapter refer only to ΔΣ modulators.[b] ΔΣ modulators rely on oversampling and feedback filtering in order to noise-shape the quantization noise away from the desired signal [19]. The aim of noise-shaping is to leave as little quantization noise as possible in the desired signal bandwidth and the rest out of it. The out-of-band noise is then filtered out by a decimation filter as will be explained herein. According to Chapter 5, we define the oversampling ratio OSR as the ratio of the sampling frequency F_s to the Nyquist frequency $F_{Nyquist}$ or $OSR \equiv F_s/F_{Nyquist}$. Typically, ΔΣ modulators, employed in wireless receivers, tend to have a higher OSR than 2. In general, oversampled ADCs have an advantage over their Nyquist counterparts due to the simplified requirements they place on antialiasing filter. While a Nyquist converter may require an antialiasing filter with a sharp transition band, thus causing phase distortion to the desired signal, an oversampled converter typically does not place the same stringent requirements on the filter. In this section, we will discuss the various aspects of ΔΣ modulators varying from analysis to architecture.

6.3.1 The basic loop dynamics

In order to understand the basic workings of oversampling converters, consider the lowpass ΔΣ converter depicted in Figure 6.30. The subscript "*a*" in this context denotes an analog signal to differentiate it from its equivalent discrete signal. The circuit is comprised of four major components namely, a summing element, an analog

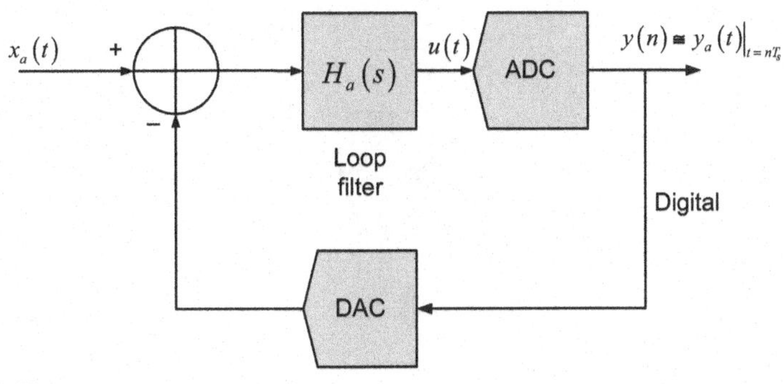

FIGURE 6.30

The basic ΔΣ loop architecture.

[b]In this chapter, ΔΣ modulators will be treated as oversampled converters. There have been techniques, however, such as the one described in [18], where ΔΣ modulators have been implemented in parallel, thus increasing the resolution of the converter and reducing its oversampling rate or completely eliminating it.

loop filter with transfer function $H_a(s) = \mathfrak{L}\{h_a(t)\}$ where $\mathfrak{L}\{\}$ denotes the Laplace operator, and an ADC and a DAC placed in the feedback portion of the loop. Assume that both ADC and DAC are clocked at the sampling frequency F_s. Given that the input signal has much narrower frequency content than the sampling frequency, the converted digital signal can faithfully, with low quantization noise in the desired band, represent the information content present in the input analog signal. It will be shown that the quantization noise increases as a function of frequency due to the highpass shaping nature of the loop [20].

In order to further illustrate the basic functionality of the $\Delta\Sigma$ converter shown in Figure 6.30, let the input signal to the loop filter be comprised of the difference between the input signal $x_a(t)$ and its reconstructed approximation $y_a(t)$. The resulting output $u(t)$ is then digitized via an ADC in order to provide a sampled version of the output signal $y(n)$. To gain further understanding of the role of the loop in the converter, let the ADC-DAC combination be represented simply by its analog output $y_a(t)$, which is the sum of the output of the loop filter $u(t)$ and the analog-equivalent quantization noise of the converter $e_q(t)$. This process is illustrated in Figure 6.31.

Taking the Laplace transform of the output of the loop filter $u(t)$, we obtain the relation

$$U(s) = H_a(s)[X_a(s) - Y_a(s)] \tag{6.51}$$

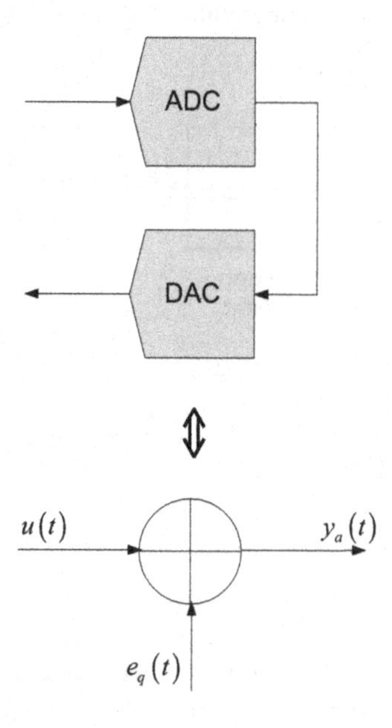

FIGURE 6.31

Simplified analog-equivalent representation of the ADC-DAC circuit.

Furthermore, the output $y_a(t)$ can be written in the Laplace domain as the sum

$$Y_a(s) = U(s) + E_q(s) \tag{6.52}$$

Substitute the value of $U(s)$ found in Eqn (6.52) into Eqn (6.51) in order to obtain the input–output relation:

$$\begin{aligned} U(s) \triangleq Y_a(s) - E_q(s) &= H_a(s)[X_a(s) - Y_a(s)] \Rightarrow Y_a(s) \\ &= \frac{H_a(s)}{1 + H_a(s)} X_a(s) + \frac{1}{1 + H_a(s)} E_q(s) \end{aligned} \tag{6.53}$$

Based on the relationship in Eqn (6.53), we will define two important transfer functions, namely signal transfer function (*STF*) defined as:

$$STF(s) = \frac{H_a(s)}{1 + H_a(s)} \tag{6.54}$$

and the noise transfer function (*NTF*) defined as:

$$NTF(s) = \frac{1}{1 + H_a(s)} \tag{6.55}$$

In Eqn (6.53), the output signal $Y_a(s)$ is the sum of the input signal $X_a(s)$ filtered by the *STF* and the quantization noise $E_q(s)$ filtered by the *NTF*. At this point, in order to simplify the analysis, assume that the loop filter $H_a(s)$ is an integrator filter, that is

$$\begin{aligned} H_a(s) &= \frac{1}{s} \\ |H_a(j\Omega)|^2 &= \frac{1}{\Omega^2} \end{aligned} \tag{6.56}$$

Then the signal transfer function can be expressed as:

$$STF(s) = \frac{H_a(s)}{1 + H_a(s)} = \frac{\frac{1}{s}}{1 + \frac{1}{s}} = \frac{1}{s + 1} \tag{6.57}$$

The magnitude-squared response of $STF(j\Omega)$ can be computed as:

$$|STF(j\Omega)|^2 = \frac{1}{\Omega^2 + 1} \tag{6.58}$$

The *STF* is a lowpass function that satisfies the lowpass filter characteristics at DC and $\Omega \to \infty$, that is

$$\begin{aligned} \lim_{\Omega \to 0} |STF(j\Omega)|^2 &= \lim_{\Omega \to 0} \left(\frac{1}{\Omega^2 + 1} \right) = 1 \\ \lim_{\Omega \to \infty} |STF(j\Omega)|^2 &= \lim_{\Omega \to \infty} \left(\frac{1}{\Omega^2 + 1} \right) = 0 \end{aligned} \tag{6.59}$$

Similarly, the noise transfer function can be expressed by substituting $H_a(s)$ in Eqn (6.56) into the expression in Eqn (6.55), and we obtain

$$NTF(s) = \frac{1}{1+H_a(s)} = \frac{1}{1+\frac{1}{s}} = \frac{s}{s+1} \tag{6.60}$$

The magnitude-squared response of the noise transfer function then becomes

$$|NTF(j\Omega)|^2 = \frac{\Omega^2}{\Omega^2+1} \tag{6.61}$$

The *NTF* is a highpass function that satisfies the characteristics of a highpass filter at DC and $\Omega \to \infty$, that is

$$\begin{aligned} \lim_{\Omega\to 0} |NTF(j\Omega)|^2 &= \lim_{\Omega\to 0}\left(\frac{\Omega^2}{\Omega^2+1}\right) = 0 \\ \lim_{\Omega\to \infty} |NTF(j\Omega)|^2 &= \lim_{\Omega\to \infty}\left(\frac{\Omega^2}{\Omega^2+1}\right) = \lim_{\Omega\to \infty}\left(\frac{1}{1+\frac{1}{\Omega^2}}\right) = 1 \end{aligned} \tag{6.62}$$

This implies that, according to Eqn (6.53), the desired signal $x_a(t) \Leftrightarrow X_a(s)$ is filtered via a lowpass filter whereas the quantization noise $e_q(t) \Leftrightarrow E_q(s)$ of the ADC is filtered by a highpass filter. In the latter case, the noise is said to be "shaped away" from the desired signal. This shaping of the quantization noise implies that its distribution is not uniform; rather, it follows a highpass function, in contrast with the quantization noise present in Nyquist ADCs where the quantization noise is uniformly distributed. This process is illustrated in Figure 6.32 using a first order $\Delta\Sigma$ converter. The input to the converter is a noise-free tone with an oversampling ratio of 8 as illustrated in Figure 6.32 (a). The ADC is simply a single-bit converter or slicer. The output of the first order $\Delta\Sigma$ converter is captured at the output of the ADC as shown in Figure 6.32 (b). The shaping effect can be noted as "pushing" the quantization noise away from the signal, thus minimizing its degradation effect and increasing the SQNR. Increasing the order of the loop pushes the noise further away from the signal and hence increases the SQNR as is shown in Figure 6.33.

The loop filter for the first order continuous-time (CT) $\Delta\Sigma$ converter is comprised of one integrator that can be implemented as an active RC or g_m-C based circuit as shown in Figure 6.34. As a matter of fact, for high-resolution CT $\Delta\Sigma$ converters, a combination of both types of integrators is used with the first integrator being an active RC while the remaining integrators are gm-C as explained in [21]. Both CT integrators realize the basic function $H(s) = 1/\tau s$ where τ is the time constant. The time constant is obtained as the product of the capacitive and resistive components of the integrator. In practical designs, the behavior of CT integrators depends on the circuit design where a certain number of design trade-offs have to be conducted. For example, active RC integrators tend to be more linear than their gm-C counterparts and can afford larger signal swing. On the other hand, gm-C integrators can operate at higher frequencies than active RC integrators [22].

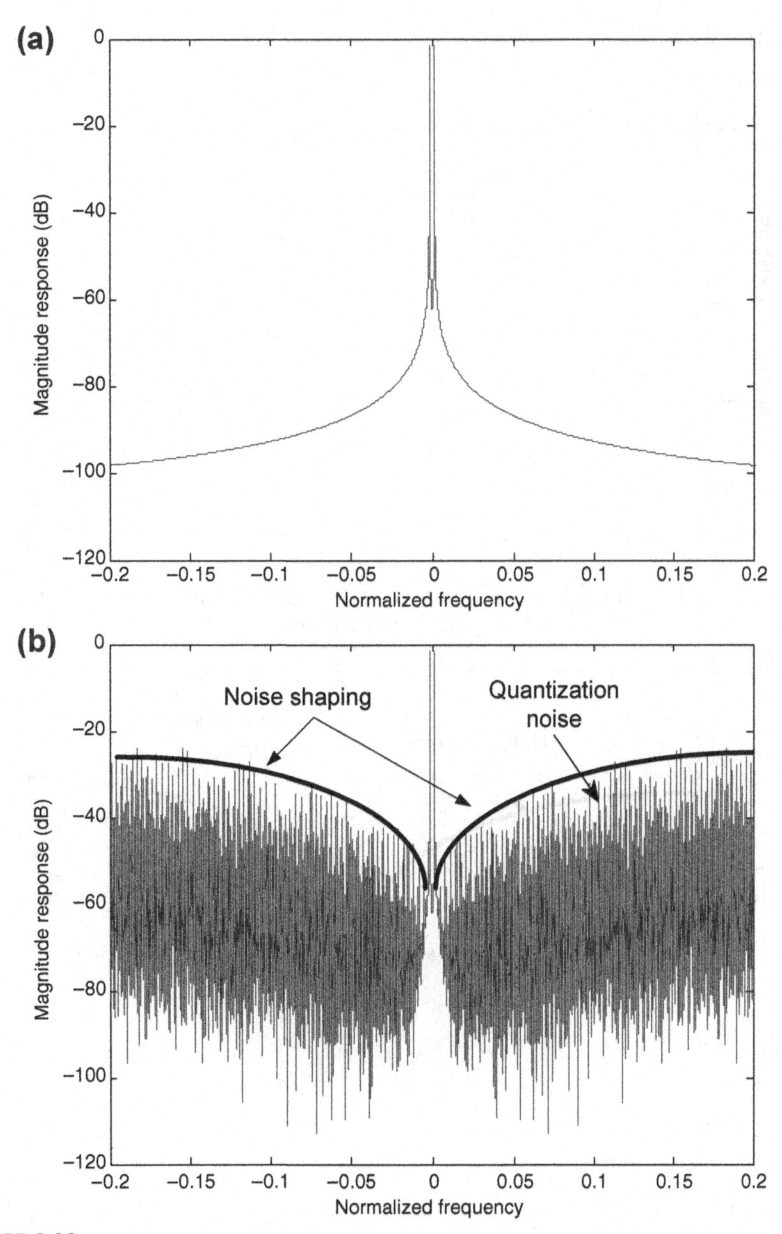

FIGURE 6.32

Quantization noise-shaping in first order ΔΣ converter: (a) single-tone input, and (b) quantization noise-shaping due to first order ΔΣ converter. (For color version of this figure, the reader is referred to the online version of this book.)

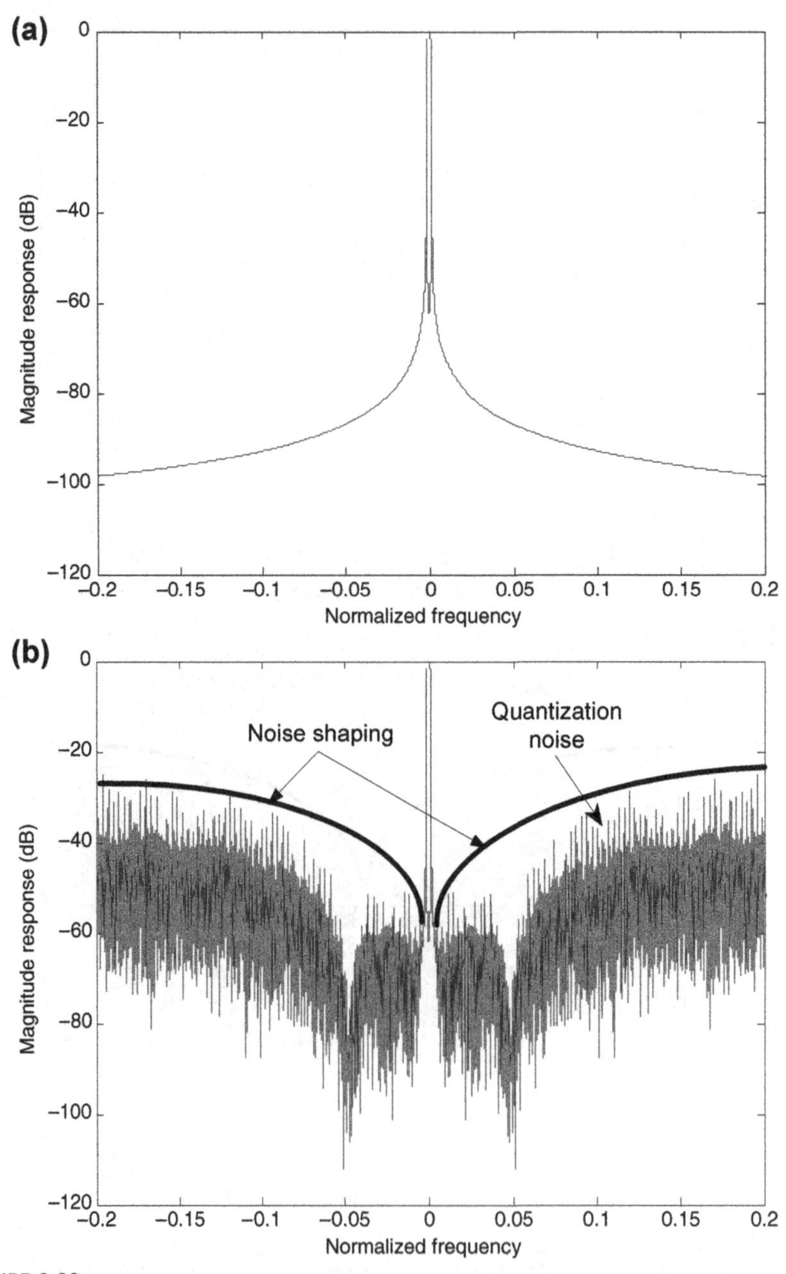

FIGURE 6.33

Quantization noise-shaping in third order $\Delta\Sigma$ converter: (a) single-tone input, and (b) quantization noise-shaping due to third order $\Delta\Sigma$ converter. (For color version of this figure, the reader is referred to the online version of this book.)

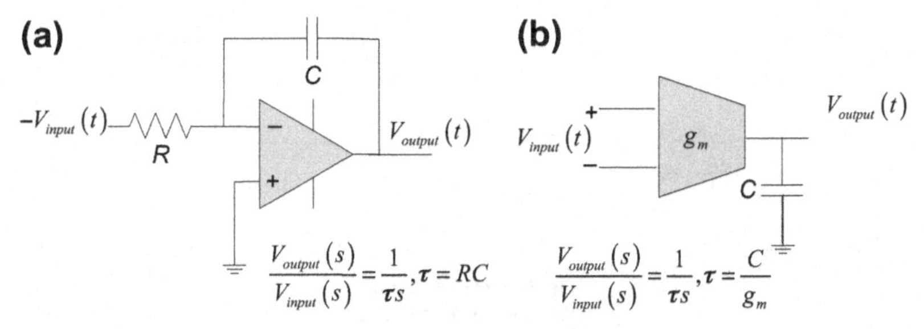

FIGURE 6.34

Basic integrator for continuous-time (CT) ΔΣ converter using: (a) active RC, and (b) gm-C.

6.3.2 Continuous-time versus discrete-time ΔΣ modulators

ΔΣ modulators come in two flavors, namely continuous-time or discrete-time (DT) modulators. Unlike the CT modulator depicted in Figure 6.30, the DT modulator depicted in Figure 6.35 first samples the analog signal via an S/H circuit. The output is then presented to the rest of the loop in discrete-sampled form. This implies that any sampling imperfections, such as sampling clock jitter, due to the S/H circuit are added to the input signal and filtered by the STF and consequently will not undergo any shaping by the NTF. In contrast, in CT-ΔΣ modulators, the sampling process

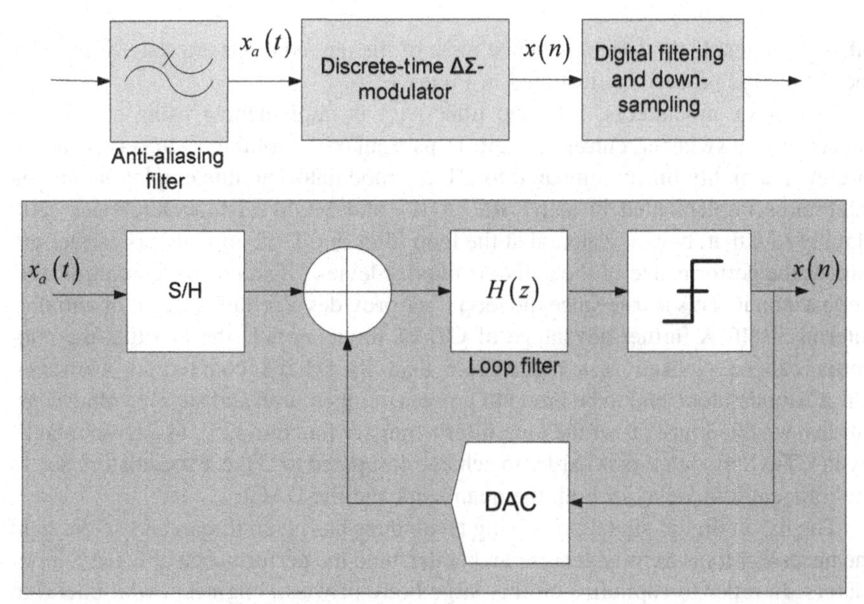

FIGURE 6.35

Basic architecture of DT-ΔΣ modulator. (For color version of this figure, the reader is referred to the online version of this book.)

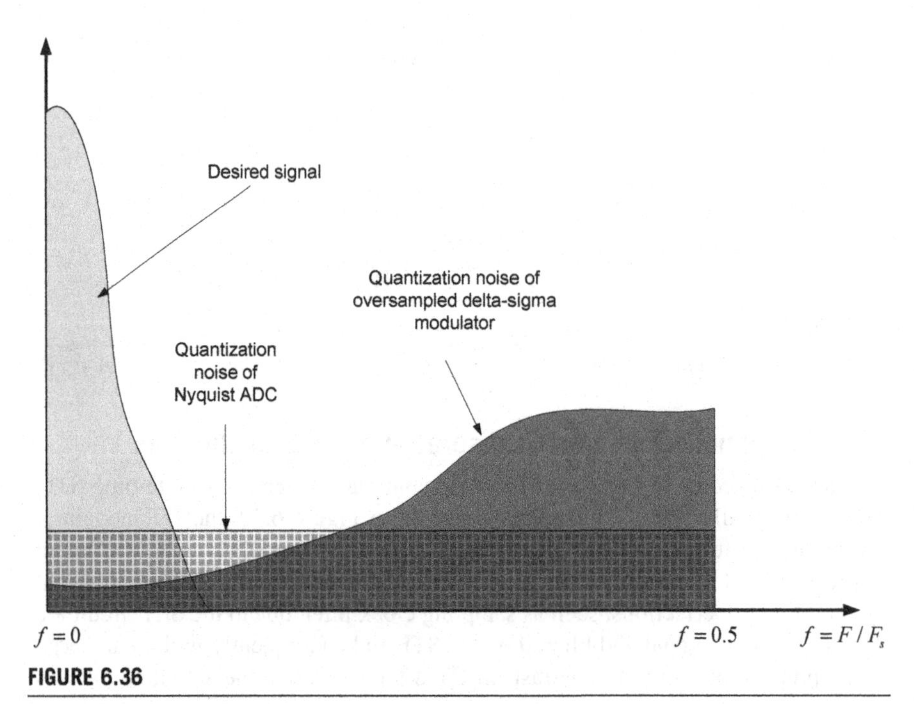

FIGURE 6.36

Noise-shaping in ΔΣ modulators.

takes place inside the loop and hence most of the degradations are shaped away by the NTF. This process is illustrated in Figure 6.36.

In DT-ΔΣ modulators, the loop filter $H(z)$ is implemented using a switched capacitor or a switched current circuit. In particular, monolithic switched capacitor circuits are highly linear compared to CT-ΔΣ modulator circuits employing analog integrators implemented in active RC, g_m-C, and MOSFET-C technologies [20]. Having said that, however, note that the loop filter in CT-ΔΣ modulators further enhances the performance of the antialias filter or levies off some of the requirements imposed on it. This is true since the loop filter provides a certain amount of antialias filtering itself. A further advantage of CT-ΔΣ modulators is the fact that the converter can be clocked at a higher rate than its DT-ΔΣ counterpart. However, CT-ΔΣ modulators tend to be prone to process, temperature, and supply voltage variations, which in turn affect the loop filter's transfer function [23]. Moreover, stability in CT-ΔΣ modulators is harder to achieve compared to DT-ΔΣ modulators due to the inherent loop delay of both the quantizers and the DACs.

The use of digital signal processing techniques has given researchers in the field the necessary tools to fully analyze and understand the performance of DT-ΔΣ modulators. In order to capitalize on this large body of work, engineers who intend to design and implement CT-ΔΣ modulators model their behavior as DT-ΔΣ modulators first. The design is then converted to continuous time and implemented. One such implementation technique requires that the DT-ΔΣ modulator has an identical impulse response as the corresponding CT-ΔΣ modulator at the sampling instant nT_s at the input to the quantizer. This, in turn, will ensure that the outputs of both loops are identical at all sampling instants in response to all input signals.

To further demonstrate this modeling process, consider the CT-ΔΣ modulator shown in Figure 6.37. The analog portion of the circuit is comprised of a DAC, an analog loop filter, and an ADC or bit slicer that all can be modeled as a discrete filter. This assumption is reasonable since the input to the analog circuit or the DAC as well as the output of the analog circuit or ADC are both discrete. Define $H(z)$ as the equivalent discrete-time filter that satisfies the relation

$$Z^{-1}\{H(z)\} \cong \mathfrak{L}^{-1}\{H_a(s)D_a(s)\}\big|_{t=nT_s} \quad \text{for all } t \tag{6.63}$$

where $D_a(s)$ is the DAC's linear impulse response and $Z^{-1}\{.\}$ is the inverse Z-transform operator. The implication of Eqn (6.63) is that the sampled response of $NTF(s)$ when sampled at nT_s is identical to the discrete response $NTF(z)$. The relationship in Eqn (6.63) implies that the discrete-time impulse response $h(n)$ of $H(z)$ is the sampled version of the convolution between the continuous-time loop filter $h_a(t)$ and the DAC linear response $d_a(t)$ at the sampling instance nT_s or

$$h(n) \cong d_a(t) * h_a(t)\big|_{t=nT_s} = \int_{-\infty}^{\infty} d_a(\tau)h_a(t-\tau)d\tau \Big|_{t=nT_s} \tag{6.64}$$

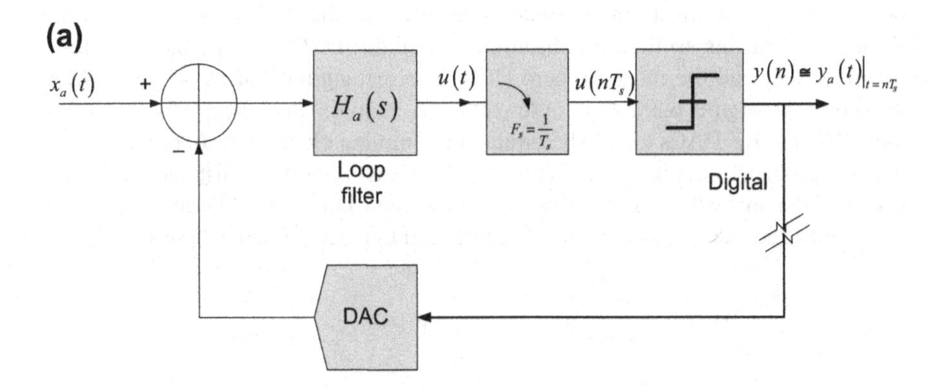

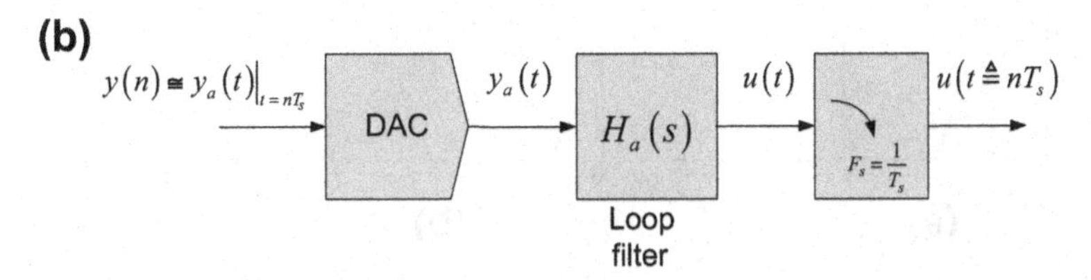

FIGURE 6.37

CT-ΔΣ modulator showing (a) closed-loop modulator, and (b) open-loop analog circuit of the modulator that can be modeled as a discrete circuit.

where the mapping between the CT and DT domains is simply the impulse invariance mapping. Expressing analog impulse response as a parallel fraction expansion as discussed in [24], we obtain

$$H_a(s) = \sum_{k=1}^{N} \frac{\alpha_k}{s - p_k} \Leftrightarrow h_a(t) = \sum_{k=1}^{N} \alpha_k e^{p_k t} \quad \text{for } t \geq 0 \tag{6.65}$$

where $\{\alpha_k(t)\}$ are the coefficients of the analog loop filter $h_a(t)$ and $\{p_k(t)\}$ are its poles. The equivalent discrete-time filter can then be shown as

$$H(z) = \sum_{k=1}^{N} \frac{\alpha_k}{1 - e^{p_k T_s} z^{-1}}, \quad \text{where } z_k = e^{p_k T_s} \quad \text{for } k = 1, 2, \cdots, N \tag{6.66}$$

Note that the mapping of the poles via the relationship $z_k = e^{p T}_{ks}$ does not imply that the zeros in the analog and digital domains conform to the same relationship, thus implying that the equivalence model produces satisfactory results only when the sampling rate is sufficiently high. Finally, in order to find the equivalent DT transfer function, we must also model the analog transfer function of the DAC. This topic is treated in the next section.

6.3.3 Linear DAC models

Thus far, according to Eqn (6.63), in order to obtain the equivalent discrete-time loop filter $H(z)$, we must obtain, in addition to $H_a(s)$, the analog transfer function of the DAC. In this section, we discuss two popular DAC models: the nonreturn to zero (NRZ) and the return to zero (RZ). The rectangular pulses of both DACs are shown in Figure 6.38. It is noteworthy to mention that the performance of both NRZ and RZ DACs tend to be limited by sampling clock jitter. In order to alleviate the problem, circuit designers resort to DACs with more sophisticated impulse responses that make them more tolerant to clock jitter. This added tolerance comes at the cost of added design complexity. The time and Laplace domain responses of both

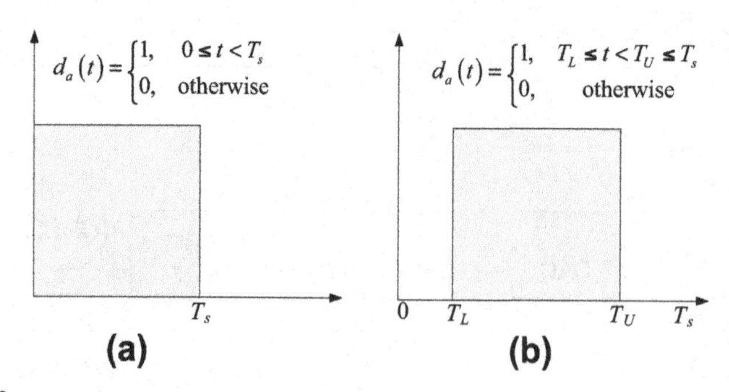

FIGURE 6.38

DAC rectangular pulse response of: (a) NRZ DAC, and (b) RZ DAC. (For color version of this figure, the reader is referred to the online version of this book.)

Table 6.2 Time and Laplace Domain Responses of NRZ and RZ DACs

DAC	Time-Domain Response	Laplace-Domain Response
NRZ	$d_{a,NRZ}(t) = \begin{cases} 1 & 0 \leq t < T_s \\ 0 & \text{otherwise} \end{cases}$	$D_{a,NRZ}(s) = \dfrac{1 - e^{-sT_s}}{s}$
RZ	$d_{a,RZ}(t) = \begin{cases} 1 & T_L \leq t < T_U \leq T_s \\ 0 & \text{otherwise} \end{cases}$	$D_{a,RZ}(s) = \dfrac{e^{-sT_L}(1 - e^{-sT_U})}{s}$

DACs are given in Table 6.2. Unlike the NRZ DAC, which remains constant over the entire sampling period, the RZ DAC's response is nonzero only between a lower bound T_L and an upper bound T_U as defined in Table 6.2. In the event where $T_L = 0$ and $T_U = T_s/2$, the DAC is simply referred to as RZ DAC. On the other hand, if $T_L = T_s/2$ and $T_U = T_s$, in this case the DAC is referred to as half-delay RZ DAC. In lowpass ΔΣ modulator designs, RZ DACs are less vulnerable to nonlinearities as clearly shown in [25,26].

Next we analyze a CT-ΔΣ modulator with an NRZ DAC in the feedback as illustrated in the linear model shown in Figure 6.39. The intent is to examine the loop and derive an equivalent DT ΔΣ modulator. Substituting the NRZ DAC transfer function given in Table 6.2 into the relationship in Eqn (6.63), we obtain the DT impulse response of $h(nT_s)$ as

$$h(t = nT_s) = Z^{-1}\{H(z)\} = L^{-1}\left\{\frac{1 - e^{sT_s}}{s} H_a(s)\right\}\Bigg|_{t=nT_s} \tag{6.67}$$

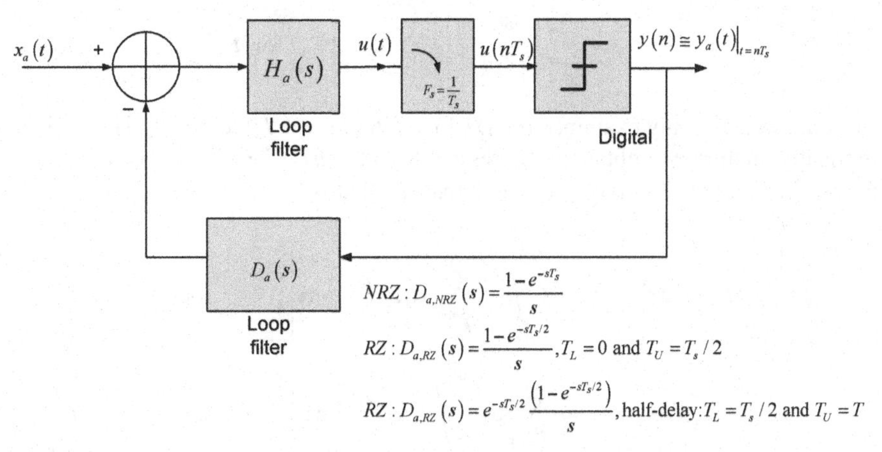

FIGURE 6.39

CT-ΔΣ modulator linear model with NRZ or RZ DAC in the feedback path.

Assume that $d_a(t)$ is a pulse with width T_s. Then expanding the relationship in Eqn (6.67) as a time-domain convolution between $h_a(t)$ and the NRZ time domain response $d_a(t)$ we obtain the pulse response of the CT domain loop as

$$\begin{aligned} h(t = nT_s) &= h_a(t) * \mathrm{d}_a(t)|_{t=nT_s} = \left.\int_{-\infty}^{\infty} h_a(t-\tau)\mathrm{d}_a(\tau)\mathrm{d}\tau\right|_{t=nT_s} \\ &= \left.\int_0^{T_s} h_a(t-\tau)\mathrm{d}\tau\right|_{t=nT_s} \end{aligned} \tag{6.68}$$

Next, substitute the analog filter expression for $h_a(t)$ found in Eqn (6.65) into Eqn (6.68), that is

$$\begin{aligned} \left.\int_0^{T_s} h_a(t-\tau)\mathrm{d}\tau\right|_{t=nT_s} &= \left.\int_0^{T_s} \sum_{k=1}^{N} \alpha_k e^{p_k(t-\tau)}\,\mathrm{d}\tau\right|_{t=nT_s} && \text{for } t > T_s \\ &= \left.\sum_{k=1}^{N} \alpha_k e^{p_k t} \int_0^{T_s} e^{-p_k \tau}\mathrm{d}\tau\right|_{t=nT_s} && \text{for } t \geq T_s \end{aligned} \tag{6.69}$$

The integral inside Eqn (6.69) can be evaluated and its value substituted into Eqn (6.69), resulting in the closed form solution

$$\begin{aligned} h(t = nT_s) &= \left.\sum_{k=1}^{N} \frac{\alpha_k e^{p_k t}}{-p_k}\left(e^{-p_k T_s} - 1\right)\right|_{t=nT_s} && \text{for } t \geq T_s \\ &= \sum_{k=1}^{N} \frac{\alpha_k e^{p_k n T_s}}{-p_k}\left(e^{-p_k n T_s} - 1\right) && \text{for } t \geq T_s \end{aligned} \tag{6.70}$$

Thus far, we have only examined the DT loop filter $h(t = nT_s)$ for $t \geq T_s$. However, it is equally important to obtain an expression for $h(t = nT_s)$ for $t < T_s$. To do so, reexamine the integral Eqn (6.69) for t less than T_s, that is

$$\begin{aligned} \left.\int_0^{0 \leq t < T_s} h_a(t-\tau)\mathrm{d}\tau\right|_{t=nT_s} &= \left.\int_0^{0 \leq t < T_s} \sum_{k=1}^{N} \alpha_k e^{p_k(t-\tau)}\,\mathrm{d}\tau\right|_{t=nT_s} && \text{for } 0 \leq t < T_s \\ &= \left.\sum_{k=1}^{N} \alpha_k \int_0^{0 \leq t < T_s} e^{p_k(t-\tau)}\mathrm{d}\tau\right|_{t=nT_s} && \text{for } 0 \leq t < T_s \end{aligned} \tag{6.71}$$

For $0 \le t < T_s$, evaluate $h(t = nT_s)$ for $n = 0$

$$h(0) \cong \left. \int_0^t \sum_{k=1}^{N} \alpha_k e^{p_k(t-\tau)} \, d\tau \right|_{\substack{t = nT_s \\ t < T_s \Rightarrow n = 0}}$$

$$= \left. \sum_{k=1}^{N} \frac{\alpha_k e^{p_k t}}{-p_k} \left(e^{-p_k t} - 1\right) \right|_{\substack{t = nT_s \\ n = 0}} \quad \text{for } 0 \le t < T_s$$

$$= \left. \sum_{k=1}^{N} \frac{\alpha_k e^{p_k n T_s}}{-p_k} \left(e^{-p_k n T_s} - 1\right) \right|_{n=0} = 0 \quad \text{for } 0 \le t < T_s \tag{6.72}$$

Note that the DT response is only valid for $n = 0$ where it was stated that the upper bound for the integral is t, which itself is bounded within the interval $0 \le t < T_s$. Within these bounds, it is clear from Eqn (6.72) that $h(0) = 0$, simply implying that the first sample in the loop pulse response is zero. This indicates the existence of a delay in the numerator of the pulse-invariant transfer function as clearly shown in [27]. Using the Z-transform, the discrete-time loop filter becomes:

$$H(z) = \sum_{n=-\infty}^{\infty} h(n) z^{-n} = \sum_{n=1}^{\infty} \left\{ \sum_{k=1}^{N} \frac{\alpha_k e^{p_k n T_s}}{-p_k} \left(e^{-p_k T_s} - 1\right) \right\} z^{-n} \tag{6.73}$$

Further manipulation of the summation around n results in the simplified expression

$$H(z) = \sum_{k=1}^{N} \left\{ \frac{\alpha_k}{-p_k} \left(e^{-p_k T_s} - 1\right) \sum_{n=1}^{\infty} \left(e^{p_k T_s} z^{-1}\right)^n \right\} = \sum_{k=1}^{N} \frac{\alpha_k}{-p_k} \frac{\left(1 - e^{p_k T_s}\right) z^{-1}}{1 - e^{p_k T_s} z^{-1}} \tag{6.74}$$

Thus far, the expression in Eqn (6.74) does not directly deal with collocated poles such as the ones encountered in second order lowpass ΔΣ modulators. For example, consider the two collocated poles:

$$p_{k,discrete}(z) = \frac{\beta_k}{\left(1 - e^{p_k T_s} z^{-1}\right)^2} \tag{6.75}$$

In Eqn (6.75), β_k is the coefficient for collocated double poles equivalent to α_k as presented in Eqn (6.70). In the CT domain, the equivalent expression for the double collocated poles is given as

$$p_{k,continuous}(s) = \beta_k \left\{ \frac{\left(1 - e^{-p_k T_s} - p_k T_s\right) \frac{s}{T_s} + p_k^2}{(s - p_k)^2 \left(1 - e^{-p_k T_s}\right)^2} \right\} \tag{6.76}$$

A close examination of Eqn (6.76) shows that there exist two collocated double poles. The work in [28] offers a complete treatment of transfer functions involving more than two collocated poles.

Next we derive an expression equivalent to Eqn (6.74) for the RZ DAC. To do so, replace the zero-order hold expression for the NRZ DAC with $D_{a,RZ}(s)$ for RZ DAC as found in Table 6.2. Note that, in the case of an RZ DAC, the integration limits in Eqn (6.69) is the interval $[0, T_s/2]$ whereas for the half-delay RZ DAC the integration limit is simply $[T_s/2, T_s]$. This in turn implies that $H(z)$ can be expressed as

$$H(z) = \sum_{k=1}^{N} \left[\frac{-\alpha_k e^{p_k T_s/2}}{p_k} \frac{\left(1 - e^{p_k T_s/2}\right) z^{-1}}{1 - e^{p_k T_s} z^{-1}} \right] \tag{6.77}$$

for the RZ case and

$$H(z) = \sum_{k=1}^{N} \left[\frac{-\alpha_k}{p_k} \frac{\left(1 - e^{p_k T_s/2}\right) z^{-1}}{1 - e^{p_k T_s} z^{-1}} \right] \tag{6.78}$$

for the half-delay RZ DAC.

Similar to Eqn (6.74), the two expressions for $H(z)$ in Eqns (6.77) and (6.78) do not address collocated poles that arise in second order $\Delta\Sigma$ modulators, for example, for the RZ and half-delay RZ cases. In this chapter, we state that for the simple RZ DAC [28], the pulse transformation can be expressed as

$$p_{k,continuous}(s) = \beta_k e^{-p_k T_s} \left\{ \frac{\left(1 - e^{-p_k T_s/2} - p_k T_s\left(1 - \frac{1}{2} e^{-p_k T_s/2}\right)\right) \frac{s}{T_s} + p_k^2 \left(1 - \frac{1}{2} e^{-p_k T_s/2}\right)}{(s - p_k)^2 \left(1 - e^{-p_k T_s/2}\right)^2} \right\} \tag{6.79}$$

In a similar vein, we state that for half-delay DAC [28], the pulse transformation can be expressed as

$$p_{k,continuous}(s) = \beta_k e^{-p_k T_s} \left\{ \frac{\left(1 - 0.5 p_k T_s - e^{-p_k T_s/2}\right) \frac{s}{T_s} + 0.5 p_k^2}{(s - p_k)^2 \left(1 - e^{-p_k T_s/2}\right)^2} \right\} \tag{6.80}$$

Obviously, similar expressions for pulse transformation can be found for higher number of collocated poles. Hence, one can obtain similar CT loop filter expressions for higher order loop analysis.

6.4 Performance analysis of $\Delta\Sigma$ modulators

In this section, the basic performance of the $\Delta\Sigma$ modulator is presented, beginning by introducing the first order modulator and then discussing higher order modulators. The section concludes by discussing MASH converter architectures.

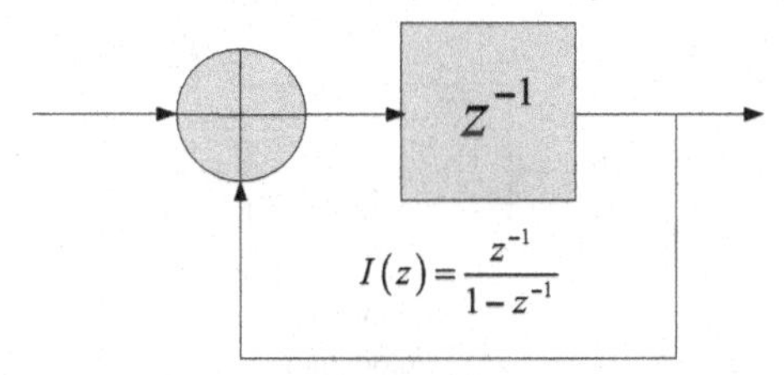

FIGURE 6.40

Loop filter comprised of single-pole integrator function for first order modulator.

6.4.1 First order ΔΣ modulators

The first order ΔΣ modulator, shown in Figure 6.42, uses a single integrator $I(z)$ as a loop, as shown in Figure 6.40, defined as

$$I(z) = \frac{z^{-1}}{1 - z^{-1}} \tag{6.81}$$

The loop filter implemented as a switched capacitor (SC) integrator, in a slightly modified form for $I(z) = C_1 z^{-1}/C_2(1 - z^{-1})$, is shown in Figure 6.41. In this analysis, we will assume the quantization noise to be additive and signal independent, thus making the loop simple to analyze and allowing us to derive key performance parameters. However, it is important to note that this straightforward linear model could lead to serious flaws in the analysis.

Let $U(z)$ and $E_q(z)$ be the output of the loop filter and additive-quantization noise, respectively, then the output of the loop $Y(z)$ can be expressed as

$$Y(z) = U(z) + E_q(z) \tag{6.82}$$

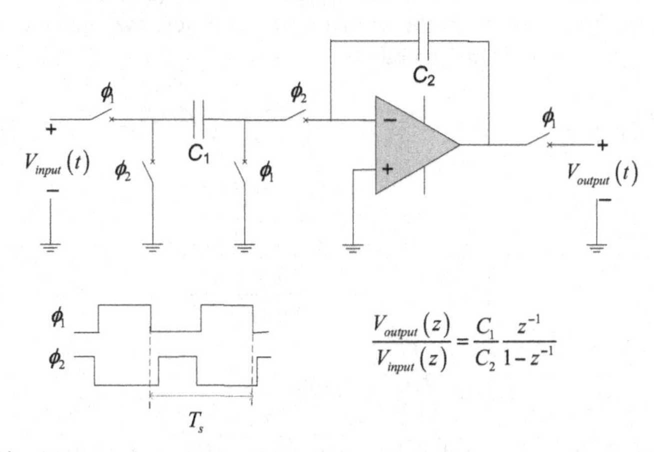

FIGURE 6.41

Noninverting SC integrator employed in DT-ΔΣ modulator. (For color version of this figure, the reader is referred to the online version of this book.)

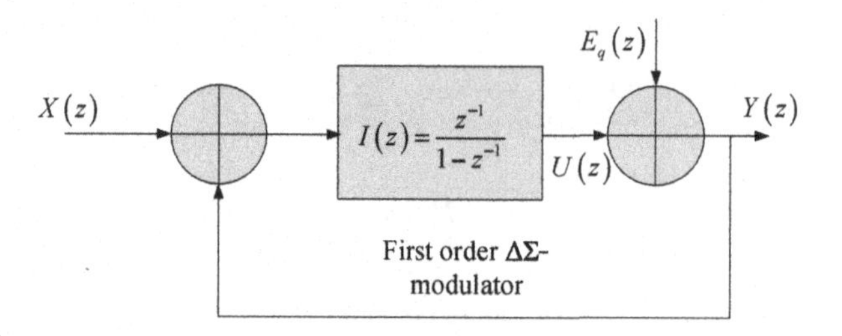

FIGURE 6.42

First order DT-ΔΣ modulator.

The output of the loop filter $U(z)$ can be further expressed in terms of the input signal $X(z)$ and the loop's output signal $Y(z)$

$$U(z) = \frac{z^{-1}}{1-z^{-1}}\{X(z) - Y(z)\} \tag{6.83}$$

Substituting Eqn (6.83) into Eqn (6.82), we obtain

$$\begin{aligned} Y(z) &= \frac{z^{-1}}{1-z^{-1}}\{X(z) - Y(z)\} + E_q(z) \\ \Rightarrow Y(z) &= z^{-1}X(z) + (1 - z^{-1})E_q(z) \\ \Rightarrow STF(z) &= z^{-1} \text{ and } NTF(z) = 1 - z^{-1} \end{aligned} \tag{6.84}$$

The relationship in Eqn (6.84) implies that the *STF*(.) is an all-pass filter in the form of a simple delay that allows the input signal to pass through the loop basically unchanged. On the other hand, the $NTF(z) = 1 - z^{-1}$ is a highpass filter that impacts the quantization noise only. The *NTF*'s magnitude and phase responses are depicted in Figure 6.43. Analytically, according to Eqn (6.84), $NTF(z)$ has a null at DC. That is, for $z = exp(j2\pi f)$, where f is the normalized frequency, the squared magnitude response of $1 - z^{-1}$ can be computed as

$$\begin{aligned} \left|NTF\left(e^{j2\pi f}\right)\Big|_{f=F/F_s}\right|^2 &= \left|1 - e^{-j2\pi f}\right|^2 = \left|e^{-j2\pi f/2}\left(e^{j2\pi f/2} - e^{-j2\pi f/2}\right)\right|^2 \\ &= \left|2je^{-j2\pi f/2}\underbrace{\frac{\left(e^{j2\pi f/2} - e^{-j2\pi f/2}\right)}{2j}}_{=\sin(\pi f)}\right|^2 \\ &= \left|2je^{-j2\pi f/2}\sin(\pi f)\right|^2 \end{aligned} \tag{6.85}$$

In this case, it is obvious according to Eqn (6.85) and Figure 6.43 that the quantization noise that exists between DC and the sampling frequency is shaped by a

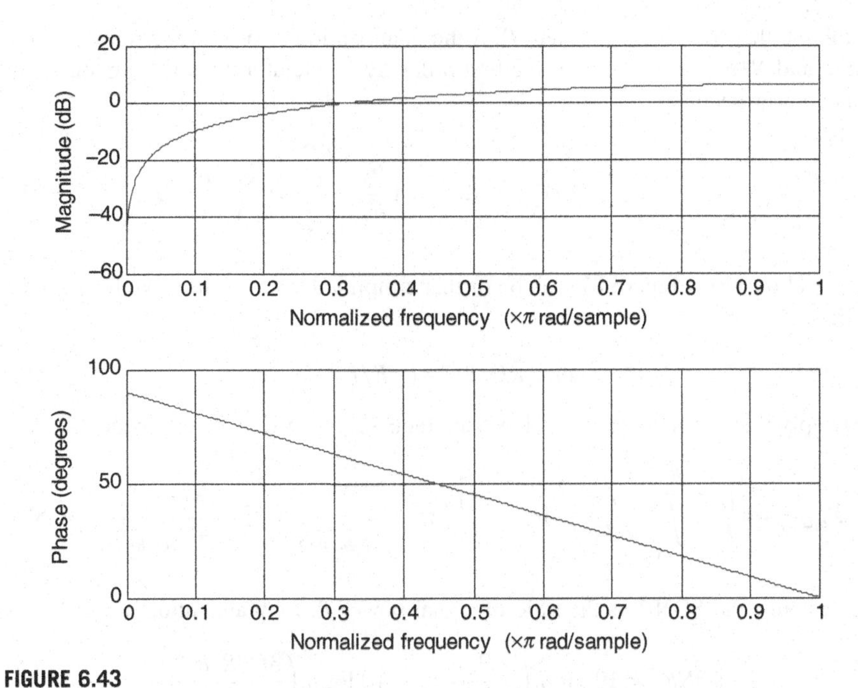

FIGURE 6.43

Magnitude and phase response versus normalized frequency of $NTF(z) = 1 - z^{-1}$. (For color version of this figure, the reader is referred to the online version of this book.)

highpass function that must be taken into account when computing the in-band degradation due to quantization noise.

The linear model for the first order ΔΣ modulator predicts that the loop is unconditionally stable. This is true provided that the quantizer can be modeled via a linear model. In reality, however, the quantizer may exhibit strong nonlinear behavior that must be accounted for. Furthermore, a DC signal input to the loop such that $|x(n)| =$ constant > 1, or even a slowly varying signal, causes the output of the integrator to *theoretically* grow without bound. In practice, however, both analog and digital implementation of the integrator will cause the output to saturate. In contrast, if the input signal is an irrational-bounded DC signal such that $|x(n)| =$ constant < 1, the output sequence of the loop becomes quasiperiodic.

In order to compute the SQNR for the first order ΔΣ modulator, we must compute the quantization noise power $P_{e,\Delta\Sigma_1}$ defined as the integral of the power spectral density $S_e(f)$ shaped by the $NTF(.)$, or

$$P_{e,\Delta\Sigma_1} = \int_{-B/2}^{B/2} S_e(F)|NTF(F)|^2 \mathrm{d}F \tag{6.86}$$

Realizing that $S_e(f) = P_e/F_s$ where P_e is the quantization error power assumed to be white and $NTF(z) = 1 - z^{-1}$ for the first order $\Delta\Sigma$ modulator, then the quantization noise power becomes

$$P_{e,\Delta\Sigma_1} = \frac{P_e}{F_s}\int_{-B/2}^{B/2} |NTF(F)|^2 dF = 4\frac{P_e}{F_s}\int_{-B/2}^{B/2} \sin^2\left(\frac{\pi F}{F_s}\right) dF \tag{6.87}$$

The relationship in Eqn (6.87) can be further simplified if we assume that for a given OSR such that

$$\sin^2(\pi F/F_s) \approx (\pi F/F_s)^2 \tag{6.88}$$

or simply that the ratio $F/F_s << 1$ is true, then $P_{e,\Delta\Sigma_1}$ in Eqn (6.87) becomes

$$P_{e,\Delta\Sigma_1} \approx 4\frac{P_e}{F_s}\int_{-B/2}^{B/2} \left(\frac{\pi F}{F_s}\right)^2 dF = \frac{4}{3}\frac{P_e}{F_s^3}\pi^2 F^3\Big|_{F=-B/2}^{F=B/2} = \frac{1}{3}\frac{P_e \pi^2}{OSR^3}\Big|_{OSR=\frac{F_s}{B}} \tag{6.89}$$

Next, define the SQNR as the ratio of signal power P_x to quantization power

$$SQNR = 10\log_{10}\left(\frac{P_x}{P_{e,\Delta\Sigma_1}}\right) = 10\log_{10}\left(\frac{3OSR^3 P_x}{P_e \pi^2}\right) \tag{6.90}$$

The relationship in Eqn (6.90) can be further expanded as

$$SQNR = 10\log_{10}\left(\frac{3}{\pi^2}\right) + 30\log_{10}(OSR) + 10\log_{10}\left(\frac{P_x}{P_e}\right) \tag{6.91}$$

Recall from Chapter 5 that

$$10\log_{10}\left(\frac{P_x}{P_e}\right) = 10\log_{10}\left(\frac{12\sigma_x^2}{\delta^2}\right) = 10\log_{10}\left(\frac{12 \times 2^{2b} \times \sigma_x^2}{V_{x,Fs}^2}\right)$$
$$= 10.8 + 6.0206b + 10\log_{10}\left(\frac{\sigma_x^2}{V_{x,FS}^2}\right) \tag{6.92}$$

Then substituting Eqn (6.92) into Eqn (6.91), the $SQNR$ can be then be expressed as

$$SQNR = 4.7712 + 6.0206b + 10\log_{10}\left(\frac{3}{\pi^2}\right) + 30\log_{10}(OSR) - PAPR \tag{6.93}$$

where again the PAPR (peak-to-average-power ratio) is defined according to Chapter 5 as

$$PAPR = 10\log_{10}\left(\frac{P_{peak}}{P_s}\right) = 10\log_{10}\left(\frac{V_{x,FS}^2}{4\sigma_x^2}\right) = 10\log_{10}\left(\frac{1}{\kappa^2}\right) \tag{6.94}$$

According to Eqn (6.93), doubling the OSR results in an increase of 9.03 dB in *SQNR*. Recall that doubling the OSR in a Nyquist converter resulted roughly in 3-dB improvement in *SQNR*. This dramatic impact on performance will be further emphasized in the coming sections.

EXAMPLE 6.3 DC INPUT SIGNALS AND LIMIT CYCLES IN ΔΣ MODULATORS

Consider the ΔΣ modulator depicted in Figure 6.44. Demonstrate that if the output of the integrator is periodic with period T, that is $u(n+T) = u(n)$ and the input signal is DC, or $x(n) = c$, then the DC input signal is a rational number.

From Figure 6.44, the input signal $x(n)$ and output signal $y(n)$ can be related to loop error signal $v(n)$ as

$$v(n) = x(n) - y(n) \tag{6.95}$$

Furthermore, the output of the integrator can be expressed in terms of $v(n)$ and $u(n)$ as

$$u(n) = v(n-1) - u(n-1) \tag{6.96}$$

Solving for $v(n-1)$ in Eqn (6.96) and equating it to $v(n-1)$ in Eqn (6.95) we obtain the simple relationship

$$x(n-1) - y(n-1) = u(n) - u(n-1) \tag{6.97}$$

Note that since $u(n)$ is periodic with period T, then $u(n+T) - u(n) = 0$. According to Eqn (6.97), we can write

$$\begin{aligned} u(T) - u(0) &= [u(T) - u(T-1)] + [u(T-1) - u(T-2)]\cdots + [u(1) - u(0)] \\ &= \underbrace{\sum_{l=0}^{T-1} x(l)}_{=Tc} - \sum_{l=0}^{T-1} y(l) \end{aligned} \tag{6.98}$$

Due to the periodic nature of $u(n)$, we can claim that

$$TC - \sum_{l=0}^{T-1} y(l) = 0 \Rightarrow C = \frac{1}{T}\sum_{l=0}^{T-1} y(l) \tag{6.99}$$

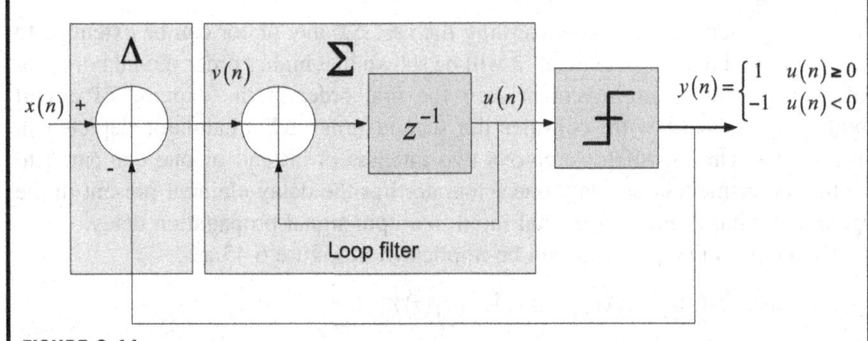

FIGURE 6.44

First order simplified ΔΣ modulator.

Continued

EXAMPLE 6.3 DC INPUT SIGNALS AND LIMIT CYCLES IN ΔΣ MODULATORS—cont'd

Recall that the output of the slicer is defined as:

$$y(n) = \begin{cases} 1 & u(n) \geq 0 \\ -1 & u(n) \langle 0 \end{cases} \tag{6.100}$$

Then, the DC input $x(n) = C$ can be expressed as the ratio

$$C = \frac{1}{T}\sum_{l=0}^{T-1} y(l) = \frac{P-Q}{T}, \ P, Q \text{ are integers} \tag{6.101}$$

Since both P and Q are integers, so is their difference $P-Q$. Hence, the ratio $(P-Q)/T$ is a rational number. The periodic sequence with period T is known as a limit cycle. A limit cycle manifests itself at the output of the modulator in the form of idle tones, which in turn depend on the DC input signal. These tones do not cause instability in the loop and could be filtered out provided they fall outside the desired signal bandwidth. The tones are present at F_s/T and any subsequent harmonics.

Next, assume that the ratio in Eqn (6.101) can be written as

$$\frac{P-Q}{T} = \frac{\lambda}{\rho}, \ \text{gcd}\ (\lambda, \rho) = 1 \ \text{and} \ \lambda < \rho \Rightarrow \lambda \ \text{and} \ \rho \text{ are relative primes} \tag{6.102}$$

The period $T = 2\rho$ if ρ or λ is even and $T = \rho$ if both ρ and λ are odd [29]. Furthermore, there are in-band and out-of-band spectral components at the output of the modulator:

$$\begin{aligned} f_{ib,n} &= \frac{n\lambda}{2\rho} F_s, n = 1, 2, \ldots \text{for in}-\text{band tones} \\ f_{ob,n} &= \frac{n(\rho-\lambda)}{2\rho} F_s, n = 1, 2, \ldots \text{for out of band tones} \end{aligned} \tag{6.103}$$

For the special case where $x = 0$, the output will oscillate between $+1$ and -1. The output spectrum in this case is a single tone situated at $F_s/2$. For a complete treatment of limit cycles, the reader should consult the work presented in [30–32]. For an irrational DC input signal, the resulting quantization noise consists of discrete tones (see [30,32]).

6.4.2 Higher order ΔΣ modulators

The ideas presented thus far concerning the first ΔΣ modulator can be extended to high order modulators. In general, it will be shown that higher order modulators provide a performance improvement over the first order at the expense of circuit complexity. To start with, consider the second order ΔΣ modulator depicted in Figure 6.45. The modulator employs two integrators instead of one compared to the first order modulator. Only one integrator has the delay element present in the signal path, thus reducing the total input to output signal propagation delay.

The input to the quantizer can be implied from Figure 6.45 as

$$\begin{aligned} U(z) &= I_1(z)I_2(z)(X(z) - Y(z)) - I_1(z)Y(z) \\ &= \left(\frac{z^{-1}}{1-z^{-1}}\right)\left(\frac{1}{1-z^{-1}}\right)(X(z) - Y(z)) - \left(\frac{z^{-1}}{1-z^{-1}}\right)Y(z) \end{aligned} \tag{6.104}$$

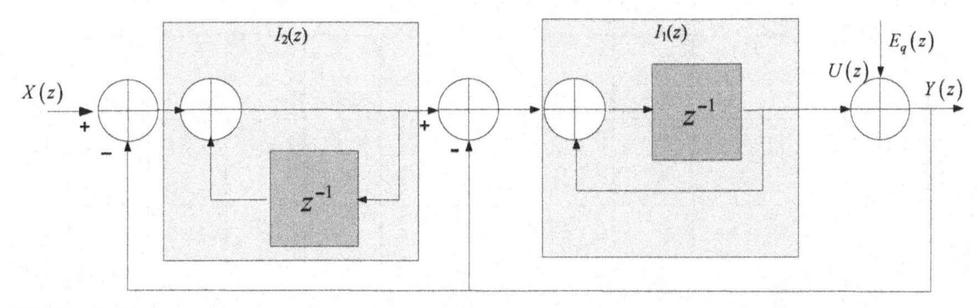

FIGURE 6.45

Second order linear model of ΔΣ modulator. (For color version of this figure, the reader is referred to the online version of this book.)

where the output signal of the modulator $Y(z)$ is related to the output of the integrators $U(z)$ as

$$Y(z) = U(z) + E_q(z) \tag{6.105}$$

Substitute the value of $U(z)$ in Eqn (6.104) into Eqn (6.105) and solving for $Y(z)$, we obtain the simplified relationship

$$Y(z) = z^{-1}X(z) + \left(1 - z^{-1}\right)^2 E(z) = STF(z)X(z) + NTF(z)E(z) \tag{6.106}$$

The $STF(.)$ in Eqn (6.106) is simply a delay as is the case for the first order loop. That is, the desired signal $x(.)$ will be unaffected by the loop. The $NTF(.)$, on the other hand, is a second order highpass function, as opposed to the $NTF(.)$ of the first order loop, which happens to be a first order highpass function. The magnitude and phase response of the $NTF(.)$ of the second order loop is depicted in Figure 6.46. Comparing the $NTF(.)$ of the first order loop depicted in Figure 6.43 to that of the second order loop in Figure 6.46, it is obvious that the latter provides more suppression of the quantization noise in the desired band, which is near DC, and more amplification of the noise outside the signal band. Therefore, it can be deduced that the second order loop performs more shaping of the quantization noise than the first order loop. The magnitude response of the second order $NTF(.)$ can be obtained after certain trigonometric manipulations, that is

$$\begin{aligned} \left|NTF\left(e^{j2\pi f}\right)\big|_{f=F/F_s}\right|^2 &= \left|\left(1 - e^{-j2\pi\frac{F}{F_s}}\right)^2\right|^2 = \left| - 4e^{-j2\pi\frac{F}{F_s}} \sin^2\left(\frac{\pi F}{F_s}\right)\right|^2 \\ &= 16 \sin^4\left(\frac{\pi F}{F_s}\right) \end{aligned} \tag{6.107}$$

Furthermore, for large OSR the sine of the argument approximates the argument, that is $\sin(\alpha) \approx \alpha$ for very small α or

$$\sin^4\left(\frac{\pi F}{F_s}\right) \approx \left(\frac{\pi F}{F_s}\right)^4 \quad \text{for large } OSR \tag{6.108}$$

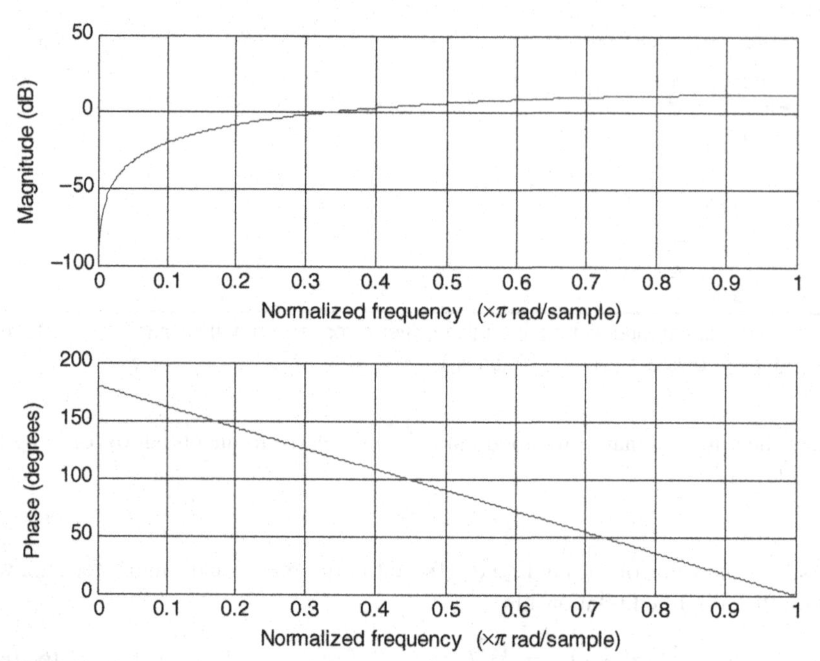

FIGURE 6.46

Magnitude and phase response of $NTF(z) = (1 - z^{-1})^2$. (For color version of this figure, the reader is referred to the online version of this book.)

The quantization noise power can then be computed according to the relationship

$$
\begin{aligned}
P_{e,\Delta\Sigma_2} &= \int_{-B/2}^{B/2} S_e(F)|NTF(F)|^2 \mathrm{d}F \\
&= 16\frac{P_e}{F_s}\int_{-B/2}^{B/2} \sin^4\left(\frac{\pi F}{F_s}\right)\mathrm{d}F \approx 16\frac{P_e}{F_s}\int_{-B/2}^{B/2}\left(\frac{\pi F}{F_s}\right)^4 \mathrm{d}F = \frac{1}{5}\frac{P_e\pi^4}{OSR^5}
\end{aligned}
\quad (6.109)
$$

Next, the relationship for the second order $\Delta\Sigma$ modulator loop is given in dB as

$$
\begin{aligned}
SQNR &= 10.8 + 6.0206b + 10\log_{10}\left(\frac{5}{\pi^4}\right) + 50\log_{10}(OSR) + 10\log_{10}\left(\frac{\sigma_x^2}{V_{x,FS}^2}\right) \\
&= 4.7712 + 6.0206b + 10\log_{10}\left(\frac{5}{\pi^4}\right) + 50\log_{10}(OSR) - PAPR
\end{aligned}
\quad (6.110)
$$

According to Eqn (6.110), doubling the sampling rate improves the *SQNR* roughly by 15 dB or 2½ bits. Second order loops, just like first order loops, also suffer from limit cycles.

Cascading N integrators in the loop increase the order of the loop to an Nth order and consequently the *SQNR* becomes

$$
\begin{aligned}
SQNR &= 10.8 + 6.0206b + 10\log_{10}\left(\frac{2N+1}{\pi^{2N}}\right) \\
&\quad + 10(2N+1)\log_{10}(OSR) + 10\log_{10}\left(\frac{\sigma_x^2}{V_{x,FS}^2}\right)
\end{aligned}
\tag{6.111}
$$

The relationship in Eqn (6.111) can be further simplified as

$$
\begin{aligned}
SQNR &= 4.7712 + 6.0206b + 10\log_{10}\left(\frac{2N+1}{\pi^{2N}}\right) \\
&\quad + 10(2N+1)\log_{10}(OSR) - PAPR
\end{aligned}
\tag{6.112}
$$

According to Eqn (6.112), doubling the OSR increases the *SQNR* by roughly $6 \times (N + 1/2)$ dBs or, in other words, increases the converter resolution by $N + 1/2$ bits. Furthermore, the *SQNR* improves roughly by 6 dBs for each additional bit in the quantizer.

The improvements according to Eqn (6.112) come at a price. For instance, increasing the order of the modulator leads to potential stability problems. The need for loop-gain scaling leads to performance degradation of the loop. Likewise, increasing the OSR implies an increase in the sampling clock and consequently an increase in the power consumption of the converter. Finally, increasing the number of bits in the quantizer implies that the multibit DAC use in the feedback must exhibit higher linearity, which ultimately leads to higher current consumption and design complexity.

6.4.3 MASH ΔΣ modulators

In this section, we discuss a popular ΔΣ modulator known as a multistage noise-shaping (MASH) modulator. MASH modulators serve to alleviate the stability issues associated with high order single-loop ΔΣ modulators. A MASH modulator is constructed simply by cascading two or more ΔΣ modulators in tandem as depicted in Figure 6.47. The individual modulators are made up of either first, second, or higher order modulators. The input to the first modulator stage is the input signal typically processed by a single-stage ΔΣ modulator. The output of each consecutive stage is the quantization error of the preceding stage as shown in Figure 6.47 where $Y_1(z)$ and $Y_2(z)$, the z-transform of $y_1(n)$ and $y_2(n)$, can be expressed simply as

$$
Y_1(z) = STF_1(z)X(z) + NTF_1(z)E_{q1}(z) \tag{6.113}
$$

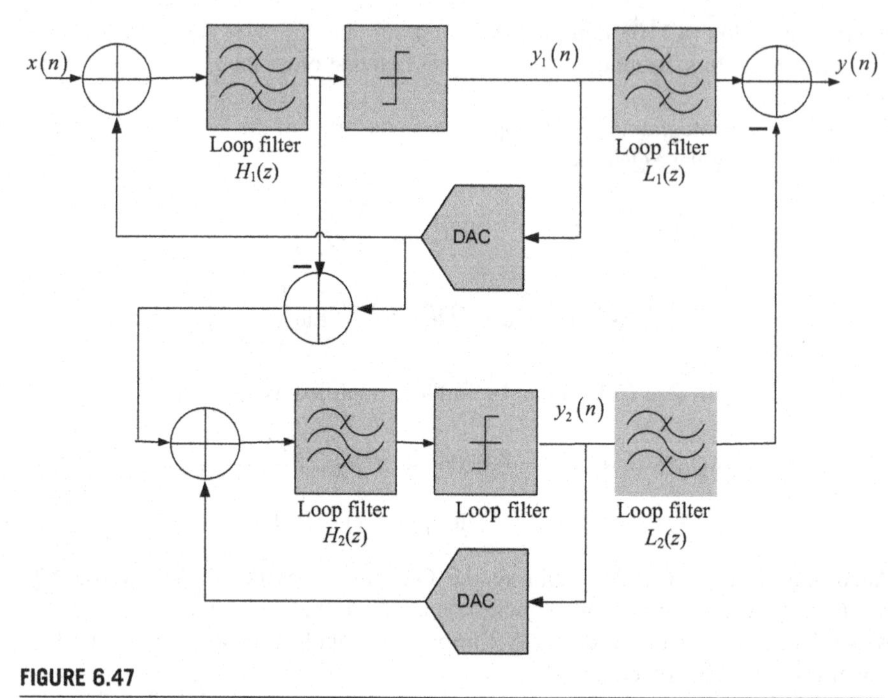

FIGURE 6.47

Two-stage MASH modulator.

and

$$Y_2(z) = STF_2(z)E_{q1}(z) + NTF_1(z)E_{q2}(z) \tag{6.114}$$

Where $E_{q1}(z)$ and $E_{q2}(z)$, $STF_1(.)$ and $STF_2(.)$, and $NTF_1(.)$ and $NTF_2(.)$ are the quantization errors, and signal and noise transfer functions due to the first and second loops consecutively. According to Figure 6.47, the output of the modulator $Y(z)$ can be written as a linear combination of the outputs of the first and second modulators

$$Y(z) = L_1(z)Y_1(z) - L_2(z)Y_2(z) \tag{6.115}$$

Furthermore, $Y(z)$ can be written in terms of the signal and noise transfer functions of the various stages by substituting Eqns (6.113) and (6.114) into Eqn (6.115) in order to obtain the output of the MASH modulator

$$\begin{aligned} Y(z) = {} & L_1(z)\{STF_1(z)X(z) + NTF_1(z)E_{q1}(z)\} \\ & - L_2(z)\{STF_2(z)E_{q1}(z) + NTF_2(z)E_{q2}(z)\} \end{aligned} \tag{6.116}$$

In this context, let $L_1(.)$ and $L_2(.)$ satisfy the following two relationships:

$$L_1(z)NTF_1(z) = L_2(z)STF_2(z) \tag{6.117}$$

and

$$L_1(z) = STF_2(z) \text{ and } L_2(z) = NTF_1(z) \tag{6.118}$$

Then, substitute Eqns (6.117) and (6.118) into Eqn (6.116) in order to express $Y(z)$ as function of signal and noise transfer functions:

$$Y(z) = STF_1(z)STF_2(z)X(z) - NTF_1(z)NTF_2(z)E_{q2}(z) \tag{6.119}$$

In order to understand the implication of Eqn (6.119), assume that both stages are comprised of first order ΔΣ modulators with

$$\begin{aligned} STF_1(z) &= STF_1(z) = z^{-1} \\ NTF_1(z) &= NTF_2(z) = 1 - z^{-1} \end{aligned} \tag{6.120}$$

The output of the MASH ΔΣ modulator, known as a 1-1 MASH modulator, is simply

$$Y(z) = z^{-2}X(z) - \left(1 - z^{-1}\right)^2 E_{q2}(z) \tag{6.121}$$

The quantization error output of the second modulator $e_{q2}(.)$ is white and uniformly distributed. The cascaded response of the MASH modulator itself performs as if the loop is second order ΔΣ modulator. As a matter of fact, it can be shown that if each stage in the MASH modulator is a second order ΔΣ modulator with

$$\begin{aligned} STF_1(z) &= STF_2(z) = z^{-2} \\ NTF_1(z) &= NTF_2(z) = \left(1 - z^{-1}\right)^{-2} \end{aligned} \tag{6.122}$$

Then the output of the MASH modulator, known as a 2-2 MASH, is theoretically equivalent to the output of a fourth order ΔΣ modulator:

$$Y(z) = z^{-4}X(z) - \left(1 - z^{-1}\right)^4 E_{q2}(z) \tag{6.123}$$

thus leading to the generalization that an *M-N-P* MASH modulator is comprised of three ΔΣ modulators of *M*, *N*, and *P* loop order each. A distinguishing difference between a MASH structure and an equivalent ΔΣ modulator of the same order is stability. For a fourth order MASH modulator, for example, the converter has to satisfy the stability of a second order ΔΣ modulator and not that of a corresponding fourth order modulator.

In practice, however, it is possible that the relationship in Eqn (6.117) is not satisfied and that

$$L_1(z)NTF_1(z) - L_2(z)STF_2(z) = F(z),\ F(z) \neq 0 \tag{6.124}$$

is true. The implication of Eqn (6.124) is the leakage of the first stage's quantization error $e_{q1}(.)$ onto the output of the MASH converter. This undesired effect adds to the converter's quantization noise and serves to degrade the overall *SQNR* of the converter. This leakage is typically caused by component mismatches that could have more profound impact on a MASH converter's performance compared to its single loop counterpart.

6.5 Nonidealities in ΔΣ modulator circuits

This section recaps the previous sections concerning circuit nonidealities and performance degradation in CT and DT ΔΣ modulators. There are a number of circuit nonidealities in ΔΣ modulator circuits that force the converter to deviate from its ideal behavior. The enormity of the error itself depends on its impact on the overall performance of the loop, specifically its impact on the *STF* and *NTF* functions. In SC implementations, for example, amplifiers, quantizers, capacitors, switches, and DACs all exhibit a certain amount of nonideality that can impact the overall performance of the modulator [33,34]. More specifically, the amplifiers are affected by large output swing, finite linear and nonlinear DC gain, gain bandwidth, and slew rate. Hysteresis, offset, and gain error impact the performance of the quantizer. Capacitors suffer from mismatches and nonlinearities. Offset, gain error, and nonlinearity can severely impact the performance of the feedback multibit DAC. Switches suffer from clock feedthrough, as discussed earlier, thermal noise, nonlinearity, and charge injection [19]. The amplifier DC gain, as well as capacitor mismatch and incomplete integrator settling error could directly impact and change the *NTF*. As mentioned in the previous section, MASH converters are particularly sensitive to finite DC gain and capacitor mismatch.

Similar to SC implementations in DT-ΔΣ modulators, finite Op-Amp DC gain for active RC implementations, integrator time constant error, circuit noise, and nonlinearities such as input voltage to current gm-C integrators affect the performance of CT-ΔΣ modulators. In addition, architectural timing errors such as quantizer metastability, excess loop delay, and clock jitter also contribute to the source of errors in the modulator. Quantizer metastability can be prevented by including latches between the quantizer and the feedback DAC [35]. The latches are clocked on opposite clock phases thus allowing each latch stage to settle properly. Obviously, the additional latches serve to introduce additional delay in the feedback loop [28].

References

[1] B. Razavi, Principles of Data Conversion System Design, John Wiley and Sons, New York, 1995.

[2] H. Kobayashi, et al., Finite aperture time and sampling jitter effects in wideband data acquisition systems, in: IEEE Conf. ARFTG Microwave Measurement, Digest 56, 38, November 2000, pp. 1–7.

[3] H. Orser, A. Gopinath, A 20 GS/s 1.2 V 0.13 μm CMOS switched cascade track-and-hold amplifier, IEEE Trans. Circuits Syst. II 57 (7) (July 2010) 512–516.

[4] L. Dai, R. Harjani, CMOS switched-Op-Amp-based sample-and-hold circuit, IEEE J. Solid-State Circuits 35 (1) (January 2000) 109–113.

[5] D.A. Johns, K. Martin, Analog Integrated Circuit Design, John Wiley and Sons, Toronto, 1997.

[6] Z. Tao, M. Keramat, A low-voltage, high-precision sample-and-hold circuit, IEEE Int. Symp. Circuits Syst. (July 2007) 433–456.

[7] W. Xu, E.G. Friedman, A CMOS Miller hold capacitance sample-and-hold circuit to reduce charge sharing effect and clock feedthrough, in: 15th International IEEE Conf. on ASIC/SOC, September 2002, pp. 92–96.

[8] K. Poulton, et al., A 6-b, 4 GS/s GaAs HBT ADC, IEEE J. Solid-State Circuits 30 (10) (October 1995) 1109–1118.

[9] R. Hagelauer, et al., A gigasample/second 5-b ADC with on-chip track and hold based on an industrial 1 μm GaAs MESFET E/D process, IEEE J. Solid-State Circuits 27 (10) (October 1992) 1313–1320.

[10] A. Matuszawa, et al., A 10-bit 30 MHz two-step parallel BiCMOS ADC with internal S/H, in: ISSCC Digest of Technical Papers, February 1990, pp. 162–163.

[11] T. Rouphael, RF and Digital Signal Processing for Software Defined Radio, Elsevier, Boston, MA, 2009.

[12] H. Veenrdrick, The behavior of flip-flops used as synchronizers and prediction of their failure rate, IEEE J. Solid-State Circuits 15 (2) (April 1980) 169–176.

[13] R. Van de Plassche, Integrated Analog-to-Digital and Digital-to-Analog Converters, Kluwer Academic Publishers, Boston, 1994.

[14] J. Li, U. Moon, Background calibration techniques for multistage pipelined ADC with digital redundancy, IEEE Trans. Circuits Syst. II 50 (9) (September 2003) 531–538.

[15] B. Brandt, J. Lutsky, A 75-mW. 10-b, 20-MSPS CMOS subranging ADC with 9.5 effective bits at Nyquist, IEEE J. Solid-State Circuits 34 (12) (December 1999) 1788–1795.

[16] U. Fiedler, D. Seitzer, A high-speed 8 bit A/D converter based on a Gray code multiple folding circuit, IEEE J. Solid-State Circuits 14 (3) (June 1979) 547–551.

[17] W. Kester, ADC Architecture VI: Folding ADCs, Analog Devices Application Note, Norwood, MA, 2008, www.analog.com.

[18] E. King, et al., A Nyquist-rate delta-sigma A/D converter, IEEE J. Solid-State Circuits 33 (1) (1998) 45–52.

[19] J. De La Rosa, Sigma-delta modulators: tutorial overview, design guide, and state of the art survey, IEEE Trans. Circuits Syst. I 58 (1) (January 2011) 1–21.

[20] I. Galton, Delta-sigma data conversion in wireless transceivers, IEEE Trans. Microwave Theory Tech. 50 (1) (January 2002) 302–315 (Invited Paper).

[21] M. Orthmans, F. Gerfers, Continuous-time Sigma-Delta A/D Conversion: Fundamentals, Performance Limits, and Robust Implementations, Springer, New York, NY, 2006.

[22] Y. Tsividis, Integrated continuous-time filter design-an overview, IEEE J. Solid-State Circuits 29 (March 1994) 166–176.

[23] K.T. Chan, K.W. Martin, Components for a GaAs delta-sigma modulator oversampled analog-to-digital converter, Proc. IEEE Int. Conf. Circuits Syst. 3 (1992) 1300–1303.

[24] A.M. Thurston, T.H. Pearce, M.J. Hawksford, Bandpass implementation of the sigma-delta A-D conversion technique, in: Proc. IEE International Conf. on Analog-to-Digital and Digital-to-Analog Conversion, September 1991, pp. 81–86. Swansea, U.K.

[25] B.P.D. Signore, D.A. Kerth, N.S. Sooch, E.J. Swanson, A monolithic 20-b delta-sigma A/D converter, IEEE J. Solid-State Circuits 25 (6) (December 1990) 1311–1316.

[26] O. Shoai, W.M. Snelgrove, Design and implementation of a tunable 40 MHz-70 MHz Gm-C bandpass ΔΣ modulator, IEEE Trans. Circuits Syst. II 44 (7) (July 1997) 521–530.

[27] O., Shoai, Continuous-time Delta-Sigma A/D Converters for High Speed Applications, PhD Dissertation, Carlton University, 1995.

[28] J.A. Cherry, W.M. Snelgrove, Continuous-time Delta-Sigma Modulators for High Speed A/D Conversion, Kluwer Academic Publisher, Boston, 1999.

[29] S. Hein, A. Zakhor, On the stability of sigma delta modulators, IEEE Trans. Signal Process. (July 1993) 2322–2348.
[30] V. Friedman, The structure of limit cycles in sigma delta modulation, IEEE Trans. Commun. (September 1981) 1316–1323.
[31] J. Candy, O. Benjamin, The structure of quantization noise from sigma-delta modulation, IEEE Trans. Signal Process. (July 1993) 2322–2348.
[32] R. Gray, Spectral analysis of quantization noise in a single-loop sigma modulator with dc input, IEEE Trans. Commun. (June 1989) 588–599.
[33] J.H. Fischer, Noise sources and calculation techniques for switched capacitor filters, IEEE J. Solid-State Circuits SSC-17 (August 1996) 742–752.
[34] C.C. Enz, G.C. Temes, Circuit techniques for reducing the effects of Op-Amp imperfections: autozeroing, correlated double sampling, and chopper stabilization, Proc. IEEE 84 (November 1996) 1584–1614.
[35] J.A. Cherry, W.M. Snelgrove, Clock jitter and quantizer metastability in continuous-time delta-sigma modulators, IEEE Trans. Circuits Syst. II 50 (January 2003) 31–37.

Further reading

[1] B. Razavi, Design of a 100-MHz 10-mW 3-V sample-and-hold amplifier in digital bipolar technology, IEEE J. Solid-State Circuits 30 (7) (July 1995) 724–730.
[2] B. Razavi, Design of sample-and-hold amplifiers for high-speed low-voltage A/D converters, IEEE Conf. Custom Integr. Circuits (1997) 59–66.

CHAPTER

Frequency Synthesis and Gain Control

7

CHAPTER OUTLINE

This chapter is divided into two main sections addressing key areas in transceiver design. The first section deals with automatic gain control. Automatic gain control adjusts the received signal strength in the receive chain, either via analog or digital gain, to a certain desired power suitable for best performance. The gain lineup is manipulated "up or down" while maintaining the best possible signal-to-noise ratio (SNR). At high input power, either due to the desired signal or blocker, the gain control algorithm may lower the low-noise amplifier (LNA) gain, for example, in order to minimize the degradation due to nonlinearity. In contrast, when the input signal power is low, the gain control algorithm may choose to boost the overall gain in

Wireless Receiver Architectures and Design. http://dx.doi.org/10.1016/B978-0-12-378640-1.00007-X

the receive chain in a manner that reduces noise figure. In Section 7.1, the purpose of automatic gain control is discussed in some detail. The architecture of a closed-loop algorithm is also discussed.

Section 7.2 is devoted to frequency synthesis-based phase-locked loop (PLL) design. The linear PLL model is discussed and performance parameters pertaining to error convergence, order and type, stability, and operating range are presented. In particular, we discuss the performance of the second and third order loop in some detail. The various blocks that constitute the PLL, such as the frequency and phase detector, loop filter, and voltage-controlled oscillator (VCO) are discussed. Finally, we focus on synthesizer architecture relevant to integer-N and fractional-N PLLs. Concerning the latter, we discuss various programmable digital counters such as the dual modulus prescalar and counters based on $\Delta\Sigma$ modulators. The performance of both architectures, due to spurs, bandwidth, and resolution, are also evaluated.

7.1 Automatic gain control

In this section, we discuss automatic gain control in the context of a closed-loop system. The algorithm is partitioned such that the gain stages are assumed to be implemented in the analog domain, whereas the detector, loop filter, and calibration algorithm are all implemented in the digital domain. Obviously, this architecture is not unique, as will be discussed in Section 7.1.3, but rather representative of common implementations of closed-loop gain control algorithms.

7.1.1 The purpose of automatic gain control in the receiver

The purpose of automatic gain control (AGC) in the receiver is to condition the total received signal, such that the desired signal can be received with acceptable SNR yielding an acceptable bit error rate (BER). In this context, a total received signal is comprised of the desired signal plus any possible interferers or blockers. Assuming that the total received signal is within the dynamic range of the receiver, then the AGC must perform any gain adjustments in order for this signal to be received within the dynamic range of the analog-to-digital converter (ADC). This process is illustrated in Figure 7.1.

The received signal strength (RSS) of the total received signal varies over time. This variation could be broadly divided into three categories. The first category is long-term RSS variation of the total received signal due to the receiver's mobility or change of radio frequency (RF) environment such as path loss. This variation takes place over the duration of many symbols and typically over the duration of many frames of symbols as shown in Figure 7.2. The first category is the only category that is effectively dealt with AGC. The second category is short-term fading such as Rayleigh fading, for example, as depicted in Figure 7.2. These changes occur relatively quickly, within a frame, for example, and can be compensated for via

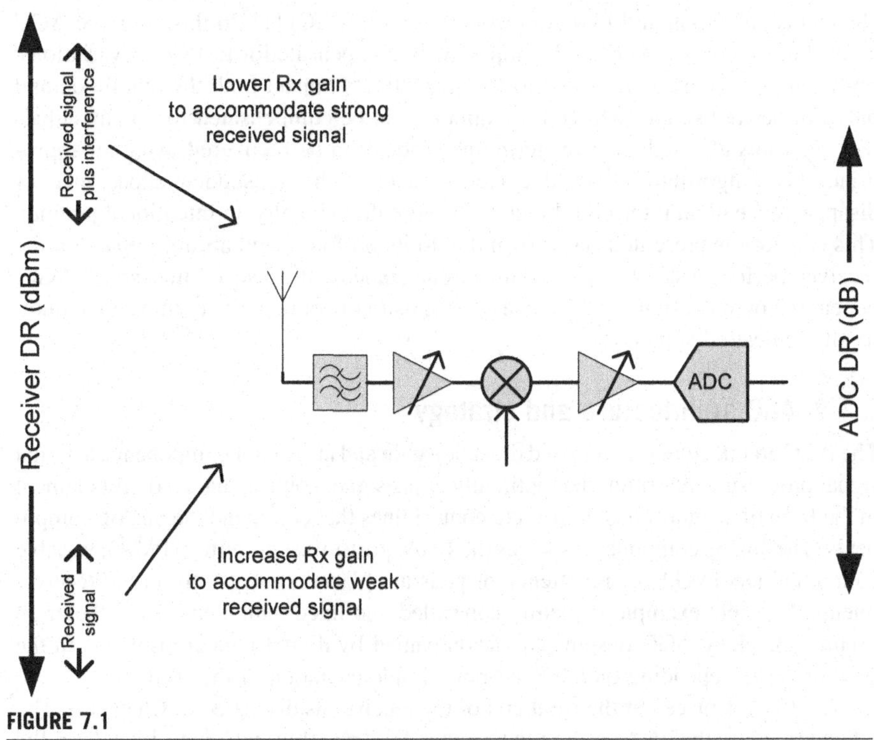

FIGURE 7.1

The purpose of automatic gain control is to adjust the gain of the receiver such that the received signal with large or small received signal strength is within the DR of the ADC. ADC, analog-to-digital converter.

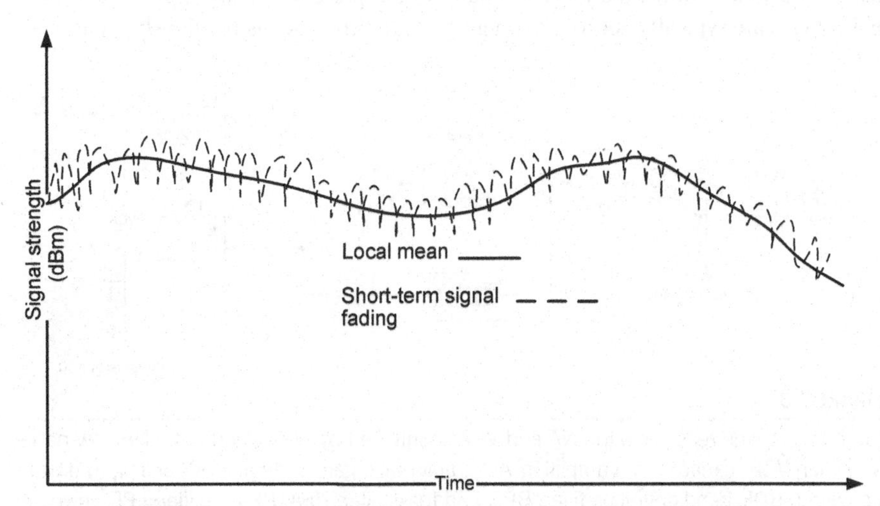

FIGURE 7.2 Short- and long-term signal strength variations in a mobile channel.

channel equalization and forward error correction (FEC) [1]. In this case, the AGC algorithm cannot be successfully employed to compensate for fast signal variations. For example, fast variations due to Doppler effects impact both the amplitude and phase of desired signal, which, for example, can be compensated for via an equalizer. Symbols affected by short-term deep fades can be recovered using an appropriate FEC algorithm. A third category deals with the sudden appearance or disappearance of an interferer, blocker, or more dramatically an intentional jammer. This category is presented on its own due to its profound and abrupt impact on the receiver performance. Under certain circumstances, the use or misuse of AGC, depending on its design strategy, can either positively or negatively impact the quality of the desired signal.

7.1.2 AGC architecture and strategy

The AGC architecture is comprised of analog gain and attenuator components, a digital signal processing algorithm that optimally adjusts gain control, and a control element in the form of a control bus or discrete control lines that command the analog components. The analog components such as the LNA, post-mixer amplifier (PMA), variable gain amplifiers (VGAs), and attenuator pads are placed strategically in the receiver lineup. A simple example depicting controlled and fixed gain elements is shown in Figure 7.3. Many AGC designs are accompanied by digital gain adjustments in the DSP or ASIC depending on the fixed-point implementation of the modem.

The LNA is placed at the front end of the receiver as discussed in Chapter 3. The intent is to place the LNA as close to the antenna as possible and hence drive down the noise figure of the receiver, maximize the desired signal's SNR, while at the same time amplify the incoming weak signal. The LNA's gain and noise figure are heavily influenced by the fabrication process used. Other controlled gain stages, such as the PMA, are placed farther down the receiver chain. In the presence of low signal power, the gain stages are typically set to maximum gain. In this case, as previously mentioned

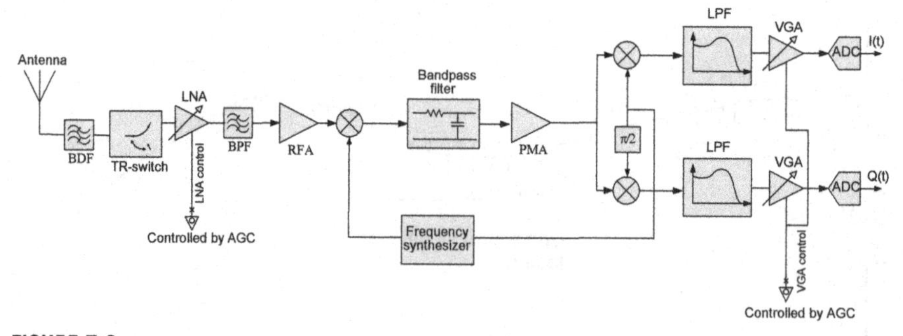

FIGURE 7.3

Superheterodyne receiver with LNA and VGAs controlled by AGC algorithm. LNA, low-noise amplifier; VGA, variable gain amplifier; AGC, automatic gain control; ADC, analog-to-digital converter; BDF, band definition filter; BPF, bandpass filter; RFA, RF-amplifier; LPF, Lowpass filter.

in Chapter 3, the noise figure of the system typically dominates the performance of the receiver. On the other hand, at high input signal power, the gain stages of the receiver are set to low. In this case, the LNA is operating at low gain or bypassed altogether. Under these circumstances, the linearity of the receiver rather than its noise figure plays a key role in influencing its performance. The designer, however, while thriving to maximize the desired signal SNR, must be mindful of the receiver's complexity and current consumption. In the case of the LNA, for example, the IIP3 is directly tied to current consumption that is a higher IIP3 point and usually implies higher current consumption in the device. Other important design parameters that impact the LNA performance, but not directly related to gain control, are reverse isolation, along with input/output return loss and impedance.

Other programmable gain amplifiers following the LNA can be at intermediate frequency (IF) or baseband. The linearity and gain of these devices depends on their location in the receive chain and the required dynamic range of the receiver. One common type of programmable amplifier is the VGA. Depending on the receiver architecture, the VGA could be placed either at IF in a dual conversion receiver for example[1] or at baseband as is always the case in a direct conversion receiver. The gain of the VGA may be continuously adjusted or adjusted in steps as is more commonly the case. In an architecture where the VGA is placed at IF, then only a single amplifier is needed. On the other hand, if the VGA is placed after the quadrature mixers at baseband as shown in Figure 7.3, then two VGA devices are needed. In this case, calibration may be necessary to ensure that the in-phase and quadrature VGA gains are within margin of error.

EXAMPLE 7.1 GAIN CONTROL STRATEGY

Consider the receiver lineup described in Table 7.1. The order of the components in the receiver lineup corresponds with lineup shown in Table 7.1. Assume for the sake of this example that the noise figure and IIP3 of the individual components is not affected by the state of their gains. Furthermore, assume that the input signal power varies between sensitivity at −95 and −25 dBm. Assume that the VGA is programmable in 2-dB steps, and that the LNA and PMA are programmable in high gain and low gain modes. The high gain of the LNA is 16.5 dB, while its low gain is 4.5 dB. Similarly, the high gain of the PMA is 10 dB, while its low gain is 2 dB. The VGA gain varies between 2 dB at low gain and 46 dB at high gain. Assume that, with the lineup given below, the signal reaches the ADC at an optimum level. Assume that the bandwidth of the desired signal is 5 MHz and the required SNR to satisfactorily demodulate the desired signal is 3 dB. All computations are performed at room temperature. Determine the gain strategy of the AGC. The intent of this example is not to cover all the corner cases of AGC strategy, such as the exact placement of the gain stages and attenuators, but to provide a simple illustrative example.

A good gain control strategy ensures that the cumulative gain of the receiver is linearly increasing or decreasing as a function of the received signal power. This in turn ensures that the received signal at the ADC has on average the same signal power level for any

Continued

[1]Note that in an architecture that employs an IF stage, the VGA may be placed either at IF or baseband depending on the design.

Table 7.1 Receiver Lineup for Example 7.1

Parameter	Band Definition Filter	Band Switch	Duplexer	LNA	Tx Notch	Mixer	PMA	Attenuator	LP Filter/ Gain	VGA	ADC
NF (dB)	0.8	1.3	2.4	1.9	1.2	10.5	9	0.8	19	14.5	40
In-band IIP3 (dBm)	50	60	40	6	10.3	30	35	28.6	15	10	40
In-band gain (dB)	−0.8	−1.3	−2.4	16.5	−0.2	−1.75	10	−0.25	9	46	0

EXAMPLE 7.1 GAIN CONTROL STRATEGY—cont'd

received signal power. Furthermore, the total SNR of the receiver as a function of cumulative gain must be monotonically increasing as the cumulative gain decreases.

With the given lineup, at high gain, the noise figure and cumulative IIP3 of the receiver is computed as shown in Table 7.2 using the methods developed in previous chapters.

The SNR due to thermal noise, the SDR due to IIP3, and the total SNR are shown in Table 7.3.

As the input signal power varies between −95 and −51 dBm, let the VGA be the only gain control that varies between 46 and 2 dB. This decision makes sense since lowering the VGA gain at the low received signal power hardly impacts the total noise figure of the receiver as depicted in Figure 7.4. In receiver design, total SNR is dominated by the receiver's noise figure at low signal level. Hence, lowering the VGA gain tends to impact the receiver's noise figure performance the least.

Note that if the signal power drops from −51 to −49 dBm, for example, the gain of the LNA or PMA would have to drop since the VGA is now at its lowest gain setting of 2 dB. Obviously, in order to maintain a monotonic increase in total SNR and not drastically increase the noise figure, the PMA gain is set to low gain or 2 dB. However, since the cumulative is only supposed to change by 2 dB, changing the gain setting in the PMA from high (10 dB) to low (2 dB) changes the cumulative gain by 8 dB. Therefore, the VGA would have to compensate by increasing its gain by 6 dB in order to maintain the linear decrease of the cumulative gain by 2 dB and uphold a monotonic increase in total SNR as graphically depicted in Figure 7.5. Note that the decrease in PMA gain from high to low is accompanied by simultaneous increase in VGA gain. That is, the gain change of both VGA and PMA has to occur before receiving any new data. Both noise figure and cumulative gain are monotonically increasing and decreasing correspondingly.

Any decrease in signal power at this point, from −45 dBm to say −43 dBm, would have to involve changing the LNA gain from high to low. Similar to the PMA case, the difference between the LNA's high gain and low gain setting is 12 dB, which is 10 dB higher than the required 2 dB step. In an actual design lineup, the degradation due to receiver nonlinearity may start to impact total SNR. In order to compensate for the 10 dB gain due to changing the LNA gain from high to low, and hence keep the cumulative gain linear, both the PMA gain and VGA gain are increased. The PMA gain is set to high or 10 dB and the VGA gain is increased from 2 to 4 dB. A similar step takes place by changing the signal power from −41 to −39 dBm. In this case, the PMA gain goes from high to low, the LNA gain stays at low, and the VGA gain goes from 2 to 8 dB. Any further increase in signal power is compensated for by decreasing the gain of the VGA until it hits its minimum gain value of 2 dB. This whole process is depicted in Figure 7.6. Note that increasing the signal power beyond this point directly impacts the performance of the ADC and may cause clipping along compression of the analog front end. Figure 7.7 illustrates the gain lineup strategy for the whole receiver chain.

7.1.3 Types of AGC algorithms

There are various flavors of AGC algorithms that could be classified into the various categories varying in complexity and performance. For example, some design categories are: open loop, closed loop, data aided, non-data-aided, zero-forcing, minimum mean square error (MMSE) gain control, or a combination thereof [2]. Non-data-aided or noncoherent AGC refers to the fact that the algorithm has no

Table 7.2 Receiver Lineup Analysis Showing Receiver Noise Figure and Cumulative IIP3

Parameter	Band Definition Filter	Band Switch	Duplexer	LNA	Tx Notch	Mixer	PMA	Attenuator	LP Filter/ Gain	VGA	ADC
NF (dB)	0.8	1.3	2.4	1.9	1.2	10.5	9	0.8	19	14.5	40
NF linear	1.202264435	1.348962883	1.73780083	1.548817	1.318257	11.22018	7.943282	1.202264435	79.43282347	28.18383	10,000
In-band IIP3 (dBm)	50	60	40	6	10.3	30	35	28.6	15	10	40
In-band IIP3 linear	100,000	1,000,000	10,000	3.981072	10.71519	1000	3162.278	724.4359601	31.6227766	10	10,000
In-band gain (dB)	−0.8	−1.3	−2.4	16.5	−0.2	−1.75	10	−0.25	9	46	0
In-band gain linear	0.831763771	0.741310241	0.57543994	44.66836	0.954993	0.668344	10	0.944060876	7.943282347	39,810.72	1
Aggregate gain linear	1	0.831763771	0.616595	0.354813	15.84893	15.13561	10.11579	101.1579454	95.4992586	758.5776	30,199,517
Aggregate gain (dB)	0	−0.8	−2.1	−4.5	12	11.8	10.05	20.05	19.8	28.8	74.8
Aggregate NF linear	1.202264435	1.621810097	2.81838293	4.385239	4.385239	5.06048	5.748859678	6.570152164	6.605987	6.605987	6.606319
Aggregate NF (dB)	0.8	2.1	4.5	6.4	6.419933	7.041917	7.594306	7.595817082	8.175754279	8.199377	8.199595
Agg Gain/IIP3 linear	0.00001	8.31764E-07	6.166E-05	0.089125	1.479108	0.015136	0.003199	0.139636836	3.01995172	75.85776	3019.952
Aggregate IIP3 linear	0.000322523										
Aggregate IIP3 (dBm)	−34.91439539										
Aggregate NF (dB)	8.199595096										

Table 7.3 SNR, SDR, and Total SNR Computed at Sensitivity for Example 7.1

Signal power (dBm)	−95
Noise floor (dBm)	−98.73786909
Signal power linear	3.16228E-10
SDR due to IIP3	60.08560461
1/SDR lin	9.80482E-07
SNR due to noise	3.737869089
1/SNR lin	0.422876052
Total SNR	3.73785902

knowledge of the received data symbols. Figure 7.8 depicts an example of a non-data-aided open-loop AGC algorithm often used entirely in the digital domain to compensate for signal power after filtering and scaling. In this case, the output of the power detector is lowpass filtered in order to obtain the average of the received envelope signal power or

$$P_r(n) = E\{I^2(n) + Q^2(n)\}; \quad P_r(n) \neq 0 \tag{7.1}$$

where $E\{.\}$ is the expectation operator. The lowpass filtering is assumed to average out the received instantaneous signal power, and hence the output of the lowpass filter is the expected value of the received signal power. The received signal is then scaled by factor $g(n) = \sqrt{P_d/P_r(n)}$, where P_d represents the desired signal power.

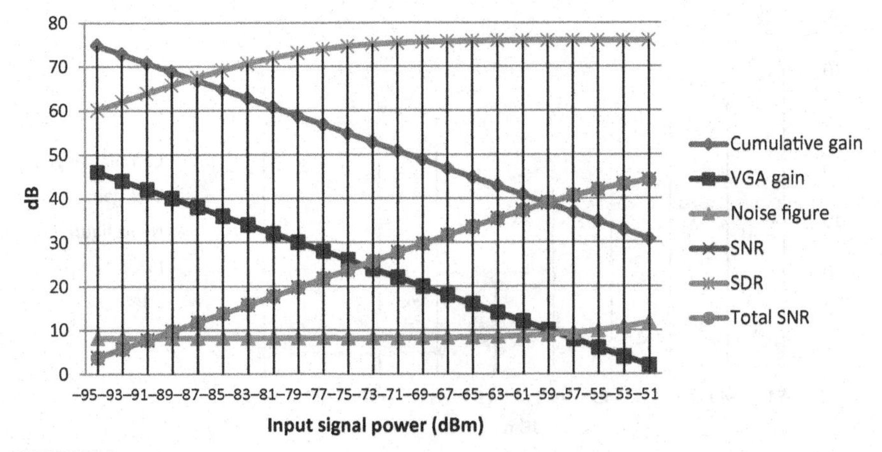

FIGURE 7.4

VGA gain versus received signal power varying between −95 and −51 dBm. VGA, variable gain amplifier, SNR, signal-to-noise ratio. (For color version of this figure, the reader is referred to the online version of this book.)

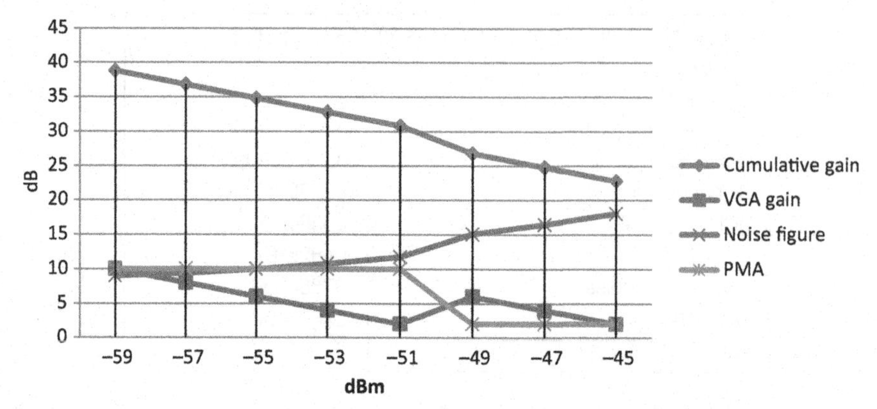

FIGURE 7.5

VGA gain and PMA gain for an input signal power varying between −59 and −45 dBm. (For color version of this figure, the reader is referred to the online version of this book.)

Once the received signal is adjusted by the scaling factor, its average power will approach P_d. In this type of AGC algorithm, the computation of $g(n)$ tends to be highly inaccurate if the input signal power varies widely. This problem is alleviated by employing a closed-loop AGC. For this reason, a non-data-aided open-loop AGC algorithm is not widely relied upon to perform automatic gain control on the received signal, especially if the algorithm is used to control the analog gain stages. Moreover, all noncoherent AGC algorithms amplify the desired signal along with

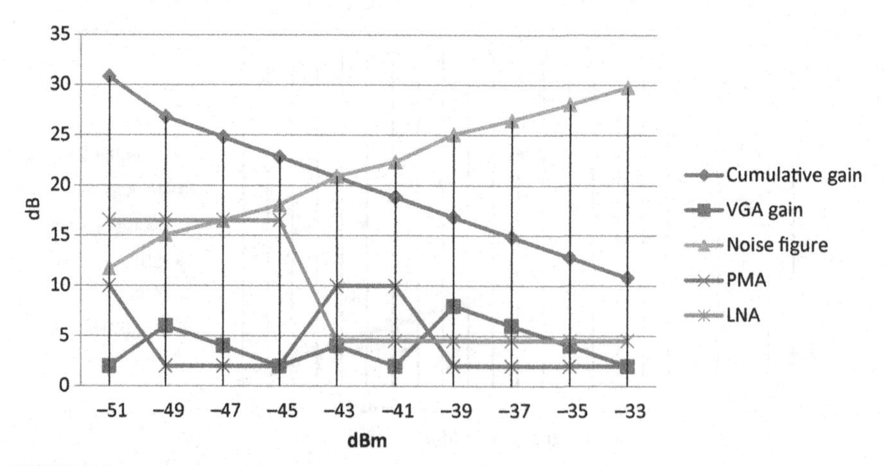

FIGURE 7.6

VGA, PMA, and LNA gain for an input signal varying between −51 and −33 dBm. VGA, variable gain amplifier; PMA, post-mixer amplifier; LNA, low-noise amplifier. (For color version of this figure, the reader is referred to the online version of this book.)

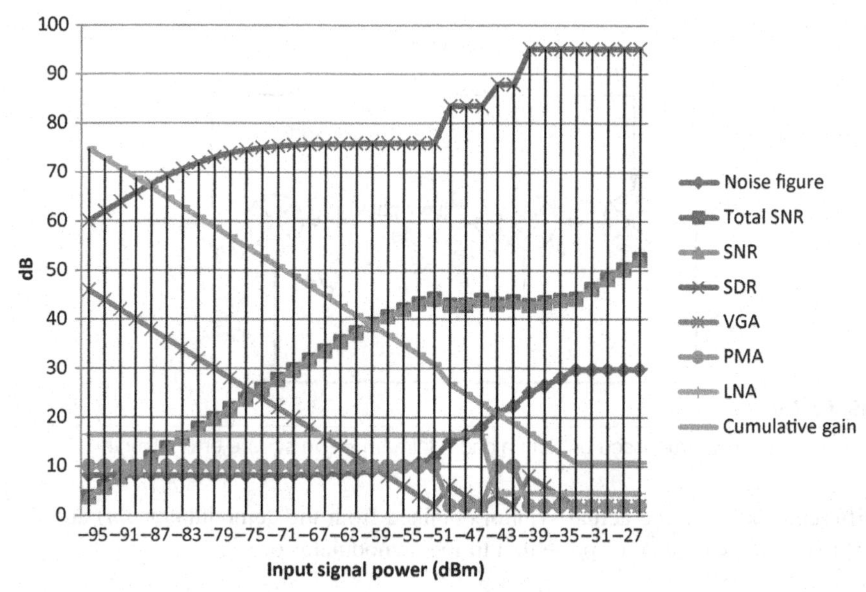

FIGURE 7.7

Gain lineup settings for the VGA, PMA, and LNA as a function of cumulative gain, SNR, SDR, and total SNR. VGA, variable gain amplifier; PMA, post-mixer amplifier; LNA, low-noise amplifier; SNR, signal-to-noise ratio. (For color version of this figure, the reader is referred to the online version of this book.)

the thermal noise present in the receiver as well as any in-band blocker or interferer. The latter problem could be mitigated by using a data-aided or coherent AGC such as a zero-forcing algorithm or a minimum mean square error gain control algorithm.

The MMSE AGC is for all intents and purposes an equalizer designed to minimize the mean square error power of $e(n)$, namely $\varepsilon \triangleq E\{e^2(n)\}$. A simplified version of this algorithm is depicted in Figure 7.9. The error signal is obtained as the

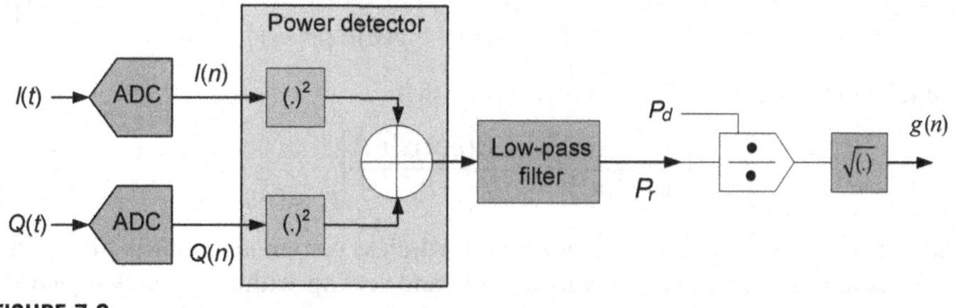

FIGURE 7.8

Non-data-aided automatic gain control with open loop. ADC, analog-to-digital converter. (For color version of this figure, the reader is referred to the online version of this book.)

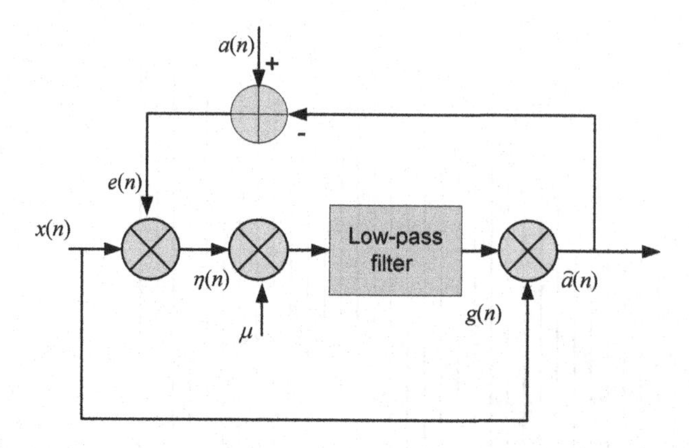

FIGURE 7.9

Automatic gain control based on data-aided minimum mean square error algorithm.

difference between the actual symbol obtained from the demodulator $a(n)$ and the approximate symbol $\widehat{a}(n)$ presented to the demodulator or

$$\begin{aligned} e(n) &= a(n) - \widehat{a}(n) \\ &= a(n) - g(n)x(n) \end{aligned} \tag{7.2}$$

Note that the data $x(n)$ as well as the symbol $a(n)$ are independent of the gain function $g(n)$. The mean square error power ε forms a quadratic function of $g(n)$ with a *unique minimum* $\overline{g} \triangleq \min\{g(n)\}$. In this case, we can express

$$\frac{\partial \varepsilon}{\partial g(n)} = 2E\left\{ e(n) \frac{\partial e(n)}{\partial g(n)}\bigg|_{e(n)=a(n)-g(n)x(n)} \right\} = -2E\{e(n)x(n)\} \tag{7.3}$$

where $\partial e(n)/\partial g(n) = -x(n)$ is true due to the independence of $x(n)$ and $a(n)$ from $g(n)$. In steady state, the loop attempts to remove any correlation between the error $e(n)$ and the received signal $x(n)$, in essence removing the DC at the input of lowpass filter, that is

$$\begin{aligned} E\{e(n)x(n)\} &= E\{[a(n) - g(n)x(n)]x(n)\} \\ &= E\{a(n)x(n)\} - g(n)E\{x^2(n)\} = 0 \end{aligned} \tag{7.4}$$

The relationship in Eqn (7.4) simply implies that

$$\overline{g} = \frac{E\{a(n)x(n)\}}{E\{x^2(n)\}} = \frac{E\{a(n)x(n)\}}{P_x}\bigg|_{P_x=E\{x^2(n)\}} \tag{7.5}$$

The utility of the MMSE AGC algorithm in wireless communications is limited to digital compensation of gain due to filtering and scaling within the modem and is rarely used to control analog gain stages of the receiver. In the next section, we will discuss a non-data-aided AGC loop frequently used in wireless receivers. We will analyze in detail the various components of the loop.

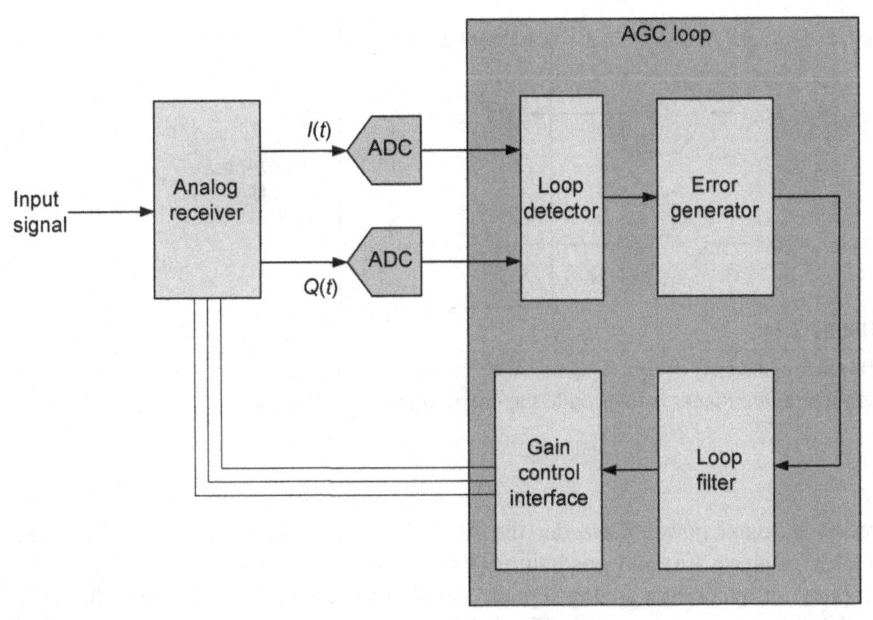

FIGURE 7.10

The conceptual AGC loop comprised of a loop detector, an error generator, a loop filter, and gain control interface. AGC, automatic gain control; ADC, analog-to-digital converter. (For color version of this figure, the reader is referred to the online version of this book.)

7.1.4 The closed-loop AGC algorithm

The AGC loop discussed in this section is commonly used in most modern wireless receivers. The loop, depicted conceptually in Figure 7.10, consists of a digital algorithm and analog gain and attenuator blocks. In this section we will concentrate on the digital signal processing portion of the loop. As stated earlier, the purpose of the AGC loop in most modern wireless communication receivers is to compensate for long-term fading and path loss effects that impact the received signal. It is not intended to correct for short-term fast fades, especially frequency selective fades that can be appropriately corrected for using certain types of equalization and FEC. The AGC loop is slow and may be updated once every slot or frame depending on the modulation, channel, error correction, and the makeup of the signaling scheme or waveform itself. The digital signal processing algorithm portion of the loop is comprised of a loop detector, error generator, loop filter, and gain control interface. In the following subsections, we will analyze the interworkings of these blocks.

7.1.4.1 AGC loop dynamics

The purpose of the AGC detector is to detect the instantaneous power of the received signal and average it over a certain period of time to obtain the average of the

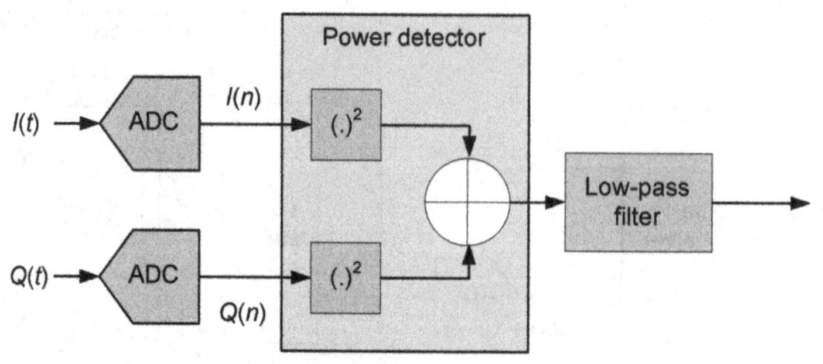

FIGURE 7.11

The automatic gain control detector. ADC, analog-to-digital converter. (For color version of this figure, the reader is referred to the online version of this book.)

received signal power. Consider the AGC detector depicted in Figure 7.11. The analog in-phase $I(n)$ and quadrature $Q(n)$ signals are the sampled versions of the in-phase or quadrature analog signals. Let the instantaneous signal power be given at the output of the summer in Figure 7.11 as:

$$r^2(n) = I^2(n) + Q^2(n) = I^2(t) + Q^2(t)\big|_{t=nT_s} \tag{7.6}$$

where n implies the sample taken at nT_s. A typical implementation of the detector is of an approximation of Eqn (7.6) or simply of the envelope $r(n)$. The most common approximation involves the minimum and maximum values of $I(n)$ and $Q(n)$ or

$$r(n) \approx \alpha \max\{|I(n)|, |Q(n)|\} + \beta \min\{|I(n)|, |Q(n)|\} \tag{7.7}$$

where α and β are constants. This algorithm is easily implemented in ASIC. Table 7.4 lists various constant values that are typically used for α and β. The performance of the detector for various parameters is shown in Figure 7.12.

In order to ensure that the AGC does not react to instantaneous power surges and recesses, the power signal is averaged further via an integrate and dump filter, for

Table 7.4 AGC Detector According to Eqn (7.7) for Various Values of α and β

α	β	$r(n)$
$\alpha = 1$	$\beta = 1/2$	$r(n) \approx \max\{\lvert I(n)\rvert, \lvert Q(n)\rvert\} + \frac{1}{2}\min\{\lvert I(n)\rvert, \lvert Q(n)\rvert\}$
$\alpha = 1$	$\beta = 1/4$	$r(n) \approx \max\{\lvert I(n)\rvert, \lvert Q(n)\rvert\} + \frac{1}{4}\min\{\lvert I(n)\rvert, \lvert Q(n)\rvert\}$
$\alpha = 1$	$\beta = 3/8$	$r(n) \approx \max\{\lvert I(n)\rvert, \lvert Q(n)\rvert\} + \frac{3}{8}\min\{\lvert I(n)\rvert, \lvert Q(n)\rvert\}$
$\alpha = 15/16$	$\beta = 15/32$	$r(n) \approx \frac{15}{16}\left[\max\{\lvert I(n)\rvert, \lvert Q(n)\rvert\} + \frac{1}{2}\min\{\lvert I(n)\rvert, \lvert Q(n)\rvert\}\right]$

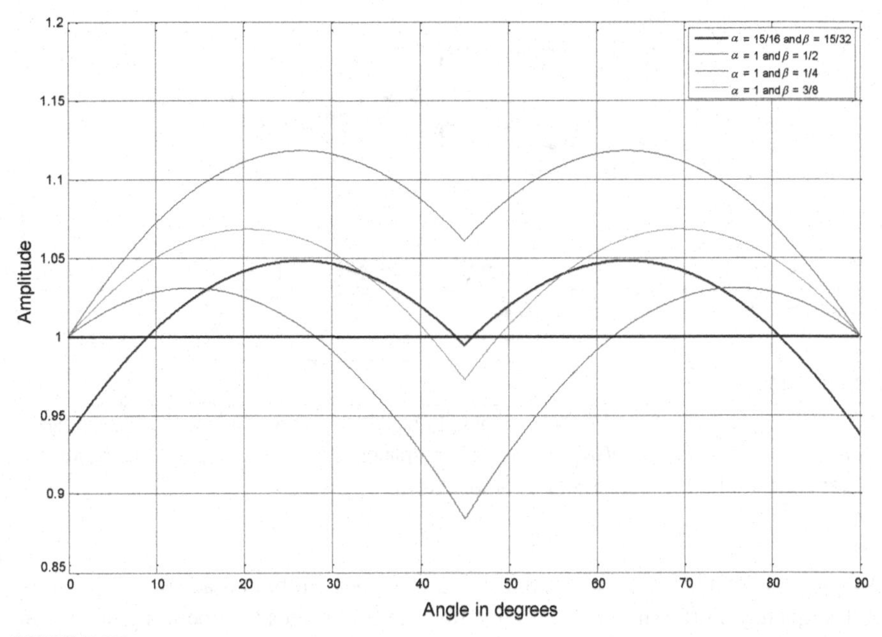

FIGURE 7.12

Performance of various automatic gain control detectors as specified in Table 7.4. (For color version of this figure, the reader is referred to the online version of this book.)

example, as shown in Figure 7.13, in order to obtain the average signal power defined as the expected value of $r^2(n)$:

$$E\{r^2(n)\} = \frac{1}{N}\sum_{n=0}^{N-1}[I^2(n) + Q^2(n)] \tag{7.8}$$

where $E\{.\}$ is the expectation operator. The output of the integrate-and-dump filter is fed into an *optional* logarithmic lookup table or function. The purpose of the logarithm is to speed up the reaction of the loop due to a small received signal power. In the presence of an in-band strong blocker, the received signal power is high and a rapid reaction of the AGC is required. The logarithmic function compresses the input signal and helps speed up the reaction of the loop to a strong signal. However, introducing the logarithm into the loop causes asymmetry in attack and decay time of the loop. Furthermore, note that any integration of error between the desired signal power and the average power of the signal is happening in dB scale and not in the linear scale, hence introducing nonlinearity into the loop. Thirdly, the quantization error at the output of the loop filter is no longer uniformly distributed. In order to ensure that the loop will converge, we introduce an exponential function of the form $10^{(.)/20}$ into the gain-control lookup table (the divide by 20 is due to voltage gain rather than power gain, which implies divide by 10). In most receiver

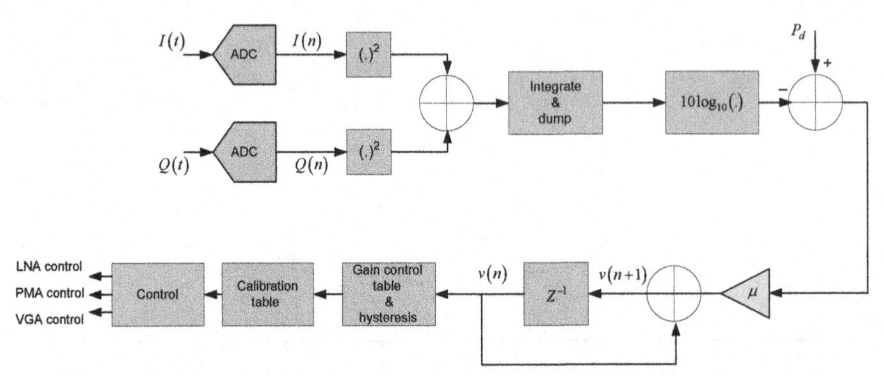

FIGURE 7.13

The automatic gain control loop. ADC, analog-to-digital converter; LNA, low-noise amplifier; PMA, post-mixer amplifier; VGA, variable gain amplifier. (For color version of this figure, the reader is referred to the online version of this book.)

designs, the AGC lookup table contains gain values calibrated across temperature and frequency. Furthermore, we assume that the gain stages are operating in a linear fashion without compression and that the all gain stages including the VGA and all gain attenuators can be increased and decreased according to some *monotonically* varying function.

Next, consider the output of the single-pole loop filter taken at the output of the delay

$$\begin{aligned} v(n+1) &= v(n) + \mu\left[P_d - 10\log_{10}\left(10^{v(n)/10}\widehat{r}^{2}(n)\right)\right] \\ v(n+1) &= v(n) + \mu\left[P_d - v(n) - 10\log_{10}\left(\widehat{r}^{2}(n)\right)\right] \end{aligned} \tag{7.9}$$

where $v(n)$ is the state at the output of the loop filter and $\widehat{r}^{2}(n)$ is the mean squared input power estimated over M samples and obtained via integration and dump as

$$\widehat{r}^{2}(n) = E\{r^2(n)\} \approx \frac{1}{M}\sum_{m=0}^{M-1}\{I^2(n-m) + Q^2(n-m)\} \tag{7.10}$$

The mean squared input power is then input to a logarithmic table, and its output is then compared to the desired set power P_d. In this particular implementation, the average mean square input power is obtained via an integrate-and-dump operation, say $h(.)$, which in turn mimics a lowpass filter or

$$\begin{aligned} H(z) &= 1 + z^{-1} + z^{-2} + \cdots + z^{-M+1} \\ &= \frac{1 - z^{-M}}{1 - z^{-1}} \end{aligned} \tag{7.11}$$

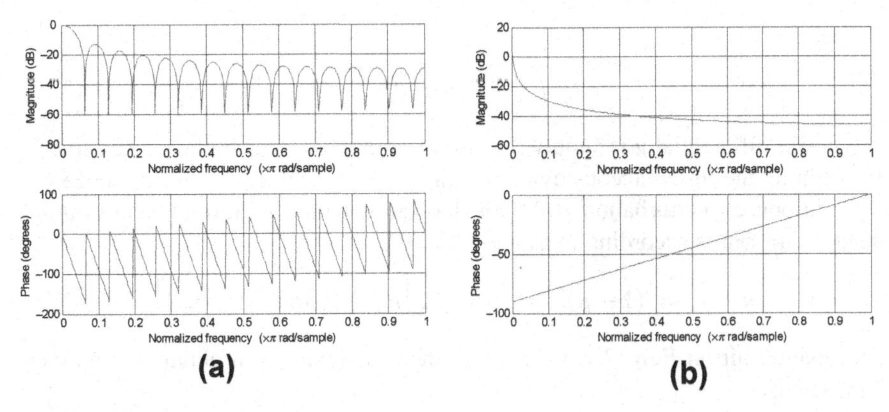

FIGURE 7.14

Magnitude and phase response of: (a) integrate-and-dump function taken over 30 samples, and (b) of mean function taken over an infinite number of samples. (For color version of this figure, the reader is referred to the online version of this book.)

The magnitude and phase response for $H(z)$ for $M = 30$ compared to the response of the actual mean $1/1 - z^{-1}$ is depicted in Figure 7.14. The integrate-and-dump filter can be replaced with a single-pole filter that resets it memory (output of delay element) on the integration boundary.

Note that the choice of a single-pole loop filter is not necessary but sufficient for the convergence of the loop. The parameter P_d is the desired set power of the AGC loop chosen in such a way that accommodates the dynamic range of the ADCs, the dynamic range of the signal and its peak power-to-average power ratio, and the dynamic range of the blocking signal. Its value is certainly dictated by the desired SNR of the received signal as well as parameters above as detailed in Chapter 5.

7.1.4.2 Steady state analysis of AGC loop

Assume that during steady state the error signal $e(n) = P_d - 10\log_{10}(10^{\nu(n)/10}\widehat{r}^2(n))$ tends to zero. In this case, the output of the loop filter becomes

$$\nu(n+1) = \nu(n) + \mu \underbrace{e(n)}_{\to 0} \approx \nu(n) \tag{7.12}$$

According to Eqn (7.9)

$$\begin{aligned}\nu(n+1) &= \nu(n) + \mu\left[P_d - \nu - 10\log_{10}\left(\widehat{r}^2(n)\right)\right] \\ &= (1-\mu)\nu(n) + \mu\left[P_d - 10\log_{10}\left(\widehat{r}^2(n)\right)\right]\end{aligned} \tag{7.13}$$

In steady state, $\nu(n+1) \approx \nu(n) = \nu$ then Eqn (7.13) becomes

$$\nu = (1-\mu)\nu + \mu\left[P_d - 10\log_{10}\left(\widehat{r}^2(n)\right)\right] \tag{7.14}$$

or simply

$$\nu = P_d - 10\log_{10}\left(\widehat{r}^2(n)\right) \tag{7.15}$$

The power differential ν is simply the equilibrium point, which plays a key role in determining the instantaneous dynamic range of the receiver. To fully analyze the loop, introduce a perturbation $\chi(n)$ to the loop around the equilibrium point ν during steady state that is according to Eqn (7.14)

$$\nu + \chi(n+1) = (1-\mu)[\nu + \chi(n)] + \mu\left[P_d - 10\log_{10}\left(\widehat{r}^2(n)\right)\right] \tag{7.16}$$

The relationship in Eqn (7.16) can be further manipulated to obtain a simplified expression:

$$\chi(n+1) = (1-\mu)\chi(n) - \mu\nu + \mu\left[P_d - 10\log_{10}\left(\widehat{r}^2(n)\right)\right] \tag{7.17}$$

Substituting Eqn (7.15) into Eqn (7.17), we obtain

$$\begin{aligned}\chi(n+1) &= (1-\mu)\chi(n) - \mu\left[P_d - 10\log_{10}\left(\widehat{r}^2(n)\right)\right] + \mu\left[P_d - 10\log_{10}\left(\widehat{r}^2(n)\right)\right]\\ &= (1-\mu)\chi(n)\end{aligned} \tag{7.18}$$

Note that the relationship Eqn (7.18) implies that

$$\chi(n+1) = (1-\mu)\chi(n) \Rightarrow \chi(n+1) = (1-\mu)^n\chi(0) \tag{7.19}$$

The stability condition imposed on the loop can be inferred from Eqn (7.19) as n approaches infinity and is

$$\lim_{n\to\infty}|\chi(n+1)| = \lim_{n\to\infty}\left|(1-\mu)^n\chi(0)\right|_{|\chi(0)|\neq 0} = 0 \Rightarrow |1-\mu| < 1 \Rightarrow \begin{cases}\mu > 0\\ \mu < 2\end{cases} \tag{7.20}$$

where the loop filter gain μ dictates both attack and decay times. The AGC loop is stable if and only if Eqn (7.20) satisfied.

To illustrate the theoretical performance of the loop, let the input to AGC consist of a single sinusoid with the AGC set point taking into account the 3-dB backoff from full scale. In this case, there is no blocker or other imperfections introduced to the simulation. Comparing the average input signal power of the sinusoid to the AGC set point requires the AGC to make a power adjustment of 65 dB. The output of the loop filter versus time or number of samples slowly and continuously adjusts the signal power as shown in Figure 7.15. The in-phase signal of the AGC as it is being continuously gain adjusted as a function of time is also shown in Figure 7.16.

7.1.4.3 Observations

In this section, we attend to a series of common issues with the AGC algorithm. A conceptual IF-receiver lineup is depicted in Figure 7.17. The purpose of this

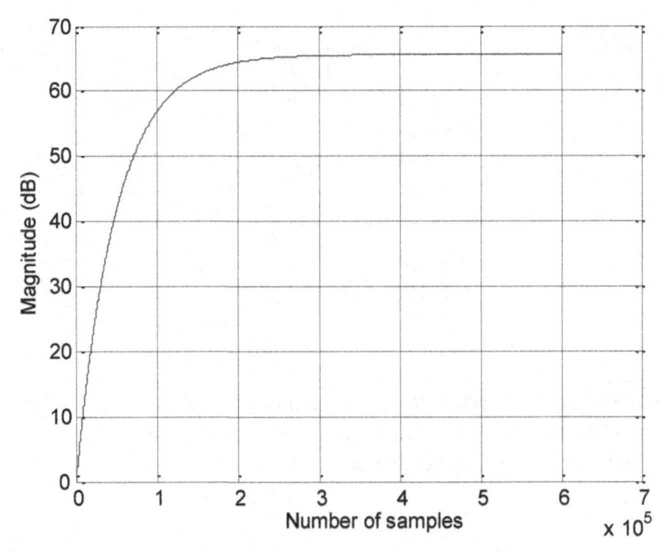

FIGURE 7.15

Output of automatic gain control loop filter as a function of time or samples. (For color version of this figure, the reader is referred to the online version of this book.)

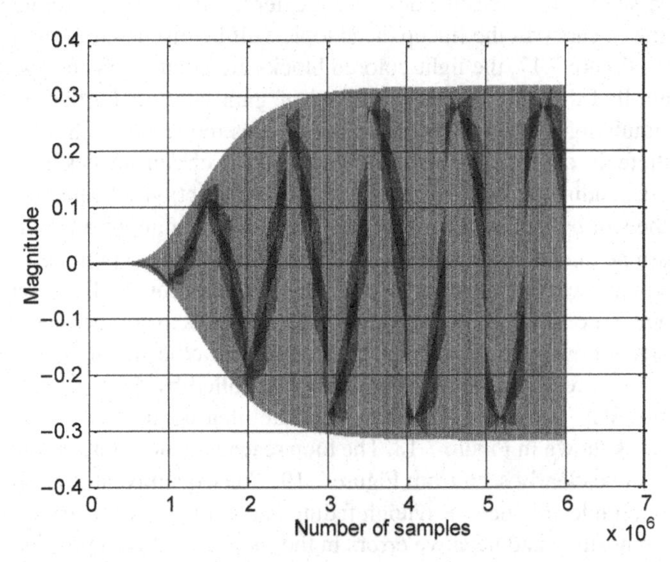

FIGURE 7.16

In-phase component of a sinusoidal signal as it is being continuously gain adjusted by automatic gain control loop. (For color version of this figure, the reader is referred to the online version of this book.)

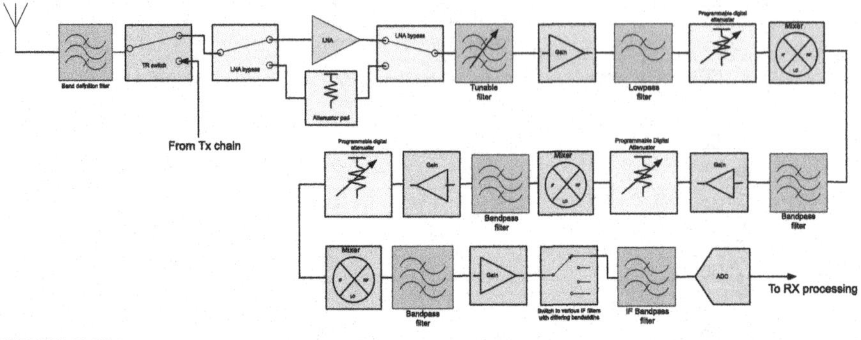

FIGURE 7.17

Conceptual lineup of an intermediate frequency-sampling receiver. LNA, low-noise amplifier; ADC, analog-to-digital converter; IF, intermediate frequency. (For color version of this figure, the reader is referred to the online version of this book.)

simulation is not to discuss the merits of the design and the lineup of the components of the receiver shown in Figure 7.17 but rather to uncover certain common problems with automatic gain control algorithms. The behavior of the AGC in Rayleigh fading channels and in the presence of a narrowband blocker is discussed.

The IF receiver presented in Figure 7.17 relies on gain stages and attenuators rather than a VGA. This gives the designer greater flexibility in managing the noise figure and the linearity in the lineup. This topic will be discussed in more detail in Chapter 8. In Figure 7.17, the light-colored blocks are attenuators and switches that can be controlled digitally for the purposes of gain control. Furthermore, in the following simulations, the AGC is run in continuous mode and without any hysteresis to illustrate the algorithm's performance under the circumstances in question. In most cases, depending on the modulation and error correction scheme, the AGC will update on the slot boundary or possibly only at the beginning of a frame.

Next, assume that the desired signal is undergoing Rayleigh two-path fading and that the average received signal strength is at −90 dBm or 25 dB lower than the AGC set point. Once the AGC is turned on, the loop starts to compensate for the difference in signal power by decreasing the attenuation in the signal path. The rate at which the gain correction is made is partially controlled by the loop-filter gain. At this point, the AGC is trying to reach steady state; that is, the error signal is trying to reach 0 dB as shown in Figure 7.18. The loop reaches steady state at roughly 2000 samples as can be clearly seen from Figure 7.19. Consequently, all the variations in the signal amplitude are due to Rayleigh fading and not the asymmetry in which the loop reacts to positive and negative errors in the loop. Note that a positive error implies that the received signal strength, including blocking signals, is less than the AGC set point whereas a negative error implies that the received signal strength is larger than the AGC set point. In steady state, the loop reacts to the changes in the signal strength due to channel variations. Note that although the AGC seems

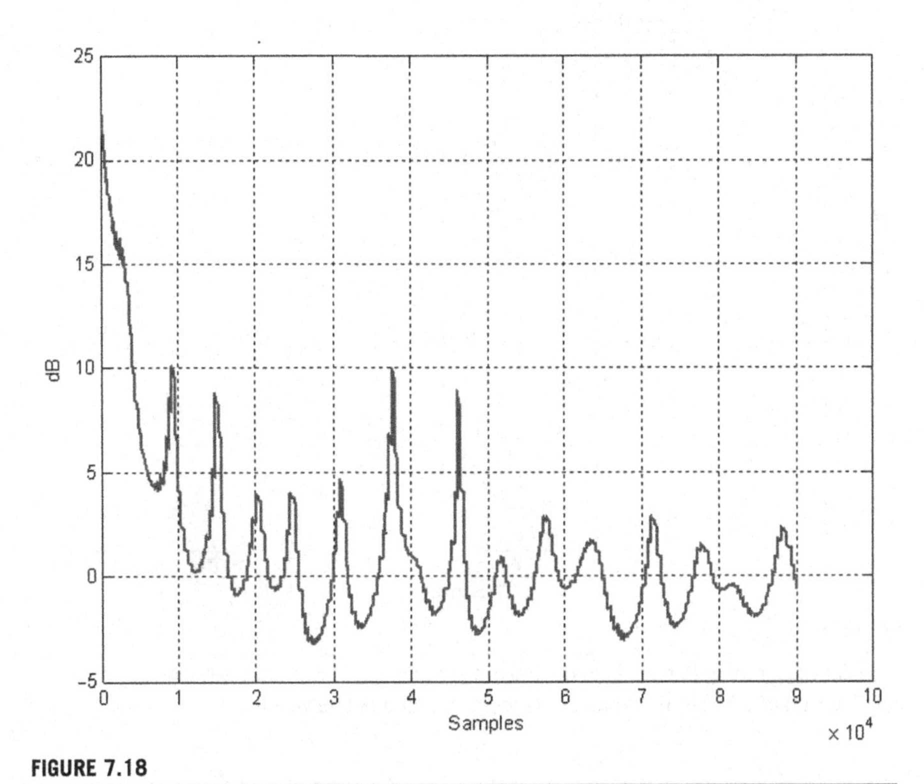

FIGURE 7.18

Error signal of automatic gain control in the presence of two-path Rayleigh fading. (For color version of this figure, the reader is referred to the online version of this book.)

to track signal variations in fading, updating the AGC in the middle of a FEC slot will only serve to degrade the performance. This is true, since most FEC schemes rely on the signal amplitude and phase within an FEC slot in order to perform error correction.

Next, the AGC performance is analyzed in the presence of a blocker whose power throughout the simulation is −70 dBm or 20 dB higher than the signal power. Furthermore, the blocker is assumed to be received along with the desired signal within the IF bandwidth of the receiver. The blocker obviously does not overlap with the desired signal's instantaneous bandwidth. In some other cases, depending on the receiver's overall filtering strategy and/or the blocking signal's frequency location relative to that of the desired signal, the blocker may be either slightly or severely attenuated by the various filters. Again, the intent is to show how the AGC loop behaves in the presence of a blocker. In order to simplify the analysis, we restrict the duty cycle of the desired signal to 100%. In the first case, we assume the duty cycle of the blocker to be 1/3 or 33.33%. The error signal as well as the blocker duty cycle are both depicted in Figure 7.20. Given the loop-filter gain chosen for this simulation, the error signal barely reaches an average of 0 dB indicating that

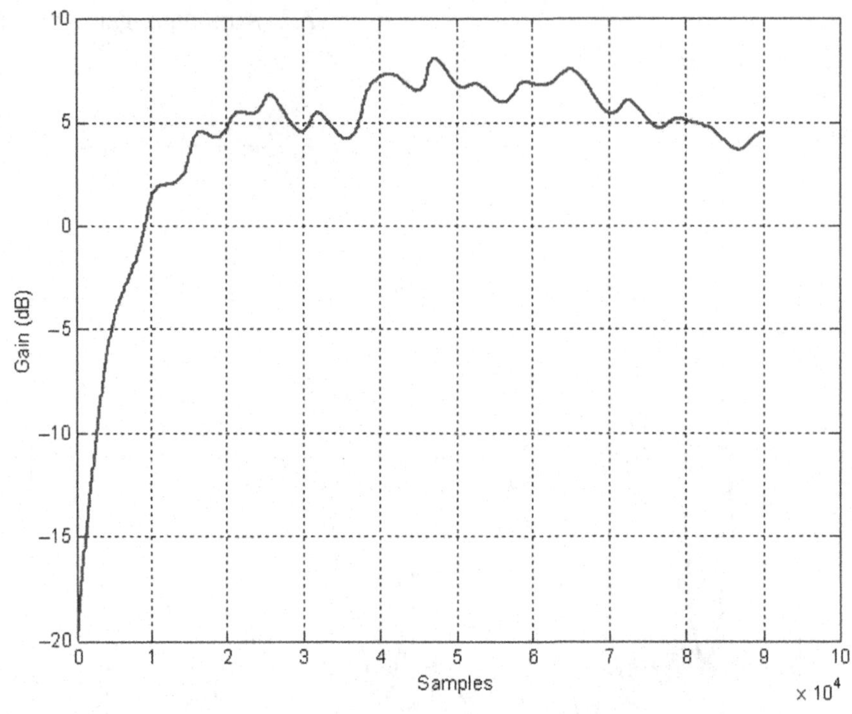

FIGURE 7.19

Cumulative gain of automatic gain control in the presence of two-path Rayleigh fading. (For color version of this figure, the reader is referred to the online version of this book.)

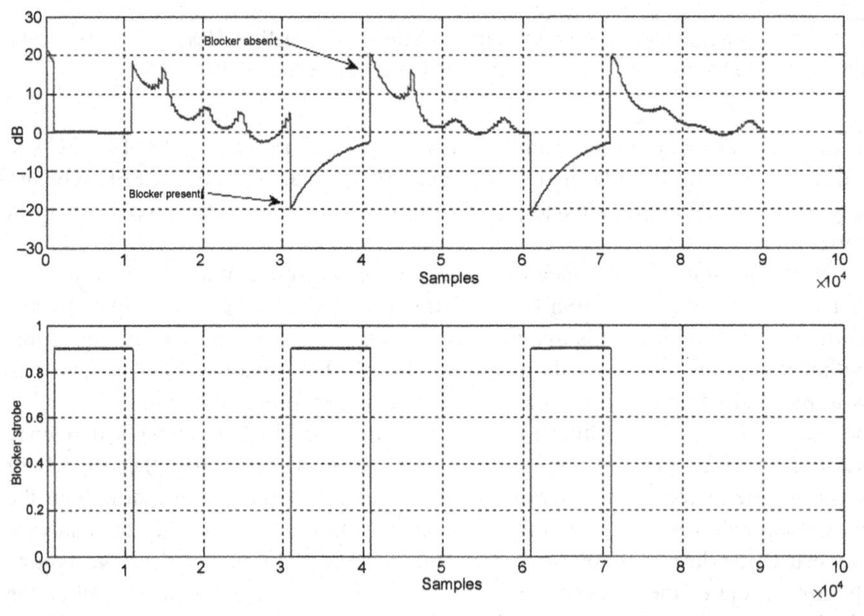

FIGURE 7.20

Error signal of automatic gain control in the presence of two-path Rayleigh fading and blocker signal with duty cycle = 1/3 (above) error signal and (below) strobe indicating presence of blocker. (For color version of this figure, the reader is referred to the online version of this book.)

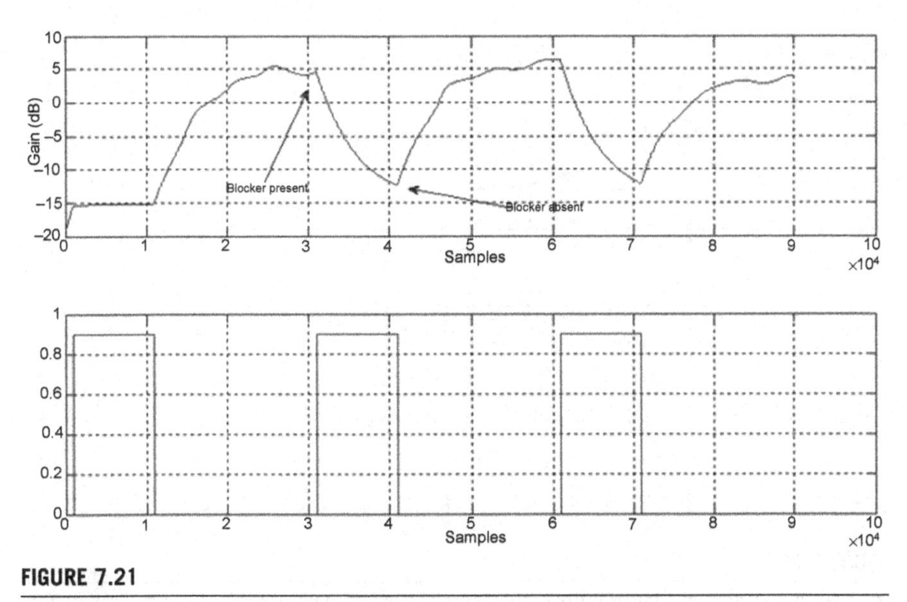

FIGURE 7.21

Gain control of automatic gain control in the presence of two-path Rayleigh fading and blocker signal with duty cycle = 1/3 (above) error signal and (below) strobe indicating presence of blocker. (For color version of this figure, the reader is referred to the online version of this book.)

the loop has reached a steady state recovering from the absence of a blocker when the blocking signal impinges back in the antenna causing the AGC error to become negative. The AGC loop reacts, slowly this time, to adjust the gain in the receiver lineup, hardly reaching steady state before the blocker disappears again causing the loop to react to the error by increasing the gain. This process can be well understood by further examining the output of the loop filter as shown in the upper portion of Figure 7.21. The top figure indicates the output of the loop filter block whereas the bottom portion of Figure 7.21 depicts the duty cycle of the blocker. In order to improve on the performance of the AGC, a faster loop response may be required at the expense of steady state performance. Next, we examine the performance of the loop for the same set of parameters with the exception of the duty cycle, which is now set at 10%. The outputs of the error signal, the loop filter, and the blocking signal duty cycle are shown in Figures 7.22 and 7.23, respectively. In this case, when the blocking signal is present, the AGC loop has barely any time to react to react before the signal blocker disappears. Again, the designer may choose a larger loop filter gain, thus speeding up the loop at the expense of steady state performance, or "tighten" the IF or baseband filters' response and allow for the FEC to deal with the bursty nature of the errors caused by the blocker. These are some examples of how to deal with gain control in the presence of the blocker. In general, the AGC algorithm seems to be simple enough, however, in practice the designer must

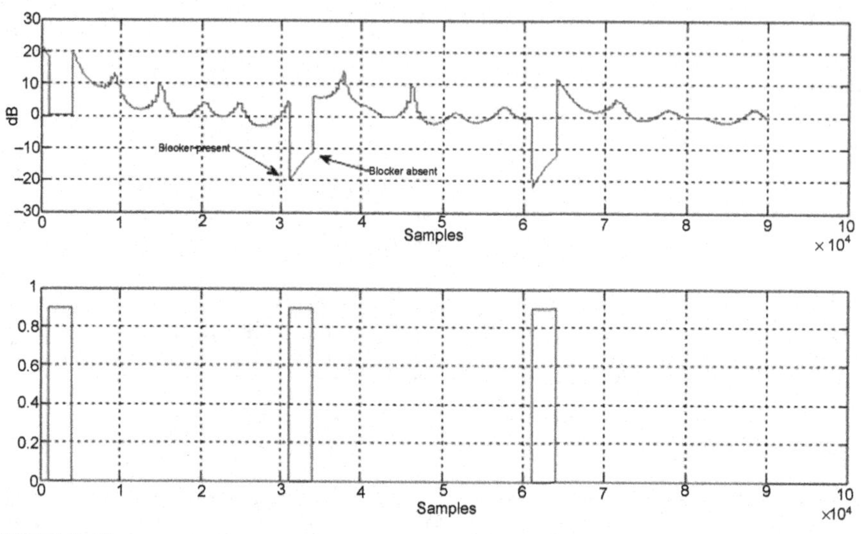

FIGURE 7.22

Error signal of automatic gain control in the presence of two-path Rayleigh fading and blocker signal with duty cycle = 1/10 (above) error signal and (below) strobe indicating presence of blocker. (For color version of this figure, the reader is referred to the online version of this book.)

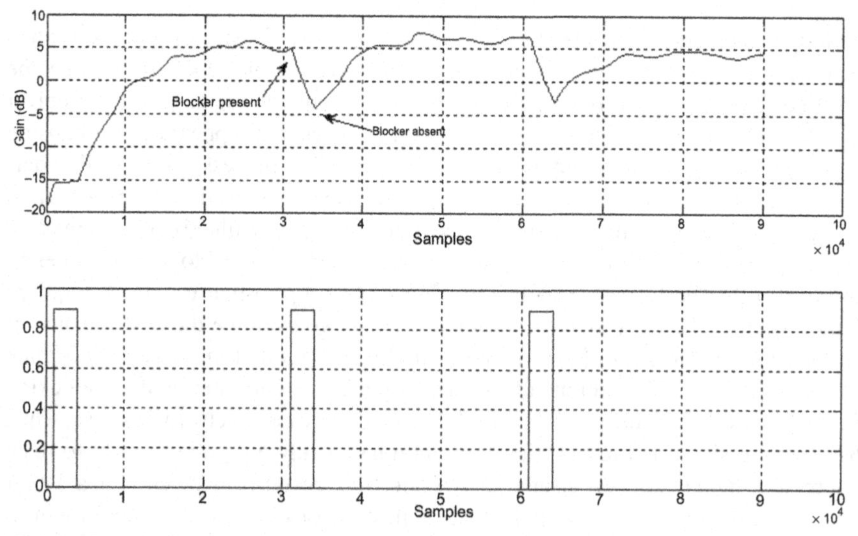

FIGURE 7.23

Gain control of automatic gain control in the presence of two-path Rayleigh fading and blocker signal with duty cycle = 1/10 (above) error signal and (below) strobe indicating presence of blocker. (For color version of this figure, the reader is referred to the online version of this book.)

understand the signaling waveform in full detail, the channel or environment where the radio will operate along with the possible blocking signals and interferers, and design the receiver lineup and the AGC loop with this information in mind.

7.2 Frequency synthesizers

Frequency synthesizers are an essential part of any wireless transceiver. Their role is to translate the desired signal from one frequency band, for example, RF, to another frequency band, for example IF, with minimal degradation. In this chapter, we will focus on frequency synthesis using PLLs. As was discussed in Chapter 3, and will be discussed in more detail herein, a PLL uses a single reference frequency to generate multiple output frequencies at the LO with accuracy and precision.

There are various types of PLL-based frequency synthesizers. An integer-N synthesizer, for example, uses a reference frequency F_{ref} to generate the output frequency F_{out} such that $F_{out} = NF_{ref}$. That is, the output frequencies of the synthesizer are generated as integer multiples of the reference frequency. An integer-N synthesizer's beauty lies in its simplicity, as compared to a fractional-N synthesizer, for example, and its low power consumption and die area. On the other hand, its shortcoming is evident in its flexibility of choosing the reference frequency F_{ref}. This limitation is overcome by the use of a fractional-N synthesizer. The output frequency of a fractional-N synthesizer is related to the reference frequency as $F_{out} = (P + \varepsilon)F_{ref}$, $\varepsilon = A/B$, where $0 \leq \varepsilon < 1$ and P, A, and B are integers.[2] This flexibility, however, comes at a price manifested in design complexity, increased power consumption and die area, and certain performance degradation. This degradation manifests itself in spurs occurring at the output spectrum of the synthesizer at regular intervals, as will be discussed following. These fractional spurs can be minimized by breaking up the periodicity of the division using a $\Delta\Sigma$ modulator for example. A $\Delta\Sigma$ modulator can be used to randomize the switching mechanism in the divider block and hence the instantaneous divider ratio. In this section, PLL-based frequency synthesizers will be studied. We first study the basics of PLL theory and its building blocks in the context of frequency synthesis and then proceed to discuss various synthesizer architectures.

7.2.1 Phase-locked loops

In Chapter 3 we discussed the role of the PLL in the context of frequency synthesis and its impact on phase noise. The emphasis was on the impact of phase noise degradation on the desired signal. In this chapter, the study of PLLs is mainly focused on the design of the loop. We first discuss the loop dynamics in general and then

[2]In order to be consistent with the notation below, we exchanged the letter N with P to denote the integer portion of the fractional-N divider.

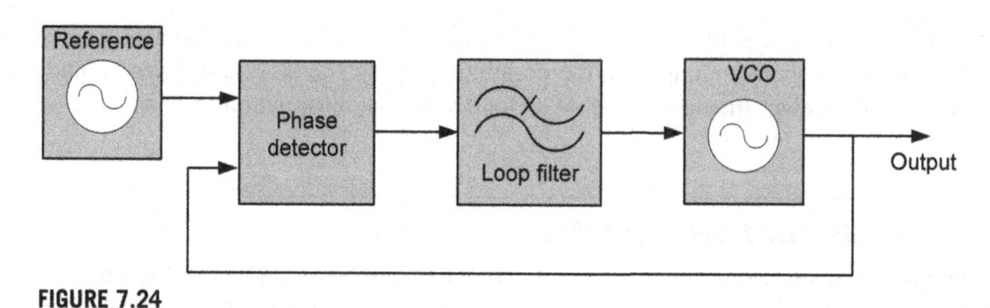

FIGURE 7.24

The basic blocks in a phase-locked loop. VCO, voltage-controlled oscillator.

proceed to the specifics. The loop order, its type, and general design parameters are studied in some detail.

A PLL, depicted in Figure 7.24, is comprised of a phase detector, a loop filter, and a VCO. The input to the PLL, say a frequency reference, is compared to the output of the VCO in the phase detector block. The output signal of the phase detector signifies the error or phase difference between the phase of the reference signal and the output signal of the VCO, possibly divided down to the level of the reference frequency. The error signal is then integrated via the loop filter whose output is presented to the VCO block. In the coming discussion, we will analyze and discuss the various aspects of PLL design. The linear PLL model is presented first followed by an analytical discussion concerning error convergence in the loop. The stability of the loop is then discussed in some detail with particular attention to first, second, and third order loops. The role of the phase detector is then discussed as well.

7.2.1.1 The linear-PLL model revisited

Consider the linear-PLL model depicted in Figure 7.25. In this model, the reference signal $\theta_i(s)$ is compared to the output phase of the VCO $\theta_o(s)$ in the phase detector

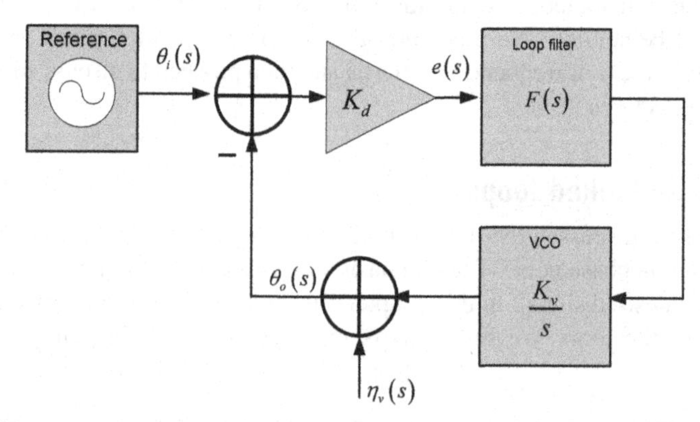

FIGURE 7.25

The linear-PLL model.

block. The phase detector is comprised of a summer block followed by a gain block K_d representing the gain of the phase detector. The analysis will be conducted in the Laplace domain. The output of the phase detector is the error signal given as

$$e(s) = K_d\{\theta_i(s) - \theta_o(s)\} \tag{7.21}$$

The VCO is comprised of an integrator with gain K_v plus additive phase noise $\eta_v(s)$. The frequency of the VCO is dictated by the output voltage of the loop filter. According to Figure 7.25, the output of the VCO $\theta_o(s)$ can be expressed in terms of the error signal $e(s)$:

$$\theta_o(s) = \frac{K_v}{s}F(s)e(s) + \eta_v(s) \tag{7.22}$$

Solving for the error signal $e(s)$ in terms of $\theta_o(s)$, we obtain

$$e(s) = \frac{s}{K_v L(s)}\{\theta_o(s) - \eta_v(s)\} \tag{7.23}$$

Substituting the value of the error function $e(s)$ found in Eqn (7.23) into Eqn (7.21), we obtain:

$$\frac{s}{K_v F(s)}\{\theta_o(s) - \eta_v(s)\} = K_d\{\theta_i(s) - \theta_o(s)\} \tag{7.24}$$

Solving for $\theta_o(s)$ in Eqn (7.24) results in the expression

$$\theta_o(s) = \frac{K_d K_v F(s)}{s + K_d K_v F(s)}\theta_i(s) + \frac{s}{s + K_d K_v F(s)}\eta_v(s) \tag{7.25}$$

Define the transfer function $H(s)$ as

$$H(s) = \frac{K_d K_v F(s)}{s + K_d K_v F(s)} \tag{7.26}$$

Then we can express

$$\begin{aligned} 1 - H(s) &= 1 - \frac{K_d K_v F(s)}{s + K_d K_v L(s)} \\ &= \frac{s}{s + K_d K_v F(s)} \end{aligned} \tag{7.27}$$

Define the open-loop gain function $G(s)$ as

$$G(s) = \frac{K_d K_v F(s)}{s} \tag{7.28}$$

Then the closed-loop transfer function can be expressed in terms of $G(s)$ by substituting Eqn (7.28) into Eqn (7.26), and we obtain:

$$H(s) = \frac{G(s)}{1 + G(s)} \quad \text{and} \quad 1 - H(s) = \frac{1}{1 + G(s)} \tag{7.29}$$

Furthermore, the relationship in Eqn (7.25) can be expressed in a simplified notation in terms of $H(s)$:

$$\theta_o(s) = H(s)\theta_o(s) + \{1 - H(s)\}\eta_v(s) \tag{7.30}$$

The transfer function $H(s)$ is known as the closed-loop transfer function and plays a pivotal role in shaping the reference and VCO phase noise (topic is addressed later).

7.2.1.2 Error convergence, order, and type of PLL

The investigation herein is pertinent to steady state analysis rather than transient analysis, that is, all transients are assumed to have subsided. Let us rewrite the error signal in Eqn (7.22) in terms of the reference signal $\theta_i(s)$ in order to obtain the error transfer function; that is, substitute Eqn (7.22) into Eqn (7.21) and solve for $e(s)/\theta_i(s)$:

$$\begin{aligned} e(s) &= K_d\{\theta_i(s) - \theta_o(s)\} \\ &= K_d\left\{\theta_i(s) - \frac{K_v}{s}F(s)e(s) + \eta_v(s)\right\} \Rightarrow \\ e(s) &= \frac{K_d s}{s + K_d K_v F(s)}\{\theta_i(s) + \eta_v(s)\} \\ &= K_d\{1 - H(s)\}\{\theta_i(s) + \eta_v(s)\} = \frac{K_d\{\theta(s) + \eta_v(s)\}}{1 + G(s)} \end{aligned} \tag{7.31}$$

Define the error transfer function as

$$\begin{aligned} \varepsilon(s) &= \frac{e(s)}{\theta_i(s)} = \frac{K_d s}{s + K_d K_v F(s)} \\ &= K_d\{1 - H(s)\} = \frac{K_d}{1 + G(s)} \end{aligned} \tag{7.32}$$

In steady state, we desire the error signal $e(t) = \mathfrak{L}^{-1}\{e(s)\}$ to approach zero at steady state. That is, provided a steady state value exists, then using the final value theorem we claim

$$\begin{aligned} \lim_{t\to\infty} e(t) &= \lim_{s\to 0} se(s) \\ &= \lim_{s\to 0} \frac{K_d s^2}{s + K_d K_v F(s)}\theta_i(s) = 0 \end{aligned} \tag{7.33}$$

The relationship in Eqn (7.33) can be satisfied depending on the choice of the loop filter as well as the input signal $\theta(t) \overset{\mathfrak{L}\{\cdot\}}{\rightleftarrows} \theta(s)$. To further elaborate, consider the elementary phase inputs:

$$\theta_i(t) = \begin{cases} \Delta\theta & \text{Phase step} \\ \Delta\omega t & \text{Frequency step} \\ \Delta\omega t^2 & \text{Frequency ramp} \end{cases} \tag{7.34}$$

The input phase functions presented in Eqn (7.34) are representative of most common inputs [3].

Consider first $\theta_i(t)$ to be a phase step $\Delta\theta$ whose s-domain equivalent is $\theta_i(t) = \Delta\theta \overset{\mathfrak{L}\{\cdot\}}{\rightleftarrows} \theta_i(s) = \Delta\theta/s$, then the condition that the loop filter must satisfy in order for the error to converge is

$$\begin{aligned}\lim_{t\to\infty} e(t) &= \lim_{s\to 0}\left\{s\frac{\Delta\theta}{s}\frac{k_d s}{s+K_d K_v F(s)}\right\}\\ &= \lim_{s\to 0}\left\{\Delta\theta\frac{k_d s}{s+K_d K_v F(s)}\right\}\\ &= \Delta\theta\frac{k_d\times 0}{0+K_d K_v F(0)}, \quad \text{iff } F(0)\neq 0\end{aligned} \tag{7.35}$$

The relationship in Eqn (7.35) implies that if the input is simply a constant phase shift, then the PLL's error will converge provided that $F(s = 0)\neq 0$, that is, for example, the loop filter $F(s)$ cannot be a highpass filter but it can be a constant, that is, $F(s) = \chi\in\mathbb{R}, \chi\neq 0$. Next, consider the input signal to be simply a frequency step $\Delta\omega t$ whose equivalent Laplace transform is $\theta_i(t) = \Delta\omega t \overset{\mathfrak{L}\{\cdot\}}{\rightleftarrows} \theta_i(s) = \Delta\omega/s^2$, then the error function of the PLL will converge provided that

$$\begin{aligned}\lim_{t\to\infty} e(t) &= \lim_{s\to 0}\left\{s\frac{\Delta\omega}{s^2}\frac{k_d s}{s+K_d K_v F(s)}\right\}\\ &= \lim_{s\to 0}\left\{\Delta\omega\frac{k_d}{s+K_d K_v F(s)}\right\}\\ &= \Delta\omega\frac{k_d}{K_d K_v F(0)} = 0 \quad \text{iff } F(0)\neq 0 \quad \text{and} \quad F(s)\\ &\qquad \text{has at least one pole at the origin}\end{aligned} \tag{7.36}$$

If only the condition $F(s = 0)\neq 0$ applies and the loop filter is a constant over all frequencies, that is, $F(s) = \chi\in\mathbb{R}, \ \chi\neq 0$, then the error $e(t)$ will not converge since

$$\begin{aligned}\lim_{t\to\infty} e(t) &= \lim_{s\to 0}\left\{s\frac{\Delta\omega}{s^2}\frac{k_d s}{s+K_d K_v F(s)}\right\}\\ &= \lim_{s\to 0}\left\{\Delta\omega\frac{k_d}{s+K_d K_v \chi}\right\}\\ &= \Delta\omega\frac{k_d}{K_d K_v \chi} = \text{constant and } \chi\neq 0\end{aligned} \tag{7.37}$$

Therefore, the relationships in Eqns (7.36) and (7.37) imply that the loop filter must have at least one pole at the origin. Finally, consider the case where the input signal is a frequency ramp, that is $\theta_i(t) = \Delta\omega t^2$ and whose Laplace transform pair is given simply as $\theta_i(t) = \Delta\omega t^2 \overset{\mathcal{L}\{\cdot\}}{\rightleftarrows} \theta_i(s) = \Delta\omega/s^3$, then the error signal of the PLL will converge provided that the loop filter satisfies the conditions:

$$\begin{aligned}
\lim_{t\to\infty} e(t) &= \lim_{s\to 0}\left\{s\frac{\Delta\omega}{s^3}\frac{k_d s}{s+K_d K_v F(s)}\right\} \\
&= \lim_{s\to 0}\left\{\frac{\Delta\omega}{s}\frac{k_d}{s+K_d K_v F(s)}\right\} \\
&= \Delta\omega\frac{k_d}{K_d K_v F(0)} = 0 \quad \text{iff } F(0)\neq 0 \quad \text{and} \quad F(s) \\
&\quad \text{has at least two poles at the origin}
\end{aligned} \tag{7.38}$$

If the conditions on the loop filter expressed in Eqn (7.38) are not met, then the error will not converge to zero at steady state. For example, for $F(s) = \chi \in \mathbb{R}, \chi \neq 0$, the error signal at steady state for $\theta_i(t) = \Delta\omega t^2$ becomes undefined:

$$\begin{aligned}
\lim_{t\to\infty} e(t) &= \lim_{s\to 0}\left\{s\frac{\Delta\omega}{s^3}\frac{k_d s}{s+K_d K_v F(s)}\right\} \\
&= \lim_{s\to 0}\left\{\frac{\Delta\omega}{s}\frac{k_d}{s+K_d K_v F(s)}\right\} \\
&= \frac{k_d}{K_d K_v \chi}\lim_{s\to 0}\left\{\frac{\Delta\omega}{s}\right\} \to \infty
\end{aligned} \tag{7.39}$$

In a similar vein, if the loop filter contains only one pole DC, for example, for $F(s) = c/s, \ c = \text{constant} \neq 0$, then

$$\begin{aligned}
\lim_{t\to\infty} e(t) &= \lim_{s\to 0}\left\{s\frac{\Delta\omega}{s^3}\frac{k_d s}{s+K_d K_v F(s)}\right\} \\
&= \lim_{s\to 0}\left\{\frac{\Delta\omega}{s}\frac{s k_d}{s+K_d K_v}\right\} \\
&= \frac{\Delta\omega k_d}{K_d K_v} \neq 0
\end{aligned} \tag{7.40}$$

According to Eqn (7.40), the error does not converge to zero.

The order of a loop is determined by the number of poles in the open-loop gain function $G(s)$. For example, if the loop transfer function $H(s) = \Upsilon/(s+\Upsilon)$, Υ is a constant, then

$$H(s) = \frac{\Upsilon}{s+\Upsilon} = \frac{\Upsilon/s}{1+\Upsilon/s} = \frac{G(s)}{1+G(s)} \Rightarrow G(s) = \frac{\Upsilon}{s} \tag{7.41}$$

The relationship in Eqn (7.41) implies that the loop is a first order loop. Note that the order of the loop can also be determined from the largest power of s in the denominator of the closed-loop transfer function $H(s)$. The type of a PLL is given as the number of perfect integrators in $G(s)$. A perfect integrator has its pole s at DC. So the loop described in Eqn (7.41) is a first order type 1 PLL. In theory, every PLL has at least one perfect integrator due to the VCO's transfer function. The type of the loop is important to applications where the steady state error performance is of paramount importance.

EXAMPLE 7.2 EQUIVALENT LOOP BANDWIDTH

Consider the first order loop $H(s) = \vartheta/s+\vartheta$, compute the loop noise equivalent bandwidth defined as:

$$B_L = \frac{1}{2\pi}\int_0^\infty |H(j\Omega)|^2 d\Omega \tag{7.42}$$

Compute the noise equivalent bandwidth of the second order loop:

$$H(s) = \frac{2\xi\Omega_n s+\Omega_n^2}{s^2+2\xi\Omega_n s+\Omega_n^2} \tag{7.43}$$

In Eqn (7.43), Ω_n is known as the natural frequency and ξ is the damping factor. According to Eqn (7.42), we need to compute:

$$\begin{aligned} B_L &= \frac{1}{2\pi}\int_0^\infty |H(j\Omega)|^2 d\Omega = \frac{1}{2\pi}\int_0^\infty \left|\frac{\vartheta}{j\Omega+\vartheta}\right|^2 d\Omega \\ &= \frac{\vartheta^2}{2\pi}\int_0^\infty \frac{1}{\Omega^2+\vartheta^2} d\Omega = \frac{\vartheta^2}{2\pi}\frac{1}{\vartheta}\tan^{-1}\left(\frac{\Omega}{\vartheta}\right)\Bigg|_0^\infty \\ &= \frac{\vartheta^2}{2\pi}\frac{1}{\vartheta}\frac{\pi}{2} = \frac{\vartheta}{4} \end{aligned} \tag{7.44}$$

The noise equivalent bandwidth of the first order loop is then dictated solely by the constant $\vartheta = K_d K_v$, which is the product of the loop detector constant and the constant due to the integrator of the VCO as shown in Figure 7.26.

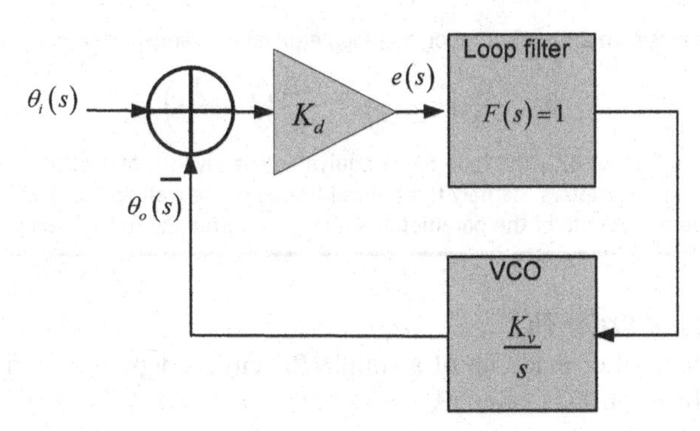

FIGURE 7.26

The first order phase-locked loop.

Continued

EXAMPLE 7.2 EQUIVALENT LOOP BANDWIDTH—cont'd

Similarly, the loop noise equivalent bandwidth of the second order loop whose transfer function is expressed in Eqn (7.43) is:

$$
\begin{aligned}
B_L &= \frac{1}{2\pi}\int_0^\infty |H(j\Omega)|^2 d\Omega \\
&= \frac{1}{2\pi}\int_0^\infty \frac{1+\left(2\xi\frac{\Omega}{\Omega_n}\right)^2}{\left[1-\left(\frac{\Omega}{\Omega_n}\right)^2\right]^2+\left(2\xi\frac{\Omega}{\Omega_n}\right)^2} d\Omega \\
&= \frac{\Omega_n}{2\pi}\int_0^\infty \frac{1+4\xi^2\left(\frac{\Omega}{\Omega_n}\right)^2}{\left(\frac{\Omega}{\Omega_n}\right)^4+2\left(2\xi^2-1\right)\left(\frac{\Omega}{\Omega_n}\right)^2+1} d\Omega
\end{aligned}
\tag{7.45}
$$

The integral in Eqn (7.45) can be divided into two subintegrals, namely

$$
B_L = \frac{\Omega_n}{2\pi}\left[\int_0^\infty \frac{1}{\left(\frac{\Omega}{\Omega_n}\right)^4+2\left(2\xi^2-1\right)\left(\frac{\Omega}{\Omega_n}\right)^2+1} d\Omega + \int_0^\infty \frac{4\xi^2\left(\frac{\Omega}{\Omega_n}\right)^2}{\left(\frac{\Omega}{\Omega_n}\right)^4+2\left(2\xi^2-1\right)\left(\frac{\Omega}{\Omega_n}\right)^2+1} d\Omega\right] \tag{7.46}
$$

According to Ref. [4], each integral in Eqn (7.46) is of the form

$$
\int_0^\infty \frac{x^{r-1}}{(x^2+\alpha)(x^2+\beta)} dx = \frac{\pi}{2}\frac{\beta^{r/2-1}-\alpha^{r/2-1}}{\alpha-\beta}\csc\left(\frac{\pi r}{2}\right) \tag{7.47}
$$

where $|\arg(\alpha)| < \pi$, $|\arg(\beta)| < \pi$, and $0 < \mathrm{Re}\{r\} < 4$.

Let

$$
\begin{aligned}
\alpha &= \left(\xi+\sqrt{\xi^2-1}\right)^2 \\
\alpha &= \left(\xi-\sqrt{\xi^2-1}\right)^2
\end{aligned}
\tag{7.48}
$$

Then after a certain amount of labor, the loop equivalent bandwidth can be obtained as

$$
B_L = \frac{\Omega_n}{2}\left(\xi+\frac{1}{4\xi}\right) = \frac{\xi\Omega_n}{2}\left(1+\frac{1}{4\xi^2}\right) \tag{7.49}
$$

According to Eqn (7.49), the loop noise equivalent bandwidth of a second order loop is dependent on two parameters, namely the natural frequency as well as the damping factor. We will examine the effects of the parameters on loop performance in the next section.

7.2.1.3 Second order PLL

Consider the loop filter made up of a simple RC circuit depicted in Figure 7.27(a) with transfer function $F(s)$ given as

$$
F(s) = \frac{1}{1+RCs} = \frac{1}{1+\tau s} \tag{7.50}
$$

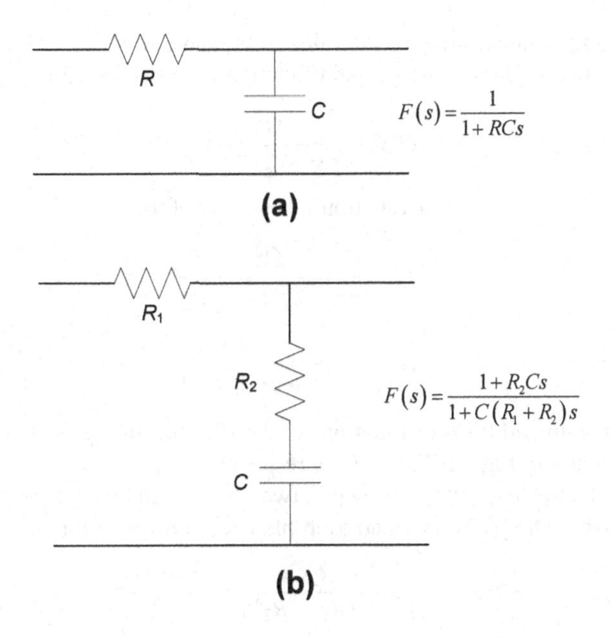

FIGURE 7.27

Second order phase-locked loop loop filters: (a) simple passive *RC* filter, and (b) active-phase lag-lead filter.

where $\tau = RC$ is the filter's time constant. According to Eqn (7.28), the open-loop gain function $G(s)$ can be expressed in conjunction with the loop filter's transfer function given in Eqn (7.50) as

$$G(s) = \frac{K_d K_v F(s)}{s} = \frac{K_d K_v \frac{1}{1+RCs}}{s} = \frac{K}{s(1+\tau s)} \tag{7.51}$$

where for simplicity we chose $K = K_d K_v$. The PLL's transfer function can then be expressed by substituting Eqn (7.51) into Eqn (7.29) in order to obtain

$$\begin{aligned} H(s) &= \frac{G(s)}{1+G(s)} = \frac{K/\tau}{s^2 + s/\tau + K/\tau} \\ 1 - H(s) &= \frac{1}{1+G(s)} = \frac{s(\tau s + 1)}{s^2 + s/\tau + K/\tau} \end{aligned} \tag{7.52}$$

Borrowing from the theory of vibrations, define the natural frequency Ω_n and damping factors ξ as

$$\begin{aligned} \Omega_n &= \sqrt{\frac{K}{\tau}} \quad \text{and} \quad \xi = \frac{\Omega_n}{2K} \\ \xi &= \frac{\sqrt{\frac{K}{\tau}}}{2K} = \frac{1}{2\Omega_n \tau} \end{aligned} \tag{7.53}$$

Then from a simple substitution of the value of Ω_n and ξ supplied in Eqn (7.53) into Eqn (7.51) we obtain the open-loop gain function $G(s)$ expressed as

$$G(s) = \frac{\Omega_n^2}{s^2 + 2\xi\Omega_n s} \tag{7.54}$$

Accordingly, the loop's transfer function takes on the form

$$\begin{aligned} H(s) &= \frac{\Omega_n^2}{s^2 + 2\xi\Omega_n s + \Omega_n^2} \\ 1 - H(s) &= \frac{s^2 + 2\xi\Omega_n s}{s^2 + 2\xi\Omega_n s + \Omega_n^2} \end{aligned} \tag{7.55}$$

The open loop gain and transfer functions of the PLL according to Eqns (7.54) and (7.55) are depicted in Figures 7.28–7.30, respectively.

Next consider the loop filter made up of two resistors and one capacitor as shown in Figure 7.27(b). The phase lag-lead loop filter has a transfer function

$$F(s) = \frac{1 + R_2Cs}{1 + C(R_1 + R_2)s} = \frac{\tau_2 s + 1}{\tau_1 s + 1} \tag{7.56}$$

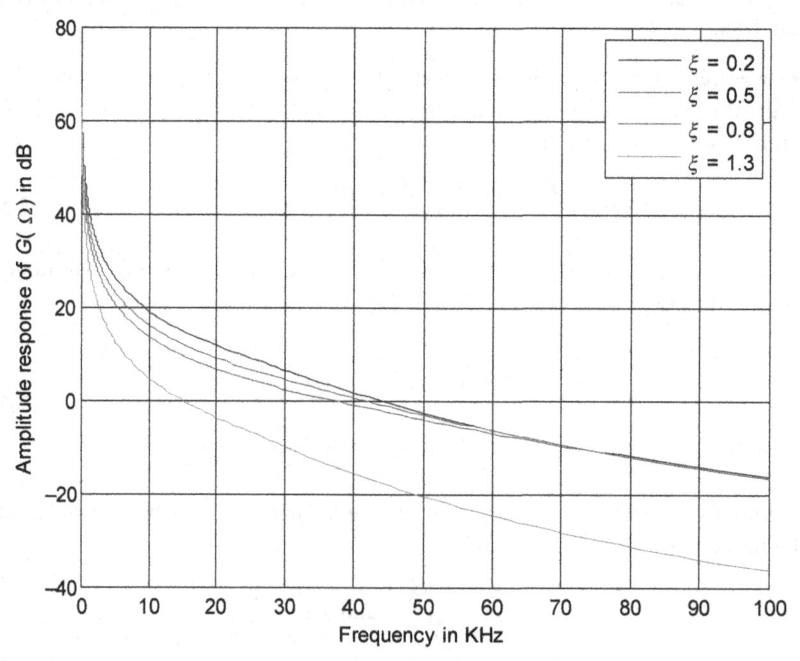

FIGURE 7.28

Amplitude response of $G(s)$ according to Eqn (7.55) for 5-KHz loop bandwidth and various damping factors ξ. (For color version of this figure, the reader is referred to the online version of this book.)

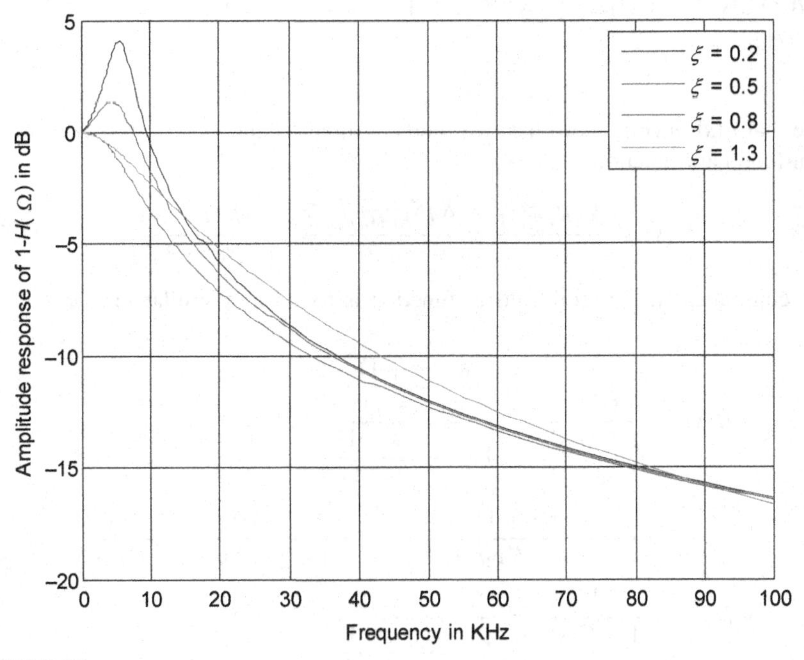

FIGURE 7.29

Amplitude response of $H(s)$ according to Eqn (7.55) for 5-KHz loop bandwidth and various damping factors ξ. (For color version of this figure, the reader is referred to the online version of this book.)

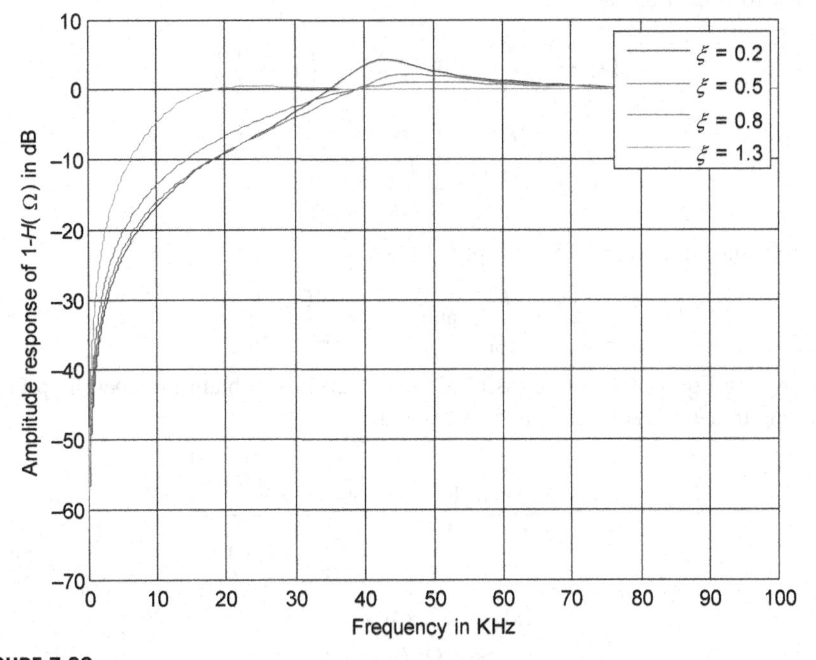

FIGURE 7.30

Amplitude response of $1-H(s)$ according to Eqn (7.55) for 5-KHz loop bandwidth and various damping factors ξ. (For color version of this figure, the reader is referred to the online version of this book.)

Note that *F*(*s*) in (Eqn 7.56) has two time constants τ_1 and τ_2 resulting more in the open-loop gain function

$$G(s) = \frac{K_d K_v F(s)}{s} = \frac{K_d K_v \frac{1+R_2 Cs}{1+C(R_1+R_2)s}}{s} = \frac{K(\tau_2 s+1)}{s(\tau_1 s+1)} \tag{7.57}$$

And consequently, the loop transfer function is found in a similar manner to Eqn (7.52):

$$H(s) = \frac{G(s)}{1+G(s)} = \frac{\dfrac{K(\tau_2 s+1)}{s(\tau_1 s+1)}}{1+\dfrac{K(\tau_2 s+1)}{s(\tau_1 s+1)}}$$

$$= \frac{K(\tau_2 s+1)}{\tau_1 s^2+s+K\tau_2 s+K} = \frac{K(\tau_2 s+1)/\tau_1}{s^2+(K\tau_2+1)s/\tau_1+K/\tau_1} \tag{7.58}$$

$$1-H(s) = \frac{1}{1+G(s)} = \frac{1}{1+\dfrac{K(\tau_2 s+1)}{s(\tau_1 s+1)}}$$

$$= \frac{s(\tau_1 s+1)}{\tau_1 s^2+s+K\tau_2 s+K} = \frac{s(\tau_1 s+1)/\tau_1}{s^2+(K\tau_2+1)s/\tau_1+K/\tau_1}$$

The definition of the natural frequency and damping factors are given in a similar manner to (Eqn 7.53) as

$$\Omega_n = \sqrt{\frac{K}{\tau_1}} \quad \text{and} \quad \xi = \frac{\Omega_n}{2}\left(\tau_2+\frac{1}{K}\right)$$

$$\xi = \frac{\sqrt{\dfrac{K}{\tau_1}}}{2}\left(\tau_2+\frac{1}{K}\right) \tag{7.59}$$

The relationship in (Eqn 7.59) simply implies

$$\tau_1 = \frac{K}{\Omega_n^2} \quad \text{and} \quad \tau_2 = \frac{2\xi}{\Omega_n}-\frac{1}{K} \tag{7.60}$$

Substituting Eqn (7.60) into Eqns (7.57) and (7.58), we obtain the open-loop gain and loop transfer functions to be expressed as

$$G(s) = \frac{K(\tau_2 s+1)}{s(\tau_1 s+1)} = \frac{K\left[\left(\dfrac{2\xi}{\Omega_n}-\dfrac{1}{K}\right)s+1\right]}{s\left(\dfrac{K}{\Omega_n^2}s+1\right)} \tag{7.61}$$

$$= \frac{(2\xi\Omega_n-\Omega_n^2/K)s+\Omega_n^2}{s(s+\Omega_n^2/K)}$$

and

$$H(s) = \frac{G(s)}{1+G(s)} = \frac{\dfrac{(2\xi\Omega_n - \Omega_n^2)s + \Omega_n^2}{s(s+\Omega_n^2/K)}}{1+\dfrac{(2\xi\Omega_n - \Omega_n^2)s + \Omega_n^2}{s(s+\Omega_n^2/K)}} = \frac{(2\xi\Omega_n - \Omega_n^2/K)s + \Omega_n^2}{s^2 + 2\xi\Omega_n s + \Omega_n^2}$$

$$1 - H(s) = \frac{1}{1+G(s)} = \frac{1}{1+\dfrac{(2\xi\Omega_n - \Omega_n^2)s + \Omega_n^2}{s(s+\Omega_n^2/K)}} = \frac{s(s+\Omega_n^2/K)}{s^2 + 2\xi\Omega_n s + \Omega_n^2}$$

(7.62)

The relationships in Eqns (7.61) and (7.62) imply that the PLL has three degrees of freedom that could manipulate its performance as opposed to the PLL described in Eqns (7.54) and (7.55). The open-loop gain and transfer functions of the PLL according to Eqns (7.61) and (7.62) are shown in Figures 7.31—7.33, respectively. With the current choice of detector and VCO gains $K = K_d K_v$, a comparison between open-loop gain and amplitude response of the PLL's transfer function

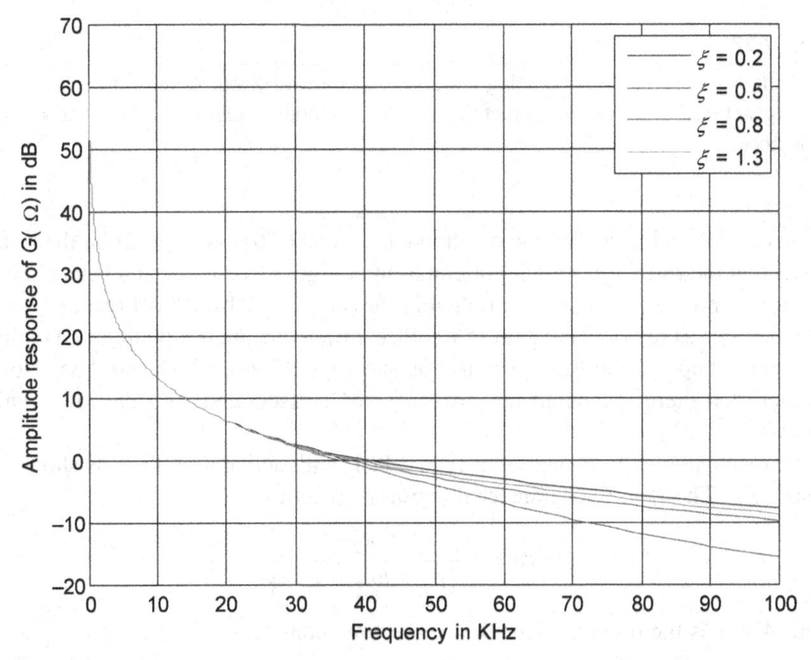

FIGURE 7.31

Amplitude response of *G*(*s*) according to Eqn (7.61) for 5-KHz loop bandwidth and various damping factors ξ. (For color version of this figure, the reader is referred to the online version of this book.)

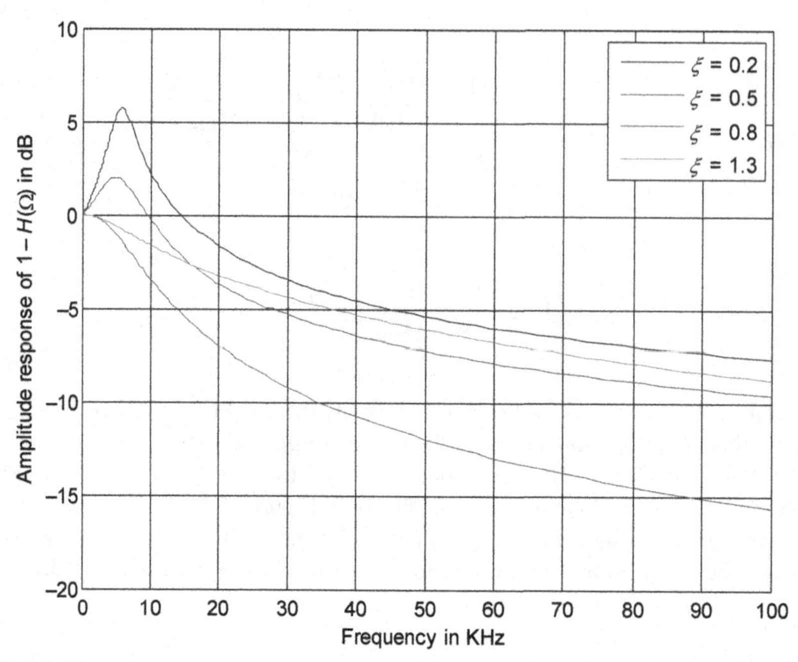

FIGURE 7.32

Amplitude response of $H(s)$ according to Eqn (7.62) for 5-KHz loop bandwidth and various damping factors ξ. (For color version of this figure, the reader is referred to the online version of this book.)

of Eqns (7.54) and (7.55) on the one hand and Eqns (7.61) and (7.62) on the other reveals that the former has higher attenuation at higher frequency and hence better rejection of noise. The impact of reducing the gain K say by 1.5 dB from its original value serves to boost the gain of the PLL's transfer function without changing the overall shape of the PLL response as shown in Figures 7.34 and 7.35. Note, however, that there is a small upward shift in Ω_n associated with the lower gain response.

Next we examine a second order type 2 PLL with active loop filter as shown in Figure 7.36. The loop filter transfer function is given as:

$$F(s) = \frac{1 + R_2Cs}{C\left(R_1 + \frac{R_1+R_2}{A}\right)s + \frac{1}{A}} \tag{7.63}$$

where $A \gg 1$ is the op-amp gain. Define the time constants τ_1 and τ_2 as

$$\begin{aligned} \tau_1 &= C\left(R_1 + \frac{R_1 + R_2}{A}\right)_{A \gg 1} \approx R_1C \\ \tau_2 &= R_2C \end{aligned} \tag{7.64}$$

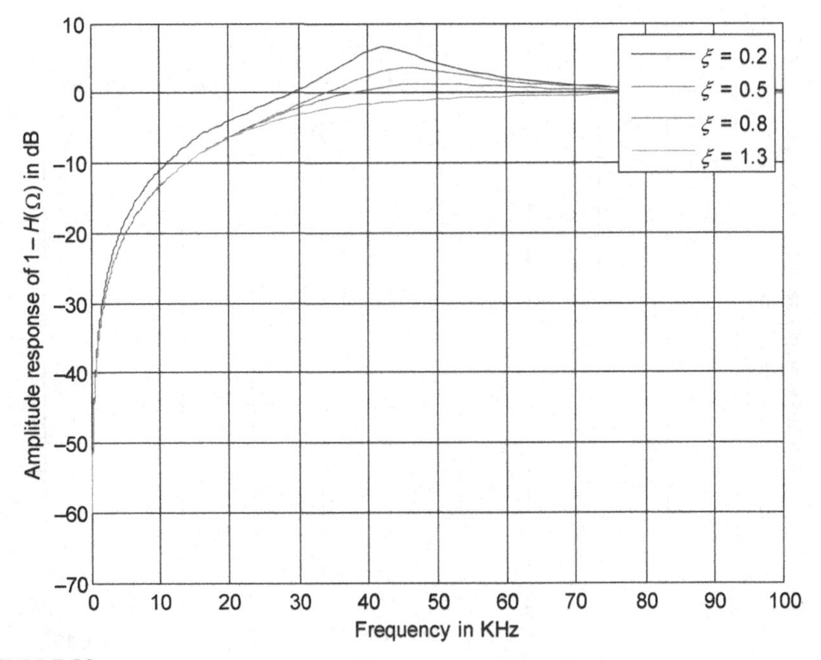

FIGURE 7.33

Amplitude response of $1 - H(s)$ according to Eqn (7.62) for 5-KHz loop bandwidth and various damping factors ξ. (For color version of this figure, the reader is referred to the online version of this book.)

Substituting the time constants given in Eqn (7.64) into Eqn (7.63), we obtain the loop filter transfer function as

$$F(s) = \frac{\tau_2 s + 1}{\tau_1 s + \frac{1}{A}} \tag{7.65}$$

According to Eqn (7.57), the open-loop gain for using an active loop filter such as Eqn (7.65) can be found as

$$\begin{aligned} G(s) &= \frac{K_d K_v F(s)}{s} = \frac{K_d K_v \dfrac{\tau_2 s + 1}{\tau_1 s + \dfrac{1}{A}}}{s} \\ &= \frac{K_d K_v (\tau_2 s + 1)}{s\left(\tau_1 s + \dfrac{1}{A}\right)} = \frac{K(\tau_2 s + 1)}{s\left(\tau_1 s + \dfrac{1}{A}\right)} \end{aligned} \tag{7.66}$$

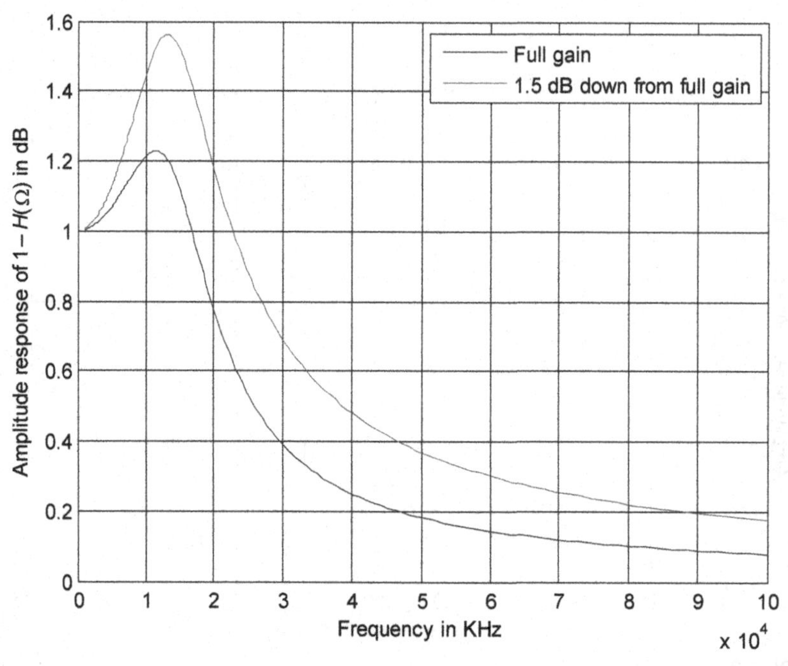

FIGURE 7.34

Amplitude response of $H(s)$ according to Eqn (7.62) for 1.5-KHz loop bandwidth $\xi = 0.2$ and various values of K. (For color version of this figure, the reader is referred to the online version of this book.)

The magnitude squared response of $G(s)$ implies that at high frequencies the gain of the loop diminishes to a gain similar to that of a first order loop $K\tau_2/\tau_1$, as can be shown

$$\lim_{\Omega\to\infty}|G(j\Omega)|^2 = \lim_{\Omega\to\infty}\left|\frac{K^2(\tau_2^2\Omega^2+1)}{\Omega^2\left(\tau_1^2\Omega^2+\dfrac{1}{A^2}\right)}\right|_{A\gg 1}$$
$$\approx \lim_{\Omega\to\infty}\left|\frac{K^2\tau_2^2\Omega^2}{\tau_1^2\Omega^4}\right| = \lim_{\Omega\to\infty}\left|\frac{K^2\tau_2^2}{\tau_1^2\Omega^2}\right| \tag{7.67}$$

This may imply that the loop exhibits the same lock-in characteristics as the first order loop at high frequencies. According to Eqn (7.67), the effective gain of the loop has been reduced to $K\tau_2/\tau_1$. The PLL transfer function for the loop filter given in Eqn (7.65) is given according to Eqn (7.52) as

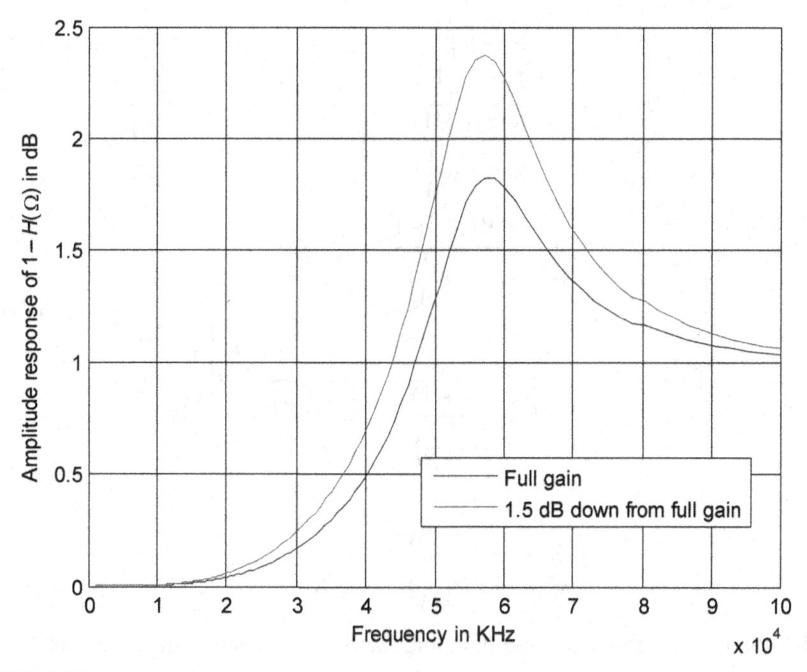

FIGURE 7.35

Amplitude response of $1 - H(s)$ according to Eqn (7.62) for 1.5-KHz loop bandwidth $\xi = 0.2$ and various values of K. (For color version of this figure, the reader is referred to the online version of this book.)

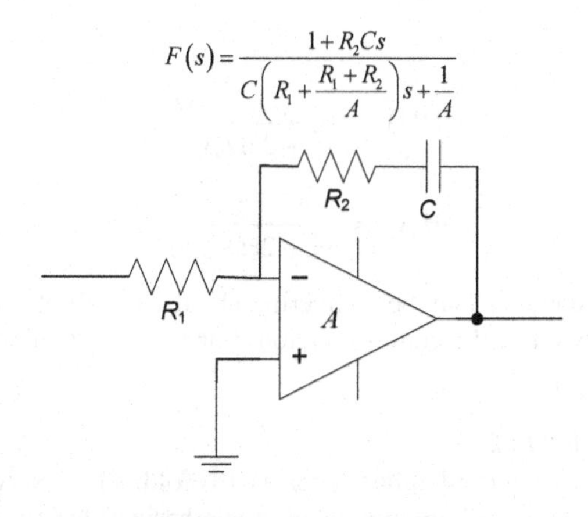

FIGURE 7.36

Active loop filter for second order type 2 phase-locked loop.

$$
\begin{aligned}
H(s) &= \frac{\dfrac{K(\tau_2 s+1)}{s(\tau_1 s+\frac{1}{A})}}{1+\dfrac{K(\tau_2 s+1)}{s(\tau_1 s+\frac{1}{A})}} = \frac{K(\tau_2 s+1)}{\tau_1 s^2+(K\tau_2+\frac{1}{A})s+K} \\
&= \frac{K(\tau_2 s+1)/\tau_1}{s^2+(K\tau_2+\frac{1}{A})s/\tau_1+K/\tau_1} \\
1-H(s) &= \frac{1}{1+\dfrac{K(\tau_2 s+1)}{s(\tau_1 s+\frac{1}{A})}} = \frac{\tau_1 s^2+\frac{1}{A}s}{\tau_1 s^2+(K\tau_2+\frac{1}{A})s+K} \\
&= \frac{s^2+\frac{1}{A\tau_1}s}{s^2+(K\tau_2+\frac{1}{A})s/\tau_1+K/\tau_1}
\end{aligned}
\tag{7.68}
$$

Define the natural frequency and damping factor for the active filter case as

$$
\Omega_n = \sqrt{\frac{K}{\tau_1}} \quad \text{and} \quad \xi = \frac{1}{2\sqrt{\frac{K}{\tau_1}}}\left(\frac{K\tau_2+\frac{1}{A}}{\tau_1}\right) = \frac{\tau_2\Omega_n}{2} \tag{7.69}
$$

The simplified forms of Eqns (7.66) and (7.68) can be obtained by substituting Eqn (7.69) to obtain

$$
G(s) = \frac{2\xi\Omega_n s+\Omega_n^2}{s^2} \tag{7.70}
$$

and

$$
\begin{aligned}
H(s) &= \frac{2\xi\Omega_n s+\Omega_n^2}{s^2+2\xi\Omega_n s+\Omega_n^2} \\
1-H(s) &= \frac{s^2}{s^2+2\xi\Omega_n s+\Omega_n^2}
\end{aligned}
\tag{7.71}
$$

At this juncture, the reader may be wondering about the stability of the loop. This topic will be addressed in the coming section. For now, we will turn our attention to the third order loop.

7.2.1.4 Third order PLL

Our interest in studying third order loops is twofold: (1) a second order loop, although the most common loop employed in synthesizer design, may not deliver the required spur attenuation desired, and (2) the presence of parasitic poles in a second order loop may unintentionally lead to a third order transfer function. It is important to reiterate, however, that most synthesizers rely on second order type 2 PLL in their designs. A third order PLL naturally adds a third pole to the loop filter. In frequency synthesis, the need for a third order PLL is the need for increased spurious suppression for the sake of spectral purity. Furthermore, third order PLLs suffer from shorter pull-in range as well as stability realization. A passive loop filter intended for a third order loop is comprised of two *RC* sections in cascade as shown in Figure 7.37(a) with loop filter transfer function given as

$$F(s) = \frac{1}{\tau_1 \tau_3 s^2 + (\tau_1 + \tau_2 + \tau_3)s + 1} \quad \text{where } \tau_1 = R_1 C_1, \tau_2 = R_1 C_2, \tau_3 = R_2 C_2 \tag{7.72}$$

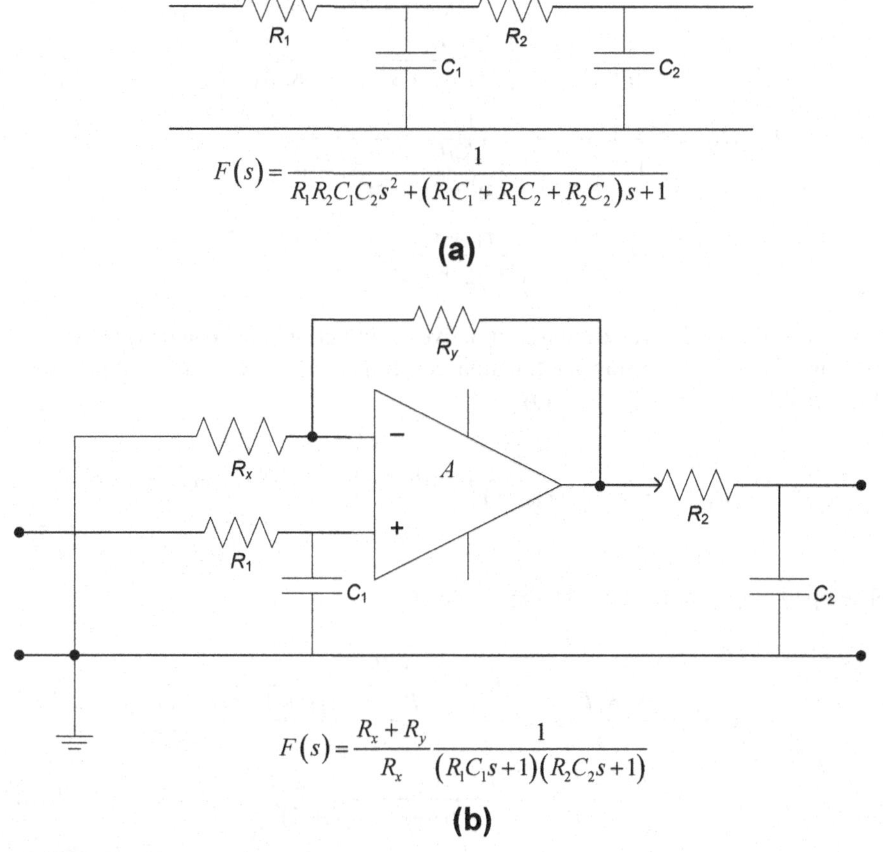

FIGURE 7.37

Loop filter for a third order loop: (a) passive filter comprised of two *RC* sections, and (b) active filter comprised of two *RC* sections.

According to Eqn (7.28), the open loop gain $G(s)$ can be expressed as

$$G(s) = \frac{K_d K_v F(s)}{s} = \frac{K_d K_v \dfrac{1}{\tau_1 \tau_3 s^2 + (\tau_1 + \tau_2 + \tau_3)s + 1}}{s} = \frac{K_d K_v}{\tau_1 \tau_3 s^3 + (\tau_1 + \tau_2 + \tau_3)s^2 + s} \tag{7.73}$$

Substituting Eqn (7.73) into Eqn (7.29), we obtain the loop transfer functions

$$H(s) = \frac{G(s)}{1+G(s)} = \frac{\dfrac{K_d K_v}{\tau_1 \tau_3 s^3 + (\tau_1 + \tau_2 + \tau_3)s^2 + s}}{1 + \dfrac{K_d K_v}{\tau_1 \tau_3 s^3 + (\tau_1 + \tau_2 + \tau_3)s^2 + s}} = \frac{K_d K_v}{\tau_1 \tau_3 s^3 + (\tau_1 + \tau_2 + \tau_3)s^2 + s + K_d K_v}$$

$$1 - H(s) = \frac{1}{1 + \dfrac{K_d K_v}{\tau_1 \tau_3 s^3 + (\tau_1 + \tau_2 + \tau_3)s^2 + s}} = \frac{\tau_1 \tau_3 s^3 + (\tau_1 + \tau_2 + \tau_3)s^2 + s}{\tau_1 \tau_3 s^3 + (\tau_1 + \tau_2 + \tau_3)s^2 + s + K_d K_v} \tag{7.74}$$

Similarly, Figure 7.37(b) displays an active amplifier supplemented with two RC sections. The resulting transfer function constitutes a loop filter for a third order PLL given as (Figures 7.38–7.40)

$$F(s) = \frac{R_x + R_y}{R_x} \frac{1}{(\tau_1 s + 1)(\tau_3 s + 1)} \quad \text{where } \tau_1 = R_1 C_1, \text{ and } \tau_3 = R_2 C_2 \tag{7.75}$$

The open loop gain $G(s)$ can be expressed as

$$G(s) = \frac{K_d K_v F(s)}{s} = \frac{K_d K_v \dfrac{R_x + R_y}{R_x} \dfrac{1}{(\tau_1 s + 1)(\tau_3 s + 1)}}{s} = \frac{K_d K_v (R_x + R_y)}{R_x} \frac{1}{s(\tau_1 s + 1)(\tau_3 s + 1)} = \Psi \frac{1}{\tau_1 \tau_3 s^3 + (\tau_1 + \tau_3)s^2 + s}\Bigg|_{\Psi = \frac{K_d K_v (R_x + R_y)}{R_x}} \tag{7.76}$$

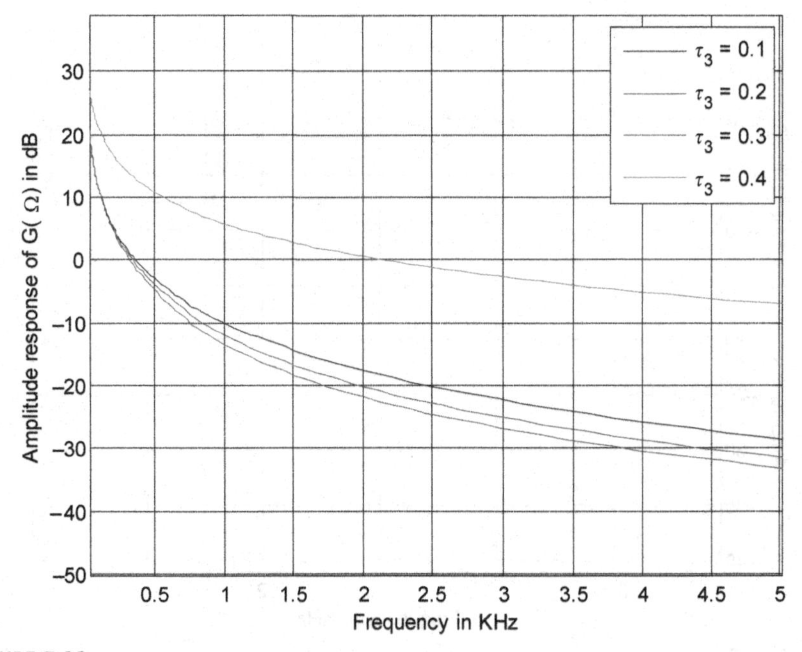

FIGURE 7.38

Open-loop gain $G(s)$ for third order PLL employing passive *RC* sections. (For color version of this figure, the reader is referred to the online version of this book.)

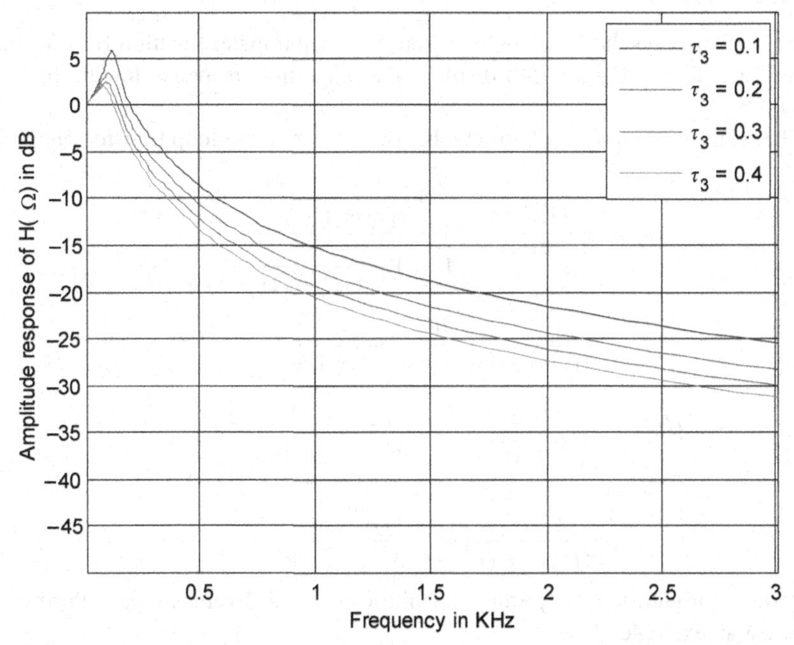

FIGURE 7.39

Magnitude response of loop transfer function $H(s)$ for third order PLL employing passive *RC* sections.(For color version of this figure, the reader is referred to the online version of this book.)

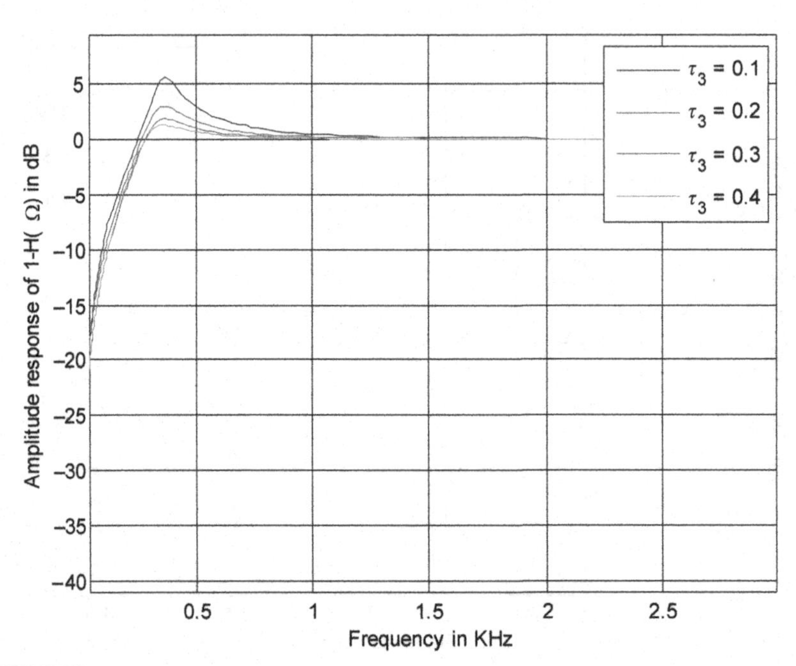

FIGURE 7.40

Magnitude response of loop transfer function $1 - H(s)$ for third order PLL employing passive *RC* sections. (For color version of this figure, the reader is referred to the online version of this book.)

Figure 7.39 displays the Magnitude response of loop transfer function H(s) for third order PLL whereas Figure 7.40 displays the magnitude response for the function 1-H(s)

The relationship in Eqn (7.76) can be used to derive the loop transfer functions:

$$\begin{aligned} H(s) &= \frac{G(s)}{1+G(s)} = \frac{\Psi \dfrac{1}{\tau_1\tau_3 s^3 + (\tau_1+\tau_3)s^2 + s}}{1+\Psi \dfrac{1}{\tau_1\tau_3 s^3 + (\tau_1+\tau_3)s^2 + s}} \\ &= \frac{\Psi}{\tau_1\tau_3 s^3 + (\tau_1+\tau_3)s^2 + s + \Psi} \\ 1 - H(s) &= \frac{1}{1+G(s)} = \frac{1}{1+\Psi \dfrac{1}{\tau_1\tau_3 s^3 + (\tau_1+\tau_3)s^2 + s}} \\ &= \frac{\tau_1\tau_3 s^3 + (\tau_1+\tau_3)s^2 + s}{\tau_1\tau_3 s^3 + (\tau_1+\tau_3)s^2 + s + \Psi} \end{aligned} \tag{7.77}$$

Next we highlight the steady state capabilities of type 2 versus type 3 third order loops via an example.

EXAMPLE 7.3 STEADY STATE PERFORMANCE OF TYPE 2 AND TYPE 3 THIRD-ORDER PLLS

The generic error transfer functions of type 2 and type 3 third order PLLs are given, respectively, in simplified form as

$$\varepsilon_{\text{type } 2}(s) = \frac{e_{\text{type } 2}(s)}{\theta_i(s)} = \frac{\Psi(\tau_1\tau_3 s^3 + \tau_1 s^2)}{\tau_1\tau_3 s^3 + \tau_1 s^2 + \Psi K\tau_2 s + \Psi K} \tag{7.78}$$

and

$$\varepsilon_{\text{type } 3}(s) = \frac{e_{\text{type } 3}(s)}{\theta_i(s)} = \frac{\Psi s^3}{\tau_1^2 s^3 + \Psi K\tau_1^2 s^2 + 2\Psi K\tau_2 s + \Psi K} \tag{7.79}$$

Determine the steady state error for the phase input signals and state your observations:

$$\theta_i(t) = \begin{cases} \Delta\theta & \text{Phase step} \\ \Delta\omega t & \text{Frequency step} \\ \Delta\omega' t^2 & \text{Frequency ramp} \end{cases} \quad \theta_i(s) = \begin{cases} \Delta\theta/s & \text{Phase step} \\ \Delta\omega/s^2 & \text{Frequency step} \\ 2\Delta\omega'/s^3 & \text{Frequency ramp} \end{cases} \tag{7.80}$$

Recall that the final value theorem, as presented in Eqn (7.33), predicts the performance of the error transfer function in steady state, that is, $\lim_{t\to\infty} e(t) = \lim_{s\to 0} se(s) = \lim_{s\to 0} s\theta_i(s)\varepsilon(s)$. For both the phase step and the frequency step inputs, it can be easily shown that the steady state error response of both loop types reaches zero, that is, for phase step

$$\lim_{t\to\infty} e_{\text{type } 2}(t) = \lim_{s\to 0} s\theta_i(s)\varepsilon_{\text{type } 2}(s) = \lim_{s\to 0}\left[s\frac{\Delta\theta}{s}\frac{\Psi(\tau_1\tau_3 s^3 + \tau_1 s^2)}{\tau_1\tau_3 s^3 + \tau_1 s^2 + \Psi K\tau_2 s + \Psi K}\right] = 0$$

$$\lim_{t\to\infty} e_{\text{type } 3}(t) = \lim_{s\to 0} s\theta_i(s)\varepsilon_{\text{type } 3}(s) = \lim_{s\to 0}\left[s\frac{\Delta\theta}{s}\frac{\Psi s^3}{\tau_1^2 s^3 + \Psi K\tau_1^2 s^2 + 2\Psi K\tau_2 s + \Psi K}\right] = 0 \tag{7.81}$$

and similarly for a frequency step

$$\lim_{t\to\infty} e_{\text{type } 2}(t) = \lim_{s\to 0} s\theta_i(s)\varepsilon_{\text{type } 2}(s) = \lim_{s\to 0}\left[s\frac{\Delta\omega}{s^2}\frac{\Psi(\tau_1\tau_3 s^3 + \tau_1 s^2)}{\tau_1\tau_3 s^3 + \tau_1 s^2 + \Psi K\tau_2 s + \Psi K}\right] = 0$$

$$\lim_{t\to\infty} e_{\text{type } 3}(t) = \lim_{s\to 0} s\theta_i(s)\varepsilon_{\text{type } 3}(s) = \lim_{s\to 0}\left[s\frac{\Delta\omega}{s^2}\frac{\Psi s^3}{\tau_1^2 s^3 + \Psi K\tau_1^2 s^2 + 2\Psi K\tau_2 s + \Psi K}\right] = 0 \tag{7.82}$$

Next we consider the case for which the input is a frequency ramp

$$\lim_{t\to\infty} e_{\text{type } 2}(t) = \lim_{s\to 0} s\theta_i(s)\varepsilon_{\text{type } 2}(s) = \lim_{s\to 0}\left[s\frac{2\Delta\omega'}{s^3}\frac{\Psi(\tau_1\tau_3 s^3 + \tau_1 s^2)}{\tau_1\tau_3 s^3 + \tau_1 s^2 + \Psi K\tau_2 s + \Psi K}\right]$$

$$= \lim_{s\to 0}\left[2\Delta\omega'\frac{\Psi(\tau_1\tau_3 s + \tau_1)}{\tau_1\tau_3 s^3 + \tau_1 s^2 + \Psi K\tau_2 s + \Psi K}\right] = \frac{2\Delta\omega'\tau_1}{K}$$

$$\lim_{t\to\infty} e_{\text{type } 3}(t) = \lim_{s\to 0} s\theta_i(s)\varepsilon_{\text{type } 3}(s) = \lim_{s\to 0}\left[s\frac{2\Delta\omega'}{s^3}\frac{\Psi s^3}{\tau_1^2 s^3 + \Psi K\tau_1^2 s^2 + 2\Psi K\tau_2 s + \Psi K}\right] = 0 \tag{7.83}$$

Continued

EXAMPLE 7.3 STEADY STATE PERFORMANCE OF TYPE 2 AND TYPE 3 THIRD-ORDER PLLS—cont'd

According to Eqn (7.83), the steady state error of a type 2 third order loop is not zero. Furthermore, it is a function of the frequency ramp $\Delta\omega'$. For large errors, the loop will not be able to retain phase coherence, thus making it very difficult to employ coherent modulation in the receiver.

7.2.1.5 PLL stability and operating range parameters

In general, a system is said to be stable if for a given bounded input signal, the system produces a bounded output signal. The transient effects of first and second order loops show that these feedback systems are *theoretically* unconditionally stable. This is true, however, provided in the second order case there are no additional poles introduced inadvertently due to certain phase detectors or possibly the use of leaky integrators. Higher order loops, on the other hand, must be designed with stability in mind since they are not inherently stable systems.

A PLL is a feedback system with open-loop transfer function $G(s)$ also known as the feedback transfer function. As a feedback system, a PLL will oscillate if

$$1 + G(s) = 0 \Rightarrow |G(j\Omega_c)|e^{j\phi_c}\big|_{\phi=\angle G(j\Omega)} = -1 \tag{7.84}$$

where Ω_c is the crossover frequency. The result in Eqn (7.84) implies simply that the magnitude $|G(j\Omega_c) = 1|$ and the phase $\phi_c = (2n+1)\pi$, $n = \cdots -2, -1, +1, +2, \cdots$. In this case, that is when $|G(j\Omega_c)| = 1$, the relationship in Eqn (7.84) can be expressed further as

$$\log[|G(j\Omega_c)|] = j[(2n+1)\pi - \phi_c] \tag{7.85}$$

According to Nyquist, a system is said to be stable if the phase margin is *positive* at crossover frequency. The phase margin Υ is obtained according to the relation

$$\Upsilon = \angle G(j\Omega_c) + \pi \tag{7.86}$$

Similarly, the gain margin can be defined as

$$\Gamma = -20\log(|G(j\Omega_c)|) \tag{7.87}$$

Asymptotic stability can then be obtained when the phase margin is greater than 0° and the gain margin is greater than 0 dB. The gain margin is a measure of how much the loop gain can increase before the system becomes unstable. Similarly, the phase margin is a measure of the phase lag function of the loop. Time delay in the loop reduces the phase margin and brings the loop closer to instability. In most cases, time delay is attributed to the digital circuits in the loop. The loop becomes unstable if the time delay increases by

$$\Delta\tau_{\max} = \frac{\Upsilon(\text{in degrees})}{\Omega_c}\frac{\pi}{180} \tag{7.88}$$

Next, we present the Hurwitz criterion [5], which is a simple method that determines whether the loop, or any feedback system, is stable or not. Assume that $G(s)$ is represented by the ratio

$$G(s) = \frac{N(s)}{s^m D(s)} \tag{7.89}$$

Then according to Eqn (7.84) $1 + G(s)$ can be expressed as

$$1 + G(s) = 1 + \frac{N(s)}{s^m D(s)} = \frac{s^m D(s) + N(s)}{s^m D(s)} \tag{7.90}$$

Define the polynomial $P(s)$ as

$$\begin{aligned} P(s) = s^m D(s) + N(s) &= \alpha_n s^n + \alpha_{n-1} s^{n-1} + \cdots + \alpha_2 s^2 \\ &\quad + \alpha_1 s + \alpha_0, \quad \alpha_n > 0 \text{ and } \alpha_i \in \mathbb{R} \end{aligned} \tag{7.91}$$

Stability of the loop can be implied from the roots of the polynomial $P(s)$. The loop is stable if real parts of the roots of $P(s)$ are negative. The polynomial is known as a Hurwitzian polynomial. To determine whether the roots are negative, we rely on computing a series of determinants. Consider the determinant

$$\begin{aligned} \Delta_1 &= \alpha_{n-1} > 0 \\ \Delta_2 &= \begin{vmatrix} \alpha_{n-1} & \alpha_n \\ \alpha_{n-3} & \alpha_{n-2} \end{vmatrix} > 0 \\ \Delta_3 &= \begin{vmatrix} \alpha_{n-1} & \alpha_n & 0 \\ \alpha_{n-3} & \alpha_{n-2} & \alpha_{n-1} \\ \alpha_{n-5} & \alpha_{n-4} & \alpha_{n-3} \end{vmatrix} > 0 \\ \Delta_n &= \begin{vmatrix} \alpha_{n-1} & \alpha_n & 0 & 0 & 0 & 0 & \cdots & 0 \\ \alpha_{n-3} & \alpha_{n-2} & \alpha_{n-1} & \alpha_n & 0 & 0 & \cdots & 0 \\ 0 & 0 & \alpha_{n-3} & \alpha_{n-2} & \alpha_{n-1} & \alpha_n & \cdots & 0 \\ 0 & 0 & 0 & 0 & \alpha_{n-3} & \alpha_{n-2} & \cdots & 0 \\ \cdots & \cdots & \cdots & \cdots & \cdots & \cdots & \cdots & 0 \\ \cdots & \cdots & \cdots & \cdots & \cdots & \cdots & \cdots & \alpha_0 \end{vmatrix} > 0 \end{aligned} \tag{7.92}$$

The polynomial is said to have all of its real parts of its roots negative if the minor determinants are positive, that is

$$\Delta_1 > 0, \Delta_2 > 0, \cdots \Delta_n > 0 \tag{7.93}$$

If the condition on the determinants as stated in Eqn (7.93) is met, the loop is said to be stable.

EXAMPLE 7.4 STABILITY OF FIRST AND SECOND ORDER PLLs

Consider the open loop gain functions of a first, second, and third order loop. Determine the stability of each and draw conclusions.

$$G(s) = \frac{\Upsilon}{s}; \quad \text{First order loop}$$
$$G(s) = \frac{\Omega_n^2}{s(1+\tau s)}; \quad \text{Second order loop} \qquad (7.94)$$
$$G(s) = \frac{K_d K_v}{\tau_1\tau_3 s^3 + (\tau_1+\tau_2+\tau_3)s^2 + s}; \quad \text{Third order loop}$$

In order to determine the stability of the various loops, we compute the determinants of $P(s)$ according to Eqn (7.92). Recall that for the various loops, we can write:

$$1+G(s) = 1+\frac{\Upsilon}{s} = \frac{s+\Upsilon}{s}; \quad \text{First order loop}$$
$$1+G(s) = 1+\frac{\Omega_n^2}{s^2+2\xi\Omega_n s} = \frac{s^2+2\xi\Omega_n s+\Omega_n^2}{s^2+2\xi\Omega_n s}; \quad \text{Second order loop}$$
$$1+G(s) = 1+\frac{K_d K_v}{\tau_1\tau_3 s^3 + (\tau_1+\tau_2+\tau_3)s^2 + s} \qquad (7.95)$$
$$= \frac{\tau_1\tau_3 s^3 + (\tau_1+\tau_2+\tau_3)s^2 + s + K_d K_v}{\tau_1\tau_3 s^3 + (\tau_1+\tau_2+\tau_3)s^2 + s}; \quad \text{Third order loop}$$

According to Eqn (7.91), the polynomial $P(s)$ is none other than the numerator of $1 + G(s)$ in Eqn (7.95), that is

$$P(s) = s+\Upsilon; \quad \text{First order loop}$$
$$P(s) = s^2+2\xi\Omega_n s+\Omega_n^2; \quad \text{Second order loop}$$
$$P(s) = \tau_1\tau_3 s^3 + (\tau_1+\tau_2+\tau_3)s^2 + s + K_d K_v; \quad \text{Third order loop} \qquad (7.96)$$
$$P(s) = \alpha_n s^n + \alpha_{n-1}s^{n-1} + \cdots + \alpha_2 s^2 + \alpha_1 s + \alpha_0; \quad \text{Arbitrary order loop}$$

Next, consider the first order loop. This is the most obvious case where the condition for stability is obvious, that is, the root of the denominator has to be in the left half plane. The loop is said to be stable if and only if $\Delta_1 = \alpha_{n-1} > 0$

$$\Delta_1 = \alpha_{n-1} = \Upsilon > 0 \qquad (7.97)$$

Next, compute the determinants of the second order loop:

$$\Delta_1 = \alpha_{n-1} = 2\xi\Omega_n > 0 \text{ since both } \xi \text{ and } \Omega_n > 0$$
$$\Delta_2 = \begin{vmatrix} \alpha_{n-1} & \alpha_n \\ \alpha_{n-3} & \alpha_{n-2} \end{vmatrix} = \alpha_{n-1}\alpha_{n-2} - \alpha_{n-3}\alpha_n = 2\xi\Omega_n^3 > 0 \qquad (7.98)$$

According to Eqn (7.98), the second order loop is also stable. As a matter of fact, both first and second order loops are unconditionally stable.

Next, we examine the third order loop. The determinants are given as:

$$\Delta_1 = \alpha_{n-1} = \tau_1+\tau_2+\tau_3 > 0$$
$$\Delta_2 = \alpha_{n-1}\alpha_{n-2} - \alpha_{n-3}\alpha_n = (\tau_1+\tau_2+\tau_3) - K_d K_v \tau_1\tau_3;$$
$$\Delta_3 = \begin{vmatrix} \alpha_{n-1} & \alpha_n & 0 \\ \alpha_{n-3} & \alpha_{n-2} & \alpha_{n-1} \\ \alpha_{n-5} & \alpha_{n-4} & \alpha_{n-3} \end{vmatrix} \qquad (7.99)$$
$$= (\alpha_{n-1}\alpha_{n-2} - \alpha_{n-3}\alpha_n)\alpha_{n-3} - \alpha_{n-1}(\alpha_{n-1}\alpha_{n-4} - \alpha_{n-5}\alpha_n)$$
$$= (\tau_1+\tau_2+\tau_3 - K_d K_v\tau_1\tau_3)K_d K_v = \Delta_2 K_d K_v; \; K_d K_v > 0$$

EXAMPLE 7.4 STABILITY OF FIRST AND SECOND ORDER PLLs—cont'd

The relationship in Eqn (7.99) does not indicate that a third order loop is unconditionally stable. As a matter of fact, stability is predicated on the fact that $\Delta_2 > 0$. As a matter of fact, loops higher than second order are not unconditionally stable and the designer must make a conscious effort when designing such loops to ensure their stability.

Finally, we very briefly discuss other important design parameters of the PLL, namely the hold-in range, the pull-in range, and the lock-in range. The hold-in range of the loop is an indicator of loop performance [6]. By definition, it is the maximum frequency difference between the input frequency Ω_i and the VCO frequency Ω_v, given as $\Delta\Omega_H = |\Omega_i - \Omega_v|$ before the PLL loses lock. In case the input frequency is close to that of the VCO, the PLL locks up simply with a phase transient [7] without cycle slipping prior to locking. Define the frequency range for which the loop acquires phase to lock without cycle slips as the lock-in range $\Delta\Omega_L$ of the loop. For a first order loop, the lock-in range is equal to the hold-in range $\Delta\Omega_H = \Delta\Omega_L|_{\text{for first order PLL only}}$. For a second order loop or higher, however, the lock-in range is always less than the hold-in range $\Delta\Omega_H < \Delta\Omega_L|_{\text{for second and higher order PLL}}$. Next, define the pull-in range of the PLL $\Delta\Omega_P$ as the frequency interval within which the loop will acquire lock after slipping a certain number of cycles. The pull-in range for second and higher order loops is bounded as

$$\Delta\Omega_L < \Delta\Omega_P < \Delta\Omega_H \tag{7.100}$$

It is important to note that the hold-in range depends on the nature of the loop detector. That is, the value of $\Delta\Omega_H$ varies depending on whether the output of the phase detector is sinusoidal or triangular, for example. Computation of the pull-in range $\Delta\Omega_P$ is even more complicated. Nonetheless, these parameters provide necessary insights into the overall performance of the loop dictated by output frequency and the modulation scheme itself.

7.2.1.6 The phase detector

The role of the phase detector is to detect the phase difference between the reference signal and the output of the VCO, which in the case of a frequency synthesizer is divided down by a frequency divider. Phase detectors can be classified as analog, sampled, or digital. Analog detectors are typically driven by sinusoidal input signals. In a simple switch, both reference and VCO output signals are multiplied and filtered in the detector and the resulting output signal is fed to the loop filter. The sampled phase detector relies on a switch connected to a capacitor. At every sampling instant, the switch connects the analog phase difference signal to a memory capacitor. The resulting error signal is then fed to the loop filter. The digital phase detector on the other hand can also be implemented in a variety of ways, with the charge pump implementation being the standard for frequency synthesizers. A block diagram of a PLL employing a charge pump (CP) is depicted in Figure 7.41. A tri-state phase/frequency detector (PFD), shown in Figure 7.42, is used along with a CP in

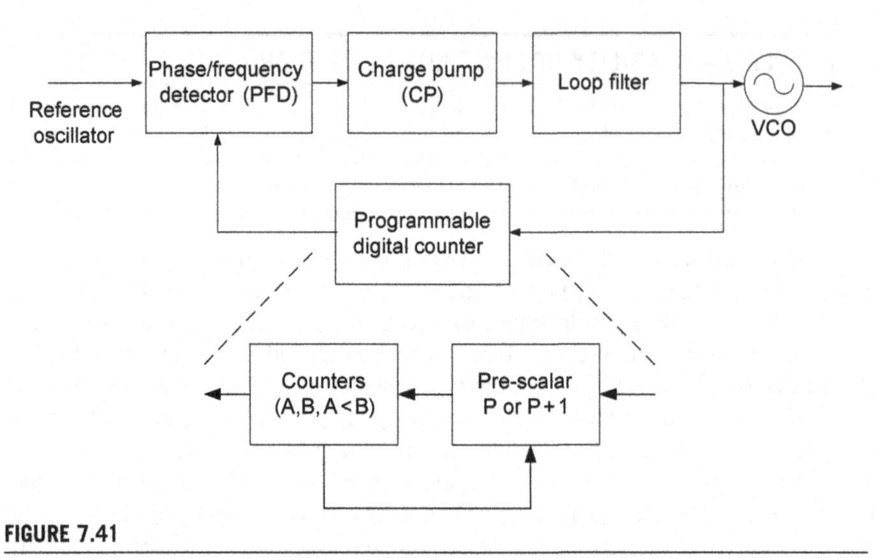

FIGURE 7.41

Charge-pump phase-locked loop used in frequency synthesis.

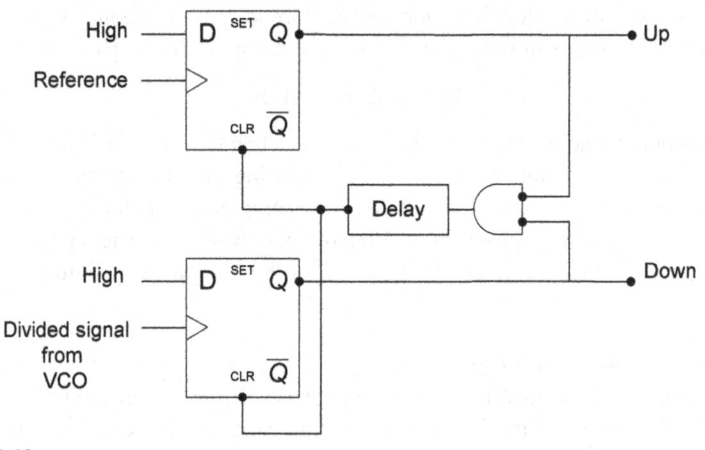

FIGURE 7.42

Tri-state phase/frequency detector used in conjunction with charge pump-phase/frequency detector.

a frequency synthesizer. The PFD as the name implies is also capable of detecting frequency errors [8,9]. An edge-triggered PFD has a linear phase detection range of $\pm 2\pi$ and is insensitive to duty cycle. The PFD is constructed with memory elements such as flip-flops. The delay element shown in the reset path is needed in order to avoid any undetectable phase difference range. The charge pump PLL is a type 2 PLL, that is, it has two poles at the origin in its open loop transfer function. A CP-PLL has a capture range that is only restricted by the VCO's tuning range.

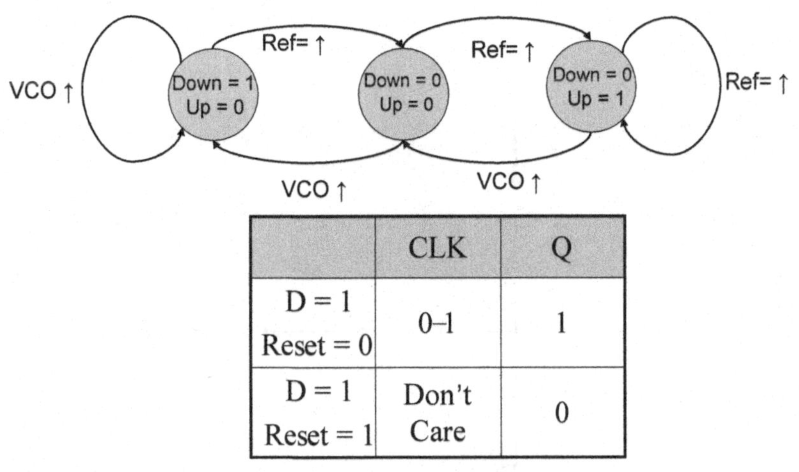

	CLK	Q
D = 1 Reset = 0	0–1	1
D = 1 Reset = 1	Don't Care	0

FIGURE 7.43

State diagram of a tri-state phase/frequency detector. VCO, voltage-controlled oscillator. (For color version of this figure, the reader is referred to the online version of this book.)

The functionality of the PFD can be illustrated via a state machine as shown in Figure 7.43 [10]. For example, the PFD output "Up" is high when the rising edge of the reference leads that of the divided VCO output. At the output of the CP, this phase difference between the reference and the divided VCO is translated into current. The CP delivers charges to capacitors in the loop filter as shown in Figure 7.44. In general, the "Up" and "Down" pulses translate into current pulses I_{CP} in the charge pump, which, in turn, change the voltage drop V_c on the loop filter impedance. Hence, as the loop filter voltage increases or decreases, the VCO frequency and phase increase or decrease accordingly.

At this juncture, given a PLL synthesizer using a CP-based PFD, the performance degradation due to reference spurs must be discussed. Reference spurs occur at the output of the synthesizer at multiples of the reference frequency, that is at $F_{out} \pm mF_{ref}, m \in \mathbb{N}$. The main culprit is periodic disturbances in the loop filter voltage caused by the CP. This is due to mismatches between the positive and negative current sources in the CP. Typically, during each reference period, the PFD turns on the positive or negative current source whenever the reference edge or divider edges occur for a given minimum duration T_D. Both current sources are turned off simultaneously T_D seconds after the latter of the two edges has occurred. The CP's output current pulse is simply the difference between the positive and negative current pulses. A significant source of nonlinear degradation can be avoided by allowing both current sources enough time to settle [11]. Having said that, however, transient and amplitude mismatches between the two current sources are expected, thus causing a certain error component in each CP pulse that is constant from period to period. After filtering the DC component of constant error pulse in the PLL, the remaining error is zero-mean periodic disturbance of the VCO's control voltage.

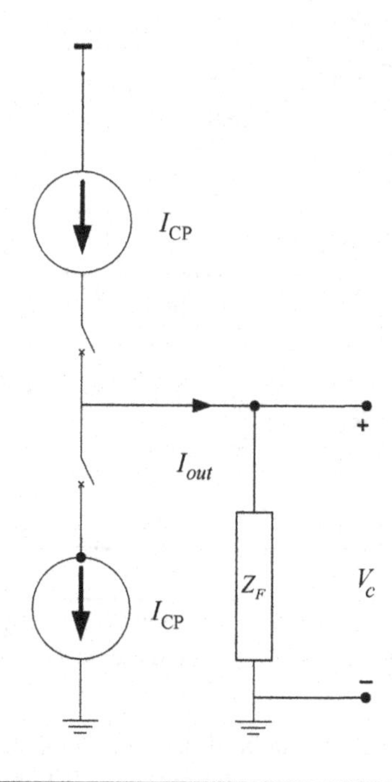

FIGURE 7.44

Charge pump feeding a loop filter.

This periodic disturbance is responsible for the occurrence of the reference spur. This periodic disturbance, however, could be removed via a sample-and-hold operation conducted between the output of the loop filter and the input of the VCO as stated in [12].

In the absence of any technique used to the remove the reference spur, the choice of both the reference frequency F_{ref} and the noise equivalent bandwidth of the loop affect the power levels of the reference spurs. In order to minimize the spur levels, the PLL's noise equivalent bandwidth must be narrowed with respect to the reference frequency such that the spurs are attenuated by the loop transfer function. Conversely, the spur levels may also be minimized by increasing the reference frequency well beyond the loop's noise equivalent bandwidth.

EXAMPLE 7.5 FREQUENCY PLANNING AND SYNTHESIS FOR MULTIBAND OFDM UWB

This example is based on the work presented in [13]. Consider a multiband (MB) ultrawideband (UWB) device that operates in the frequency bands between 3.1 and 10.6 GHz, including the UNII band for the sake of this example, as shown in Figure 7.45. According to Ref. [14], the signal is modulated over OFDM with instantaneous channel bandwidth of 528 MHz. Furthermore, assume that the signal is divided into five band groups. Furthermore, assume that the first band group centered at 3690 MHz will be used

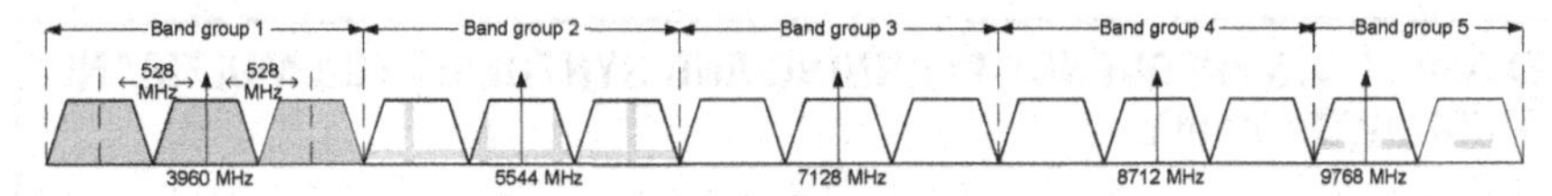

FIGURE 7.45

Frequency band groups for multiband–ultrawideband concerning Example 7.5.

EXAMPLE 7.5 FREQUENCY PLANNING AND SYNTHESIS FOR MULTIBAND OFDM UWB—cont'd

exclusively by UWB whereas all other band groups are optional. In this example, the transceiver architecture is based on direct conversion with synthesizer based on an integer-N PLL. The switching time between bands within a band group is 9.47 ns. The bands are spaced 528 MHz apart with center frequencies governed by the relationship: $F_c = (2904 + 528n)$ MHz.

In order to facilitate hopping between bands, a 528 MHz tone must be made available by the synthesizer. In the design of this synthesizer, you are constrained to use divide-by-2 and divide-by-3 circuits only in addition to MUXs and mixers.

1. What are some of the possible VCOs that can be used by the PLL? Construct a frequency tree. Restrict the VCO to be less than 15 GHz.
2. Choose 6.336 GHz as the VCO frequency, using the constraints above, derive the center frequencies of the various band groups, namely: BG1-3960 MHz, BG2-5544 MHz, BG3-7128 MHz, BG4-8712 MHz, and BG5-10296 MHz.
3. Based on the work done on the previous section, generate the rest of the center frequencies for the remaining frequency bands.
4. Show a block diagram detailing the synthesizer architecture.

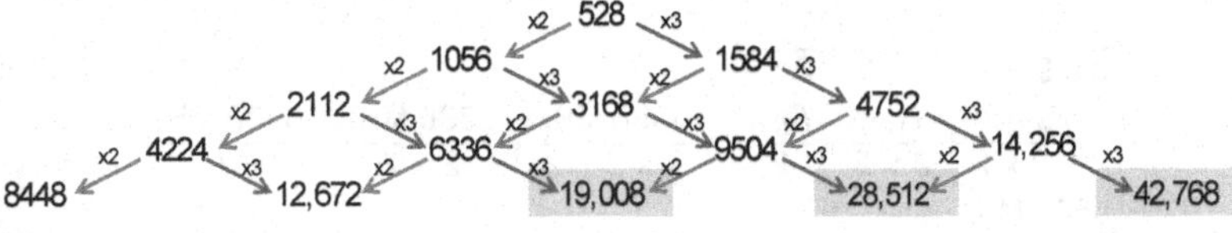

FIGURE 7.46

Frequency tree of possible voltage-controlled oscillator frequencies up to 15 GHz. The shaded frequencies are above 15 GHz and will not be considered. (For color version of this figure, the reader is referred to the online version of this book.)

1. Since the 528-MHz tone must be made available from all VCO frequencies and that only divide-by-2 and divide-by-3 circuits may be used, then the choice of the VCO frequency F_{VCO} must be divisible by products of 2 and 3 s. To do so, constitute all the products of 2 and 3 s of 528 MHz up to 15 GHz as shown in Figure 7.46. The shaded frequencies are products of 528 MHz that are above 15 GHz. For example, the frequency 19.008 GHz is the product of 528 MHz according to the relation

$$\begin{aligned} 19{,}008 \text{ MHz} &= 3 \times 6336 \text{ MHz} = 3 \times 3 \times 2112 \text{ MHz} \\ &= 2 \times 3 \times 3 \times 1056 \text{ MHz} = 2 \times 2 \times 3 \times 3 \times 528 \text{ MHz} \end{aligned} \tag{7.101}$$

By the same process, an acceptable VCO frequency of 6.336 GHz is generated according to Eqn (7.101) as 6336 MHz = 2 × 2 × 3 × 528 MHz.

Continued

EXAMPLE 7.5 FREQUENCY PLANNING AND SYNTHESIS FOR MULTIBAND OFDM UWB—cont'd

2. First express the ratio of center frequency to VCO frequency as

$$\frac{F_c}{F_{VCO}} = \frac{3960\ \text{MHz}}{6336\ \text{MHz}} = 0.625 = \frac{5}{8} \tag{7.102}$$

The center frequency of band group 1 is 3960 MHz generated using sums, differences, and multiply-by-2 and multiply-by-3 of 528 MHz. That is, express 5/8 as the sum

$$\frac{F_c}{F_{VCO}} = \frac{3960\ \text{MHz}}{6336\ \text{MHz}} = 0.625 = \frac{5}{8} = \frac{1}{8} + \frac{4}{8} = \frac{1}{2 \times 2 \times 2} + \frac{1}{2}$$

$$F_c = \left(\frac{1}{2 \times 2 \times 2} + \frac{1}{2}\right) F_{VCO} = \frac{F_{VCO}}{8} + \frac{F_{VCO}}{2}$$

$$\Rightarrow F_c = \left(\frac{1}{2 \times 2 \times 2} + \frac{1}{2}\right) \times 6336\ \text{MHz} = 792\ \text{MHz} + 3168\ \text{MHz} = 3960\ \text{MHz} \tag{7.103}$$

In a similar manner, in order to produce the center frequency 5544 MHz at the UNII band, consider the ratio

$$\begin{aligned}\frac{F_c}{F_{VCO}} &= \frac{5544\ \text{MHz}}{6336\ \text{MHz}} = 0.875 = \frac{7}{8} = \frac{1}{8} + \frac{1}{4} + \frac{1}{2} \\ &= \frac{1}{2 \times 2 \times 2} + \frac{1}{2 \times 2} + \frac{1}{2}\end{aligned} \tag{7.104}$$

Or simply

$$\begin{aligned}\frac{F_c}{F_{VCO}} &= \frac{5544\ \text{MHz}}{6336\ \text{MHz}} = 0.875 = \frac{7}{8} = \frac{1}{8} + \frac{1}{4} + \frac{1}{2} \\ &= \frac{1}{2 \times 2 \times 2} + \frac{1}{2 \times 2} + \frac{1}{2} \\ F_c &= \frac{F_{VCO}}{2 \times 2 \times 2} + \frac{F_{VCO}}{2 \times 2} + \frac{F_{VCO}}{2} = \frac{6336\ \text{MHz}}{2 \times 2 \times 2} + \frac{6336\ \text{MHz}}{2 \times 2} + \frac{6336\ \text{MHz}}{2} \\ &= 5544\ \text{MHz}\end{aligned} \tag{7.105}$$

Following the thought process presented above, we can produce the center frequency of all the band-groups as shown in Table 7.5.

Table 7.5 Multiband–Ultrawideband Band Group Center Frequencies and Synthesis Equations

Band Group	Center Frequency	Synthesis
1	3960 MHz	$\frac{F_{VCO}}{2} + \frac{F_{VCO}}{8}$
2	5544 MHz	$\frac{F_{VCO}}{2} + \frac{F_{VCO}}{4} + \frac{F_{VCO}}{8}$
3	7128 MHz	$F_{VCO} + \frac{F_{VCO}}{8}$
4	8712 MHz	$F_{VCO} + \frac{F_{VCO}}{4} + \frac{F_{VCO}}{8}$
5	10296 MHz	$F_{VCO} + \frac{F_{VCO}}{2} + \frac{F_{VCO}}{8}$

EXAMPLE 7.5 FREQUENCY PLANNING AND SYNTHESIS FOR MULTIBAND OFDM UWB—cont'd

3. Generating the remaining frequency bands is now relatively simple. Based on the center frequencies presented in Table 7.5, the center frequency of each may be generated by adding or subtracting $F_{VCO}/12 = 528$ MHz. For example, in order to generate the frequencies 3432 and 4488 MHz in band group 1, we use the synthesis equation for 3960 MHz and add or subtract 512 MHz, that is

$$\frac{F_{VCO}}{2} + \frac{F_{VCO}}{8} = 3960 \text{ MHz}$$
$$\Rightarrow \frac{F_{VCO}}{2} + \frac{F_{VCO}}{8} - \frac{F_{VCO}}{12} = 3960 \text{ MHz} - 528 \text{ MHz} = 3432 \text{ MHz} \tag{7.106}$$
$$\text{and} \quad \frac{F_{VCO}}{2} + \frac{F_{VCO}}{8} + \frac{F_{VCO}}{12} = 3960 \text{ MHz} + 528 \text{ MHz} = 4488 \text{ MHz}$$

4. The block diagram detailing the synthesizer architecture is shown in Figure 7.47.

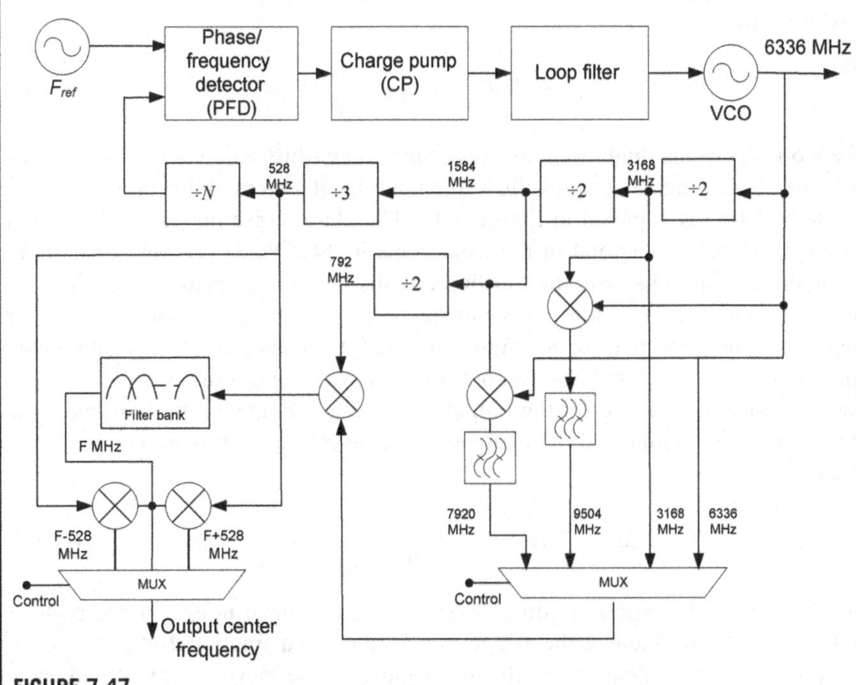

FIGURE 7.47

Ultrawideband synthesizer architecture as shown in Ref. [13].

7.2.2 Fractional-*N* frequency synthesis

Unlike integer-*N* synthesizers, the resolution of fractional-*N* synthesizers is not directly tied to the frequency reference of the loop. Their advantage over their integer-*N* counterparts is in their added design flexibility. Fractional-*N* synthesizers are divided into two categories: classical, which employ simple divider circuits such as the dual modulus prescalar with possible dithering and phase interpolation to minimize performance degradation, and $\Delta\Sigma$-based fractional-*N* synthesizers. We will discuss the pros and cons of each implementation following.

7.2.2.1 Programmable digital counter

In Chapter 3, we made a passing mention of the programmable digital counter as a divide-by-*N* block in order to analyze phase noise. In this section, we will address a popular divider that can achieve a fractional division ratio. In particular, the dividing circuit under discussion is comprised of a dual modulus prescalar with divide ratio of P or $P+1$ and programmable pulse swallowing divider where A and B are both integers as shown in Figure 7.41.[3] In this case, the total number of VCO cycles before the counters return to their preset state, as explained in Example 7.6, is simply $N_{\text{Total}} = (P+1)A + (B-A)P = BP + A$ VCO cycles resulting in an average frequency divide ratio of $N_f = P + A/B$. Consequently, the output frequency of the synthesizer is [8]

$$F_{out} = N_f F_{ref} = \left(P + \frac{A}{B}\right) F_{ref} \tag{7.107}$$

The workings of the dual modulus prescalar will be illustrated via Example 7.6 as presented following. By way of illustration, a typical dual modulus prescalar for a divide by 64/65 is depicted in Figure 7.48. The phase noise impact of the digital prescalar is best understood in the time domain [14]. Noise present at the input of the digital flip-flops or in the flip-flop circuits themselves tend to alter the triggering instant of the flip-flops, thus causing time jitter. In an asynchronous divider chain, the time jitter tends to accumulate through the chain, thus implying that the phase noise itself accumulates. In contrast, in a synchronous divider, the last flip-flop is mostly responsible for the output phase noise. In theory, we can relate the jitter in the time domain to the changes in phase in the circuits according to the relation

$$\Delta\theta = 2\pi F_{signal} v_{noise} \left.\frac{dt}{dv}\right|_{t=\text{trigering instant}} \tag{7.108}$$

where F_{signal} is the signal frequency, v_{noise} is the voltage noise. The derivative in Eqn (7.108) is taken at the triggering instant. In a fractional-*N* synthesizer, the power spectral density of the prescalar's phase noise is amplified as a

[3]We will expound on the "pulse swallowing" capability of the dual modulus prescalar in Example 7.6.

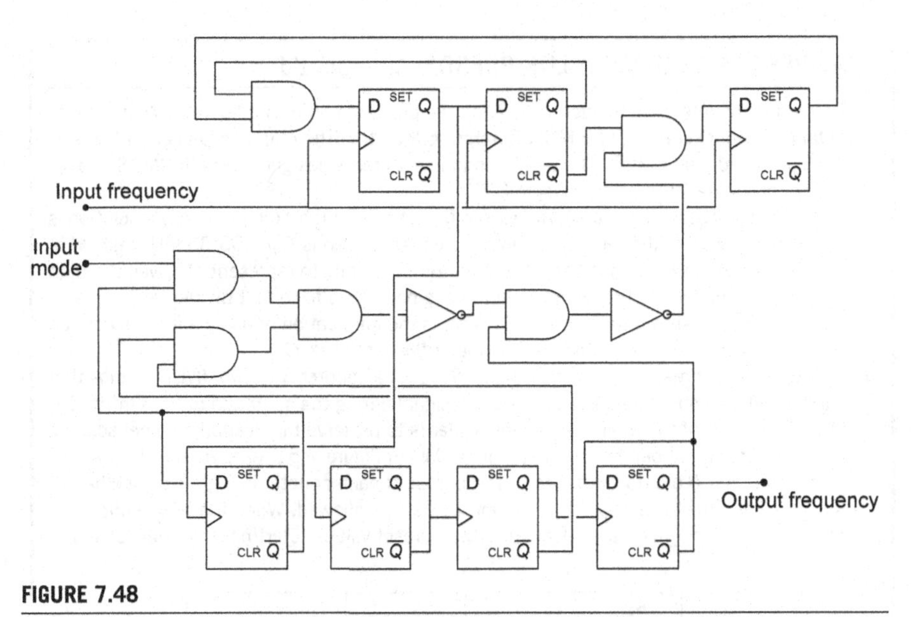

FIGURE 7.48

Dual modulus prescalar with divide by 64/65 circuit [14].

function of the square of the divider ratio due to the feedback structure of the PLL, and consequently it could have a significant impact on the synthesizer's performance [14].

EXAMPLE 7.6 DUAL MODULUS PRESCALAR

This is a rhetorical straightforward example that demonstrates the need for and working of the dual modulus prescalar in a frequency synthesizer. Consider the integer-N frequency synthesizer implemented for a system-on-a-chip (SOC) design as presented in Figure 7.49. Assume that the reference frequency used is $F_{ref} = 5$ MHz. The desired frequency output of the synthesizer is at the ISM band or 2.4 GH to be in a ZigBee application.

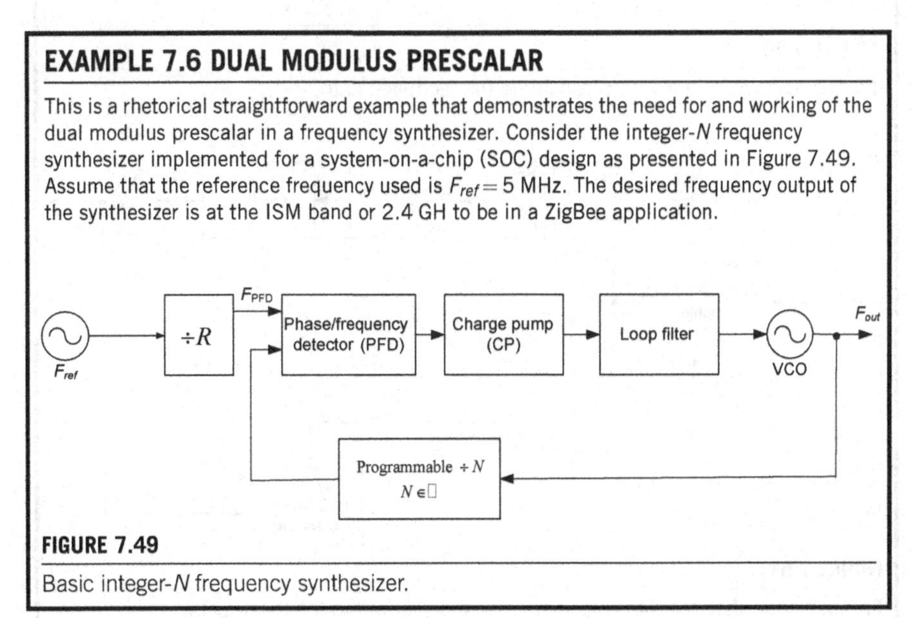

FIGURE 7.49

Basic integer-N frequency synthesizer.

Continued

EXAMPLE 7.6 DUAL MODULUS PRESCALAR—cont'd

1. Determine the reference divider ratio R and the programmable counter value N in order to obtain a channel spacing of 5 MHz. Determine R and N. The chip is designed in 1.8 V 0.18 μm epi-digital CMOS. The programmable counter is designed in both CMOS and power-consuming current mode logic (CML).
2. Next, assume that for some unknown reason the requirement for the frequency resolution is 10 KHz instead of 1 MHz and the reference frequency used is $F_{ref}=33.33$ MHz instead of 5 MHz. Determine R, N, and b the required number of bits to represent N. Given the high input VCO frequency to the counter implies the need for a fixed first divider $\div P$ in the feedback path as shown in Figure 7.50. What is the implication of using a fixed divider on the minimum resolution or channel spacing of the synthesizer?
3. A dual modulus prescalar, shown in Figure 7.51, is a counter with two division ratios that can be switched spontaneously via a control signal during the operation of the synthesizer [15]. The intent of the dual modulus prescalar is to preserve the desired channel spacing while preserving the benefit of using a prescalar. In Figure 7.51, assume that $B \geq A$ and that both A and B are down counters. The output of each counter is high, except when it reaches 0 and then it is low. No underflow below 0 is allowed. When the $\div B$ counter reaches 0, both counters are reloaded to their preset values. Starting from fully loaded

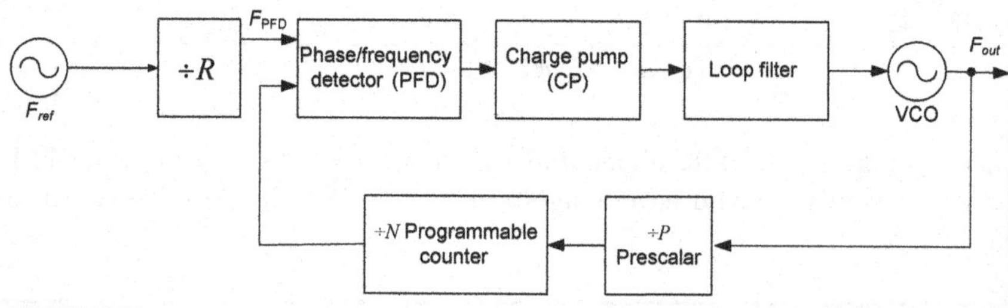

FIGURE 7.50

Frequency synthesizer with prescalar in the feedback path.

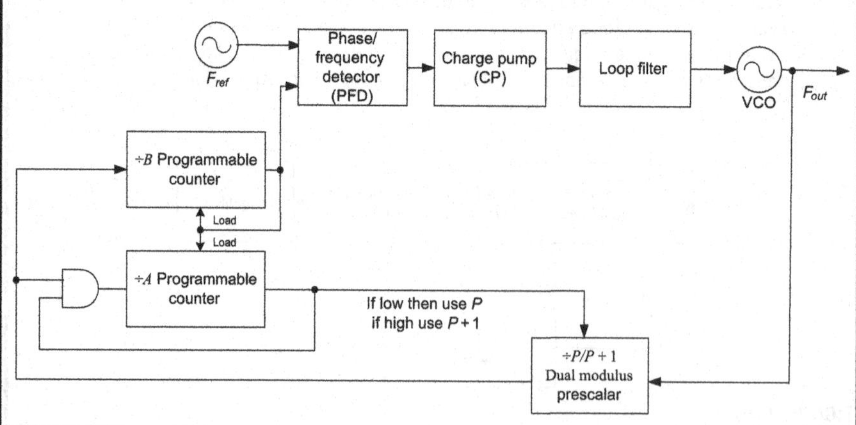

FIGURE 7.51

Frequency synthesizer with dual-modulus prescalar in the feedback path.

EXAMPLE 7.6 DUAL MODULUS PRESCALAR—cont'd

counters, determine the number of VCO cycles until the same logic state is reached. Express the output frequency in terms F_{ref}. Note that in part of the example, we have removed reference frequency divider $\div R$. Explain why the dual modulus prescalar overcomes the limitation on the frequency resolution seen with the fixed prescalar architecture.

In order to obtain the desired frequency resolution of F_{PFD} of 5 MHz, the reference signal must be divided down by R in order to obtain the input signal to the frequency detector. This signal is compared against the output signal of the synthesizer with frequency F_{out} divided down by N and then compared with the input signal to the PLL. That is, it is desired that steady state frequency of the divided VCO output $\widehat{F}_{PFD} = F_{out}/N$ is equal to F_{PFD} or

$$F_{PFD} = \widehat{F}_{PFD} \Rightarrow \frac{F_{ref}}{R} = \frac{F_{out}}{N} = 5\text{ MHz}$$

or

$$R = \frac{F_{ref}}{F_{PFD}} = \frac{5\text{ MHz}}{5\text{ MHz}} = 1 \qquad (7.109)$$

$$N = \frac{F_{out}}{\widehat{F}_{PFD}} = \frac{2.4\text{ GHz}}{5\text{ MHz}} = 480$$

In this implementation there is no need for a reference divider. On the other hand, the number of bits required to implement the programmable counter is simply [15]

$$b = \lceil \log_2(N) \rceil = \lceil \log_2(480) \rceil = \lceil 8.9 \rceil = 9\text{ bits} \qquad (7.110)$$

where $\lceil x \rceil = ceiling(x)$.

The input to the frequency divider is the output of the VCO that is 2.4 GHz. At such a high input frequency, the programmable counter is typically split into two sections. The first section of the programmable counter may be implemented in CML to bring the frequency down to a level manageable by CMOS for low frequency division. Note that the power consumption of CML is higher than that of CMOS, however, it has a higher operating frequency [16]. The programmable counter can be designed using two divide by 2/3 cells implemented in CML and six divide by 2/3 cells implemented in CMOS [17]. A divide by 2/3 topology using L cells implies that the output signal time period $\Upsilon_{div-out}$ can be expressed in the input signal time period Υ_{div-in} according to the relation

$$\Upsilon_{div-out} = \left(2^L + 2^{L-1}\rho_L + \cdots 2\rho_2 + \rho_1\right)\Upsilon_{div-in} \qquad (7.111)$$

where ρ_l, $l \in \{1, \cdots L\}$ is a control bit. When ρ_l is "high" or 1, the lth divide by 2/3 cell divides the input frequency by 3. In contrast, when ρ_l is "low" or 0 the lth divide by 2/3 cell divides the input frequency by 2.

Next, we address the second part of the question and compute R and N in the same manner shown previously. In this case,

$$F_{PFD} = \widehat{F}_{PFD} \Rightarrow \frac{F_{ref}}{R} = \frac{F_{out}}{N} = 10\text{ KHz} \qquad (7.112)$$

Continued

EXAMPLE 7.6 DUAL MODULUS PRESCALAR—cont'd

This implies that the reference divider ratio and the value of the divided by N counter are:

$$R = \frac{F_{ref}}{F_{PFD}} = \frac{33.333 \text{ MHz}}{10 \text{ KHz}} \approx 3334$$
$$N = \frac{F_{out}}{\widehat{F}_{PFD}} = \frac{2.4 \text{ GHz}}{10 \text{ KHz}} = 240,000 \tag{7.113}$$

The number of bits required to implement the counter in the feedback loop is simply

$$b = \lceil \log_2(N) \rceil = \lceil \log_2(240,000) \rceil = \lceil 17.8726 \rceil = 18 \text{ bits} \tag{7.114}$$

Again, the divide by N block cannot be fully programmable at 2.4 GHz for an 18-bit counter due to the limited speed of CMOS devices. In this case, a prescalar in the form of a fixed counter is used to precede the programmable counter. However, using a fixed counter impacts the resolution or channel spacing achieved by the synthesizer. In other words, according to Figure 7.50, the modified relationship analogous to Eqn (7.112) is

$$\frac{F_{ref}}{R} = \frac{F_{out}}{\underbrace{N}_{\text{programmable}} \times \underbrace{P}_{\text{fixed}}} \Rightarrow F_{out} = \frac{N(PF_{ref})}{R} \tag{7.115}$$

The fixed prescalar factor in Eqn (7.115) implies that the frequency resolution is no longer 10 KHz but rather $P \times 10$ KHz. That is, it is no longer possible to generate frequencies at the output of the VCO at multiples of 10 KHz. In order to overcome this limitation, a dual-modulus prescalar is typically used.

Next, we turn our attention to the third and final part of this problem. According to Figure 7.51, both $\div A$ and $\div B$ counters start to count down, but since $A \leq B$, the $\div A$ counter will reach 0 first and its output will go low. That is, for A-instances, the prescalar $\div P/P+1$ will be in $\div P+1$ mode for a total of $(P+1)A$ VCO cycle before the output of the $\div A$ counter goes low. At this point, there remain a total of $(B-A)P$ VCO cycles before both the $\div A$ and $\div B$ counters reset to their maximum count values. During the transition between $\div P+1$ and $\div P$ operation of the dual modulus prescalar, one pulse is removed and is said to be swallowed, hence the terminology of a swallowing divider. So the total number of VCO cycles before the $\div A$ and $\div B$ counters return to the same logic state is

$$N_{Total} = (P+1)A + (B-A)P = BP + A \tag{7.116}$$

According to Eqn (7.116), the output frequency of the VCO is no longer a simple multiple of F_{ref} and P as was the case with the fixed prescalar but rather a multiple of F_{ref} as it will become more obvious in the ensuing discussion. The fractional divide ration is then somewhere between P and $P+1$. More precisely, the average divide ratio, taken over B reference periods, can be expressed as

$$N_f = \frac{N_{Total}}{B} = P + \frac{A}{B}, \quad 0 \leq \frac{A}{B} \leq 1, \quad A, B \in \mathbb{N} \tag{7.117}$$

The integer B is generally referred to as the fractional modulus and the ratio $1/B$ is referred to as the fractionality of the synthesizer. The fractional modulus is chosen such that the channel spacing is simply the ratio

$$\text{Channel spacing} \triangleq F_{PFD} = \frac{F_{ref}}{B} \tag{7.118}$$

EXAMPLE 7.6 DUAL MODULUS PRESCALAR—cont'd

The frequency F_{ref}/B is also known as the beat-note. The frequency at the output of the VCO can then be expressed in terms of the input frequency simply as

$$F_{out} = N_f F_{ref} = \left(P + \frac{A}{B}\right) F_{ref} \tag{7.119}$$

Now recall that the limitation on $\div A$ and $\div B$ is such that $A \in \{0, 1, \cdots A_{max}\}$, that is, A can be any number between 0 and A_{max}, and $B \geq A$[4] implies that the minimum value for Eqn (7.116) is simply

$$N_{Total,\min} = BP \tag{7.120}$$

and

$$B \geq A_{max} \text{ or } B_{\min} = A_{max} \tag{7.121}$$

Furthermore, the relationship in Eqn (7.116) implies that the values that the $\div A$ counter can take are bounded by P or

$$A \in \{0, 1, \cdots P-1\} \Rightarrow A_{max} = P - 1 \Rightarrow B_{\min} = P - 1 \tag{7.122}$$

By way of an example, assume that the prescalar is one of the ratios 9/10 or 16/17. In this case, the smallest realizable division ratios are

$$\begin{aligned} P = 9 : N_{Total,\min} &= B_{\min} P = (P-1)P = 8 \times 9 = 992 \\ P = 16 : N_{Total,\min} &= B_{\min} P = (P-1)P = 15 \times 16 = 240 \end{aligned} \tag{7.123}$$

In certain scenarios where the application requires an extended frequency range of the synthesizer, the designer may resort to a quad-modulus prescalar, for example.

[4]In order to best illustrate the workings of the dual modulus prescalar, we have allowed the relationship between the integers A and B to be such that BAA. In practice, however, a fractional-N synthesizer invariably implies that $B > A$.

7.2.2.2 Spurs in dual modulus fractional-N frequency synthesizers

As mentioned in the introduction of this section, and according to Figure 7.51, fractional spurs in the PLL occur at multiples of the beat-note frequency, resulting in sidebands around the output frequency of the PLL. Theoretically, the spurs that are most troublesome are the ones that are closest to the channel spacing F_{ref}/B. Typically, the power of a fractional spur is large if it is located within the loop noise equivalent bandwidth. Fractional spurs with frequencies higher than the loop bandwidth are attenuated by the loop filter. Hence, the power of a fractional spur can be reduced simply by reducing the loop bandwidth. On the other hand, a fractional spur may be reduced by choosing a high enough frequency reference F_{ref} such that the spurs occur outside of the loop bandwidth. Therefore, the choice of the loop bandwidth, the reference frequency, and the synthesizer resolution somewhat dictate the level of suppression of the fractional spurs.

By way of illustration, consider a fractional-N PLL operating in the range 2.11−2.155 GHz. The channel spacing is 1.4 MHz. Assume that the reference frequency is 40 MHz and the loop bandwidth is 20 KHz. The first five fractional spurs shown in Figure 7.52 are enumerated versus their respective amplitudes as shown in

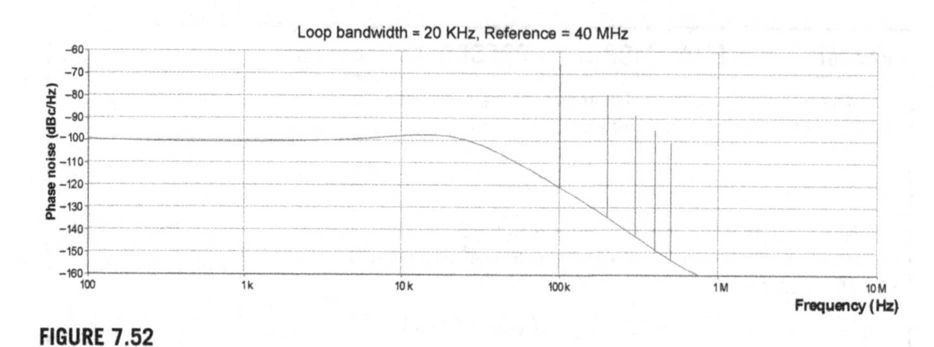

FIGURE 7.52

Spurs of a fractional-*N* synthesizer with 40-MHz reference frequency and 20-KHz loop bandwidth. (For color version of this figure, the reader is referred to the online version of this book.)

Table 7.6. Keeping the reference frequency at 40 MHz but increasing the loop bandwidth from 20 to 50 KHz, as depicted in Figure 7.53, increases the amplitude of the spurs across the board as can be deduced from Table 7.6. Naturally, the spur locations remained the same since the reference frequency and the divide ratio stayed the same. Next, we compare the spurs due to two reference frequencies, namely 18 and 40 MHz using the same PLL. The first apparent observation is that the locations of the fractional spurs have changed due to the change in the reference frequency as noted in Figure 7.54 and enumerated in Table 7.7. Furthermore, note that decreasing the frequency reference from 40 to 18 MHz while maintaining the same loop bandwidth of 20 KHz shows that the level of the various fractional spurs is higher due to decreased attenuation by the PLL's lowpass transfer function.

Next we focus our attention on the generation mechanism of the fractional spurs and on the various ways by which they could be minimized. In this analysis, we will follow the approach taken in Ref. [18]. According to Eqn (7.117), the dual modulus

Table 7.6 Spurs of a Fractional-*N* PLL Using Loop Bandwidths of 20 and 50 KHz, Respectively, and Employing a 40-MHz Reference

Frequency	Loop Bandwidth = 20 KHz, Frequency Reference = 40 MHz Spur Level (dBc)	Loop Bandwidth = 50 KHz, Frequency Reference = 40 MHz Spur Level (dBc)
100 KHz	−65.3	−49.6
200 KHz	−79.2	−61.2
300 KHz	−88.3	−68.8
400 KHz	−95.1	−74.5
500 KHz	−101	−79.2

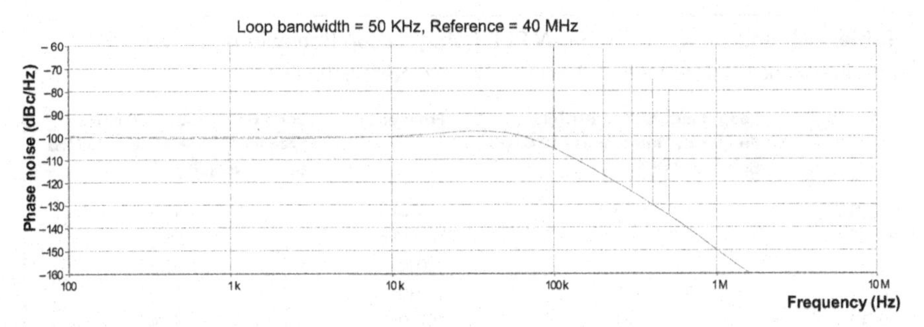

FIGURE 7.53

Spurs of a fractional-N synthesizer with 40-MHz reference frequency and 50-KHz loop bandwidth. Note that the amplitudes of the spurs are higher than those shown in Figure 7.52. (For color version of this figure, the reader is referred to the online version of this book.)

prescalar sets the divide ratio to P for A reference periods and to $P+1$ for $B-A$ reference periods, thus resulting in N_f as the fractional divide ratio. Although N_f is the correct fractional divide ratio on the average, the instantaneous divide ratio is typically in error. The resulting phase error sequence is periodic and repeats itself every B-cycles. The spurs contained in the output signal of the phase detector are all harmonics of the beat-note with the dominant spurs being F_{ref}/B and $(A/B)F_{ref}$. The spurs manifest themselves as discrete tones in the PLL's output spectrum. During steady state, the phase difference between the reference signal and the divided VCO signal implies that in theory

$$\frac{1}{F_{ref}} - \frac{N_f}{F_{out}} = 0 \tag{7.124}$$

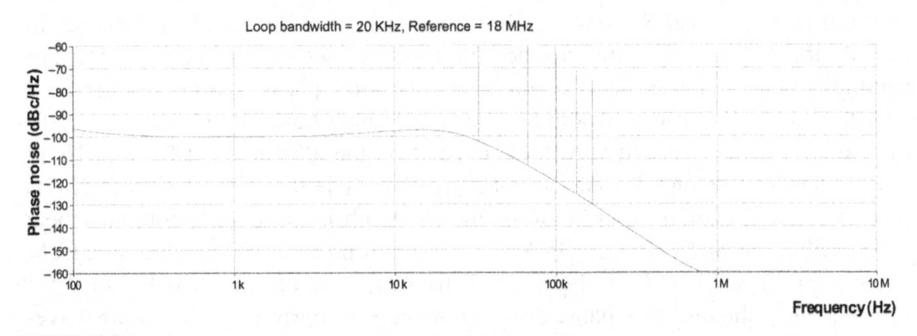

FIGURE 7.54

Spurs of a fractional-N synthesizer with 18-MHz reference frequency and 20-KHz loop bandwidth. (For color version of this figure, the reader is referred to the online version of this book.)

Table 7.7 Spurs of a Fractional-*N* PLL Using a Loop Bandwidth of 20 KHz and Frequency References of 40 and 18 MHz, Respectively

Frequency	Loop Bandwidth = 20 KHz, Frequency Reference = 40 MHz Spur Level (dBc)	Frequency	Loop Bandwidth = 20 KHz, Frequency Reference = 18 MHz Spur Level (dBc)
100 KHz	−65.3	33.3 KHz	−47.2
200 KHz	−79.2	66.7 KHz	−58
300 KHz	−88.3	100 KHz	−65.3
400 KHz	−95.1	133 KHz	−70.9
500 KHz	−101	167 KHz	−75.4

Consequently, the instantaneous phase error accumulating each cycle can be expressed in the form

$$\Delta\theta(n) = \Delta\theta(n-1) + 2\pi F_{ref}\left(\frac{1}{F_{ref}} - \frac{P+q(n)}{F_{out}}\right),\ q(n)\in\{0,1\}\forall n = 1,\cdots B \tag{7.125}$$

where $q(n)$ is a sequence of 0's and 1's comprised of A 1's and $B-A$ 0's. Substituting the value of F_{out} according to Eqn (7.119), the relationship in Eqn (7.125) can be further expressed as

$$\begin{aligned}\Delta\theta(n) &= \Delta\theta(n-1) + 2\pi\left(1 - \frac{P+q(n)}{P+\frac{A}{B}}\right)\\ \Delta\theta(n) &= \Delta\theta(n-1) + 2\pi\left(\frac{A-Bq(n)}{BP+A}\right)\end{aligned} \tag{7.126}$$

At this point, it is instructive to examine the sequence $q(n)$ for the dual modulus prescalar over B instances. By the very nature of the dual modulus prescalar, all the 0's and 1's are grouped together as shown in Figure 7.55. For simplicity's sake, let $\Delta\theta(0)=0$, then the resulting phase error over the next B instances is a discrete sawtooth as shown in Figure 7.56. The maximum amplitude and frequency of said waveform depend on the allocation of 0's and 1's in $q(n)$. When the 0's and 1's are grouped together as is the case of dual modulus prescalar, the maximum amplitude of the sawtooth is large with dominant beat spectra of F_{ref}/B and $(A/B)F_{ref}$. In Ref. [10], Egan presents an accumulator technique wherein the 0's and 1's of $q(n)$ are distributed evenly as shown in Figure 7.55. Consequently, the resulting phase error sequence is comprised of a sawtooth waveform whose maximum amplitude is smaller than that produced by the dual modulus prescalar.

Finally, in order to determine the spectral content of the signal at the output of the CP, we examine the continuous-time current pulse $I_{CP}(t)$. According to Ref. [18], CP

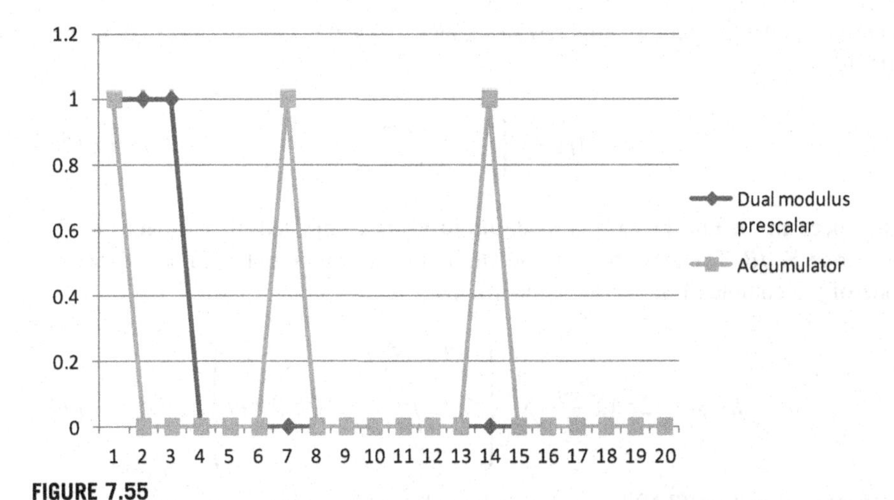

FIGURE 7.55

The sequence $q(n)$ for $B=20$ for the dual modulus prescalar and accumulator method discussed in Ref. [10].

output current is a pulse width modulated signal whose duty cycle is in accordance with the relative phase error:

$$I_{CP}(t) = 2\pi K_d \sum_{n=1}^{B}\left[u\left(t-\frac{n}{F_{ref}}\right) - u\left(t-\left(\frac{n}{F_{ref}}+\frac{\Delta\theta(n)}{2\pi F_{ref}}\right)\right)\right] \tag{7.127}$$

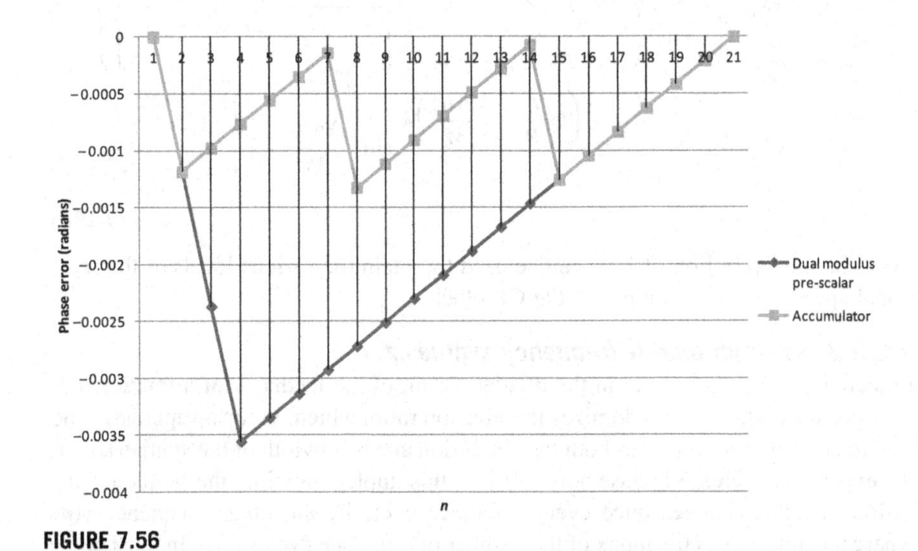

FIGURE 7.56

Phase error sequences due to dual modulus prescalar and accumulator method discussed in Ref. [10].

where K_d is the CP gain in amperes/2π radians and $u(t)$ is the step function defined simply as

$$u(t) = \begin{cases} 0 & t < 0 \\ 1 & \text{otherwise} \end{cases} \tag{7.128}$$

The spectrum of Eqn (7.127) is made up of spurs comprised of harmonics of the beat-tone F_{ref}/B. The amplitude of the kth harmonic can be obtained as the magnitude of the complex Fourier series coefficient:

$$I_{\mathrm{CP},k} = 2\pi K_d \frac{F_{ref}}{B} \sum_{n=1}^{B} \left\{ \int_{n/F_{ref}}^{n/F_{ref} + \frac{\Delta\theta(n)}{2\pi F_{ref}}} e^{-j2\pi \frac{F_{ref}}{B} kt} dt \right\} \tag{7.129}$$

The integral in Eqn (7.129) can be evaluated, thus resulting in a more simplified expression of $I_{\mathrm{CP},k}$

$$\begin{aligned} I_{\mathrm{CP},k} &= 2\pi K_d \frac{F_{ref}}{B} \sum_{n=1}^{B} \left\{ e^{-j2\pi \frac{n}{B} k} \left[\frac{1 - e^{-j\frac{\Delta\theta(n)}{M} k}}{jn} \right] \right\} \\ &= 2\pi K_d \frac{F_{ref}}{B} \sum_{n=1}^{B} \left\{ \frac{2}{n} e^{-j\left(2\pi \frac{n}{B} + \frac{\Delta\theta(n)}{2M}\right)k} \left[\frac{e^{-j\frac{\Delta\theta(n)}{2M} k} - e^{-j\frac{\Delta\theta(n)}{2M} k}}{2j} \right] \right\} \\ &= 2\pi K_d \frac{F_{ref}}{B} \sum_{n=1}^{B} \left\{ \frac{2}{n} e^{-j\left(2\pi \frac{n}{B} + \frac{\Delta\theta(n)}{2M}\right)k} \sin\left(\frac{\Delta\theta(n)}{2M} k\right) \right\} \end{aligned} \tag{7.130}$$

The relationship in Eqn (7.130) can be used to obtain the various levels of the fractional spurs seen at the output of the CP block.

7.2.2.3 ΔΣ fractional-N frequency synthesizer

Embedding a ΔΣ modulator in the divider circuit of the feedback of a fractional-N PLL is a technique that randomizes the division ratio, which, in turn, suppresses the fractional spurs and increases both the resolution and bandwidth of the synthesizer at the expense of increased phase noise [19]. In this implementation, the frequency division modulus changes once every reference cycle by an integer sequence $y(n)$ where n in this case is the index of the number of reference cycles [20]. In most common implementations, $y(n)$ is generated by quantizing a certain digital constant $\propto$. Recall from Chapter 6 that for a second order ΔΣ modulator, for example, as shown in Figure 7.57, the error signal $e_q(n)$ is related to the input $x(n) \cong \alpha$ and the output $y(n)$ according to the relation

$$\begin{aligned} y(n) &= \alpha + e_q(n) - 2e_q(n-1) + e_q(n-2) \\ Y(z) &= \alpha + \left(1 - 2z^{-1} + z^{-2}\right)E_q(z) \end{aligned} \tag{7.131}$$

where the error sequence $e_q(.)$ has been shaped by the highpass function

$$NTF(z) = 1 - 2z^{-1} + z^{-2} \tag{7.132}$$

For a "white-quantization distributed" error sequence $e_q(n)$, the resulting error is shaped by the highpass transfer function $NTF(z)$. In steady state, the VCO frequency is a multiple of the average division ratio $P + y(n)$, namely

$$F_{\text{out}} = (P + y(n))F_{\text{ref}} \tag{7.133}$$

This implies that the divided output frequency of the VCO $F_{out}/(P + y(n))$ is equal to the input reference frequency F_{ref} only *on average*. The instantaneous error can be expressed according to Eqn (7.133) as

$$\text{Instantaneous frequency error} = F_{ref}\left[y(n) - \alpha|_{n \text{ designates the instant } nT}\right] \tag{7.134}$$

As in the case of the dual modulus fractional-N PLL discussed previously, the highpass-shaped frequency error is lowpass filtered by the loop filter, which suppresses the high frequency content of the input signal as it presents its output to the VCO. Therefore, lowering the bandwidth of the loop filter will automatically imply lowering the phase noise of the ΔΣ-PLL synthesizer. It turns out that the resolution of this type of fractional-N synthesizer is limited only by the resolution of the

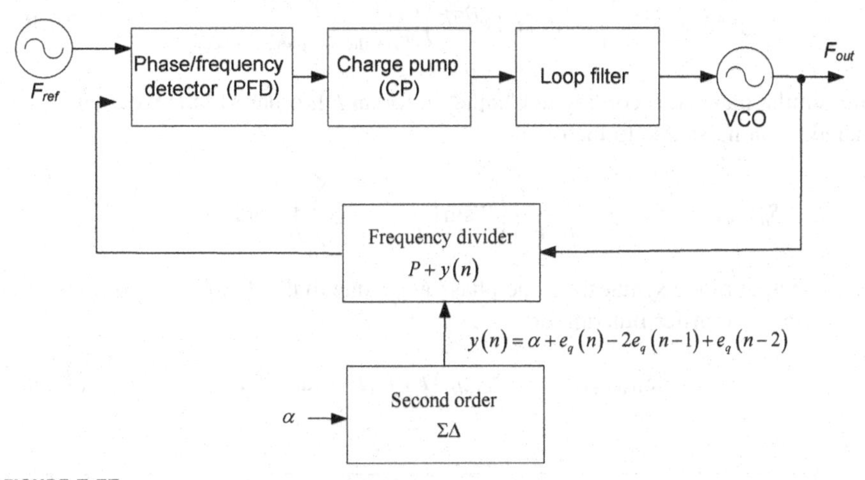

FIGURE 7.57

Fractional-N phase-locked loop employing second order ΣΔ modulator.

$\Delta\Sigma$ modulator. Although it is beyond the scope of this text, $\Delta\Sigma$ fractional-N synthesizers have been used effectively in the implementation of polar transmitters by replacing the digital constant-input to the PLLα by a slowly varying digital-phase sequence $m(n)$ representing the transmitted data sequence, thus improving the efficiency of the power amplifiers and lowering the power consumption of the overall wireless transmitter. This is true provided that the data sequence $m(n)$ has significantly lower bandwidth than the PLL itself.

There are some challenges that limit the utility of the $\Delta\Sigma$-PLL synthesizer, namely the impact of quantization noise on phase noise performance as well as the spurs present due to the output of the oversampled modulator itself. For a small $\propto$, the spurs occur within the passband of the loop filter and hence experience very little, if any, suppression by it. The quantization noise itself may fold in and cause a significant increase in close-in phase noise at the output of the synthesizer. A further source of errors is due to modulus-dependent divider delays [20]. The delay itself depends on the value of $y(n)$ and occurs during the $(n+1)$th reference cycle where the rising edge generated by the divider circuit happens a certain delay after $(N+y(n))$ VCO cycles [21].

The quantization noise due to the $\Delta\Sigma$ modulator can be mapped to PLL's output phase noise. The phase noise due to the divider is given according to Chapter 6 and Ref. [22] as[5]:

$$S_{divider}(F) = \left(\frac{F_{ref}}{P \times F/F_s} S_{\Delta\Sigma}(F/F_s)\right)^2 \Bigg|_{F_s \text{ is the sampling frequency}} \quad \text{rad}^2/\text{Hz} \tag{7.135}$$

where root mean square spectral density of the modulator's quantization noise is given as

$$S_{\Delta\Sigma}(F) = \frac{1}{\sqrt{12F_{ref}}}\left|NTF\left(e^{j2\pi\frac{F}{F_s}}\right)\right|_{F_s \text{ is the sampling frequency}} \tag{7.136}$$

In a similar manner, according to Chapter 7, for an Lth-order MASH modulator, the quantization noise due to the divider is

$$S_{divider}(F) = \frac{F_{ref}}{(P \times F/F_s)}\left(2\sin\left(\frac{\pi F}{F_{ref} \times F_s}\right)\right)^{2L} \text{rad}^2/\text{Hz} \tag{7.137}$$

At the output of the synthesizer, the phase noise due to the divider circuit is shaped by the loop's transfer function or

$$S_{output}(F) = S_{divider}(F)|H(F)|^2 \text{rad}^2/\text{Hz} \tag{7.138}$$

[5]The notation for the $\Delta\Sigma$ modulator has slightly changed from that of Chapter 4 in order to stay consistent with the notation of this chapter.

Within the passband of the loop's transfer function, the close-in phase noise due to the divider can be approximated according to the relationship in Eqn (7.138) as Ref. [19]

$$S_{output}(F) = P^2 S_{divider}(F)\text{rad}^2/\text{Hz} \tag{7.139}$$

Again, the major advantage of using a $\Delta\Sigma$ modulator in the feedback of a fractional-*N* frequency synthesizer versus a dual modulus prescalar is suppression of the fractional spurs. Various $\Delta\Sigma$ modulator topologies may be used to design the divider circuits as long as the design is appropriate for high frequency operation and exhibits stable DC-input ranges that meet the performance requirements of the intended application.

References

[1] J.D. Gibson, The Mobile Communications Handbook, second ed., CRC in cooperation with IEEE Press, Boca Raton, FL, 1999.

[2] J.W.M. Bergmans, Digital Baseband Transmission and Recording, Kluwer Academic Publishers,, Boston, MA, 1996.

[3] H. Meyr, G. Ascheid, Synchronization in Digital Communications, vol. 1, John Wiley and Sons, New York, 1990.

[4] I.S. Gradshteyn, I.M. Ryzhik, Table of Integrals, Series and Products, Academic Press, New York, NY, 1980.

[5] C. Chen, Linear System Theory and Design, Holt, Reinhart, and Winston, New York, NY, 1984.

[6] F.M. Gardner, Phaselock Techniques, second ed., Wiley, New York, NY, 1979.

[7] G.C. Hsieh, J.C. Hung, Phase-locked loop techniques – a survey, IEEE Trans. Ind. Electron. 43 (December 1996) 609–615.

[8] B. Razavi, RF Microelectronics, Prentice Hall, New York, NY, 1998.

[9] W. Rhee, Design of high performance CMOS charge pumps in phase locked loops, in: Proc. IEEE Int. Symp. Circuits and Systems (ISCAS), vol. 1, May 1999. Orlando, FL.

[10] W.F. Egan, Frequency Synthesis by Phase Lock, second ed., John Wiley and Sons, New York, NY, 2000.

[11] B. Rezavi, Design of Analog CMOS Integrated Circuits, McGraw-Hill, 2001.

[12] A. Swaminathan, K. Wang, I. Galton, A wide-bandwidth 2.4 GHz ISM band fractional-*N* PLL with adaptive phase noise cancellation, IEEE J. Solid-State Circuits 42 (No. 12) (December 2007) 2639–2650.

[13] A. Barta, et al., Multi-band OFDM Physical Layer Proposal for IEEE 802.15 Task Group 3a, IEEE, Piscataway, NJ, March 2004. IEEE P802.15–03/268r3-TG3a.

[14] B. De Muer, M. Steyaert, CMOS Fractional-N Synthesizers, Kluwer Academic Publishers, Boston, MA, 2003.

[15] Analog Devices Tutorial, Fundamentals of Phase Lock Loops (PLLs), MT-086 Tutorial.

[16] A. Shinmyo, M. Hashimoto, H. Onodera, Design and measurement of 6.4 Gbps 8:1 multiplexer in 0.18 μm CMOS process, in: ASP-DAC 2005: Proceedings of the 2005 Conference on Asia South Pacific Design Automation, 36, 2005, 9–10.

[17] C.S. Vaucher, et al., A family of low-power truly modular programmable dividers in standard 0.35-μm CMOS technology, IEEE J. Solid-State Circuits 35 (July 2000) 1039–1045,.

[18] D. Butterfield, B. Sun, Prediction of fractional-N spurs for UHF PLL frequency synthesizers, in: IEEE MTT-S Symposium on Technologies for Wireless Applications Tech. Dig., February 1999, pp. 29–34.

[19] B. Miller, B. Conley, A multiple modulator fractional divider, in: 44th Annual Symposium on Frequency Control, 1990, pp. 559–568.

[20] S. Pamarti, Digital techniques for integrated frequency synthesizers: a tutorial, IEEE Commun. Mag. (April 2009) 126–133.

[21] S. Pamarti, L. Janson, I. Galton, A wideband 2.4-GHz delta-sigma fractional-*N* frequency synthesizer for DCS-1800, IEEE J. Solid-State Circuits 39 (No. 1) (January 2004) 49–62.

[22] T. Riley, M. Copeland, T. Kwasnewski, Delta-sigma modulation in fractional-*N* frequency synthesis, IEEE J. Solid-State Circuits 28 (May 1993) 553–559.

Further reading

[23] C. Mishra, et al., Frequency planning and synthesizer architectures for multiband OFDM UWB radio, IEEE Trans. Microwave Tech. 53 (No. 12) (December 2005) 3744–3756.

CHAPTER

8

Receiver Architectures

CHAPTER OUTLINE

The choice of receiver architecture is dictated mainly by six high-level parameters: performance, implementation complexity, size, number of external components, power consumption, and cost [1]. The relative importance of each of these parameters, albeit some go hand and hand, has varied over the years depending on IC technology and the underlying wireless application. From a signal processing perspective,

Wireless Receiver Architectures and Design. http://dx.doi.org/10.1016/B978-0-12-378640-1.00008-1

the receiver can be thought of as a system divided into the following subsystems: antenna, analog front-end, analog intermediate frequency (IF) and baseband, data conversion, frequency generation, and digital baseband. In the receiver, the antenna is a transducer system that transfers electromagnetic energy into electrical or magnetic energy (see Chapter 1.) The antenna's gain, directivity, frequency bands of operation, and the antenna's form factor are some of the key parameters defining its performance. The analog front-end is comprised of filters, amplifiers, switches, and mixers. It plays a key role in setting the receiver's sensitivity and linearity. The analog baseband is also comprised of filters and gain stages. The channel selectivity and interference mitigation are predominantly set by the analog low-pass or IF filters. The impact of phase noise on the receiver performance becomes apparent at this stage. The analog baseband is followed by the data conversion subsystem. In Chapter 6, two types of analog-to-digital conversion schemes were discussed: Nyquist converters and oversampling converters. The parameters of interest, at this stage, range from converter resolution and sampling rate to converter nonlinearity, and jitter performance. The frequency conversion block is designed with a certain frequency plan in mind to facilitate the down-conversion (or up-conversion) of signals from radio frequency (RF) to baseband or RF to IF. In Chapter 7, frequency generation was discussed in some detail. Finally, the last block is the digital portion of the receiver where the signal after analog-to-digital conversion is demodulated and synchronization is performed. At this stage, the signal may undergo further equalization and decoding before the data packets are sent to MAC and higher link layers.

This chapter is divided into five sections. Sections 8.1–8.3 discuss the three most common wireless receiver architectures: direct-conversion, superheterodyne, and low IF. Section 8.4 addresses the typical driving requirements of a wireless receiver and how they impact a certain architecture. Section 5 contains some appendices.

8.1 Direct-conversion receiver

The intent of this section is to present an overview of the direct-conversion or zero-IF receiver[1] architecture depicted in Figure 8.1. This receiver architecture is by far the most common architecture used in low-power applications due to its simplicity and scalability. In the coming sections, we will review the signal flow through the various stages of the receiver and discuss the architecture's pros and cons.

8.1.1 The direct conversion architecture

Zero IF implies that the image frequency matches the frequency of the desired received signal. Direct-conversion receivers, unlike superheterodyne receivers,

[1]In this chapter, zero-IF and direct-conversion will be used interchangeably.

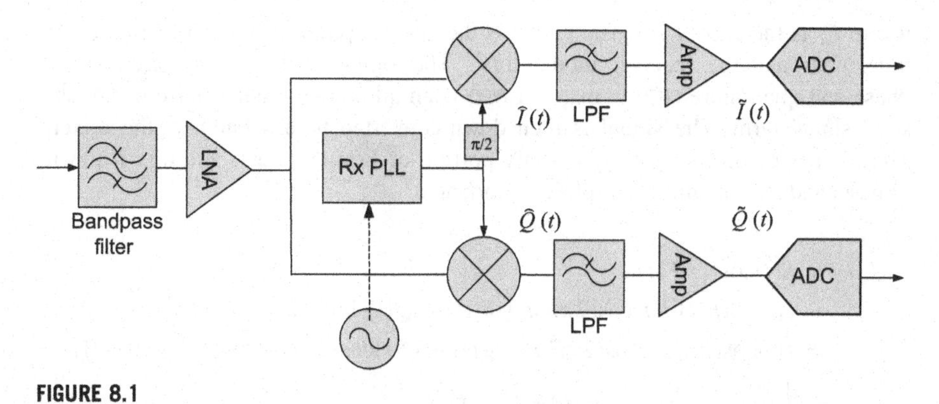

FIGURE 8.1

The conceptual direct-conversion receiver.

for example, do not require image rejection, thus avoiding the use of high-Q external image-reject filters. The architecture itself is simple, with very low component count and good performance. It is particularly suitable for broadband signals with moderate performance specifications, high integration, and multi-band multistandard operation. Nonetheless, the architecture has certain constraints and disadvantages with varying impact on design and performance that the system and circuit designers must be aware of. Until recently, these limitations kept the direct-conversion architecture lagging behind the superheterodyne architecture [2].

In Figure 8.1, the signal impinging on the antenna is filtered by a bandpass filter, typically known as the band-definition filter (BDF). The intent of this filter is to band limit the signal and noise received to within the desired frequency band of operation. Furthermore, the BDF will protect the receiver front end from signals and spurs occurring outside of its band and prevent the circuits from saturation in the presence of a strong out-of-band signal. The incoming signal is then amplified by the low noise amplifier (LNA), which in turn ensures a low system noise figure required for proper demodulation of the desired signal, especially near sensitivity. The signal at this point is at RF and can be expressed simply as[2]

$$S(t) + N(t) = \underbrace{I(t)\cos(2\pi F_c t + \theta(t)) + Q(t)\sin(2\pi F_c t + \theta(t))}_{S(t)} + N(t) \qquad (8.1)$$

[2]This is one of two alternate representations where the quadrature component is added to the in-phase component. However, the most common representation is the one where the quadrature component is subtracted from the in-phase component, thus representing the real component of the transmit signal. We chose the former for simplicity.

where F_c is the carrier frequency, $\theta(t)$ is the general phase difference between the transmitter and the receiver modulators.[3] The signals $I(t)$ and $Q(t)$ are the in-phase and quadrature (IQ) terms, and $N(t)$ is an all-encompassing noise or nondesired signal term. The signal is then down-converted to baseband by the mixer. That is, if we consider the signal-only portion of Eqn (8.1) $S(t)$ and mix it down to baseband, we obtain the in-phase component

$$\begin{aligned}\widehat{I}(t) &= S(t)\cos(2\pi F_c t)\\ &= \cos(2\pi F_c t)[I(t)\cos(2\pi F_c t + \theta(t)) + Q(t)\sin(2\pi F_c t + \theta(t))]\\ &= I(t)\cos(2\pi F_c t)\cos(2\pi F_c t + \theta(t)) + Q(t)\cos(2\pi F_c t)\sin(2\pi F_c t + \theta(t))\\ &= \frac{1}{2}I(t)[\cos(\theta(t)) + \cos(4\pi F_c t + \theta(t))]\\ &\quad + \frac{1}{2}Q(t)[\sin(\theta(t)) + \sin(4\pi F_c t + \theta(t))]\end{aligned} \tag{8.2}$$

The quadrature component can be obtained in a likewise manner as

$$\begin{aligned}\widehat{Q}(t) &= S(t)\sin(2\pi F_c t)\\ &= \sin(2\pi F_c t)[I(t)\cos(2\pi F_c t + \theta(t)) + Q(t)\sin(2\pi F_c t + \theta(t))]\\ &= I(t)\sin(2\pi F_c t)\cos(2\pi F_c t + \theta(t)) + Q(t)\sin(2\pi F_c t)\sin(2\pi F_c t + \theta(t))\\ &= \frac{1}{2}I(t)[\sin(\theta(t)) + \sin(4\pi F_c t + \theta(t))]\\ &\quad + \frac{1}{2}Q(t)[\cos(\theta(t)) - \cos(4\pi F_c t + \theta(t))]\end{aligned} \tag{8.3}$$

The IQ signals undergo further filtering via two lowpass filters. The purpose of these filters is twofold: first, to filter any undesired signals or interferers within the band that originate outside the receiver or from within the receiver itself; and second, to filter out the high frequency components of Eqn (8.8). That is, after filtering, the ideal IQ components, ignoring the effects of the amplifiers, become

$$\begin{aligned}\tilde{I}(t) &\approx \frac{1}{2}I(t)\cos(\theta(t)) + \frac{1}{2}Q(t)\sin(\theta(t))\\ \tilde{Q}(t) &\approx \frac{1}{2}I(t)\sin(\theta(t)) + \frac{1}{2}Q(t)\cos(\theta(t))\end{aligned} \tag{8.4}$$

---

[3] In order to simply illustrate the workings of the direct-conversion receiver, the received signal as well as the receiver itself are assumed to be ideal.

In the event where $\theta(t)$ is negligible, then $\cos(\theta(t)) \approx 1$ and $\sin(\theta(t)) \approx 0$ and the relationship in Eqn (8.5) can be further simplified as

$$\tilde{I}(t) \approx \frac{1}{2} I(t)\cos(\theta(t))$$
$$\tilde{Q}(t) \approx \frac{1}{2} Q(t)\cos(\theta(t)) \quad (8.5)$$

Of course, throughout this discussion we ignored the effect of nonidealities for the sake of simplifying the analysis and illustration of the working of the receiver.

8.1.2 Receiver performance

In this section, the pros and cons of the direct-conversion receiver architecture and its overall performance will be discussed.

8.1.2.1 Advantages of direct-conversion receiver

As mentioned earlier, the main attraction of the direct-conversion architecture for wireless devices is its low cost, small size, and low power consumption. In terms of component counts, it uses the smallest number of external components as compared to superheterodyne or low IF. The signal is mixed directly from RF to baseband. In most cases, this architecture requires very high-Q voltage-controlled oscillators (VCOs). The use of a single LO to perform the signal down-conversion limits the degradation due to phase noise compared to architectures that employ dual or triple down-conversion. The local oscillator (LO) itself has quadrature outputs for proper modulation (transmit) and demodulation (receive). Compared to the superheterodyne architecture that requires a mixer to down-convert the signal from RF to IF followed by other mixers that mix the signal to baseband, direct conversion requires only a quadrature mixer with very high linearity that mixes the signal from RF to baseband. The high linearity requirement is put in place in order to minimize the distortion impacting the signal due to the mixer and the LNA. Both mixer and LNA are exposed to the wide RF spectrum of the band definition filter whose bandwidth is as wide as the signaling band. The linearity requirement for the mixer could be alleviated by providing additional filtering before the mixer. Unlike the superheterodyne and low-IF architectures, the direct-conversion architecture does not suffer from image rejection since the desired signal is down-converted directly from RF to baseband. Consequently, no image reject filter is needed and the LNA does not need to drive a 50 Ω load as is the case, for example, with the superheterodyne architecture.

Following the mixer, and in the event where the architecture requires the use of a premixer filter, the signal is further amplified by a postmixer amplifier (not shown in Figure 8.1). In practice, this filter is needed to lessen the impact of out-of-band blockers and interferers in order to prevent the desensitization of the received signal. The quadrature mixers are followed by integrated monolithic lowpass filters with varying complexity depending on the desired performance of the receiver. The filters are followed by amplifiers, most likely voltage gain amplifiers (VGAs), and two analog-to-digital converters (ADCs) for the IQ components. Although direct-conversion receivers require two ADCs, both converters require slower clocks and lower resolution and number of bits than say a single ADC performing bandpass sampling operating at IF as is the case of the superheterodyne IF-sampling receiver. The power consumption for both converters, in most cases, is also lower than a single ADC performing bandpass sampling at IF.

8.1.2.2 Disadvantages of direct conversion receiver

Direct-conversion receivers suffer from various sources of degradations chief amongst them are DC signals due to self-mixing, IQ imbalance, second-order nonlinearity, and flicker noise. We will address each of these sources of degradation individually.

8.1.2.2.1 LO leakage, large blockers, and DC-offset signals

The LO signal may be radiated or conducted inadvertently through any number of paths to the mixer's RF input port as shown in Figure 8.2. The LO signal mixes with itself to produce an unwanted DC signal (or DC offset) at baseband. In certain designs, the signal may even reach the input of the LNA, get amplified and then self-mix down to baseband. This may particularly be true in a design where the LNA, mixer, and LO are all integrated on the same substrate. In this case, substrate coupling, bond-wire radiation, ground bounce, and capacitive and magnetic coupling are all mechanisms that lead to poor isolation.

LO self-mixing can be explained mathematically as follows. Assume that the LO signal is expressed as $LO_I(t) = \gamma\cos(\Omega_c t)$ for the in-phase component and $LO_Q(t) = \gamma\sin(\Omega_c t)$ for the quadrature component where γ is the amplitude. After the LO leakage signal gets amplified by the LNA, its amplitude and phase characteristics undergo a certain variation resulting in the signal $LO_{leakage}(t) = g\cos(\Omega_c t + \delta)$ where

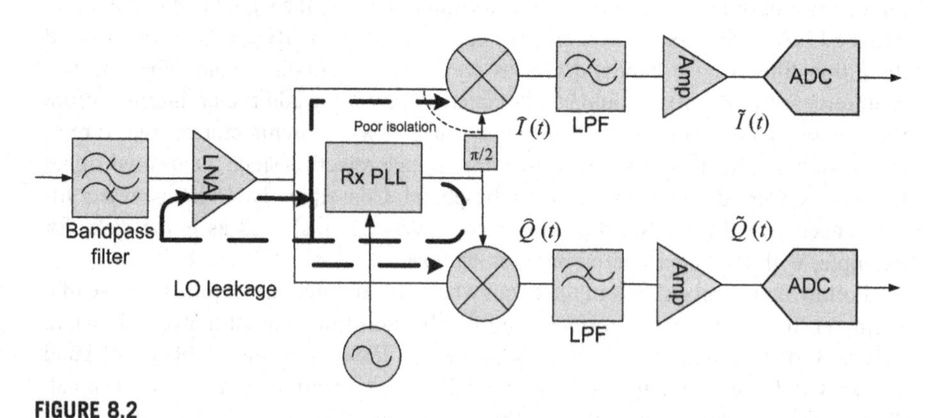

FIGURE 8.2

Effect of LO leakage on direct-conversion receiver performance.

g is the amplitude and δ is the phase of the LO leakage signal. Basically, after mixing the LO signal with a phase and amplitude modified version of itself, we obtain the IQ components at the output of the quadrature mixer as

$$\begin{aligned} LO_{leakage}(t)LO_I(t) &= g\cos(\Omega_c t+\delta)\gamma\cos(\Omega_c t) = \frac{g\gamma}{2}[\cos(2\Omega_c t+\delta)+\cos(\delta)] \\ LO_{leakage}(t)LO_Q(t) &= g\cos(\Omega_c t+\delta)\gamma\cos(\Omega_c t) = \frac{g\gamma}{2}[\sin(2\Omega_c t+\delta)+\sin(\delta)] \end{aligned} \tag{8.6}$$

Once the signals pass through the baseband filters, the high frequency components are removed and the resulting DC signals ensue:

$$\begin{aligned} LO_{leakage}(t)LO_I(t)\Big|_{\substack{\text{filtered by} \\ h_I(t)}} &\approx \hbar_Q\frac{g\gamma}{2}\cos(\delta) \\ LO_{leakage}(t)LO_Q(t)\Big|_{\substack{\text{filtered by} \\ h_Q(t)}} &\approx \hbar_I\frac{g\gamma}{2}\sin(\delta) \end{aligned} \tag{8.7}$$

where $\hbar_I$ and $\hbar_Q$ are the respective DC gains of the IQ baseband filters $h_I(.)$ and $h_Q(.)$. With the exception of $\delta = \pi/4$, the DC offset values on the IQ component ADCs are different.

Similarly, a strong blocker or interferer, compared to the desired signal, impinging on the antenna undergoes amplification by the LNA, and then due to poor forward isolation couples onto the LO and mixes itself to baseband as illustrated in Figure 8.3. Another possible degradation, of lesser impact, occurs when a certain amount of LO power due to poor reverse isolation propagates through

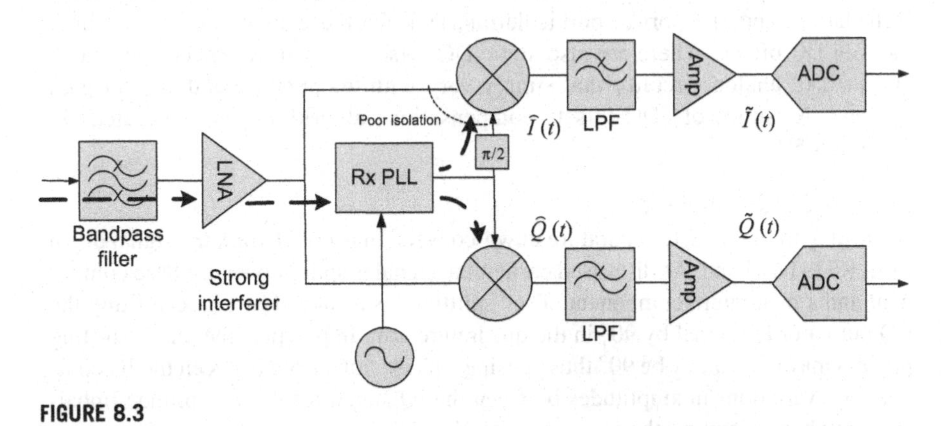

FIGURE 8.3

Impact of strong interferer or blocker on direct-conversion receiver performance.

the antenna. The signal travels through the mixer, LNA, unattenuated by the band-pass filters, and finally radiates through the antenna causing interference to close-by receivers operating in the same band.

8.1.2.2.2 Even order nonlinearity and DC-offset signals

Even order nonlinearity was discussed in some detail in Chapter 4. It was shown that a signal that undergoes any second order nonlinearity distortion will generate a signal at DC. For example, if the signal the input signal is $x(t) = A(t)\cos(\Omega_c t)$, where $A(t)$ is the information bearing signal, undergoes a second order nonlinearity such that $y(t) = \beta_1 x(t) + \beta_2 x^2(t)$, then the output signal has a signal component at DC according to the relation

$$\begin{aligned} y(t) &= \beta_1 x(t) + \beta_2 x^2(t)\big|_{x(t)=A(t)\cos(\Omega_c t)} = \underbrace{\beta_1 A(t)\cos(\Omega_c t)}_{\text{linear term}} + \underbrace{\beta_2 A^2(t)\cos^2(\Omega_c t)}_{\text{non-linear term}} \\ &= \underbrace{\beta_1 A(t)\cos(\Omega_c t)}_{\text{linear term}} + \underbrace{\frac{1}{2}\beta_2 A^2(t)}_{\text{DC term}} + \underbrace{\frac{1}{2}\beta_2 A^2(t)\cos(2\Omega_1 t)}_{\text{Modulated second harmonic term}} \end{aligned} \tag{8.8}$$

The modulated signal at DC or baseband $A^2(t)$ has twice the bandwidth as the signal $A(t)$. The DC component may be caused by even order linearity occurring primarily at the mixer output or in any of the analog blocks at baseband. As mentioned in Chapter 4, the second order nonlinearity in the mixer is characterized by the input-referred second order intercept point (IIP2) and its corresponding second order intermodulation product (IM2) where the power of the former is related to the power of the latter as $P_{IM_2,dB} = 2P_{i,dBm} - P_{IIP2,dBm}$. Degradation due to second order nonlinearity can be significantly reduced by employing a balanced mixer. Recall that there are three classifications of mixers: unbalanced, single-balanced, and double-balanced. The difference between balanced and unbalanced is the latter's superior port-to-port isolation. Thus far, these offsets are known time-varying DC offsets. There are also static DC offsets caused by process mismatch and drift of analog circuitry that simply vary with temperature and current gain setting. A suite of DC-offset compensation algorithms is presented in Section 8.5.1.

8.1.2.2.3 IQ imbalance

According to Figure 8.1, quadrature down-conversion is used to mix the signal down from RF to baseband. As discussed earlier, the signal is split into an in-phase component and a quadrature component. The "splitting" is achieved by phase shifting the LO output or RF signal by 90° in the quadrature arm. In practice, the phase-shifting process may not exactly be 90° thus causing a phase imbalance between the IQ components. Variations in amplitudes between the IQ signal result in amplitude imbalance, further corrupting the incoming signal.

To further understand IQ imbalance, consider the received signal

$$S(t) = I(t)\cos(\Omega_c t) + Q(t)\sin(\Omega_c t) \tag{8.9}$$

The IQ components are added instead of subtracted to simplify the mathematics. Furthermore, consider the LO outputs mixing the signal at the IQ paths

$$\begin{aligned} LO_I(t) &= 2(1-\alpha)\cos\left(\Omega_c t - \frac{\theta}{2}\right) \\ LO_Q(t) &= 2(1+\alpha)\sin\left(\Omega_c t + \frac{\theta}{2}\right) \end{aligned} \tag{8.10}$$

In Eqn (8.10), α is the amplitude imbalance and θ is the phase imbalance, and the factor of 2 is used to simplify the math. The input of the in-phase component to the ADCs can then be expressed as

$$S(t)LO_I(t) = 2(1-\alpha)\cos\left(\Omega_c t - \frac{\theta}{2}\right)\{I(t)\cos(\Omega_c t) + Q(t)\sin(\Omega_c t)\} \tag{8.11}$$

Expanding the relationship in Eqn (8.11), we obtain

$$\begin{aligned} S(t)LO_I(t) &= 2(1-\alpha)\left\{I(t)\cos(\Omega_c t)\cos\left(\Omega_c t - \frac{\theta}{2}\right) + Q(t)\sin(\Omega_c t)\cos\left(\Omega_c t - \frac{\theta}{2}\right)\right\} \\ &= \frac{2(1-\alpha)}{2}\left\{I(t)\cos\left(2\Omega_c t - \frac{\theta}{2}\right) + I(t)\cos\left(\frac{\theta}{2}\right)\right. \\ &\quad \left. + Q(t)\sin\left(2\Omega_c t - \frac{\theta}{2}\right) + Q(t)\sin\left(\frac{\theta}{2}\right)\right\} \end{aligned} \tag{8.12}$$

After filtering, the terms that occur at twice the carrier frequency are considerably attenuated, and hence their impact on the demodulated baseband signal is negligible. Hence, the relationship in Eqn (8.12) becomes

$$I'(t) = S(t)LO_I(t) = (1-\alpha)\left\{I(t)\cos\left(\frac{\theta}{2}\right) + Q(t)\sin\left(\frac{\theta}{2}\right)\right\} \tag{8.13}$$

In a similar manner, the product due to mixing the received signal with the LO quadrature component is

$$S(t)LO_Q(t) = 2(1+\alpha)\sin\left(\Omega_c t + \frac{\theta}{2}\right)\{I(t)\cos(\Omega_c t) + Q(t)\sin(\Omega_c t)\} \tag{8.14}$$

Further expansion of the relationship in Eqn (8.14) implies

$$
\begin{aligned}
S(t)LO_Q(t) &= 2(1+\alpha)\left\{I(t)\cos(\Omega_c t)\sin\left(\Omega_c t+\frac{\theta}{2}\right)+Q(t)\sin(\Omega_c t)\sin\left(\Omega_c t+\frac{\theta}{2}\right)\right\} \\
&= \frac{2(1+\alpha)}{2}\left\{I(t)\sin\left(2\Omega_c t+\frac{\theta}{2}\right)+I(t)\sin\left(\frac{\theta}{2}\right)\right. \\
&\qquad \left.+\,Q(t)\cos\left(2\Omega_c t+\frac{\theta}{2}\right)+Q(t)\cos\left(\frac{\theta}{2}\right)\right\}
\end{aligned}
\tag{8.15}
$$

Similar to the in-phase component case after filtering of the second harmonic as provided in Eqn (8.13), we obtain

$$
Q'(t) = S(t)LO_Q(t) = \frac{2(1+\alpha)}{2}\left\{I(t)\sin\left(\frac{\theta}{2}\right)+Q(t)\cos\left(\frac{\theta}{2}\right)\right\} \tag{8.16}
$$

Rewrite Eqn (8.13) and Eqn (8.16) in matrix form as

$$
\begin{bmatrix} I'(t) \\ Q'(t) \end{bmatrix} = \underbrace{\begin{bmatrix} (1-\alpha)\cos\left(\frac{\theta}{2}\right) & (1-\alpha)\sin\left(\frac{\theta}{2}\right) \\ (1+\alpha)\sin\left(\frac{\theta}{2}\right) & (1+\alpha)\cos\left(\frac{\theta}{2}\right) \end{bmatrix}}_{\Psi} \begin{bmatrix} I(t) \\ Q(t) \end{bmatrix} \tag{8.17}
$$

In order to restore the amplitude and phase imbalance, provided they are the same across the band of interest, we must compute the inverse of the matrix Ψ, that is

$$
\begin{aligned}
\Psi^{-1} &= \frac{\Psi^H}{\det(\Psi)} = \left.\frac{\begin{bmatrix} (1+\alpha)\cos\left(\frac{\theta}{2}\right) & -(1-\alpha)\sin\left(\frac{\theta}{2}\right) \\ -(1+\alpha)\sin\left(\frac{\theta}{2}\right) & (1-\alpha)\cos\left(\frac{\theta}{2}\right) \end{bmatrix}}{\underbrace{(1-\alpha^2)}_{(1-\alpha)(1+\alpha)}\cos(\theta)}\right|_{\substack{\theta\neq\pm\pi/2 \\ \alpha\neq\pm1}} \\
&= \begin{bmatrix} K_1 & K_3 \\ K_4 & K_2 \end{bmatrix}
\end{aligned}
\tag{8.18}
$$

where K_1, K_2, K_3, and K_4 are constants. A correction technique for IQ imbalance is depicted in Figure 8.4. The compensation can be done in the digital domain provided there is sufficient resolution for the algorithm to be realized with high fidelity. An IQ imbalance compensation algorithm is presented in the appendix. Finally, to gain a "visual" understanding of the impact of IQ imbalance on the signal constellation, consider a QPSK signal as depicted in Figure 8.5. The noisy signal is presented

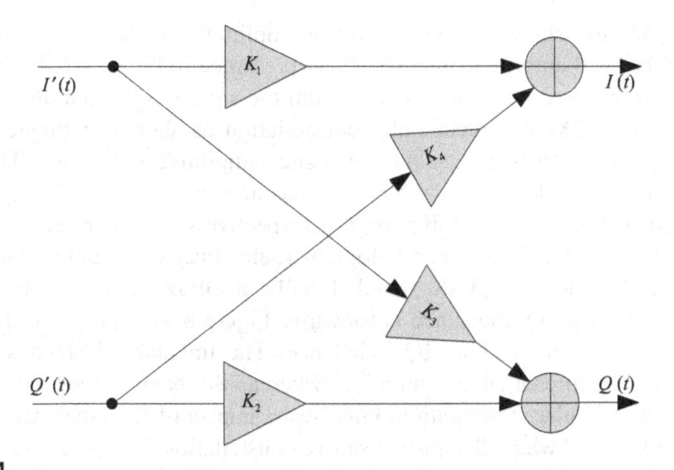

FIGURE 8.4

In-phase and quadrature imbalance compensation technique.

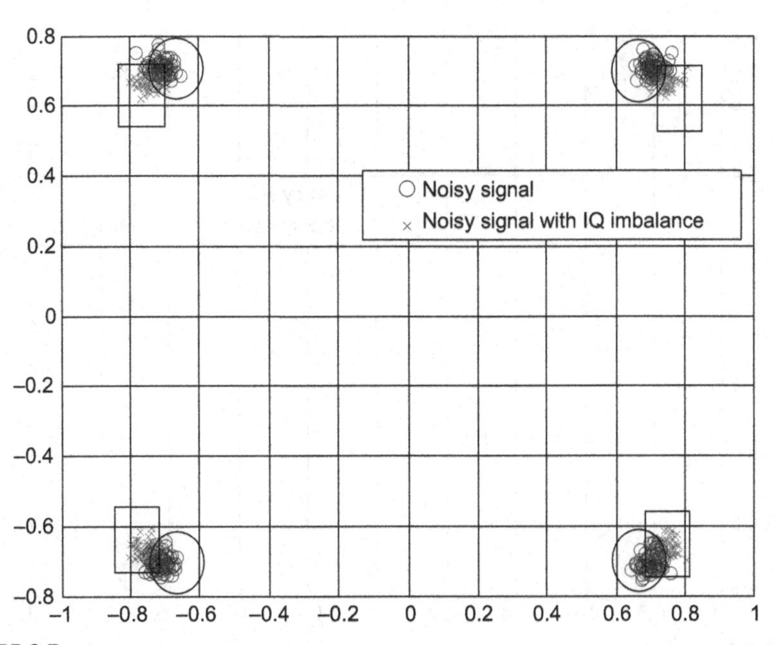

FIGURE 8.5

QPSK signal depicted with amplitude imbalance of 1 dB and no phase imbalance. The samples due to the signal with no amplitude imbalance are within the circles, whereas the samples due to the signal with amplitude imbalance are within the squares. (For color version of this figure, the reader is referred to the online version of this book.)

along with the same signal exhibiting 1 dB of amplitude imbalance. The simulation is repeated for 2-dB amplitude imbalance and presented in Figure 8.6. Note that the imbalanced signal's constellation departs from the square constellation familiar to the reader for QPSK to a rectangular constellation as shown in Figure 8.5 and even more pronounced in Figure 8.6 for higher amplitude imbalance. The signal constellations for QPSK signals with phase imbalance of 30° and 45°, respectively, are depicted in Figure 8.7 and Figure 8.8, respectively. The constellation of the phase-imbalanced signals is a parallelogram rather than the square constellation associated with a normal QPSK signal. Finally, a QPSK signal exhibiting both amplitude and phase IQ imbalance is shown in Figure 8.9. Figure 8.10 depicts an 8-PSK signal with and without IQ imbalance. The imbalanced signal's samples are placed on an ellipsoidal constellation, whereas the balanced signal's samples are placed on a circular constellation. Finally, the impact of IQ imbalance is shown for a 16-QAM signal where the normal square constellation for a perfectly balanced signal has been replaced with a parallelogram signal constellation for the IQ-imbalanced signal as shown in Figure 8.11. In all cases, IQ imbalance bears degradation on the signal's signal-to-noise ratio (SNR), and hence it is desired that its effects be minimized or calibrated out.

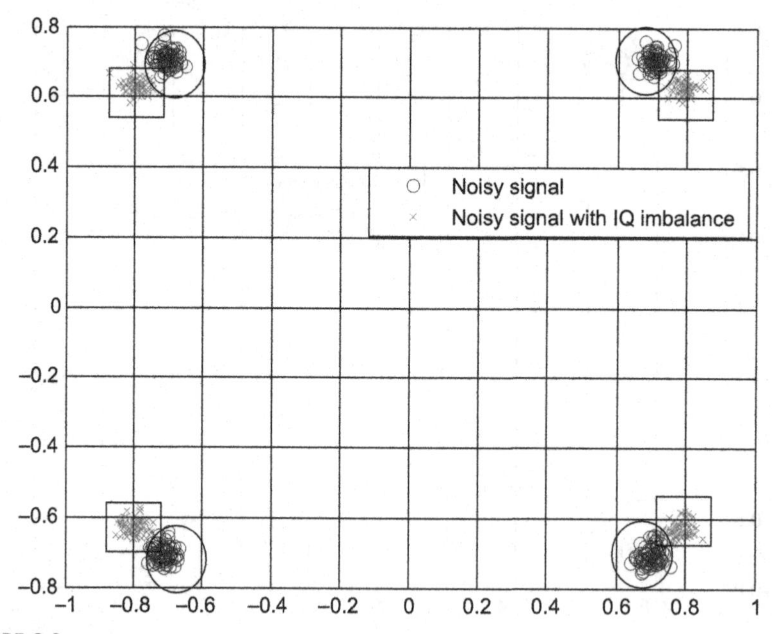

FIGURE 8.6

QPSK signal depicted with amplitude imbalance of 2 dB and no phase imbalance. The samples due to the signal with no amplitude imbalance are within the circles, whereas the samples due to the signal with amplitude imbalance are within the squares. (For color version of this figure, the reader is referred to the online version of this book.)

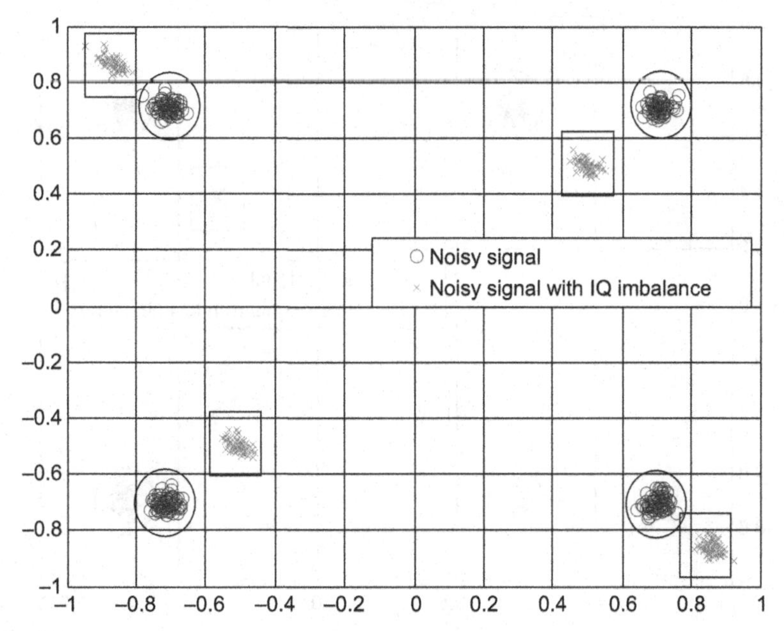

FIGURE 8.7

QPSK signal depicted with phase imbalance of 30° and no amplitude imbalance. The samples due to the signal with no amplitude imbalance are within the circles, whereas the samples due to the signal with amplitude imbalance are within the squares. (For color version of this figure, the reader is referred to the online version of this book.)

8.1.2.2.4 Low frequency noise

In direct-conversion receivers, the signal is converted directly from RF to baseband. Hence, the impact of low frequency flicker or 1/f noise, especially on narrowband signals with less 1 MHz of bandwidth, can be significant. This is true since flicker noise occurs directly in band of the desired signal. The impact of this noise degradation is particularly severe for MOS devices. Flicker noise is generated due to additional electron-energy states that exist on the boundaries of Si and SiO_2, which catches and releases electrons from the channel. This electron trapping and releasing phenomenon is slow, thus resulting in noise that is mostly concentrated at low frequencies. In a MOSFET transistor, it is fair to assume that the device transfer function can be modeled as

$$H(F) = \frac{\upsilon}{\sqrt{F}} \tag{8.19}$$

where υ is a constant that embodies the device characteristics. The flicker noise power can then be approximated over a certain frequency range as

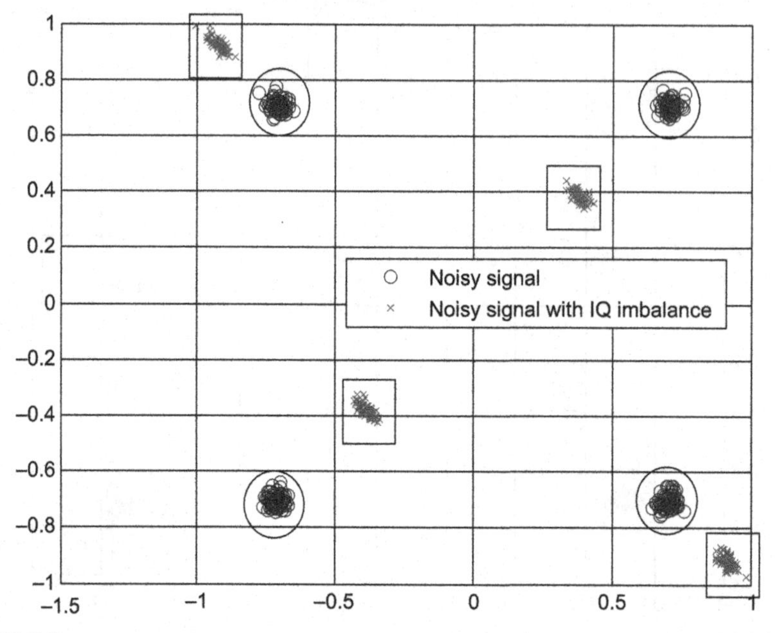

FIGURE 8.8

QPSK signal depicted with phase imbalance of 45° and no amplitude imbalance. The samples due to the signal with no amplitude imbalance are within the circles, whereas the samples due to the signal with amplitude imbalance are within the squares. (For color version of this figure, the reader is referred to the online version of this book.)

$$\int_{\substack{\text{Frequency}\\\text{range}}} \frac{\gamma^2}{F} dF = \gamma^2 \ln(F)\Big|_{\substack{\text{Frequency}\\\text{range}}} \tag{8.20}$$

where the constant in Eqn (8.20) can be approximated as $\gamma^2 = K/(WLC_{ox})$ where K is a constant that depends on the process, W is the width of the gate, L is the length of the gate, and C_{ox} is the oxide capacitance. Thus far, the discussion has centered on flicker noise within one device. However, each transistor in the analog baseband receiver contributes to $1/f$ noise. That is, circuitry in the mixer, amplifiers, and low-pass analog filters all contribute to the total noise.

8.2 Superheterodyne receiver

The superheterodyne architecture is widely used in devices where high performance and receiver (and transmitter) flexibility is desired. In this section, we will discuss

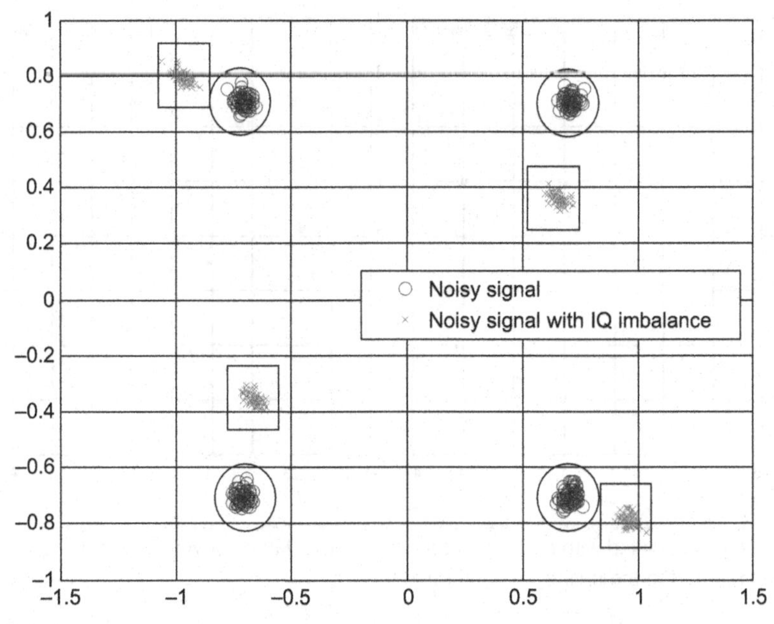

FIGURE 8.9

QPSK signal depicted with phase imbalance of 30° and amplitude imbalance of 3 dB. The samples due to the signal with no amplitude imbalance are within the circles, whereas the samples due to the signal with amplitude imbalance are within the squares. (For color version of this figure, the reader is referred to the online version of this book.)

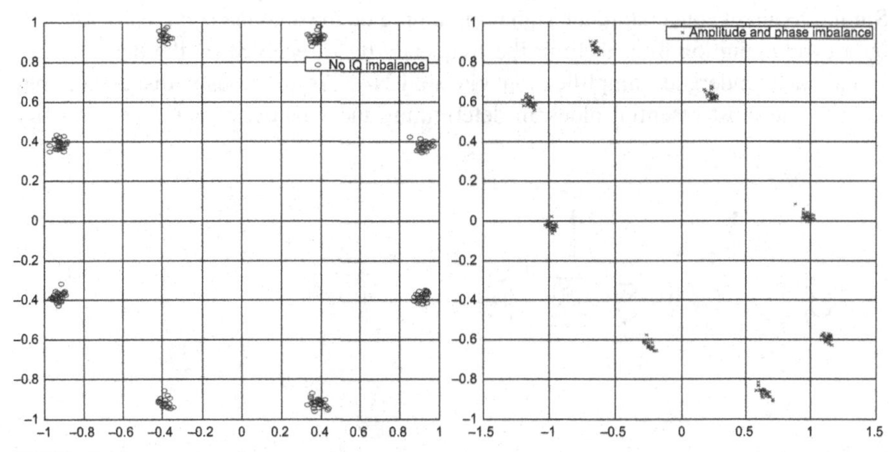

FIGURE 8.10

8-PSK signal depicted with and without IQ imbalance. (For color version of this figure, the reader is referred to the online version of this book.)

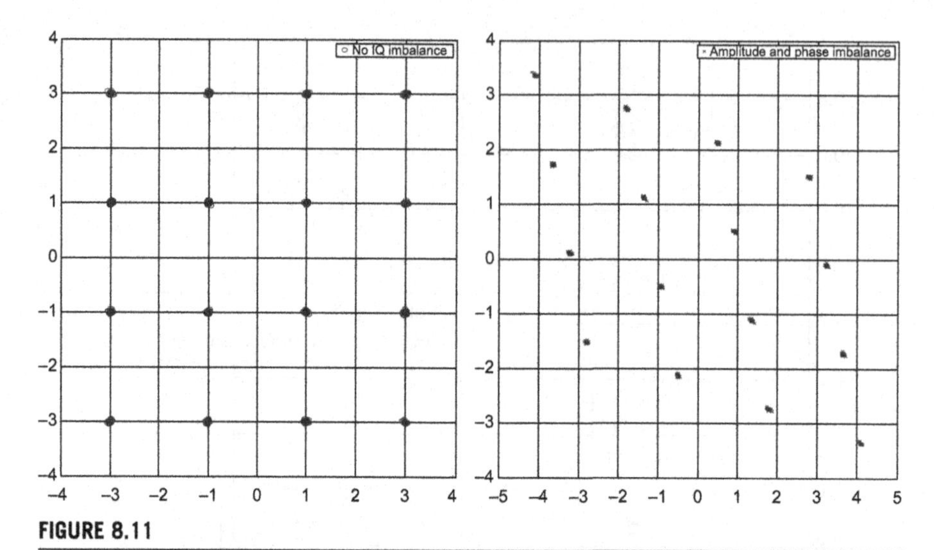

FIGURE 8.11

16-QAM signal depicted with and without IQ imbalance. (For color version of this figure, the reader is referred to the online version of this book.)

the traditional superheterodyne architecture along with the IF-sampling receiver used mostly in software defined radios. The pros and cons of this architecture will be discussed.

8.2.1 The superheterodyne architecture

A conceptual view of the superheterodyne receiver is depicted in Figure 8.12. Similar to direct conversion, the signal impinging on the antenna is typically filtered by a band definition filter to limit the frequency band received by the device. The signal then undergoes amplification via an LNA. As previously discussed, the LNA is the most essential block in determining the sensitivity of the receiver by

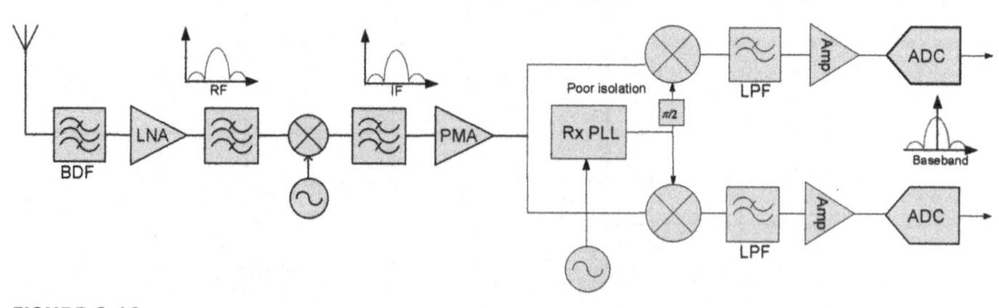

FIGURE 8.12

The superheterodyne receiver. (For color version of this figure, the reader is referred to the online version of this book.)

possessing high gain and low noise figure. The bandpass filter following the LNA plays a major role in image rejection. Any signal centered at the image frequency will be attenuated by this filter. Recall from Chapter 4 that the image frequency is centered at $F_c \pm 2F_{IF}$. Ideally, we desire this filter to have minimal insertion loss in the signal path while significantly attenuating the image noise or image signal. Both objectives can be accomplished simultaneously provided that the desired signal's center frequency is situated far enough from the image frequency, which consequently implies a significantly large IF.[4] A large IF, however, has consequences on the design of the IF filter.

The received signal is then down-converted to IF after the first mixer. The resulting signal is centered at IF, which happens to be located at either the sum of the RF and the LO frequencies or at the difference between the RF and LO frequencies. The signal at IF is then filtered in order to minimize the degradation impact of interferers and signals due to nonlinear artifacts within the band. This filter, often times referred to as the channel selection filter, is neither intended to minimize the impact of close-in blockers situated very close to the center frequency nor interference due to adjacent channels. The latter is achieved by filtering at baseband by the analog lowpass IQ filters. The channel selection filter, however, is intended to reject certain interferers and blockers as well as certain degradations due to the receiver itself. In order to simplify the design of this filter, while at the same time achieve the specified performance, it is desired to have a low IF with the aim of lowering the filter's Q. This is in direct contrast to the requirement of the image reject filter, which we prefer to have a high Q. Lowering the IF to alleviate the design of the IF channel selection filter would imply greater loss in the image reject filter. Again, recall from Chapter 4 that image reject mixers may also be used to perform certain image rejection.

After the channel filter, the desired signal is then amplified by a postmixer amplifier (PMA) before it undergoes another frequency conversion from IF to baseband. The quadrature mixer is followed by amplifier stages intermingled with filtering stages in order to preserve the linearity and selectivity of the system or simply by amplifiers such as a VGAs and analog lowpass filters. In addition to performing channel selection, the lowpass filters on the IQ paths act as antialiasing filters that limit the amount of out-of-band energy that folds over onto the desired signal band.

The superheterodyne architecture is also popular in modern software-defined radios in the form of an IF-sampling receiver architecture as shown in Figure 8.13. In this case, the signal is sampled at IF and the final IF-to-baseband down-conversion takes place in the digital domain. Depending on the chosen IF frequency, the desired signal may fall into any one of the Nyquist zones as shown via example in Figure 8.14. According to Figure 8.13, unlike the traditional superheterodyne architecture shown in Figure 8.12, the IF-sampling architecture performs bandpass sampling at IF with a single ADC. The requirements imposed on this single ADC are

[4] In certain cases, for a sufficiently large IF, the band definition filter may partially attenuate the signal at the image frequency thus performing limited image rejection.

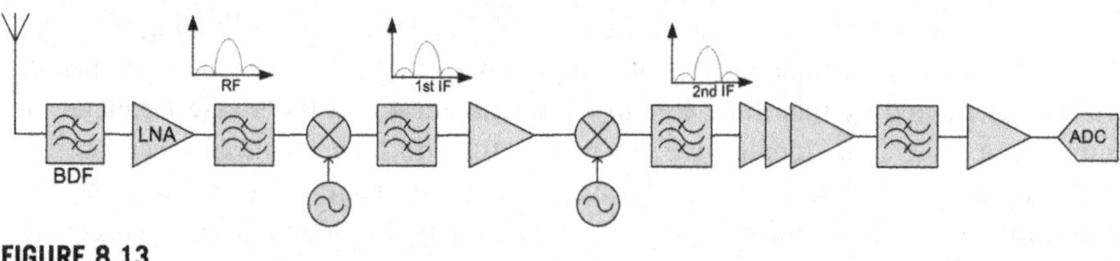

FIGURE 8.13

IF-sampling dual-conversion receiver. (For color version of this figure, the reader is referred to the online version of this book.)

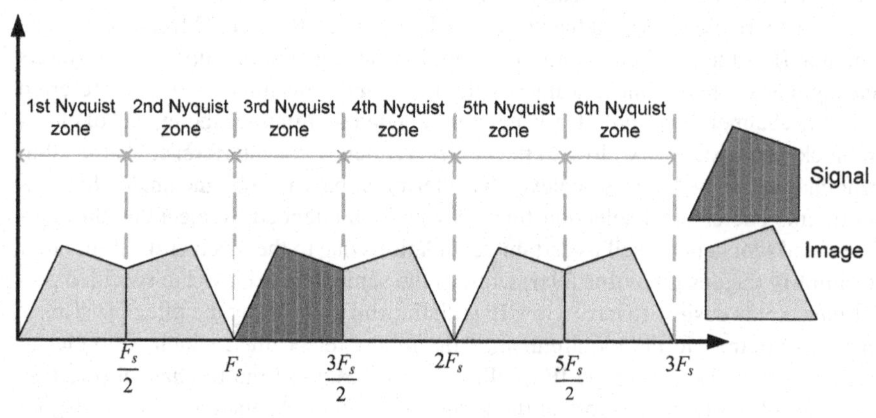

FIGURE 8.14

Desired signal (in third Nyquist zone) and spectral images. (For color version of this figure, the reader is referred to the online version of this book.)

extremely stringent. The antialiasing filter is a bandpass filter implemented in today's technology as a surface acoustic wave (SAW) filter or similar technology. SAW filters are attractive since they can provide less than 1-dB ripple in the passband and over 60 dB of rejection in the stopband depending on the bandwidth. The need to minimize triple transient effects, however, has led to higher insertion losses.

8.2.2 Receiver performance

Similar to the direct-conversion receiver architecture, in this section we will discuss the advantages and disadvantages of the superheterodyne architecture. The architecture of the traditional superheterodyne receiver is presented, followed by a very brief presentation of the IF-sampling receiver, which also falls into the superheterodyne receiver category. The pros and cons of this architecture will also be discussed.

8.2.2.1 Advantages of superheterodyne receiver

The superheterodyne receiver distinguishes itself from other receiver architectures by its overall high performance characteristics. The architecture is suitable for all modulation schemes with narrow or broad bandwidths. Excellent selectivity[5] and sensitivity are traits of the superheterodyne architecture, with perhaps, selectivity being its most distinguishing feature. Unlike direct conversion, where bandpass filtering takes place only at RF, in the superheterodyne receiver bandpass filtering could take place progressively at lower center frequencies at first and second IF. This trait of the architecture affords it higher selectivity and consequently better performance, especially in RF congested bands. These performance figures are attained by properly choosing the IF and the filters throughout the receive chain. The dominant culprit in degrading adjacent channel selectivity in this case, especially for narrowband channels, is phase noise. Phase noise is typically more dominant at the first LO due to the fact that its frequency is higher than that of the second LO. However, as mentioned earlier, choosing an IF filter with sufficiently high rejection can significantly mitigate the problem. For this reason, the first LO is designed with low phase-noise performance in mind, which may entail an external tank circuit. As a consequence, designing the receiver with high channel selectivity simply implies better signal sensitivity. In most common superheterodyne receivers, the majority of the gain stages are placed at IF. The filtering at RF effectively degrades the out-of-band interferers and allows the IF gain stages to be designed with lower dynamic range requirements, which in essence permits the design of the amplifiers to be more stable with higher gain and lower current than comparable gain stages at RF.

Unlike direct conversion, DC offset is not a major concern in superheterodyne receivers. This is especially true in the IF-sampling receiver depicted in Figure 8.13. Another significant advantage of IF-sampling receivers is its resilience to $1/f$-noise and analog IQ mismatch degradations. In general, superheterodyne receivers are suitable for multimode multiband signaling schemes. The receivers, as well as the transmitters, tend to be highly versatile in supporting a wide variety of wireless standards and modulation schemes. Due to the fixed IF, the backend analog components deliver outstanding performance with comparatively mild constraints.

8.2.2.2 Disadvantages of superheterodyne receiver

The superheterodyne receiver suffers from certain drawbacks. Most of these parameters have been previously discussed and analyzed in detail in previous chapters. The intent of this section is to briefly restate the problems associated with this architecture and consequently state the impact on the performance.

---

[5]Recall that selectivity is defined as the ability of the receiver to reject unwanted blockers and interferers situated in adjacent channels.

8.2.2.2.1 Integration, complexity, and power consumption

In today's technology, the superheterodyne architecture necessitates the use of external or off-chip components, making it less favorable for integration. The choice of high IF makes it very difficult to integrate the IF filter monolithically. The image reject filter, a passive bulky device, is also placed off chip and driven as a 50 Ω load at the output of the LNA [1]. The superheterodyne receiver is more complex than the direct-conversion receiver. Unlike direct conversion, dual conversion requires careful frequency planning, which in turn dictates the specification of the backend analog components of the receiver. Key receiver parameters such as the noise figure and the IIP3 require the gain blocks to be placed strategically in the receive path, and hence a significant number of design iterations are needed to arrive at an acceptable performance suitable for consistency in manufacturing. In terms of component count, and due to its superb and versatile capabilities, the superheterodyne receiver tends to have many more components than the direct-conversion receiver. Furthermore, the superheterodyne receiver is less conservative in terms of power consumption than the direct-conversion or low-IF receivers. These parameters all indicate that the superheterodyne receiver architecture is less amenable for low-cost wireless solutions. For this reason, the superheterodyne today remains a favorite in low-volume high-performance devices such as base stations, radars, and software defined radio.

8.2.2.2.2 Half-IF and image rejection

The ½ IF problem was discussed in detail in Chapter 4. Recall that the degradation is due to fourth order nonlinearity manifesting itself as second order nonlinearity. More precisely, any interference occurring at $F_c \pm \frac{1}{2}F_{IF}$ will overlap with the desired signal after mixing. In this case, the design trade centered on the filtering strategy, the choice of the IF, and linearity. This is an interesting trade; it entails the classical balancing act between design complexity and performance degradation. Certain knowledge of the level of interference at ½ IF will dictate the design parameters. Recall that the ½ IF problem is nonexistent for direct-conversion receivers since the IF for this architecture is zero.

The image frequency centered at $F_c \pm 2F_{IF}$ will *mix* on top of the desired frequency after down-conversion. Any blocker or interferer situated at the image frequency will cause significant degradation to the desired signal unless it is sufficiently attenuated. In the absence of any offending signal at the image frequency, image noise is then the sole cause of degradation. Several mitigation schemes including filtering and image reject mixers were also discussed in Chapter 4.

8.3 Low-IF receiver

The low-IF architecture is a compromise architecture that features some of the advantages, and inadvertently some of the disadvantages, of the superheterodyne and direct-conversion architectures.

8.3.1 The low-IF architecture

Similar to direct-conversion architecture, the received signal band is filtered by the band definition filter and then amplified by the LNA as shown in Figure 8.15. The signal is then down-converted from RF to a low IF via a quadrature converter. The IF can be as low as half the channel spacing or several multiples of it. After down-conversion, the image frequency is filtered via complex analog polyphase filters at both the IQ channels.

To gain a theoretical understanding of the complex polyphase filter, consider the filter's transfer function

$$H(s) = H_1(s) + jH_2(s) \tag{8.21}$$

where $H_1(s)$ and $H_2(s)$ are the real and imaginary part of $H(s)$. The transfer function $H(s)$ is complex since its frequency response is not symmetrical around DC [3,4]. Assume that the complex signal input to the complex polyphase filter is $I_{Input}(s) + jQ_{Input}(s)$, then the IQ signals at the output of the polyphase filter are given as the complex signal

$$\begin{aligned}\{I_{Input}(s) + jQ_{Input}(s)\}H(s) &= \{I_{Input}(s) + jQ_{Input}(s)\} \times \{H_1(s) + jH_2(s)\} \\ &= \{I_{Input}(s)H_1(s) - Q_{Input}(s)H_2(s)\} \\ &\quad + j\{I_{Input}(s)H_2(s) + Q_{Input}(s)H_1(s)\}\end{aligned} \tag{8.22}$$

According to Eqn (8.22), the output IQ signals of the polyphase filter are

$$\begin{aligned}I_{Output}(s) &= I_{Input}(s)H_1(s) - Q_{Input}(s)H_2(s) \\ Q_{Output}(s) &= I_{Input}(s)H_2(s) + Q_{Input}(s)H_1(s)\end{aligned} \tag{8.23}$$

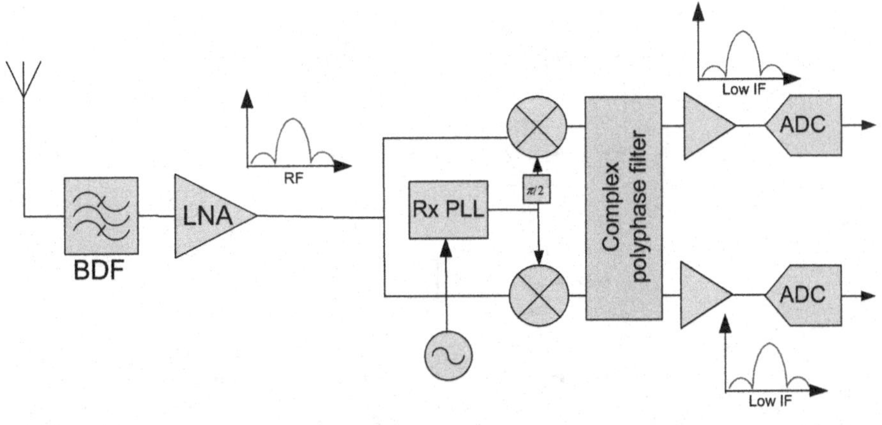

FIGURE 8.15

Low-IF receiver with complex polyphase filter. (For color version of this figure, the reader is referred to the online version of this book.)

The realization of the complex filter is shown in Figure 8.16. Given a single-stage RC network to design the polyphase filter, the filter's complex transfer function given in Eqn (8.21) becomes

$$H(s) = H_1(s) + jH_2(s) = \frac{\Upsilon\Omega_p}{s+\Omega_p} + j\frac{\Upsilon s}{s+\Omega_p} \tag{8.24}$$

where Υ is the gain and Ω_p is the pole frequency, which in this case determines the rejected frequency of the image. According to Eqn (8.24), $H_1(s)$ is a lowpass function whereas $H_2(s)$ is a highpass filter. It turns out that the desired signal at negative frequency falls within the passband of $H(-j\Omega)$ whereas the image signal at positive frequency is attenuated by the null of $H(j\Omega)$ as shown in Figure 8.17. Note that, depending on the frequency plan, the filter can be synthesized to have its passband at positive frequency and its rejected frequency at negative frequency. The image reject ratio (IRR) based on a single-stage RC network complex polyphase filter is given as the ratio

$$IRR = \frac{|H(s=j\Omega)|}{|H(s=-j\Omega)|} = \left|\frac{\Omega_p - \Omega}{\Omega_p + \Omega}\right| \tag{8.25}$$

In theory, the poles of $H_1(s)$ and $H_2(s)$ are perfectly matched and the *IRR* is zero for $\Omega = \Omega_p$. If, on the other hand, the poles of $H_1(s)$ and $H_2(s)$ are mismatched in amplitude and frequency, then $H(s)$ can be expressed as

$$H(s) = \frac{\Upsilon\Omega_p\left(1-\frac{\Delta\Upsilon}{2\Upsilon}\right)\left(1-\frac{\Delta\Omega_p}{2\Omega_p}\right)}{s+\Omega_p\left(1-\frac{\Delta\Omega_p}{2\Omega_p}\right)} + j\frac{\Upsilon\left(1+\frac{\Delta\Upsilon}{2\Upsilon}\right)s}{s+\Omega_p\left(1+\frac{\Delta\Omega_p}{2\Omega_p}\right)} \tag{8.26}$$

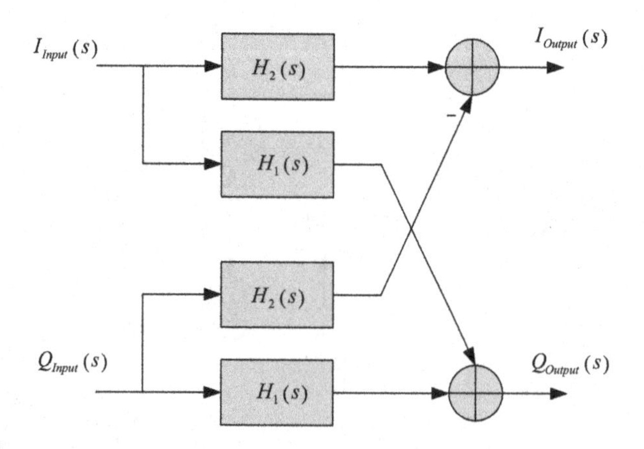

FIGURE 8.16

Block diagram of complex single-stage polyphase filter. (For color version of this figure, the reader is referred to the online version of this book.)

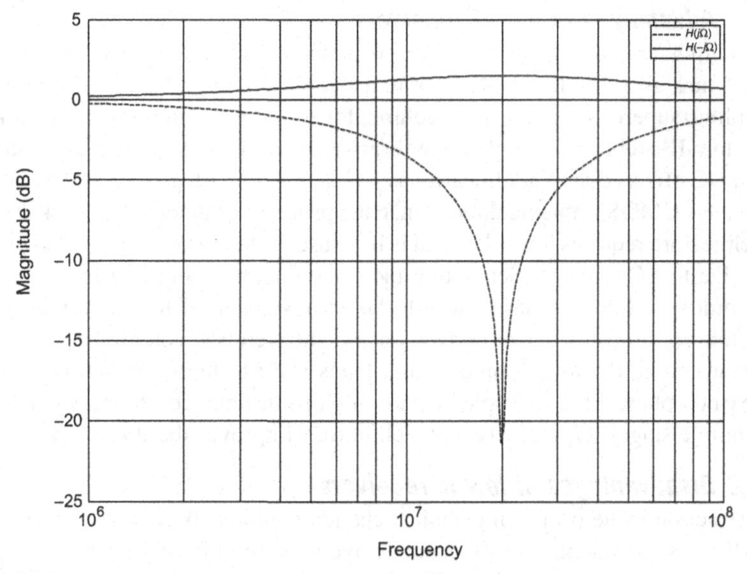

FIGURE 8.17

Magnitude response of single-stage complex polyphase filter: $H(j\Omega)$ and $H(-j\Omega)$. (For color version of this figure, the reader is referred to the online version of this book.)

The parameters $\Delta\Upsilon$ and $\Delta\Omega_p$ denote the mismatch in amplitude and frequency from the theoretical amplitude Υ and frequency Ω_p values [3]. In this case, the null due to $|H(s = j\Omega)|$ and the slight peak due to $|H(s = -j\Omega)|$ will shift in different directions and no longer "line up". In this case, the image rejection due to the complex polyphase filter is degraded from theory in proportion to $\Delta\Upsilon$ and $\Delta\Omega_p$. Therefore, we conclude that the amount of image rejection of a polyphase filter is limited by the imbalance created by component imperfections in the circuit.

A polyphase filter could be comprised of several stages in order to achieve sufficient image rejection. The resulting broadband polyphase filter possesses an image rejection theoretically obtained as the product of the image rejection due to the various stages. In practice, however, the overall image rejection is limited by the circuit imperfections amongst the various stages.

8.3.2 Receiver performance

As mentioned earlier, the low-IF receiver architecture combines some of the benefits of direct-conversion and superheterodyne architectures and inevitably some of their disadvantages. This makes it a unique architecture advantageous to use with certain modulation schemes given today's technology. These advantages and disadvantages will be discussed in the following sections.

8.3.2.1 Advantages of low-IF receivers

The sampling rate of the data converters could be as low as half the channel spacing,[6] thus allowing the ADCs to run at a low-sampling clock compared to the IF-sampling superheterodyne architecture, for example. Sampling at IF implies that the low-IF architecture, unlike the direct-conversion architecture, does not suffer from DC offsets due to nonlinearity as well as $1/f$ noise degradation. The latter is attractive for CMOS implementation since the process is inundated by flicker noise. The architecture requires lower IP2 and is less susceptible to noise generated by self-mixing due to LO leakage. Separating the desired signal from interference due to second order nonlinearity has proven to be an advantage of low-IF versus direct-conversion architecture under certain hostile interference conditions. Similar to direct conversion, the low-IF architecture lends itself to high integration since the complex polyphase filter is implemented on-chip. In practice, the polyphase filter uses multiple stages, typically two, which in turn improves the IQ matching.

8.3.2.2 Disadvantages of low-IF receivers

Image rejection is the principal technical challenge in low-IF receivers. The choice of the IF, at low frequency, prevents any image rejection filtering from taking place at RF. In most cases, the polyphase filter is designed to minimize adjacent and alternate channel interference, thus making the filter design more complex and inadvertently more power consuming. Proper choice of the IF frequency, however, can place the image in the adjacent channel. Moreover, in order to discriminate between the IQ signals, the I and Q outputs have to be processed as a complex pair. Having said that, the utility of the polyphase filter is limited by the balance accuracy between the IQ signals. Unlike direct conversion, the ADCs in low-IF architecture have to operate at IF, thus implying stricter requirements on the converters. Finally, second order distortion can result in serious in-band channel interference. In most practical implementations, the low-IF architecture has been limited to somewhat narrowband applications for the reasons cited above.

8.4 Typical driving requirements

The intent of this section is to briefly review the typical requirements that drive the design and architecture of a wireless receiver. For more details on any of these parameters, the reader is encouraged to revisit the earlier chapters where each topic is discussed in greater detail.

8.4.1 Sensitivity

The receiver's sensitivity is the lowest received signal power at the antenna at which the signal can be decoded satisfactorily. The sensitivity of the receiver is dictated by

[6] In GSM, for example, the channel bandwidth is 200 KHz, and hence the IF frequency could be as low as 100 KHz.

its noise figure, the bandwidth of the signal, and the carrier-to-noise ratio (CNR) at which the desired signal can be decoded with acceptable error rate. The noise figure of the receiver is influenced mostly by the analog front end, namely the band-definition filter, the LNA, and the first mixer. Incidentally, the LNA and the first mixer also bear the burden of setting a certain linearity performance of the receiver.

In superheterodyne receivers, the sensitivity takes into account the analog front end, the IF stage, and the analog baseband stage. The latter stages have lesser impact on sensitivity compared with the analog front end. Despite the higher component count, when compared to the direct-conversion or low-IF architectures, a well-designed superheterodyne receiver could reliably meet the sensitivity of the signaling waveform. The use of high-end passive front-end components coupled with a high gain low-noise LNA yields very consistent performance in the lab as well as products in the field. On the other hand, the sensitivity of direct-conversion receivers is mostly similar to that of low-IF receivers. Compared with the superheterodyne receiver, both the direct-conversion receiver and the low-IF receiver have lower component count in the path. Both architectures are mostly dependent on the noise figure and the gain of the LNA to set the sensitivity level of the receiver. In most cases, however, the superheterodyne receiver boasts better sensitivity performance than both the direct-conversion and the low-IF receivers.

8.4.2 Selectivity

Receiver selectivity, thoroughly addressed in Chapter 4, is the ability of a receiver to isolate the desired signal at a given frequency from other blockers and interferers situated at all other frequencies. The selectivity of a receiver is determined at all stages of the receiver, namely RF, IF, and baseband. Phase noise, spurs, linearity in general, and filtering all contribute to the final selectivity of the receiver. Ideally, the filters must be designed to pass the desired signal with minimal degradation while effectively filtering out all unwanted signals.

In superheterodyne receivers, adjacent channel selectivity is mostly impacted by phase noise. Any blocker or interferer within the passband of the band-definition filter or image reject filter is down-converted by the first LO. The first LO has the higher impact in terms of phase noise than the second LO, for example, since it operates at higher frequency. The mixing of the first LO signal with the interferer will transfer the LO phase noise unto the interferer, and the resulting signal is the convolution of the interfering signal with the LO. The danger then is the spillover of phase noise from the down-converted signal into the desired signal's band, thus degrading the received signal's SNR. The problem is exacerbated for narrowband signals with smaller channel spacing. The filtering at IF tends to be much narrower than the filtering at RF, and it occurs after the first mixer. Hence the impact of the in-band phase noise due to the first LO may not be removed simply by filtering.

In direct-conversion receivers, on the other hand, adjacent channel rejection is levied entirely on the baseband IQ filters. A well-designed baseband filter, however, can sufficiently attenuate the adjacent and alternate channels. Another limiting selectivity factor in direct conversion is the in-band phase noise whose performance almost entirely depends on the VCO.

Despite the architectural similarities between low-IF and direct-conversion receivers, the selectivity of the former is greatly affected by the performance of the complex polyphase filter. The limited rejection of the polyphase filter has grave implications on adjacent and alternate channel rejection and consequently the selectivity of the receiver. As mentioned earlier, a practical polyphase filter is limited to two stages. In this case, the implication of adjacent channel phase noise on the receiver performance is much more critical than that of direct conversion. Therefore, it may be necessary to employ a VCO with better phase noise characteristics than would be normally employed by direct-conversion receivers. This in turn implies higher complexity and form factor of the VCO with added device and integration cost. Like the two previous architectures, as would be the case with any architecture, in-band phase noise is also a limiting factor to the selectivity of low-IF receivers.

8.4.3 Image rejection

Recall from the discussion in Section 8.1.1 that the IF of a direct-conversion receiver is zero, and hence the receiver performance is not affected by image signal. In contrast, both the superheterodyne and low-IF architectures suffer from image rejection. In superheterodyne, the image reject filter is at RF. The filter can deal effectively with image rejection since twice the IF is set at a medium or high frequency offset from the filter's center frequency. That is, image rejection depends on the filter rejection at $F_c + 2F_{IF}$. The trade-off then in choosing the IF is between effective image rejection filtering or interference suppression. High IF implies that the image reject filter will better suppress the image but allow more interference into the signal path. Lowering the IF, on the other hand, narrows the image reject filter's response and thus allows it to suppress more potential blockers and interferes at the cost of allowing higher degradation due to image noise. The impact of image noise, however, can be alleviated by using image-reject mixers using a Hartley or Weaver architecture as discussed in earlier chapters.

In low-IF receivers, however, the image rejection is executed solely by the complex polyphase filter that is situated at IF and not at RF. In this case, the IF could be as low as half the channel spacing and the filter may not provide sufficient image rejection to minimize degradation to the desired signal. In this case, an image-reject mixer may be used to lessen the degradation to the signal.

8.4.4 Frequency planning and generation

Frequency planning is dictated solely by the RF for direct-conversion and low-IF receivers where using a single mixing step, the signal is down-converted to baseband or low IF. The difficulty resides in frequency generation (i.e., fractional phase-locked loop (PLL), frequency dividers, and multipliers, etc.) and the number of VCOs. The same issues and challenges are also present in superheterodyne receivers albeit to varying degrees of difficulty. On the other hand, frequency planning is an art form for superheterodyne receivers. Mixers at RF and IF tend to generate many spurious and intermodulation products related in frequency to the RF and IF signals and thus could fall in the desired signal band degrading its SNR. Therefore, careful frequency planning that takes into account the RF environment, the desired signal performance, and the receiver's lineup, including the performance of the various blocks, is fundamental to the overall performance of the receiver. A further challenge in receiver design is the RF synthesizer. Its design continues to be a demanding and time-consuming task. The classical trade-off between phase noise and tuning range is relevant to all architectures. Power dissipation also plays a key role in frequency generation, especially in prescalers.

8.4.5 Linearity

Linearity encompasses many topics in the receiver as discussed in previous chapters. Third order and to some extent second order nonlinearity are dominant concerns in receiver design. However, compared to direct-conversion and low-IF, the superheterodyne architecture typically operates at higher IIP2 and IIP3 values. This, coupled with lower sensitivity values, provides the superheterodyne receiver with wider spur-free dynamic range of operations. On the other hand, the effects of second order nonlinearity on the direct-conversion architecture are more dramatic than they are the other two architectures. Notably, direct conversion is vulnerable to DC offsets and even order distortions. Therefore, ensuring that the receiver has higher system IIP2 helps alleviate both problems.

8.5 Appendices

8.5.1 Appendix A: DC offset compensation algorithms

DC offset or DC bias is a common source of degradation in direct-conversion receivers. Under certain circumstances, due to poor port-to-port isolation, DC offset is the result of self-mixing of the LO signal with itself or with a strong blocker signal impinging on the antenna. Furthermore, certain design nonidealities in the data converters can cause considerable DC bias at their respective outputs and thus significantly degrade the desired signal's SNR. DC bias can also occur in the digital signal processing (DSP) due to truncation of the digital words. In this chapter, we are only concerned with DC offsets that are external to the DSP and due to imperfection in the analog circuits or ADCs.

The impact of DC offset on the demodulation of the signal is multifold. For example, having different DC bias on the IQ paths causes IQ imbalance in the received signal. DC offset signals can also impact certain synchronization algorithms and digital or hybrid (analog/digital) loops as well as algorithms that rely on zero-crossing techniques for detection and demodulation. For instance, DC offsets may impact the performance of an automatic frequency control (AFC) loop if the bias is not removed prior to frequency estimation. Even in a DC-free modulation scheme, such as frequency shift-keying (FSK) and orthogonal frequency division multiplexing (OFDM), a significant external DC bias to the ADCs can minimize the resolution of the converters without directly impacting the modulated signal. DC offset compensation algorithms can be divided into two main categories: real-time algorithms operating while demodulating the received signal and non-real-time algorithms where the DC offset is removed at the factory or at power-up prior to demodulation. DC offsets that originate in the analog domain can be dealt with via DC-offset compensation algorithms that operate either in the analog domain, digital domain, or both.

Figure 8.18 presents an example of typical DSP blocks that follow the ADC converters in direct-conversion architecture. After digitization, the IQ signals undergo DC-bias estimation and removal via the DC-offset compensation (DCOC) block. The simplest (DCOC) methods used to remove DC bias via DSP implementation only rely on simple filtering schemes. The intent is to estimate the average or expected value of the received signal $E\{x(n)\}|_{t=nT_s}$, where $E\{.\}$ is the expected value operator, and subtract it from $x(n)$ in order to obtain a DC-free signal.

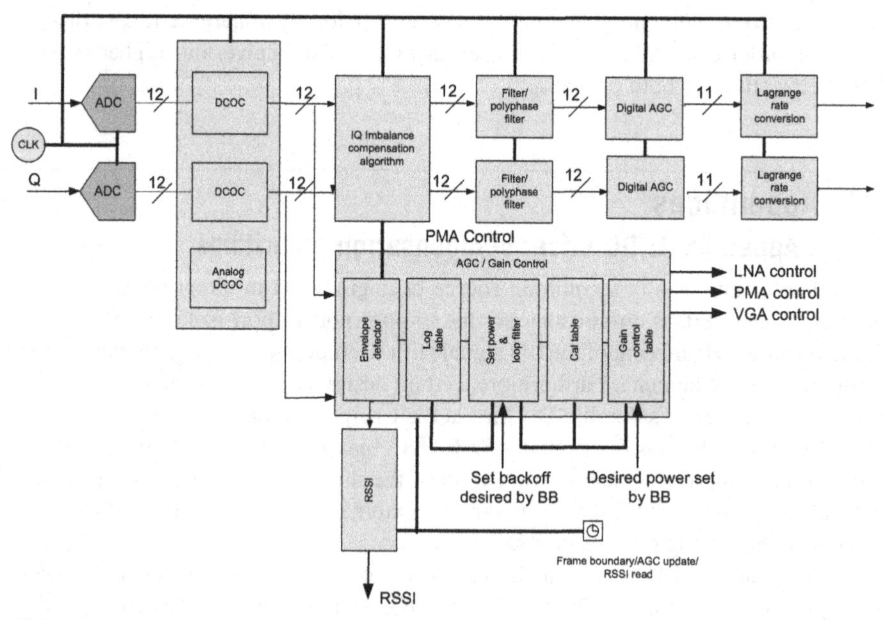

FIGURE 8.18

Baseband processing of in-phase and quadrature (IQ) signals illustrating the basic DCOC, IQ-imbalance, and AGC algorithms along with certain filters and rate conversion blocks. (For color version of this figure, the reader is referred to the online version of this book.)

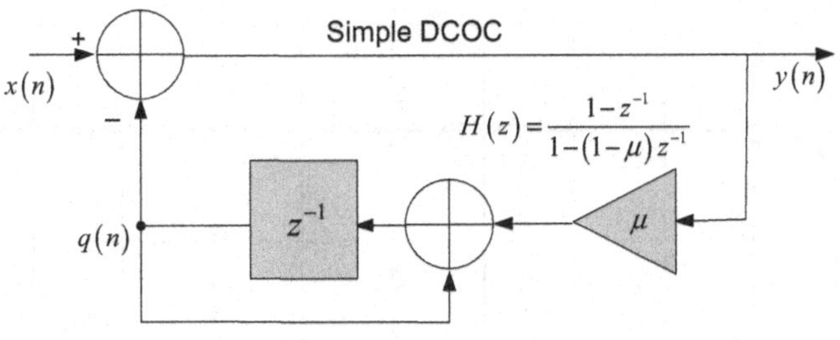

FIGURE 8.19

Single-pole IIR filter used as a DCOC algorithm to remove DC bias.

Consider the DCOC filter applicable to either the in-phase or quadrature component as depicted in Figure 8.19. The input/output relationship of the feedback filter in relation to the output $y(n)$ is

$$\frac{Q(z)}{Y(z)} = \frac{\mu z^{-1}}{1 - z^{-1}} \Leftrightarrow q(n) - q(n-1) = \mu y(n-1) \tag{8.27}$$

The input/output relationship between the original signal $x(n)$ and the corresponding DC-free output signal $y(n)$ is given according to the relationship

$$H(z) = \frac{Y(z)}{X(z)} = \frac{1 - z^{-1}}{1 - (1-\mu)z^{-1}}; \quad |\mu| \leq 1 \tag{8.28}$$

The frequency response in Eqn (8.28) is that of a highpass filter as depicted in Figure 8.20. The sharpness of the filter depends on the gain of the loop filter μ. For the sake of illustration, consider a white Gaussian signal with certain DC offset as shown in Figure 8.21. The output of the DCOC filter, also depicted in Figure 8.21, shows that the resulting signal is free of any DC bias. This is a very simple yet effective algorithm. Its main drawback is the fact that its response is very slow. In fixed point implementation, the output signal requires more bits than the input signal to preserve its accuracy due to the feedback. Truncating the output signal, however, results in quantization noise that, depending on the number of bits and the required SNR, may or may not be acceptable. A modified version of the algorithm presented above adds a delta quantizer in the feedback as shown in Figure 8.22 [5]. This algorithm tends to be more effective than the previous algorithm when it comes to removing the DC offset at the output of a coarse ADC. In this case, the DC bias is further removed by shaping the error away from DC due to the highpass response brought about by the Delta quantizer.

Another simple method of removing DC offsets via digital filtering is to employ a moving average filter that estimates the mean of the signal and subtracts it from the

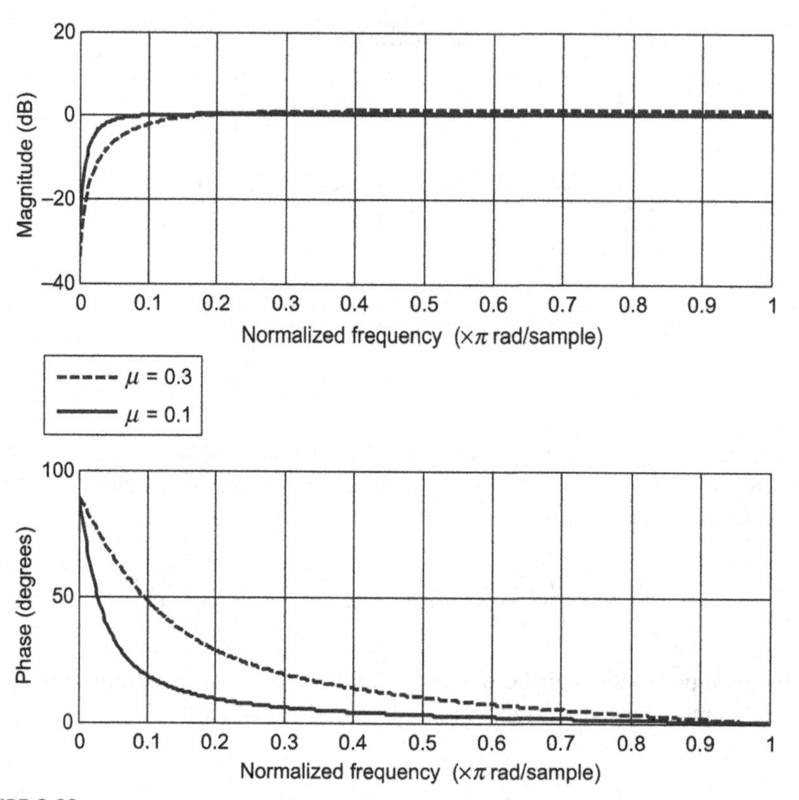

FIGURE 8.20

Frequency response of DCOC algorithm according to the filter's transfer function given in Eqn (8.28). (For color version of this figure, the reader is referred to the online version of this book.)

signal as shown in Figure 8.23. The moving average filter operating on the input signal acts as a delay-and-accumulate operation expressed as

$$\frac{1}{M}\sum_{l=1}^{M} x(n-l) \Leftrightarrow \frac{1}{M}\frac{z^{-1}-z^{-M}}{1-z^{-1}}X(z) = S(z)X(z) \tag{8.29}$$

The frequency and phase response of a 20-tap moving average finite impulse response (FIR) filter realizing $S(z)$ is depicted in Figure 8.24. The overall input/output transfer function is given according to the relation

$$\begin{aligned} Y(z) &= z^{-M}X(z) - \frac{1}{M}\frac{z^{-1}-z^{-M}}{1-z^{-1}}X(z) \\ \frac{Y(z)}{X(z)} &= H(z) = \frac{-z^{-1}+(M+1)z^{-M}-Mz^{-M-1}}{M(1-z^{-1})} \end{aligned} \tag{8.30}$$

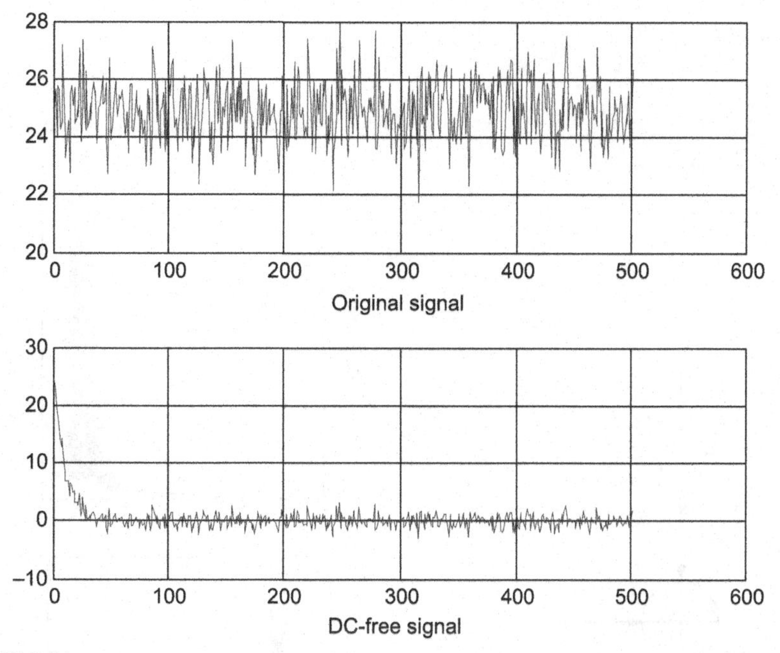

FIGURE 8.21

Original input signal and DC-free output signal of DCOC depicted in Figure 8.20 for $\mu = 0.1$. (For color version of this figure, the reader is referred to the online version of this book.)

The frequency and phase response of the DCOC realized using Eqn (8.30) for a 20- and 40-tap moving average filter is depicted in Figure 8.25. As one might expect, increasing the number of filter taps sharpens the null at DC. The advantage of having a sharper null implies better DC-bias estimation and removal. The disadvantage of having more filter taps is longer settling time. This process is illustrated by passing white noise with a certain DC bias, as shown in Figure 8.26, through the filter with varying number of taps. The output of the filter is shown in Figure 8.27 for 10, 20, and 40 taps. Figure 8.27 further depicts the performance of the DCOC utilizing a single-pole IIR filter alongside the DCOC with the moving average filter. The DCOC seems to converge faster; however, the linearity of the phase is not preserved. Another disadvantage of the moving average-based DCOC is latency. In most cases, significant latency in the data path is not tolerated. There are other types of DCOC algorithms that can be employed, varying from the very exotic to the adaptive type. These DCOC algorithms will not be discussed in this chapter.

So far, we have addressed DC-bias removal in the digital domain only. As mentioned earlier, despite utilizing an effective DCOC algorithm, the resolution of the ADCs will inevitably suffer. In the event where the DC offset severely impacts the resolution of the ADCs, then it must be removed in the analog lineup of the receiver. One common technique that is employed in removing DC offset is to

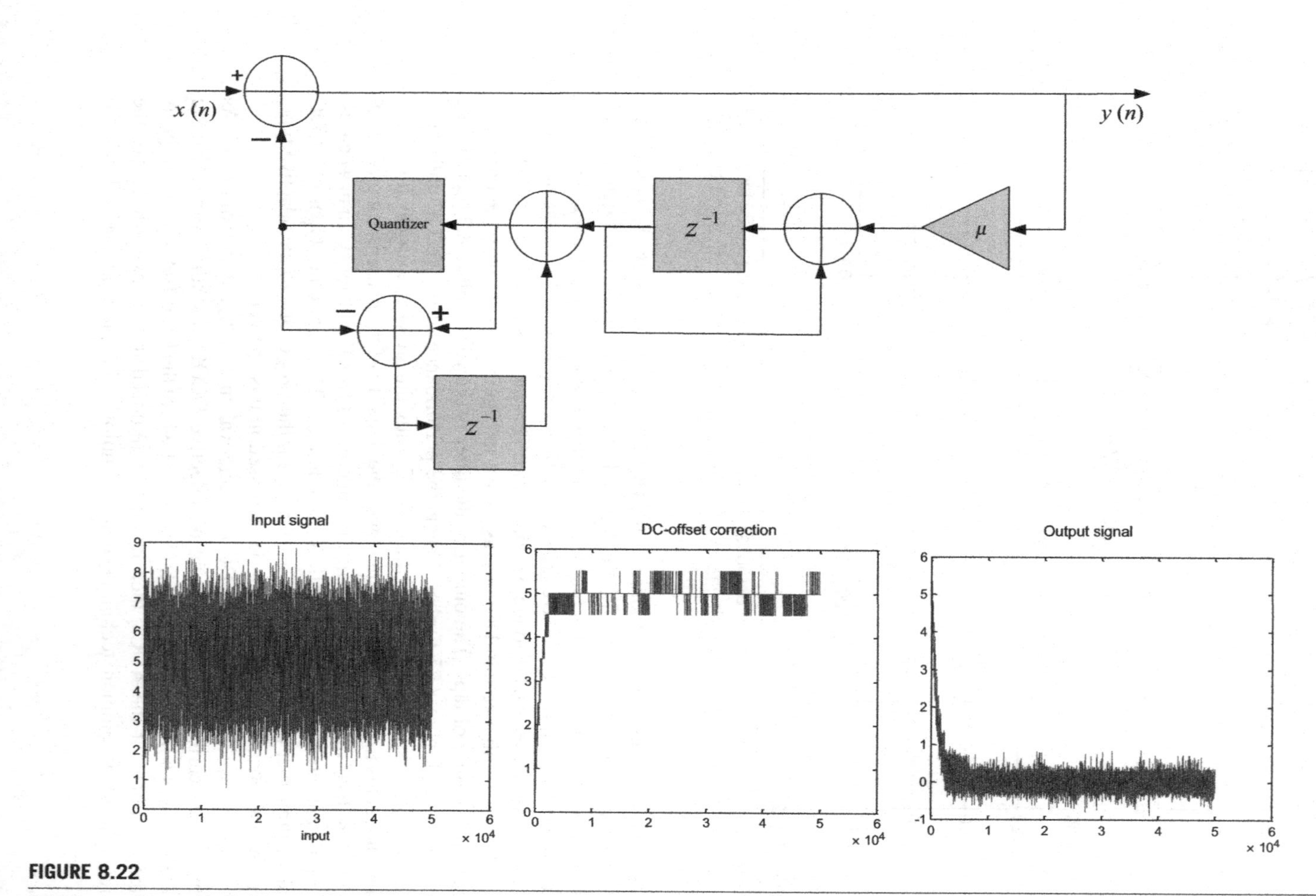

FIGURE 8.22

DCOC algorithm with Delta quantizer in the feedback. (For color version of this figure, the reader is referred to the online version of this book.)

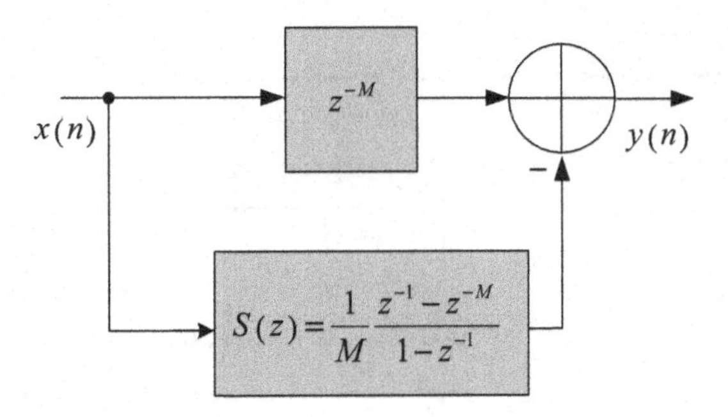

FIGURE 8.23

DCOC algorithm with a moving average implementation.

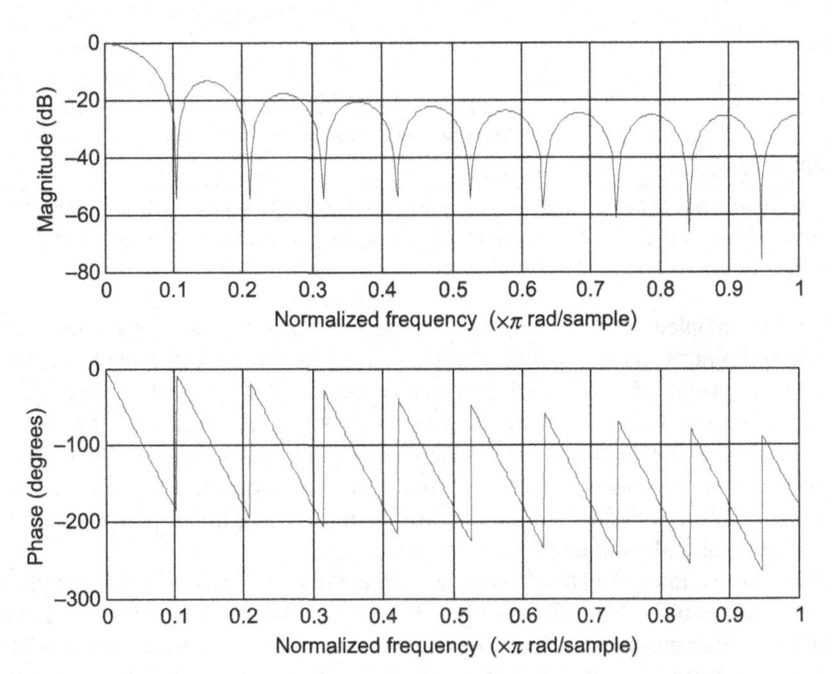

FIGURE 8.24

Frequency and phase response of a 20-tap moving average filter. (For color version of this figure, the reader is referred to the online version of this book.)

AC couple the data path [6]. AC coupling in essence implies highpass filtering the incoming down-converted signal. The implicit problem in this approach is the corner frequency of the filter itself. A high corner frequency may degrade the modulated signals with high energy content near DC. Another technique is to measure the DC offset with the antenna connector tied to the ground and the device under test completely shielded from external RF signals. In this case, the DC offset, provided

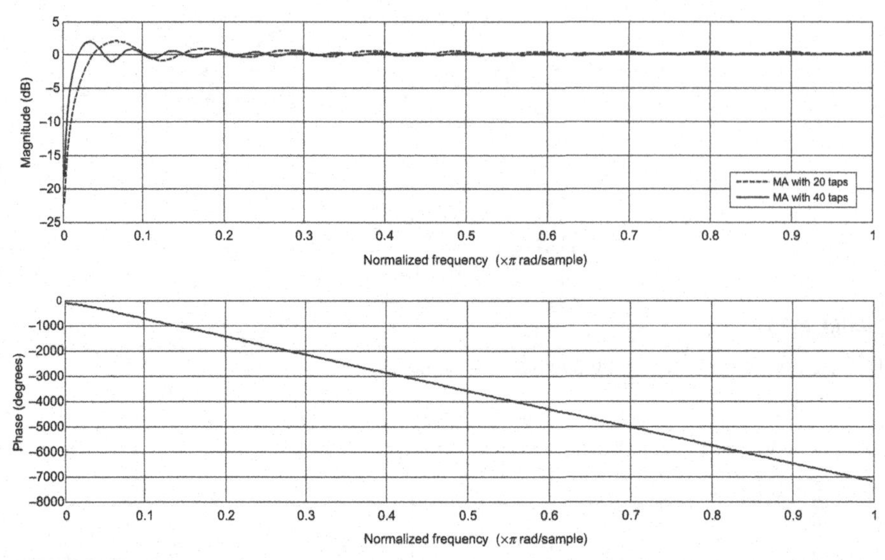

FIGURE 8.25

Frequency and phase response of $H(z)$ for a DCOC with 20- and 40-tap moving average filter in the feedforward path. (For color version of this figure, the reader is referred to the online version of this book.)

it does not completely overwhelm the converters, is measured at baseband using the DCOC techniques above, but the DC offset is then stored in memory and then subtracted via current DACs before digital conversion. In most practical systems, this is typically done at various gain stages where the DC offset is removed via current steering DACS strategically placed in the receiver path [7] as shown in Figure 8.28. In this case, one gain stage is placed before the filter and the other one after the filter. A buffer amplifier is placed before the ADC. All three amplifiers contribute to the DC offset seen at the output of the converters.

In order to remove the DC offset at the output of the converters, set the gain of the amplifier stages before and after the filters to low gain. The DC bias due to the buffer amplifiers is then measured and removed via the current DAC. The gain of the postfilter amplifiers is then set to high and again the DC offset due to these amplifiers is measured and removed. Finally, the process is repeated by setting the gain of the prefilter amplifiers to high and the DC offset is then measured and removed. Note that the strategy in removing the DC offset starts with the amplifiers closest to the converters and then moving up the chain to the gains closest to the mixer. In this way, the DC offset is almost always measurable at the output of the converters and subtracted via the current DACs.

8.5.2 Appendix B: IQ-imbalance compensation algorithm

In this appendix, we offer a simple IQ-imbalance compensation algorithm that serves to calibrate out the degradation impact in the digital domain. First, consider

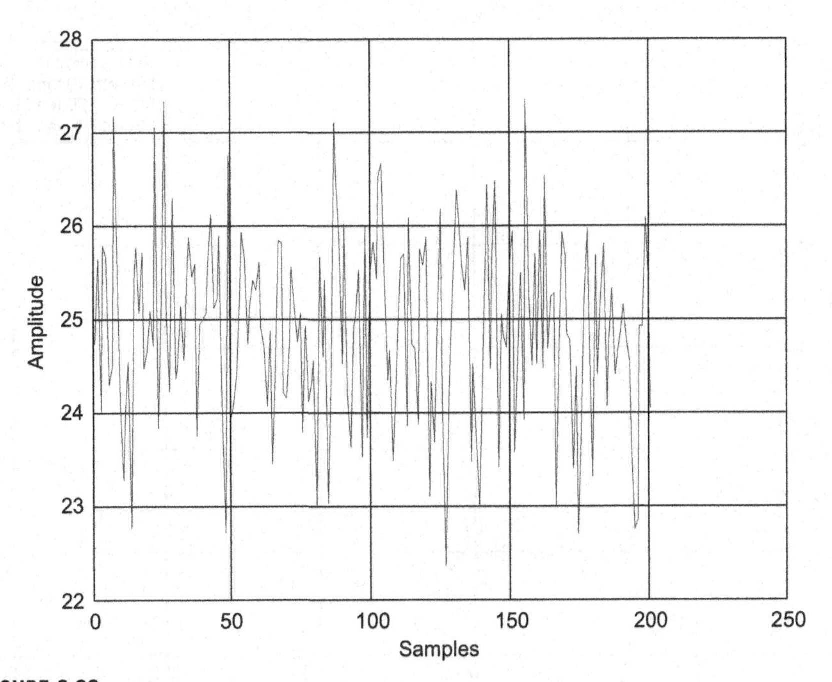

FIGURE 8.26

White Gaussian noise with DC offset. (For color version of this figure, the reader is referred to the online version of this book.)

the conceptual direct-conversion transmitter depicted in Figure 8.29. The output of the transmitter is coupled onto a square-law detector. The output of the square-law detector is then filtered and passed to the digital baseband after data conversion. The linear portion of the detector is assumed to be negligible. In this appendix, we take the liberty of mixing notation between analog and digital signals in order to simplify the mathematics.

Assume the transmit signal at the output of the IQ mixers to be

$$S(t) = 2A\cos(\omega t)\cos\left(\Omega_c t - \frac{\theta}{2}\right) + 2B\sin(\omega t)\sin\left(\Omega_c t + \frac{\theta}{2}\right) \tag{8.31}$$

[7]where in this case the IQ components transmitted from baseband IQ are simply a tone expressed as[8]

$$I(t) = \cos(\omega t) \text{ and } Q(t) = -\sin(\omega t) \tag{8.32}$$

[7]Recall that with IQ components balanced, the output of the transmitter can be theoretically expressed as $S(t) = I(t)\cos(\Omega_c t) - Q(t)\sin(\Omega_c t)$.

[8]Figure 8.29 shows a different excitation signal as will be discussed in the receive IQ-imbalance calibration.

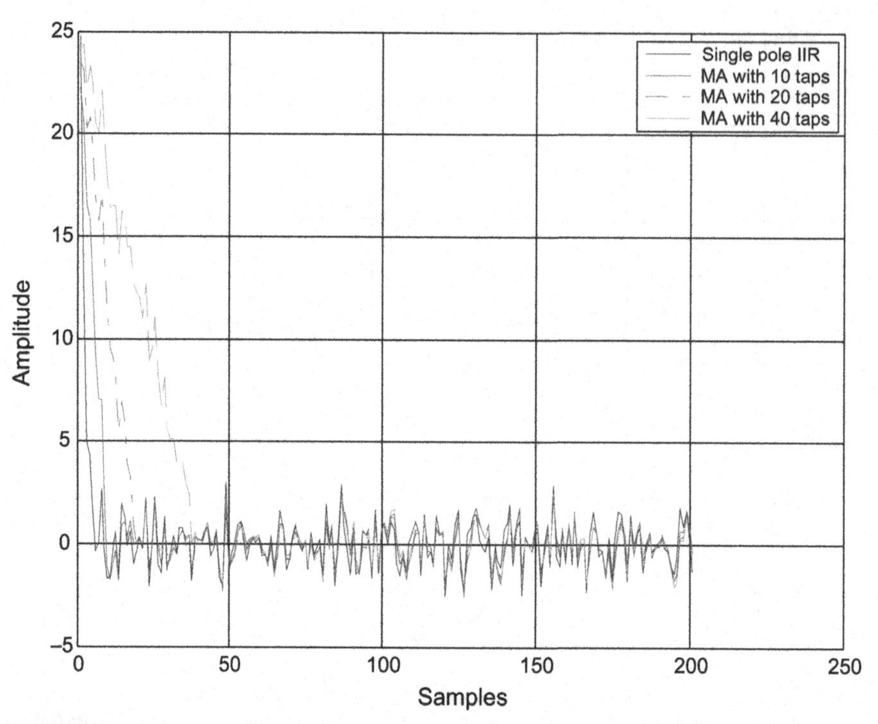

FIGURE 8.27

Output of DCOC using single-pole IIR and MA filter with 10, 20, and 40 taps. (For color version of this figure, the reader is referred to the online version of this book.)

The constants A and B Eqn (8.31) are the IQ amplitude imbalance defined as

$$A = 1 - \alpha \text{ and } B = 1 + \alpha \tag{8.33}$$

and θ is the phase imbalance. After squaring, the relation in Eqn (8.31) becomes

$$\begin{aligned} S^2(t) &= 4A^2\cos^2(\omega t)\cos^2\left(\Omega_c t - \frac{\theta}{2}\right) + 4B^2\sin^2(\omega t)\sin^2\left(\Omega_c t + \frac{\theta}{2}\right) \\ &\quad + 8AB\cos(\omega t)\cos\left(\Omega_c t - \frac{\theta}{2}\right)\sin(\omega t)\sin\left(\Omega_c t + \frac{\theta}{2}\right) \end{aligned} \tag{8.34}$$

The first term in Eqn (8.34) can be further expressed as

$$\begin{aligned} 4A^2\cos^2(\omega t)\cos^2\left(\Omega_c t - \frac{\theta}{2}\right) &= A^2\{\cos(2\omega t) + 1\}\{\cos(2\Omega_c t - \theta) + 1\} \\ &= A^2\left\{1 + \cos(2\omega t) + \cos(2\Omega_c t - \theta) + \frac{1}{2}\cos(2\Omega_c t \right. \\ &\quad \left. + 2\omega t - \theta) + \frac{1}{2}\cos(2\Omega_c t - 2\omega t - \theta)\right\} \end{aligned} \tag{8.35}$$

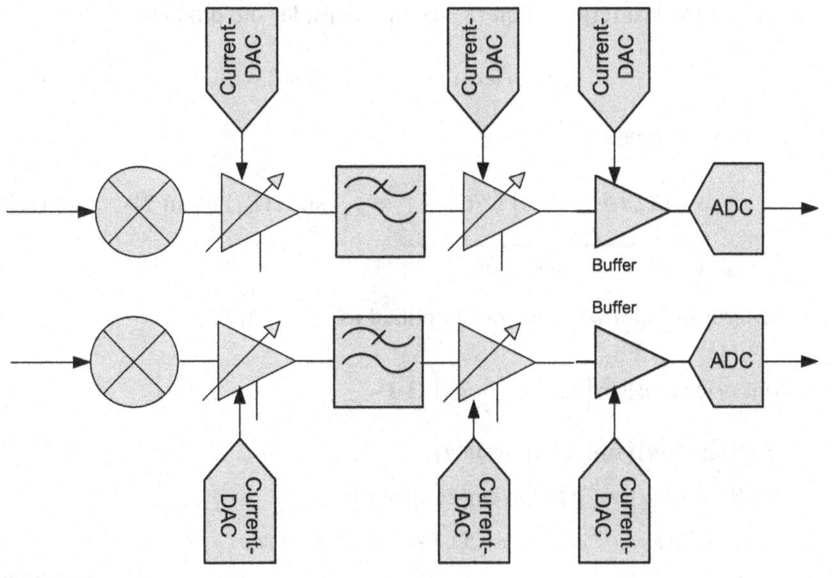

FIGURE 8.28

Conceptual diagram of the postmixer analog baseband receiver with current steering DACs used to remove DC offsets at various gain stages.

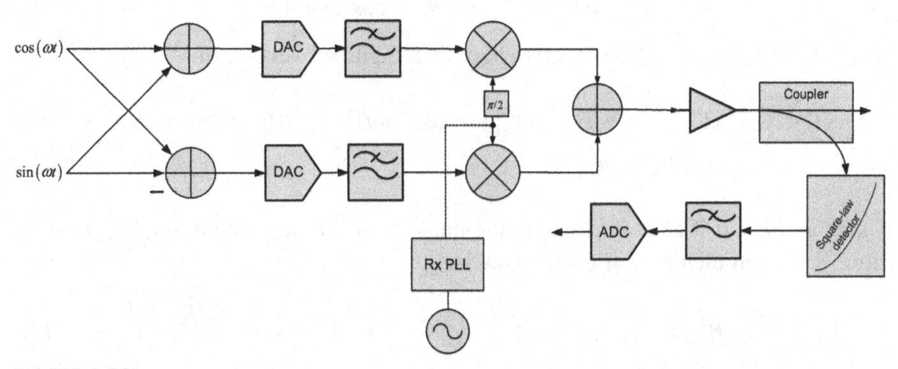

FIGURE 8.29

Conceptual direct-conversion transmitter with coupler and square-law detector. (For color version of this figure, the reader is referred to the online version of this book.)

After analog lowpass filtering, the signal due to the first term in Eqn (8.34) and expressed in Eqn (8.35) becomes

$$\widehat{T}_1(t) = A^2\{1 + \cos(2\omega t)\} \tag{8.36}$$

In a similar manner, the second term Eqn (8.34) can be manipulated in the same fashion as the first term Eqn (8.35) and then filtered to obtain

$$\widehat{T}_2 = B^2\{1 - \cos(2\omega t)\} \tag{8.37}$$

Finally, given the last term in Eqn (8.34), first consider the product

$$\sin(\omega t)\cos(\omega t) = \frac{1}{2}\sin(2\omega t) \tag{8.38}$$

Next, express the product

$$\underbrace{\sin\left(\Omega_c t + \frac{\theta}{2}\right)\cos\left(\Omega_c t - \frac{\theta}{2}\right)}_{\sin\alpha\cos\beta = \frac{1}{2}\{\sin(\alpha+\beta)+\sin(\alpha-\beta)\}} = \frac{1}{2}\{\sin(2\Omega_c t) + \sin\theta\} \tag{8.39}$$

Then the term in Eqn (8.34) can be simplified as

$$\begin{aligned} &8AB\sin(\omega t)\cos(\omega t)\sin\left(\Omega_c t + \frac{\theta}{2}\right)\cos\left(\Omega_c t - \frac{\theta}{2}\right) \\ &= 2AB\sin(2\omega t)\{\sin(2\Omega_c t) + \sin\theta\} \\ &= 2AB\{\sin(2\omega t)\sin(2\Omega_c t) + \sin(2\omega t)\sin\theta\} \\ &= AB\{\cos(2\Omega_c t - 2\omega t) - \cos(2\Omega_c t + 2\omega t) + \cos(2\omega t - \theta) - \cos(2\omega t + \theta)\} \end{aligned} \tag{8.40}$$

After analog lowpass filtering, the term in Eqn (8.40) becomes

$$\widehat{T}_3 = AB\{\cos(2\omega t - \theta) - \cos(2\omega t + \theta)\} \tag{8.41}$$

The resulting signal received at baseband is a digitized version of

$$\begin{aligned} \widehat{S}(t) = {} & A^2\{1 + \cos(2\omega t)\} + B^2\{1 - \cos(2\omega t)\} + AB\{\cos(2\omega t - \theta) \\ & - \cos(2\omega t + \theta)\} \end{aligned} \tag{8.42}$$

Using an FFT, if it already exists in the modem, or a running sum, the DC term in Eqn (8.42) can then be estimated. That is,

$$K = A^2 + B^2 = (1-\alpha)^2 + (1+\alpha)^2 = 2 + 2\alpha^2 \Rightarrow \alpha = \left|\frac{\sqrt{K-2}}{2}\right| \tag{8.43}$$

Next, consider the DC-free signal of Eqn (8.42) obtained via a highpass filter

$$\begin{aligned} \widehat{S}(t)\Big|_{DC-free} &= (A^2 - B^2)\cos(2\omega t) + AB\left\{\underbrace{\cos(2\omega t - \theta) - \cos(2\omega t + \theta)}_{\sin\alpha\sin\beta = \frac{1}{2}[\cos(\alpha-\beta)-\cos(\alpha+\beta)]}\right\} \\ &= -4\alpha\cos(2\omega t) + (1-\alpha^2)\sin(2\omega t)\sin(\theta) \end{aligned} \tag{8.44}$$

The intent then is to estimate the phase imbalance θ from Eqn (8.44). Modulating Eqn (8.44) with $\sin(2\omega t)$ we obtain

$$\begin{aligned}\sin(2\omega t)\times\widehat{S}(t)\Big|_{DC-free} &= -4\alpha\underbrace{\cos(2\omega t)\sin(2\omega t)}_{\sin(4\omega t)=2\cos(2\omega t)\sin(2\omega t)} +(1-\alpha^2)\underbrace{\sin^2(2\omega t)}_{\left\{\frac{1-\cos(4\omega t)}{2}\right\}}\sin(\theta)\\ &= -2\alpha\sin(4\omega t)-\frac{(1-\alpha^2)}{2}\cos(4\omega t)\sin(\theta)\\ &\quad+\frac{(1-\alpha^2)}{2}\sin(\theta)\end{aligned} \tag{8.45}$$

Using a running sum, for example, to obtain the DC portion ρ of Eqn (8.45), we obtain

$$\rho = \frac{(1-\alpha^2)}{2}\sin(\theta)\Rightarrow\theta = \arcsin\left(\frac{2\rho}{1-\alpha^2}\right) \tag{8.46}$$

Having obtained an estimate of the phase and amplitude imbalance, we can then use this data to calibrate the IQ imbalance in the transmit path.

Next, we discuss the IQ-imbalance calibration of the receiver. This can be done with external signals using a signal generator, for example, or via loop back in the radio itself. The latter is an attractive but expensive feature and at times not so easy to implement.

Assume the transmit excitation signal to be IQ-balanced generated from the digital baseband as

$$\begin{aligned}I(t) &= \cos(\omega t)+\sin(\omega t)\\ Q(t) &= \cos(\omega t)-\sin(\omega t)\end{aligned} \tag{8.47}$$

where $I(t)$ and $Q(t)$ are the IQ components. Again, the reader is advised that we will take immense liberties in mixing between the analog and digital domains in discussing this technique in order to simplify the mathematics.

The output of the analog transmitter according to Figure 8.29 is

$$\begin{aligned}S_{IQ-\text{balanced}}(t) &= I(t)\cos(\Omega_c t)-Q(t)\sin(\Omega_c t)\\ &= \cos(\Omega_c t)\{\cos(\omega t)+\sin(\omega t)\}-\sin(\Omega_c t)\{\cos(\omega t)-\sin(\omega t)\}\\ &= \cos((\Omega_c-\omega)t)-\sin((\Omega_c-\omega)t)\end{aligned} \tag{8.48}$$

On the receiver side, mixing the signal $S_{IQ\text{-balanced}}(t)$ with the imbalanced IQ Los, we obtain

$$\begin{aligned} S_{IQ-\text{balanced}}(t)LO_I(t) &= 2(1-\alpha)\cos\left(\Omega_c t - \frac{\theta}{2}\right)\{\cos((\Omega_c - \omega)t) - \sin((\Omega_c - \omega)t)\} \\ &= (1-\alpha)\left\{\cos\left(2\Omega_c t - \omega t - \frac{\theta}{2}\right) + \cos\left(\omega t - \frac{\theta}{2}\right)\right. \\ &\quad \left. - \sin\left(2\Omega_c t - \omega t - \frac{\theta}{2}\right) + \sin\left(\omega t - \frac{\theta}{2}\right)\right\} \end{aligned} \tag{8.49}$$

After analog lowpass filtering, the input to the in-phase ADC is

$$I'(t) = (1-\alpha)\left\{\cos\left(\omega t - \frac{\theta}{2}\right) + \sin\left(\omega t - \frac{\theta}{2}\right)\right\} \tag{8.50}$$

Similarly, after mixing the signal $S_{IQ\text{-balanced}}(t)$ can be expressed as

$$\begin{aligned} S_{IQ-\text{balanced}}(t)LO_Q(t) &= 2(1+\alpha)\sin\left(\Omega_c t + \frac{\theta}{2}\right)\{\cos((\Omega_c - \omega)t) - \sin((\Omega_c - \omega)t)\} \\ &= (1+\alpha)\left\{\sin\left(2\Omega_c t - \omega t + \frac{\theta}{2}\right) + \sin\left(\omega t + \frac{\theta}{2}\right)\right. \\ &\quad \left. + \cos\left(2\Omega_c t - \omega t + \frac{\theta}{2}\right) - \cos\left(\omega t + \frac{\theta}{2}\right)\right\} \end{aligned} \tag{8.51}$$

After analog lowpass filtering on the receiver quadrature side, the resulting signal at the input to the ADC is

$$Q'(t) = (1+\alpha)\left\{-\cos\left(\omega t + \frac{\theta}{2}\right) + \sin\left(\omega t + \frac{\theta}{2}\right)\right\} \tag{8.52}$$

Next, after digitizing, compute the average

$$\begin{aligned} \Gamma &= \frac{1}{N}\sum_{n=0}^{N-1} Q'^2(nT_s) - I'^2(nT_s) \\ &= \frac{1}{N}\sum_{n=0}^{N-1}\left[(1+\alpha)\left\{-\cos\left(\omega nT_s + \frac{\theta}{2}\right) + \sin\left(\omega nT_s + \frac{\theta}{2}\right)\right\}\right]^2 \\ &\quad - \left[(1-\alpha)\left\{\cos\left(\omega nT_s - \frac{\theta}{2}\right) + \sin\left(\omega nT_s - \frac{\theta}{2}\right)\right\}\right]^2 \end{aligned} \tag{8.53}$$

where T_s is the sampling period. For a large number of samples N, further expansion of Eqn (8.53) implies

$$\begin{aligned} \Gamma &\approx (1+\alpha)^2 - (1-\alpha)^2 = 4\alpha \\ \alpha &\approx \frac{\Gamma}{4} \end{aligned} \tag{8.54}$$

Next, consider the product

$$\begin{aligned} I'^2(t)Q'^2(t) &= [(1-\alpha)(1+\alpha)]^2 \left\{ \cos\left(\omega t - \frac{\theta}{2}\right) + \sin\left(\omega t - \frac{\theta}{2}\right) \right\}^2 \\ &\quad \left\{ -\cos\left(\omega t + \frac{\theta}{2}\right) + \sin\left(\omega t + \frac{\theta}{2}\right) \right\}^2 \\ &= \left(1-\alpha^2\right)^2 \left\{ 1 + 2\cos\left(\omega t - \frac{\theta}{2}\right) \sin\left(\omega t - \frac{\theta}{2}\right) \right\} \\ &\quad \left\{ 1 - 2\cos\left(\omega t + \frac{\theta}{2}\right) \sin\left(\omega t + \frac{\theta}{2}\right) \right\} \\ &= \left(1-\alpha^2\right)^2 \{1 + \sin(2\omega t - \theta)\}\{1 - \sin(2\omega t + \theta)\} \end{aligned} \tag{8.55}$$

A further expansion of Eqn (8.55) can be obtained as

$$\begin{aligned} I'^2(t)Q'^2(t) &= \left(1-\alpha^2\right)^2 \{1 + \sin(2\omega t - \theta) - \sin(2\omega t + \theta) \\ &\qquad - \sin(2\omega t - \theta)\sin(2\omega t + \theta)\} \\ &= \left(1-\alpha^2\right)^2 \left\{ 1 + \sin(2\omega t - \theta) - \sin(2\omega t + \theta) \right. \\ &\qquad \left. - \frac{\cos(2\theta) + \cos(4\omega t)}{2} \right\} \end{aligned} \tag{8.56}$$

Averaging the signal in Eqn (8.56) over N samples, we obtain after simplification

$$\begin{aligned} \frac{1}{N}\sum_{n=0}^{N-1} I'^2(t)Q'^2(t) &= \frac{1}{N}\left(1-\alpha^2\right)^2 \left\{ 1 + \sin(2\omega t - \theta) - \sin(2\omega t + \theta) \right. \\ &\qquad \left. - \frac{\cos(2\theta) + \cos(4\omega t)}{2} \right\} \\ &= \frac{1}{N}\left(1-\alpha^2\right)^2 \left\{ 1 - \frac{\cos(2\theta)}{2} \right\} \approx \frac{1}{N}\left(1 - \frac{\Gamma^2}{16}\right)^2 \left\{ \frac{2-\cos(2\theta)}{2} \right\} \\ &= \Upsilon \end{aligned} \tag{8.57}$$

The phase imbalance can then be computed based on Eqn (8.57) as

$$\cos(2\theta) = 2 - \frac{2N\Upsilon}{\left(1 - \frac{\Gamma^2}{16}\right)^2} \Rightarrow \theta = \frac{1}{2}\cos^{-1}\left\{2 - \frac{2N\Upsilon}{\left(1 - \frac{\Gamma^2}{16}\right)^2}\right\} \tag{8.58}$$

The relationship in Eqn (8.58) is a reasonably fair approximation of the phase imbalance. The algorithm is simple to implement and execute either on the bench or in real time in loopback mode.

$$S(t) = I(t)\cos(\Omega_c t) - Q(t)\sin(\Omega_c t)$$

References

[1] B. Razavi, Architectures and circuits for RF-CMOS receivers, IEEE Custom Integr. Conf. (1998) 393–400.

[2] A. Abidi, Direct-conversion radio transceivers for digital communications, IEEE J. Solid State Circuits 30 (12) (December 1995).

[3] C. Chou, C. Wu, The design of wideband and low-power CMOS active polyphase filter and its application in RF double-conversion receivers, IEEE Trans. Circuits Systems-1 52 (5) (May 2005) 825–833.

[4] S.H. Galal, H.F. Ragaie, M.S. Tawfik, RC sequence asymmetric polyphase network for RF integrated transceivers, IEEE Trans. Circuit Syst. Analog Digital Signal Process 47 (1) (January 2000) 18–27.

[5] R.G. Lyons, Understanding Digital Signal Processing, second ed., Prentice Hall Professional Technical Reference, Upper Saddle River, NJ, 2009.

[6] B. Razavi, Design considerations for direct-conversion receivers, IEEE Trans. Circuit Systems-II 44 (6) (June 1997).

[7] R. Xie, et al., Analog baseband for WCDMA implementation by using digital-controlled DC offset cancellation, 10th IEEE Int. Conf. Solid-State Integr. Circuit Technol. (ICSICT) (2010) 439–441.

Index

Note: Page numbers followed by f indicate figures; t, tables; b, boxes.

E

F

G

H

I

S

Printed in the United States
By Bookmasters